VECTOR QUANTIZATION AND SIGNAL COMPRESSION

THE KLUWER INTERNATIONAL SERIES
IN ENGINEERING AND COMPUTER SCIENCE

COMMUNICATIONS AND INFORMATION THEORY

Consulting Editor:
Robert Gallager

Other books in the series:

Digital Communication. Edward A. Lee, David G. Messerschmitt
ISBN: 0-89838-274-2

An Introduction to Cryptology. Henk C.A. van Tilborg
ISBN: 0-89838-271-8

Finite Fields for Computer Scientists and Engineers. Robert J. McEliece
ISBN: 0-89838-191-6

An Introduction to Error Correcting Codes With Applications.
Scott A. Vanstone and Paul C. van Oorschot
ISBN: 0-7923-9017-2

Source Coding Theory. Robert M. Gray
ISBN: 0-7923-9048-2

Switching and Traffic Theory for Integrated Broadband Networks.
Joseph Y. Hui
ISBN: 0-7923-9061-X

Advances in Speech Coding, Bishnu Atal, Vladimir Cuperman and
Allen Gersho
ISBN: 0-7923-9091-1

Coding: An Algorithmic Approach, John B. Anderson and Seshadri Mohan
ISBN: 0-7923-9210-8

Third Generation Wireless Information Networks, edited by Sanjiv Nanda
and David J. Goodman
ISBN: 0-7923-9128-3

VECTOR QUANTIZATION AND SIGNAL COMPRESSION

by

Allen Gersho
University of California, Santa Barbara

Robert M. Gray
Stanford University

KLUWER ACADEMIC PUBLISHERS
Boston/Dordrecht/London

Distributors for North America:
Kluwer Academic Publishers
101 Philip Drive
Assinippi Park
Norwell, Massachusetts 02061 USA

Distributors for all other countries:
Kluwer Academic Publishers Group
Distribution Centre
Post Office Box 322
3300 AH Dordrecht, THE NETHERLANDS

Library of Congress Cataloging-in-Publication Data

Gersho, Allen.
        Vector quantization and signal compression / by Allen Gersho,
Robert M. Gray.
        p.   cm. -- (Kluwer international series in engineering and
computer science ; SECS 159)
        Includes bibliographical references and index.
        ISBN 0-7923-9181-0
        1. Signal processing--Digital techniques. 2. Data compression
(Telecommunication) 3. Coding theory.   I. Gray, Robert M., 1943-
. II. Title. III. Series.
TK5102.5.G45   1991
621.382'2--dc20   91-28580
                                                                  CIP

This book was prepared with LaTeX and reproduced by Kluwer from camera-ready
copy supplied by the authors.

Printed on acid-free paper.

Printed in the United States of America

to Roberta & Lolly

Contents

Preface

Herb Caen, a popular columnist for the *San Francisco Chronicle*, recently quoted a Voice of America press release as saying that it was reorganizing in order to "eliminate duplication and redundancy." This quote both states a goal of data compression and illustrates its common need: the removal of duplication (or redundancy) can provide a more efficient representation of data and the quoted phrase is itself a candidate for such surgery. Not only can the number of words in the quote be reduced without losing information, but the statement would actually be enhanced by such compression since it will no longer exemplify the wrong that the policy is supposed to correct. Here compression can streamline the phrase and minimize the embarassment while improving the English style. Compression in general is intended to provide efficient representations of data while preserving the essential information contained in the data.

This book is devoted to the theory and practice of signal compression, i.e., data compression applied to *signals* such as speech, audio, images, and video signals (excluding other data types such as financial data or general-purpose computer data). The emphasis is on the conversion of analog waveforms into efficient digital representations and on the compression of digital information into the fewest possible bits. Both operations should yield the highest possible reconstruction fidelity subject to constraints on the bit rate and implementation complexity.

The conversion of signals into such efficient digital representations has several goals:

- to minimize the communication capacity required for transmission of high quality signals such as speech and images or, equivalently, to get the best possible fidelity over an available digital communication channel,

- to minimize the storage capacity required for saving such information in fast storage media and in archival data bases or, equivalently, to get the best possible quality for the largest amount of information stored in a given medium,

- to provide the simplest possible accurate descriptions of a signal so as to minimize the subsequent complexity of signal processing algorithms such as classification, transformation, and encryption.

In addition to these common goals of communication, storage, and signal processing systems, efficient coding of both analog and digital information is intimately connected to a variety of other fields including pattern recognition, image classification, speech recognition, cluster analysis, regression, and decision tree design. Thus techniques from each field can often be extended to another and combined signal processing operations can take advantage of the similar algorithm structures and designs.

During the late 1940s and the 1950s, Claude Shannon developed a theory of source coding in order to quantify the optimal achievable performance trade-offs in analog-to-digital (A/D) conversion and data compression systems. The theory made precise the best possible tradeoffs between bit rates and reproduction quality for certain idealized communication systems and it provided suggestions of good coding structures. Unfortunately, however, it did not provide explicit design techniques for coding systems and the performance bounds were of dubious relevance to real world data such as speech and images. On the other hand, two fundamental ideas in Shannon's original work did lead to a variety of coder design techniques over time. The first idea was that purely digital signals could be compressed by assigning shorter codewords to more probable signals and that the maximum achievable compression could be determined from a statistical description of the signal. This led to the idea of noiseless or lossless coding, which for reasons we shall see is often called entropy coding.

The second idea was that coding systems can perform better if they operate on vectors or groups of symbols (such as speech samples or pixels in images) rather than on individual symbols or samples. Although the first idea led rapidly to a variety of specialized coder design techniques,the second idea of coding vectors took many years before yielding useful coding schemes. In the meantime a variety of effective coding systems for analog-to-digital conversion and data compression were developed that performed the essential conversion operation on scalars, although they often indirectly coded vectors by peforming preprocessing such as prediction or transforming on the signal before the scalar quantization. In the 1980s vector coding or *vector quantization* has come of age and made an impact on the technology of signal compression. Several commercial products for speech and video coding have emerged which are based on vector coding ideas.

This book emphasizes the vector coding techniques first described by Shannon, but only developed and applied during the past twelve years.

To accomplish this in a self-contained fashion, however, it is necessary to first provide several prerequisites and it is useful to develop the traditional scalar coding techniques with the benefit of hindsight and from a unified viewpoint. For these reasons this book is divided into three parts. Part I provides a survey of the prerequisite theory and notation. Although all of the results are well known, much of the presentation and several of the proofs are new and are matched to their later applications. Part II provides a detailed development of the fundamentals of traditional scalar quantization techniques. The development is purposefully designed to facilitate the later development of vector quantization techniques. Part III forms the heart of the book. It is the first published in-depth development in book form of the basic principles of vector quantizers together with a description of a wide variety of coding structures, design algorithms, and applications.

Vector quantization is simply the coding structure developed by Shannon in his theoretical development of source coding with a fidelity criterion. Conceptually it is an extension of the simple scalar quantizers of Part II to multidimensional spaces; that is, a vector quantizer operates on vectors intsead of scalars. Shannon called vector quantizers "block source codes with a fidelity criterion" and they have also been called "block quantizers." Much of the material contained in Part III is relatively recent in origin. The development of useful design algorithms and coding structures began in the late 1970s and interest in vector quantization expanded rapidly in the 1980s. Prior to that time digital signal processing circuitry was not fast enough and the memories were not large enough to use vector coding techniques in real time and there was little interest in design algorithms for such codes. The rapid advance in digital signal processor chips in the past decade made possible low cost implementations of such algorithms that would have been totally infeasible in the 1970s.

During the past ten years, vector quantization has proved a valuable coding technique in a variety of applications, especially in voice and image coding. This is because of its simple structure, its ability to trade ever cheaper memory for often expensive computation, and the often serendipitous structural properties of the codes designed by iterative clustering algorithms. As an example of the desirable structural properties of vector quantizers, suitably designed tree-structured codes are nested and are naturally optimized for progressive transmission applications where one progressively improves a signal (such as an image) as more bits arrive. Another example is the ability of clustering algorithms used to design vector quantizers to enhance certain features of the original signal such as small tumors in a medical image.

In many applications the traditional scalar techniques remain dominant and likely will remain so, but their vector extensions are finding a steadily increasing niche in signal compression and other signal processing applications.

This book grew out of the authors' long standing interests in a wide range of theoretical and practical problems in analog-to-digital conversion and data compression. Our common interest in vector quantization and our cooperation and competition date from the late 1970s and continue to the present. Combined with our pedagogical interests, writing this book has been a common goal and chore for years. Other responsibilities and a constant desire to add more new material made progress slower than we would have liked. Many compromises have finally led to a completed book that includes most of the originally intended contents and provides a useful reference and teaching text, but it is less than the perfect treatment of our fantasies. Many interesting and useful techniques omitted here have emerged in the ever widening research literature of vector quantization. We apologize to the authors of such works for this omission. It was not possible to do justice to the entire research literature in the field and the book has already reached a length that stretches the ability of Kluwer Academic Publishers to be able to produce this volume at a reasonable price.

We cannot claim to be the first to write a book devoted to signal compression, That honor goes to the excellent text by Jayant and Noll, *Digital Coding of Waveforms*[195]. Their book and ours have little in common, however, except for the common goal of analyzing and designing A/D conversion and signal compression systems. Our emphasis is far more on vector quantization than was theirs (although they had one of the first basic treatments of vector quantization published in book form). We spend more time on the underlying fundamentals and basic properties and far more time on the rich variety of techniques for vector quantization. Much has happened since 1984 when their book was published and we have tried to describe the more important variations and extensions. While the two books overlap somewhat in their treatment of scalar quantization, our treatment is designed to emphasize all the ideas necessary for the subsequent extensions to vector quantizers. We do not develop traditional techniques in the detail of Jayant and Noll as we reserve the space for the far more detailed development of vector quantization. For example, they provide an extensive coverage of ADPCM, which we only briefly mention in presenting the basic concepts of DPCM. We do not, however, skimp on the fundamentals. Our treatment of entropy coding differs from Jayant and Noll in that we provide more of the underlying theory and treat arithmetic codes and Ziv-Lempel codes as well as the standard Huffman codes. This is not, however, a text devoted to entropy coding and the reader is referred to the cited references for further details.

Synopsis

Part I

Part I contains the underlying theory required to explain and derive the coding algorithms. Chapter 1 introduces the topic, the history, and further discusses the goals of the book. Chapter 2 provides the basic stochastic and linear systems background. Chapter 3 treats sampling, the conversion of a continuous time waveform into a discrete time waveform or sequence of sample values. Sampling is the first step in analog-to-digital conversion. Chapter 3 treats both the traditional one-dimensional sampling of a waveform and two-dimensional sampling used to convert two-dimensional image intensity rasters into a rectangular array of image pixels. Chapter 4 presents the basics of prediction theory with an emphasis on linear prediction. Prediction forms an essential component of many coding algorithms and the basic theory of prediction provides a guide to the design of such coding methods.

Part II

The traditional scalar coding techniques are developed in Part II. Chapters 5 through 8 treat analog-to-digital conversion techniques that perform the essential coding operation on individual symbols using simple scalar quantization and Chapter 9 treats entropy coding and its combination with quantization. Chapters 5 and 6 focus on direct coding of scalars by quantization, in Chapter 7 prediction is used prior to quantization, and in Chapter 8 linear transformations on the data are taken before quantization. Chapter 5 treats the basics of simple scalar quantization: the performance characteristics and common high resolution approximations developed by Bennett. Chapter 6 describes the optimality properties of simple quantizers, the structure of high-resolution optimal quantizers, and the basic design algorithm used throughout the book to design codebooks, the algorithm developed by Stuart Lloyd of Bell Laboratories in the mid 1950s. Chapters 7 and 8 build on scalar quantizers by operating on the signal before quantization so as to make the quantization more efficient. Such pre-processing is intended to remove some of the redundancy in the signal, to reduce the signal variance, or to concentrate the signal energy. All of these properties can result in better performance for a given bit rate and complexity if properly used. Chapter 7 concentrates on predictive quantization wherein a linear prediction based on past reconstructed values is removed from the signal and the resulting prediction residual is quantized. In Chapter 8 vectors or blocks of input symbols are transformed by a simple linear and orthogonal transform and the resulting transform coefficients are quantized. The issues of the optimal transform and bit allocation among the scalar quantizers are treated. The discrete-cosine transform is briefly covered and the basic concepts and performance capability of sub-band coding are presented briefly.

Part III

Part III is a detailed exposition of vector quantization fundamentals, design algorithms, and applications. Chapters 10 and 11 extend the fundamentals of scalar quantization of Chapters 5 and 6 to vectors. Chapter 10 provides the motivation, definitions, properies, structures, and figures of merit of vector quantization. Chapter 11 develops the basic optimality properties for vector quantizers and extends the Lloyd clustering algorithm to vectors. A variety of design examples to random processes, speech waveforms, speech models, and images are described and pursued through the subsequent chapters. Chapter 12 considers the shortcomings in terms of complexity and memory of simple memoryless, unconstrained vector quantizers and provides a variety of constrained coding schemes that provide reduced complexity and better performance in trade for a tolerable loss of optimality. Included are tree-structured vector quantization (TSVQ), classified vector quantizers, transform vector quantizers, product codes such as gain/shape and mean-residual vector quantizers, and multistage vector quantizers. Also covered are fast search algorithms for codebook searching nonlinear interpolative coding, and hierarchical coding.

Chapters 13 and 14 consider vector quantizers with memory, sometimes called recursive vector quantizers or feedback vector quantizers. Chapter 13 treats the extension of predictive quantization to vectors, predictive vector quantization (PVQ). Here vector predictors are used to form a prediction residual of the original input vector and the resulting residual is quantized. This chapter builds on the linear prediction theory of Chapter 4 and develops some vector extensions for more sophisticated systems. Chapter 14 treats finite-state vector quantization (FSVQ) wherein the encoder and decoder are finite-state machines. Like a predictive VQ, a finite-state VQ uses the past to implicitly predict the future and use a codebook matched to the likely behavior. Unlike a predictive VQ, a finite-state VQ is limited to only a finite number of possible codebooks. Design algorithms and examples are provided for both coding classes.

Chapter 15 is devoted to tree and trellis encoding systems. These systems have decoders like those of predictive and finite-state vector quantizers, but the encoders are allowed to "look ahead" into the future before making their decision as to which bits to send. At the cost of additional delay, such coding methods can provide improved performance by effectively increasing the input vector size while keeping complexity managable. Variations on the design algorithms of Chapters 13 and 14 which are matched to such look-ahead coders are considered.

Chapter 16 treats adaptive vector quantizers wherein the codebooks are allowed to change in a slow manner relative to the incoming data rate so as to better track local statistical variations. Both forward and backward adaptation are treated and simple gain and mean adaptive systems are described. More complicated adaptive coding techniques such as residual excited linear prediction (RELP) and code excited linear prediction (CELP) are also described.

Chapter 17 is devoted to variable-rate coding, vector quantizers that can use more bits for active signals and fewer for less active signals while preserving an overall average bit rate. Such coding systems can provide a significantly better tradeoff between bit rate and average distortion, but they can be more complex and can require buffering if they are used in conjunction with fixed rate communication links. The performance improvement often merits any such increase in complexity, however, and the complexity may in fact be reduced in applications that are inherently variable rate such as storage channels and communication networks. Much of Chapter 17 consists of taking advantage of the similarities of variable-rate vector quantizers and decision trees for statistical pattern classification in order to develop coder design algorithms for unbalanced tree-structured vector quantizers. Methods of growing and pruning such tree-structured coders are detailed. As vector quantizers can be used in conjunction with entropy coding to obtain even further compression at the expense of the added complication and the necessity of variable-rate coding, the design of vector quantizers specifically for such application is considered. Such entropy-constrained vector quantizers are seen to provide excellent compression if one is willing to pay the price. The techniques for designing variable-rate vector quantizers are shown to provide a simple and exact solution to the bit allocation problem introduced in Chapter 8 and important for a variety of vector quantizer structures, including classified and transform vector quantizers.

Instructional Use

This book is intended both as a reference text and for use in a graduate Electrical Engineering course on quantization and signal compression. Its self-contained development of prerequisites, traditional techniques, and vector quantization together with its extensive citations of the literature make the book useful for a general and thorough introduction to the field or for occasional searches for descriptions of a particular technique or the relative merits of different approaches.

Both authors (and several of our colleagues at other universities) have used the book in manuscript form as a course text for a one quarter course in quantization and signal compression. Typically in these courses much of Part I is not taught, but left as a reference source for the assumed prerequisites of linear systems, Fourier techniques, probability and random processes. Topics such as sampling and prediction are treated in varying degree depending on the particular prerequisites assumed at the different schools. Individual topics from Part I can be covered as needed or left for the student to review if his/her background is deficient. For example, the basics of linear prediction of one vector given another is treated in Chapter 4 and can be reviewed during the treatment of predictive vector quantization in Chapter 13. Part II is usually covered in the classroom at a reasonably fast pace. The basic development of scalar quantization is used as background to vector quantization. Predictive quantization and delta modulation are summarized but not treated in great detail. Chapter 8 is presented at a slower pace because the now classical development of transform coding, sub-band coding, and bit allocation should be part of any compression engineers toolbox. Similarly Chapter 9 is treated in some detail because of the important role played by entropy coding in modern (especially standardized) data compression systems. The coverage is not deep, but it is a careful survey of the fundamentals and most important noiseless coding methods.

Part III is typically the primary focus of the course. Chapters 10 through 12 are usually covered carefully and completely, as are the basic techniques of Chapter 13. The time and effort spent on the final three chapters depend on taste. Chapter 15 is typically only briefly covered as a variation on the techniques of Chapters 13 and 14. Chapter 14 can be summarized briefly by skipping the finite-state machine details such as the differences between labeled-state and labeled-transition systems. Finite state vector quantizers have been successfully applied in image coding and the open-loop/closed-loop design approach of the "omniscient" system is a useful design approach for feedback systems. Chapter 16 can be covered superficially or in depth, depending on the available time and interest.

Chapter 17 treats some of the newest and best vector quantizers from the viewpoint of performance/complexity tradeoffs and provides the only description in book form of the application of classification tree design techniques to the design of vector quantizers. This material is particularly useful if the intended audience is interested in classification and pattern recognition as well as data compression.

Each chapter provides at least a few problems for students to work through as a valuable aid to fully digesting the ideas of the chapter. The chapters having the most extensive set of problems are usually those chapters that are covered in depth in academic courses. Almost all of the problems in the book have appeared in homework assignments at the authors' universities.

Acknowledgements

While so many researchers have over the years contributed to the theory and techniques covered in this book, the name of one pioneer continually arises in this book, Stuart P. Lloyd. His now classical paper on optimal scalar quantization, which first appeared as an internal memorandum at AT&T Bell Laboratories in 1956 and was subsequently published in 1982 [219] contained the key algorithm that is the basis of so many of the design techniques treated in this book. It is a pleasure to acknowledge his significant and pervasive contribution to the discipline of vector quantization.

Numerous collegues, students, and former students at University of California, Santa Barbara and Stanford University and elsewhere have performed many of the simulations, provided many of the programs and figures, and given us a wealth of comments and corrections through the many revisions of this book during the past several years. Particular thanks go to Huseyin Abut, Barry Andrews, Ender Ayanoğlu, Rich Baker, Andrès Buzo, Geoffrey Chan, Pao Chi Chang, Phil Chou, Pamela Cosman, Vladimir Cuperman, Yariv Ephraim, Vedat Eyuboğlu, Dave Forney, Bob Gallager, Jerry Gibson, B. Gopinath, Smita Gupta, Amanda Heaton, Maurizio Longo, Tom Lookabaugh, Nader Moayeri, Nicolas Moreau, Dave Neuhoff, Mari Ostendorf, Erdal Paksoy, Antonio Petraglia, Eve Riskin, Mike Sabin, Debasis Sengupta, Ogie Shentov, Jacques Vaisey, Shihua Wang, Yao Wang, Siu-Wai Wu, and Ken Zeger, who also provided the program used to generate the Voronoi figure on the cover. We also acknowledge the financial support of several government agencies and industries for the research that led to many of design algorithms, applications, and theoretical results reported here. In particular we would like to thank the Air Force Office of Scientific Research, the National Science Foundation, the National Cancer Institute of the National Institutes of Health, the National Aeronautics and Space Administration, the California MICRO Program, Bell Communications Research, Inc., Bell-Northern Research Ltd, Compression Labs, Inc., Eastman Kodak Company, ERL, Inc., Rockwell International, and the Stanford University Information Systems Laboratory Industrial Affiliates Program. Thanks also to Blue Sky Research of Portland, Oregon, for their help in providing a beta release of their Textures^{TM} implementation of LaTeX for the Apple Macintosh computer. Of course, preparation of this book would not have been possible without the academic environment of our institutions, the University of California, Santa Barbara, and Stanford University.

Allen Gersho   Robert M. Gray
Goleta, California   La Honda, California

VECTOR QUANTIZATION AND SIGNAL COMPRESSION

Chapter 1

Introduction

1.1 Signals, Coding, and Compression

The purpose of this book is to provide the reader with a clear understanding of the principles and techniques underlying the large variety of systems, either already in use or yet to be designed, that transform signals such as speech, images, or video into a compressed digital representation for efficient transmission or storage. Before embarking on this venture, it is appropriate to first introduce and clarify the basic notions of a signal and signal coding or compression.

Basic Terminology

The word *signal* usually refers to a continuous time, continuous amplitude waveform. We will view a signal in its more general sense as a function of time, where time may be continuous or discrete and where the amplitude or values of the function may be continuous or discrete and may be scalar or vector-valued. Thus by a *signal* we will mean a sequence or a waveform whose value at any time is a real number or real vector. Sometimes a *signal* also refers to a picture or an image which has an amplitude that depends on two spatial coordinates instead of one time variable; or it can also refer to a moving image where the amplitude is a function of two spatial variables and a time variable. The word *data* is sometimes used as a synonym for signal, but more often it refers to a sequence of numbers or, more generally, vectors. Thus data can often be viewed as a discrete time signal. During recent years, however, the word *data* has been increasingly associated in the popular literature with the discrete or digital case, that is, with discrete time and discrete amplitude signals. We shall usually use *data* in its original sense, but will point out those occasions where the

1

narrower sense is assumed. We shall adopt the term *signal* to describe the input to the systems considered in this book and by a *source* or *information source* we mean the mechanism or device which produces the signals. The nature of the source will not be important; it could be the acoustic output of musical instruments or of the human vocal tract, the measuring device connected to a thermocouple producing electrical waveforms, the output of a microphone, a modem, a terminal, a computer, a video camera, and so on.

A signal may have continuous time (be a waveform) or discrete time (be data). These two possibilities are intimately related since a common source of discrete time signals is the sampling of continuous time signals, that is, a sequence of numbers or vectors which are values or samples of a waveform at sequence of times. An example is sampled human ·speech, formed by first converting the acoustic wave into a continuous electrical waveform and then sampling the resulting waveform to form a sequence of real numbers. A more complicated vector data sequence is a sequence of two dimensional image rasters, where each vector consists of a rectangular array of nonnegative real numbers giving the intensity of the image at each point or pixel in the raster. This example can be considered as two-dimensional sampling of a continuous index two-dimensional image signal.

A signal may also have a continuous or discrete amplitude. If the signal has both continuous time and continuous amplitude, it is called a continuous signal or an *analog* signal. The speech waveform exemplifies a continuous or analog signal. If the signal has both discrete time and discrete amplitude, then we say that it is a discrete signal or a *digital signal* (or is data in the narrow sense). An example of discrete or digital data is a sequence of ASCII (American Standard Code for Information Interchange) characters used to represent the English alphabet, digits, punctuation, and a few control codes. The 128 ASCII characters can be represented by seven dimensional binary vectors, or an eight dimensional binary vector (one byte) when the extra binary symbol is included as a parity check for errors in the remaining seven symbols. Obviously, one could also use three dimensional octal or decimal vectors. Transforming the ASCII sequence into one of these alternative representations is an example of *coding* data. Note that if we represent an ASCII data stream as a sequence of binary vectors, then the resulting data stream can be considered either as a sequence of binary vectors (looking at the data one vector at a time) or as a sequence of binary scalars (looking at the data one symbol at a time). Hence whether data consists of scalars or vectors is often a matter of interpretation.

Signals must often be communicated over a digital communication channel, such as a modem or packet network, or stored in a digital channel, such as a standard computer memory or an audio compact disc (CD). It may be, however, that either the original signal is analog or that it is digital but not in an appropriate form for the transmission or storage in the given channel, e.g., octal data and a binary channel, an analog speech signal and a digital channel, or a 64 kb/s digitized speech signal and a 9.6 kb/s channel. In this case the original signal must be transformed or *coded* into a form suitable for the channel. This general operation will be referred to as *signal coding*, although other names will be used for different special cases. In the cases where the data rate must be reduced for transmission, the operation is often called *signal compression*.

A specific coding scheme, a rule or a mapping that specifies how source symbols or groups of such symbols are transformed into or represented by a new set of symbols is called a *code* or a *coding system*. The word *code* is more commonly used by information theorists and is usually associated with a specific table, a dictionary listing the pairs of input/output words or symbols. The term *coding system* is more often used by engineers who are designing practical coding algorithms or hardware implementations that typically involve a complicated interconnection of building-blocks.

Information Theory and Vector Quantization

In the literature of information theory and communication theory, signal coding is called *source coding*, a terminology due originally to Shannon in his classic development of information theory, the mathematical theory of communication [289] [290]. Shannon originated the distinction between source coding, the conversion of arbitrary signals into an efficient digital representation, and *channel coding* or error control coding, the coding of signals so as to permit reliable communication over noisy channels.

In his development of information theory, Shannon also provided theoretical foundations for two other issues of particular importance to signal coding. First, he showed that one could construct nearly optimal communication systems by separately designing source codes for the source and error correcting codes for the channel. Hence a book such as this one that focuses only on the source or signal coding aspects, does not inherently mean a loss of generality. An effective overall system can always be constructed by cascading a good signal coding system (as described herein) with a good error control system (such as those described in Blahut [34], Weldon and Peterson [328], or Lin [214].) While such a separation of coding functions may not necessarily provide the best overall coding system from the point of view of minimum complexity, at least one can be assured that their is no fundamental obstacle prohibiting nearly optimal performance in this fashion. In fact, most practical communication systems today are based on this separation. One important example of this is the U.S. Telephone Industry

Association's IS-54 standard for digital cellular telephones where quite sophisticated source coding and channel coding were separately designed and cascaded together into a low bit-rate digital voice transmission system for a channel with very high rate of transmission errors.

Shannon's second relevant idea was the development of a very particular kind of source coding system: a block source code which maps consecutive, nonoverlapping segments (or blocks or vectors) of the input signal into consecutive, nonoverlapping blocks of channel symbols (usually assumed to be binary vectors). The mapping does not depend on previous actions of the encoder or decoder (i.e., it is memoryless with respect to past signal vectors) and follows a particular rule. The decoder for the block source code is assumed to also map blocks (now binary) into blocks or vectors of reproductions of the original input block.

Shannon assumed the existence of a distortion measure quantifying the (lack of) quality between a given input and reproduction. As Shannon's optimality criterion was the minimization of the average distortion subject to a rate constraint on the code, he required that the source encoder operate in a minimum distortion or nearest-neighbor fashion, that is, given an input block or vector, the encoder must select the binary codeword which when decoded yields the reproduction that has the minimum distortion with the input with respect to all possible reproductions. Shannon called such a block code a *source code subject to a fidelity criterion*, but a code of this type can also be called a *vector quantizer* and the operation of this code can be called *vector quantization* since it is a natural generalization to vectors of a simple quantizer: It is a mapping of real vectors (an ordered set of signal samples) into binary vectors using a minimum distortion rule.

The theory has largely developed under the flag of source coding subject to a fidelity criterion, but during recent years most of the work on actual coding system design and implementation has opted for the shorter and more practical sounding name of *vector quantization*. In fact, the transition from the terminology of "block source code" to the name "vector quantization" coincides with the transition from nonconstructive theoretical studies and the constructive design of algorithms and hardware for compression of real signals. As this book emphasizes the design rather than the theory, we have chosen to follow the trend and include vector quantization in the title. We use it in a more general sense, however, than that of the original Shannon formulation. Like Shannon, virtually all of the vector quantizers considered here will have a minimum distortion mapping as the heart of the encoder. Unlike Shannon, however, we will not always require that the coders map consecutive blocks in an independent fashion. In other words, the encoder and decoder may have memory as do predictive codes and finite state codes. Since the subject of signal coding is based on many established techniques that are not modeled on Shannon's block source code, we view vector quantization as a subset, albeit an important subset, of the study of signal coding or compression. For this reason, the first part of this book provides the foundation of established signal compression methods before introducing the ideas of vector quantization.

Shannon theory shows that the optimal achievable performance in the sense of minimizing average distortion for a constrained rate can be achieved in many communication systems with memoryless block codes. The theory does not, however, consider the issue of implementation complexity which can be a critical issue in most applications. Also not considered in the theory is coding delay, the delay incurred by the signal by the process of encoding and decoding, which is also a critical factor in some applications. The theory does not negate the well known fact that better performance can often be achieved for a given rate (or comparable performance at a lower rate) and a given complexity limit by using coding systems with memory. For example, delta modulation can provide superior performance to ordinary scalar quantization or PCM (pulse coded modulation) for certain sources.

Lossy Coding

If the original signal is analog, signal coding necessarily results in a loss of quality because an arbitrary continuous signal cannot be represented by discrete data in a recoverable manner. If the original signal is discrete, coding may or may not represent the signal in an invertible fashion, depending on the alphabets of the original and coded signal. For example, if a data source produces 300 ASCII characters per second, then we could code this data stream into a binary data stream provided the binary stream contained at least $7 \times 300 = 2100$ bits per second (b/s), e.g., we could use a standard 2400 b/s modem. Without some form of clever processing, however, we could not use a binary sequence of any fewer symbols per second without making errors.

These simple examples point out a principal goal of signal compression: the replacement of signals by a digital representation (discrete time and discrete amplitude) of a given transmission rate or storage capacity in such a manner that the original signal can be recovered as accurately as possible from the digital representation. In most of this book the focus will be on *lossy compression* where the original signal cannot be precisely recovered from the compressed representation.

In general, if the signal coding system includes the conversion from analog to digital, the scheme is inevitably lossy. By suitable design, the loss as indicated by a fidelity or distortion measure, can be suitably controlled and is not a deterrent to the many advantages of digital representation of signals. In addition to conversion into a form suitable for transmission or storage in a digital medium, signal compression has other uses. Digital information can be more reliably communicated over noisy channels than analog information thanks to techniques from detection (such as matched filters and correlation receivers) and error correcting coding (such as simple parity check codes). Hence if overall performance is the goal, it may be better to suffer some controlled distortion in signal coding in order to ensure very accurate communication of the digitized sequence. Digitizing signals permits signal regeneration in digital repeaters, allowing long-haul communication with negligible line losses. Digital data are easier to encrypt in a secure fashion than are analog data. Finally, most modern signal processing chips require digital inputs.

There are a variety of special cases of signal coding which have developed more or less independently in practice although they are closely related in theory. As the terminology used is not entirely standard, we introduce the most important special cases together with some of the names encountered in the literature.

Noiseless Coding

If the original signal is digital and can be perfectly reconstructed from the coded signal, then the signal coding is called *noiseless coding* or *lossless coding* or, for reasons to be explored in detail later, *entropy coding*. Noiseless coding results, for example, if we code an ASCII sequence into 7 bit binary vectors. Noiseless coding is often required in some systems, e.g., in coding binary computer programs for storage or transmission: A single bit in error can have disastrous consequences. If the noiseless coding results in a digital sequence with a smaller communications rate or storage rate than the original signal, then the noiseless coding is called *noiseless data compression*. The word *noiseless* is sometimes dropped (especially in the computer literature), but we shall retain it to avoid confusion with the more general meaning of data compression which does not imply perfect reconstruction. Noiseless data compression is also referred to as *data compaction*. The goal in noiseless data compression (and the reason for its name) is to minimize the communication or storage capacity required for the coded data. An example of noiseless data compression of the ASCII code is the Morse code representation (in binary numbers instead of dots and dashes) for the letters instead of arbitrary dimension seven binary vectors, that is, use short vectors for more likely letters and long vectors for unlikely ones. On the average fewer binary symbols will be needed in this case and the necessary communication or storage capacity will be reduced.

Noiseless coding is, of course, what every customer would like since it provides a flawless reproduction of the original. Unfortunately, it is often impossible because of rate constraints in communication or storage channels. For example, one can never perfectly communicate an arbitrary real number (say π, for example) with only a finite number of bits. If the number of bits per second or per sample available in a communication link or storage medium is smaller than the number generated by the information source (infinite in the case of analog data), then simple translation of symbols cannot yield noiseless codes.

Analog-to-Digital Conversion and Signal Compression

In the best of all possible worlds, one would always like to recreate perfectly the original signal and hence have a noiseless code, but we have seen that this is not possible if the original source produces analog or continuous data. Since continuous data cannot be perfectly recovered from a discrete representation, a general goal is to minimize the loss of fidelity as much as possible within the implementation cost constraints, that is, to be able to recreate a reproduction of the original signal from the digital representation that is as faithful a reproduction as possible. Such conversion from continuous or analog signals to digital representations is called either *analog-to-digital conversion* or *data compression*, depending on the transmission or storage rate or bandwidth of the digital representation. While there is no hard distinction, signal coding is usually called *data compression* if the rate in bits per second of the digital representation or the bandwidth of a modulated version of the digital representation is smaller than the bandwidth of the original signal. It is called *analog-to-digital conversion* or A/D conversion or ADC if the input is analog and the rate or bandwidth of the digital representation is larger than that of the original signal. In our view, the terminology *signal compression* is a better description than *data compression* since the original signal may have continuous time, although the term *data compression* has been commonly used, likely since most data compression systems for analog signals first form a data sequence by sampling the signal and then perform the compression.

Mathematically the conversion of analog data to a digital sequence can always be viewed as "compression" in the sense that data drawn from a continuous or large finite alphabet or set are replaced by a sequence of data drawn from a relatively small finite alphabet. That is, the number of *bits* or binary digits that must be communicated or stored in order to reproduce the original data has been reduced from a large or possibly infinite number to a relatively small number. Data compression in this general sense is not always equivalent to bandwidth compression, however. If, for example,

analog speech is encoded via pulse code modulation (PCM) which first
samples and then quantizes the samples, then the bandwidth required is
increased from approximately 4 KHz to over 60 KHz, far too large to fit
on an ordinary telephone channel.   Here the mathematical compression
of an infinity of bits into a finite number does not intuitively correspond
to *compression* since the bandwidth required to communicate the coded
signal has greatly increased. If, however, the resulting digital data stream
is further compressed (as it often is), that compression will usually result in
overall bandwidth reduction since in most digital signaling schemes fewer
bits per second translates directly into fewer Hertz.

Digital Signal Compression

Signal coding that is data compression but is neither noiseless nor A/D
conversion is also common. For example, a standard speech digitizer might
first convert human speech into a high rate digital sequence by sampling at
8000 Hz and then quantizing each sample with 12 bits, yielding a data sequence of 96,000 b/s. If this digital sequence is to be communicated over a
standard 9600 b/s modem, it is clear that a further compression is required:
one must reduce the number of bits by 10 to 1! Such two-step compression
is common since the first step of high rate A/D conversion can be accomplished with very fast, simple, and small devices and the compression to low
rate can then be done entirely digitally. Another example of compression
of digital data is rounding: reducing the precision of finite precision real
numbers.

1.2   Optimality

There are two principal design goals for any system for communication or
storage:

- the performance should be as good as possible, and

- the system should be as simple (or cheap) as possible.

The first goal has always been the easiest to handle mathematically and it
has gained in importance relative to the second goal as modern technology
has made increasingly sophisticated algorithms amenable to straightforward
implementation. The two goals are clearly at odds since performance is
usually obtainable only at the expense of additional system complexity and
cost.

   The performance of a signal compression system is generally assessed
by some measure of the quality or fidelity of reproduction achieved (or

equivalently by the degree of degradation or distortion incurred) at a given bit rate. Often the rate is constrained by a particular application and the design goal is to achieve some quality objective while maintaining a tolerable complexity. For the same signal source, there can be a wide range of quality levels that are acceptable for different applications. Thus, for example, speech coding systems have been designed for rates that range from as low as 300 b/s to as high as 32 kb/s. Generally the complexity increases as the rate is lowered for a particular source.

The theory of source coding attempts to describe the structure and performance of systems which are *optimal* in the sense that they maximize the performance for a given rate or minimize the rate for a required performance. The theory generally does not address the important issue of complexity, but in this book we pay particular attention to the issue of complexity. Vector quantization in particular is a technique that can quickly lead to astronomic complexity if we ignore this practical constraint. Instead we devote much attention to alternatives and structural constraints. There are many ways of measuring complexity that are important in particular applications (e.g., the number of multiplies in an encoder corresponds to more time if done in software and more area if done in hardware), usually some simple measure of the number of operations. For example the number of multiplies is often taken as an indicator of complexity.

A natural measure of performance in a signal coding system is a quantitative measure of distortion or, alternatively, fidelity. Hence signal coding systems are considered optimal if they minimize an average or maximum distortion subject to some constraints on coder structure. Alternatively, if a specific level of performance is required, the design goal might be to provide the simplest or cheapest possible system with the given constraints having such performance. Generally, we focus on minimizing average distortion subject to some constraints in attempting to optimize a particular coding structure. The constraints on the coding system usually are related to how complicated it is to implement. While theoretically optimal performance is rarely critical for practical design, it is obviously desirable as a general goal to provide the best performance possible with reasonable effort subject to the given design constraints. One has to trade off the complexity of a system with the resulting performance, balancing cost and fidelity.

For many applications, the ultimate measure of fidelity is determined by the perceived quality by the human viewer or listener. Thus, often we try to allow more general distortion measures into the design methodology that might embody some perceptually meaningful characteristics into the objective performance assessment. This is generally quite difficult and in some case we have to rely on a simple but mathematically tractable measure, the most common of which is the mean squared error.

This book is devoted to the development and analysis of design algorithms for signal coding schemes. Many of the basic coding structures along with the optimal performance bounds that are occasionally quoted for comparison are derived from the abstract theory of such systems, especially from Shannon's rate-distortion theory [289] [290] and the high-resolution quantizer theory of Bennett, Widrow, and others [30] [332] [333]. Such theoretical bounds serve two functions: First, an unbeatable performance bound provides an absolute yardstick for comparison with real data compression systems. If the optimal performance is not good enough to meet system or customer requirements, then no effort should be wasted trying to design a real system with the given parameters since it too will be unacceptable. Either the constraints must be changed or inferior performance accepted. Conversely, if a real system yields performance close to the theoretical optimum, then perhaps no money or time should be invested in trying to obtain the little improvements that are possible. (If the system has performance that is better than the optimum, then one should definitely suspect the accuracy of the mathematical model used to derive the optimum!) The second use of such results is indicated by the word *achievable*: In principle there must exist codes that do nearly as well as the performance indicated by the theoretical optimal achievable performance. Thus, for example, if a given system yields performance much worse than the theoretical optimum, then perhaps improvements can be obtained with more clever coding techniques. This application, however, has long proved the more difficult of the two. While some simple special cases of the high-resolution quantizer bounds are developed in the first two sections, the general and complete development is not presented in this text for reasons of emphasis and space. A survey of source coding theory intended to provide theoretical support for this text may be found in Gray [159] and treatments of rate-distortion theory for block codes or vector quantizers may be found in Berger [32] and Gallager [134].

Why Signal Compression?

Since a major goal of data compression is to minimize the communication or storage rate and hence the bandwidth or storage capacity required for a given communication system, one might suspect that the need for data compression is obviated by wideband technologies such as optical fiber communication networks, packet and digital radio, and optical storage disks. While it is true in some examples that available bandwidth lessens the need for complex data compression, compression can still be important for a variety of reasons: More data channels can be multiplexed in wideband systems by using data compression to reduce the bandwidth required of each channel. Some data sources such as high definition television (HDTV) and multispectral data can generate data in such high volumes that any transmission media, including optical fiber, can be quickly overwhelmed. The emerging trend towards multi-media oriented personal computers and workstations has also led to an increasing demand to store speech, music, images, and video in compressed form on magnetic or optical disks.

Another factor that arises in communications is "facilities exhaust" where a given transmission channel gradually has its spare capacity exhausted as traffic increases with time. Once there is no residual capacity, the cost of installing an entirely new transmission facility is far greater than the cost of sophisticated terminal equipment which can perform efficient data compression and thereby greatly increase the effective capacity in a cost-effective manner.

One area where bandwidth is severely limited is in radio channels which are essential for a large range of important applications such as mobile telephones or the ultimate personal telephone carried by each user. The radio spectrum will always be a precious resource and as a result an increasing fraction of data compression applications are directed to radio communications. The widespread acceptance of cellular telephones has motivated extensive studies of both speech compression and digital modulation to increasingly squeeze more and more channels into a given bandwidth allocation. New wireless technologies, such as the long term plan for universal cordless personal telephones, are expected to lead to continuing research in speech compression for many years to come. Also, wireless image and video technologies are also of increasing interest today. In particular, terrestrial and satellite broadcasting of digital high definition television is now an exciting area of signal compression technology.

Speech recognition systems can have larger vocabularies for given memory sizes by reducing the memory required for each template. Data recording equipment can store more important information by storing low rate compressed or reduced data instead of high rate raw data. For example, a 10:1 compression can mean an order of magnitude difference in the capacity of an optical storage disk.

In general, resources such as bandwidth obey a corollary to Parkinson's Law: Resource use will expand to meet the resources available. Hence there will always be advantages to be gained from data compression in terms of efficient use of bandwidth and storage capacity, even though the cost of bandwidth and storage capacity continues to drop.

Signal compression can also play a valuable role in combination with systems for signal processing. If compression is done before some form of digital signal processing such as pattern recognition, enhancement, or smoothing, the subsequent signal processing in principle becomes simpler because of the reduced number of bits required to represent the input signals. In addition, portions of such signal processing can be incorporated into the compression system so as to ease the computational burden. Here the key issue becomes the preservation of the information necessary to the application rather than the signal-to-noise ratio of the compression system.

Finally, it is important to note the massive efforts taking place today to adopt national and international standards for signal compression. These activities certainly demonstrate the widespread recognition of the need for signal compression in many diverse applications where just a few years ago compression was not considered viable. Examples are the new CCITT standard algorithm, LD-CELP, for 16 kb/s speech coding, the JPEG (Joint Photographic Experts Group) standard for still image compression, the MPEG (Moving Pictures Experts Group) standard for audio/video compression for storage applications, the FCC evaluations of candidates for a new high definition terrestrial broadcast standard expected to be based on digital compression, and many others.

1.3   How to Use this Book

This book is intended to serve the needs of both a university text and an engineering reference. Students reading this book as a part of a course will usually have a well-defined curriculum provided by their instructor. Instructors should refer to the preface for suggestions on how to use this book in the classroom. Here, we provide some comments for the individual reader using this book as a reference.

For the independent reader, there are several approaches for learning needed aspects of signal compression and vector quantization. Engineers with a general background in coding, who may already be familiar with such topics as DPCM or transform coding, may want to proceed directly to Part III to study the subject of vector quantization. Chapters 10, 11, and 12, in that order, provide the core of this subject. The remaining chapters offer more advanced or specialized aspects of VQ that may be read in any order, although occasionally the text may refer back to prior chapters for some related details.

Readers with little or no prior exposure to signal coding may choose to begin with Part II where the fundamental concepts of quantization, predictive quantization (such as DPCM and delta modulation), transform and sub-band coding may be found. Also in Part II is a chapter on entropy coding, covering the main ideas of all the important techniques for noiseless coding, often included as a building block of lossy signal compression systems.

Some readers will find it helpful to refer to Part I when they encounter unfamiliar mathematical tools while reading Parts II or III. Part I provides the basic tools of probability and random processes In addition, it covers the ideas of sampling of a continuous time signal or image and a comprehensive presentation of the theory of linear prediction which is of such widespread importance in signal compression. Of course, the more industrious reader can systematically read Part I before proceeding onward to obtain a solid review of the foundation material for the rest of the book.

1.4 Related Reading

There are several books that are useful supplements to this one. Many of the papers referred to herein may be found in the collections edited by Davisson and Gray [95], Jayant [194], Swaszek [304], and Abut [2]. Of particular relevance is the extensive compendium of reprints of published papers on vector quantization available in the IEEE reprint collection *Vector Quantization*, edited by H. Abut [2].

The first comprehensive treatment of waveform coding is the book by Jayant and Noll [195]. Their book provides an extensive coverage of popular techniques for coding of speech and images and includes many practical details of the coding methods that we cover in Part II from a conceptual viewpoint as well as detailed treatment of sampling and analog-digital conversion. Vector quantization and vector coding are treated very briefly, but there have not yet been complete treatments in book form. Tutorial surveys of vector quantization may be found in the papers by Cuperman and Gersho [139], Gray [153], Makhoul, Roucos, and Gish [229], and Nasrabadi and King [243]. O'Shaughnessy [247] has a good treatment of vector quantization applied to speech coding and Rabbanni and Jones [256] provide a nice overview of image compression using traditional methods and vector quantization with extensive examples.

An extensive treatment of an important special case of vector quantization, lattice codes, can be found in Conway and Sloane [79]. A thorough treatment of noiseless coding may be found in Storer [303].

Part I

Basic Tools

Chapter 2

Random Processes and Linear Systems

2.1   Introduction

The language of signal coding is primarily that of random processes and linear systems. Although signal coding systems are inherently nonlinear, the tools and techniques of linear systems play an important part in their analysis. In a mathematical model of a communication system in general and a signal coding system in particular, it is usually assumed that the data to be communicated (or stored, compressed, processed, ...) is produced by a random process. Where this is not the case, there would be no need to communicate the signal—it would already be known to the potential user. In addition to the original signal, the system itself may have random components such as the addition of random noise or the introduction of digital errors in a random fashion.

If the system is modeled as a random process, then the performance of a system can be measured by its average or expected behavior, e.g., by such fidelity criteria as average mean squared error, by the ratio of average signal power to average error power (signal-to-noise ratio or SNR), or by the average probability of symbol error. There are, however, two types of averages. *Expectations* or *probabilistic averages* or *statistical averages* are sums or integrals weighted by a probability mass, density, or cumulative distribution function. *Time averages* or *sample averages* are formed by summing or integrating actual sample values produced by the process and then normalizing over time. While the short term behavior of such time averages is random, in many situations the long term behavior converges to something that is not random. Expectations are most useful for the-

17

ory, e.g., for computing averages based on a random process model. One can study the overall effect of design changes on a system by computing expectations of certain key variables in a mathematical model of the system. This is useful both in analysis and optimization of such systems. In practice, however, it is the time average that is measured. It is therefore important to categorize conditions under which the two averages are the same and to develop means of handling systems for which they are not. For example, when is a computed probability of error the same as the long term relative frequency of errors?

Relating expectations and long term time averages requires an understanding of stationarity and ergodic properties of random processes, properties which are somewhat difficult to define precisely, but usually have a simple intuitive interpretation. These issues are not simply of concern to mathematical dilettantes. For example, stationarity can be violated by such commonly occurring phenomena as the addition of a transient to an otherwise stationary process, yet sample averages may still converge in a useful way. An even more striking example is that of human speech: there has been much argument in the literature as to whether human speech is stationary and ergodic and whether or not the theory of such processes is applicable to the design and analysis of speech coding systems. We argue that a modest understanding of those properties will show when the theory is or is not likely to apply, regardless of whether or not speech is in fact stationary or ergodic in the strict mathematical sense.

In this chapter we survey some of the key ideas and examples from the theory of random processes. The triple goals of the chapter are:

- to introduce notation used throughout the book,

- to describe the applicability of certain of the basic results of random processes, and

- to describe the most popular examples of random processes for the theory and practice of signal compression.

The reader is assumed to be familiar with the general topics of probability and random processes and the chapter is intended primarily for reference and review. A more extensive treatment of basic random processes from a similar point of view may be found, e.g., in [164].

2.2 Probability

A *probability measure* is an assignment of real numbers to subsets (called "events") of a *sample space* Ω which represents the space of possible outcomes of an experiment. Typical sample spaces include discrete spaces for modeling coin flips and die tosses, the real line $\mathcal{R} = (-\infty, \infty)$ for modeling a single outcome of a measurement, real-valued vectors, and sequences and waveforms. The probability measure must conform to certain rules or axioms to ensure a useful theory or calculus of probability. These axioms formalize the intuitive requirements that

- probabilities are nonnegative,

- the probability of the entire space Ω is 1 (a normalization), and

- the probability of the union of a collection of disjoint events is the sum of the probabilities of the individual events.

These properties are inherent in the two most common methods of computing probabilities: integration and summation. Suppose, for example, that $\Omega = \mathcal{R}$, the real line. A probability measure P in this case can be defined in terms of a real valued function f defined on $\mathcal{R}$ with the following properties:

$$f(r) \geq 0, \text{ all } r \in \mathcal{R},$$

$$\int_{-\infty}^{\infty} f(r)\,dr = 1.$$

Then the set function P defined by

$$P(F) = \int_F f(r)\,dr \qquad (2.2.1)$$

is a probability measure and the function f is called a *probability density function* or *pdf* since it is nonnegative and it is integrated to find probability.

We admit in passing that much is being swept under the rug here. This definition will work only if the integrals make sense, which will restrict the sets or events considered since it is not possible to integrate over all sets. Furthermore, to ensure that the axioms of probability are satisfied one must take the integral in the general Lebesgue sense, rather than in the Riemann sense familiar to most engineers. These difficulties will not cause problems in this text, however, and interested readers are left to pursue them on their own. When we wish to make an informal probability statement without actually specifying a probability measure P, we will often use the notation $\Pr(F)$ as an abbreviation of the English statement "the probability that event F occurs."

Some of the more common pdf's on $\mathcal{R}$ are listed below. They are given in terms of the real-valued parameters m, $b > a$, $\lambda > 0$, and $\sigma > 0$.

The uniform pdf: $f(r) = 1/(b-a)$ for $r \in [a, b] = \{r : a \leq r \leq b\}$.

The exponential pdf: $f(r) = \lambda e^{-\lambda r}$; $r \geq 0$.

The doubly exponential or Laplacian pdf: $f(r) = \frac{\lambda}{2} e^{-\lambda |r|}$; $r \in \mathcal{R}$.

The Gaussian pdf: $f(r) = (2\pi\sigma^2)^{-1/2} e^{-(r-m)^2/2\sigma^2}$; $r \in \mathcal{R}$.

Another common construction of a probability measure arises when all of the probability sits on a discrete subset of the real line (or any other sample space). Suppose that Ω consists of a finite or countably infinite collection of points. (By countably infinite we mean a set that can be put into one-to-one correspondence with the nonnegative integers, e.g., the nonnegative integers, the integers, the even integers, the rational numbers.) Suppose further that we have a function p defined for all points in Ω which has the following properties:

$$p(\omega) \geq 0 \text{ all } \omega \in \Omega.$$

$$\sum_{\omega \in \Omega} p(\omega) = 1.$$

Then the set function P defined by

$$P(F) = \sum_{\omega \in F \cap \Omega} p(\omega)$$

is a probability measure and the function p is called a *probability mass function* or *pmf* since one adds the probability masses of points to find the overall probability. That P defined as a sum over a pmf is indeed a probability measure follows from the properties of sums.

It is common practice in engineering to use the Dirac delta function in order to retain the use of the pdf for a discrete probability distribution. With this convention, it is necessary to consider the sample space Ω as the entire real line $\mathcal{R}$ although the probability mass is limited to the discrete subset Ω' of the real line. Then the pdf $f(r)$ for describing the appropriate probability measure for the real line can be written as

$$f(r) = \sum_{r_i \in \Omega'} \delta(r - r_i) p(r_i)$$

It is usually best to avoid the clumsy notation of this approach and use a pmf instead. The Dirac delta approach is primarily useful when one must handle mixed distributions containing both continuous and discrete components. Even then, it is usually avoidable.

Some of the more common pmf's are listed below. The pmf's $p(\omega)$ are specified in terms of parameters p, n and λ where p is a real number in $[0,1]$, n is a positive integer and λ is a positive real number.

The binary pmf: $\Omega = \{0, 1\}$. $p(1) = p$, $p(0) = 1 - p$.

The uniform pmf: $\Omega = \{0, 1, \cdots, n - 1\}$. $p(k) = 1/n$; $k \in \Omega$.

The geometric pmf: $\Omega = \{1, 2, \cdots\}$. $p(k) = p(1 - p)^{k-1}$, $k \in \Omega$.

The Poisson pmf: $\Omega = \{0, 1, 2, \cdots\}$. $p(k) = \lambda^k e^{-\lambda}/k!$, $k \in \Omega$.

Thus one can define probability measures on the real line by integrating pdf's or by summing pmf's.

Another probability function of interest that is defined in both the continuous and discrete case is the *cumulative distribution function* or *cdf* as follows: Given a probability measure P on the real line (or some subset thereof), define the cdf $F(r)$ by

$$F(r) = P(\{\omega : \omega \leq r\}),$$

that is, $F(r)$ is simply the probability of the set of values less than r. If P is defined by a pdf f, then

$$F(r) = \int_{-\infty}^{r} f(x)\, dx.$$

The fundamental theorem of calculus implies that

$$f(r) = \frac{dF(r)}{dr},$$

that is, the pdf is the derivative of the cdf. Alternatively, if the cdf is differentiable, then the pdf exists and the probability measure of events can be found by integrating the pdf. In the discrete case, the pmf can be found as the differences of the cdf, e.g., if Ω is the set of integers, then $p(k) = F(k) - F(k - 1)$.

The above ideas are easily extended to sample spaces consisting of real valued vectors rather than just real numbers, that is, to Euclidean vector spaces. Suppose that $\Omega = \mathcal{R}^n$, the space of all vectors $\mathbf{x} = (x_0, x_1, \cdots, x_{n-1})$ with $x_i \in \mathcal{R}$. A function f defined on $\mathcal{R}^n$ is called a *joint pdf* or an *n-dimensional pdf* or simply a *pdf* if

$$f(\mathbf{x}) \geq 0, \ \mathbf{x} \in \mathcal{R}^n,$$

$$\int_{\mathbf{x} \in \mathcal{R}^n} f(\mathbf{x})\, d\mathbf{x} = 1,$$

where the vector integral notation is shorthand for

$$\int_{x_0 \in \mathcal{R}} \int_{x_1 \in \mathcal{R}} \cdots \int_{x_{n-1} \in \mathcal{R}} f(x_0, x_1, \cdots, x_{n-1})\, dx_0\, dx_1 \cdots dx_{n-1}.$$

Then the set function P defined by the integral

$$P(G) = \int_G f(\mathbf{x})\, d\mathbf{x},$$

where $G \subset \mathcal{R}^n$, is a probability measure (subject to considerations of existence like those of the scalar case). In a similar manner we can consider pmf's for vectors.

Given a probability measure P defined on k-dimensional Euclidean space $\mathcal{R}^n$, a *joint cdf* or *n-dimensional cdf* or simply a *cdf* is defined by

$$F(\mathbf{r}) = P(\{\omega : \omega_1 \leq r_1, \omega_2 \leq r_2, \ldots, \omega_n \leq r_n\}).$$

We can completely describe a probability measure on Euclidean space by giving the cdf (or in special cases a pdf or pmf). This lets us describe two of the most common and important examples of probability measures on vector spaces: the product measure and the Gaussian measure.

Suppose that we have a a collection of cdf's $\{F_i(r), \; r \in \mathcal{R}\}$; $i = 0, \cdots, n - 1$. They may be identical or different. Define an n-dimensional cdf $F^n(\mathbf{r})$ by

$$F(\mathbf{r}) = \prod_{i=0}^{n-1} F_i(r_i).$$

(It is straightforward to verify that F is indeed a cdf.) A cdf of this form is called a *product cdf* and the resulting probability measure is called a product probability measure. If the original cdf's F_i are defined by densities f_i, then the above is equivalent to saying that the n-dimensional pdf f is given by

$$f(\mathbf{r}) = \prod_{i=0}^{n-1} f_i(r_i).$$

Similarly, if the original cdf's F_i are defined by pmf's p_i, then the above is equivalent to saying that the n-dimensional pmf p is given by

$$p(\mathbf{r}) = \prod_{i=0}^{n-1} p_i(r_i).$$

Thus any of the scalar pdf's or pmf's previously defined can be used to construct product measures on vector spaces.

The second important example is the multidimensional Gaussian pdf. Here we describe the n-dimensional pdf which in turn implies the probability measure and the cdf. A pdf f on $\mathcal{R}^n$ is said to be *Gaussian* if it has the following form:

$$f(\mathbf{x}) = \frac{e^{-\frac{1}{2}(\mathbf{x}-\mathbf{m})^t \mathbf{\Lambda}^{-1}(\mathbf{x}-\mathbf{m})}}{\sqrt{(2\pi)^n \det \mathbf{\Lambda}}},$$

where the superscript t stands for "transpose", $\mathbf{m} = (m_0, \cdots, m_{n-1})^t$ is a column vector, and $\mathbf{\Lambda}$ is an $n \times n$ square symmetric matrix, e.g., $\mathbf{\Lambda} = \{\Lambda(i,j);\ i = 0, \cdots, n-1;\ j = 0, \cdots, n-1\}$ and $\Lambda(i,j) = \Lambda(j,i)$, for all i,j. Equivalently, $\mathbf{\Lambda}^t = \mathbf{\Lambda}$. We also require that $\mathbf{\Lambda}$ be positive definite so that for any nonzero vector $\mathbf{y}$ the quadratic form $\mathbf{y}^t \mathbf{\Lambda} \mathbf{y}$ is positive; that is,

$$\mathbf{y}^t \mathbf{\Lambda} \mathbf{y} = \sum_{i=0}^{n-1} \sum_{j=0}^{n-1} y_i \Lambda_{i,j} y_j > 0.$$

A standard result of matrix theory ensures that if $\mathbf{\Lambda}$ is positive definite, then the inverse exists and the determinant is not 0 and hence the above pdf is well defined.

For reasons that will be seen later the vector $\mathbf{m}$ is called the *mean* and the matrix $\mathbf{\Lambda}$ is called the *covariance* of the distribution.

2.3 Random Variables and Vectors

A random variable can be thought of as a real-valued measurement on a sample space: an experiment produces some outcome ω and an observer then measures some function of ω, say $X(\omega)$. For example, if ω is a real vector, X could be its Euclidean length, the minimum coordinate value, the third component, and so on. The basic idea is that if ω is produced in a random fashion as described by a probability measure on the original sample space, then X should inherit a probabilistic description, thereby giving us a new probability measure which describes the possible outcomes of X.

Mathematically, a random variable X is just a mapping from the sample space into the real line: $X : \Omega \to \mathcal{R}$; that is, X assigns a real number to every point in the sample space. Calculus can be used to find a probabilistic description of X given the underlying probabilistic description P on Ω. This is the basic *derived distribution* technique of probability theory. The fundamental formula of derived distributions states that for a given event of the form $X \in F$, we can compute the probability of this "output event" in terms of the original probability as $P(\{\omega :\ X(\omega) \in F\})$. This leads to the definition of the *distribution* of the random variable P_X as the induced probability measure defined by

$$P_X(G) = P(\{\omega : X(\omega \in G)\}); \quad \text{all events } G.$$

Thus, for example, if the probability measure on the original space is described by a pdf f, then

$$P_X(G) = \int_{r:\, X(r)\in G} f(r)\, dr.$$

If the probability measure on the original space is described by a pmf p, then analogously

$$P_X(G) = \sum_{k:\, X(k)\in G} p(k).$$

If the distribution P_X is specified by a cdf F_X of the random variable X corresponding to a pmf p_X or a pdf f_X, then

$$P_X(G) = \int_{x\in G} f_X(x)\, dx$$

or

$$P_X(G) = \sum_{x\in G} p_X(x)$$

depending on whether X has a pdf or takes on discrete values.

Note that if $\Omega = \mathcal{R}$, then a trivial random variable is $X(r) = r$, the identity function. By this method all the pdf's and pmf's described for the real line can be considered as equivalent to distributions of random variables and the resulting random variables share the name of the probability functions. For example, we speak of a Gaussian random variable as one with a distribution induced by the Gaussian pdf. If the distribution P_X of a random variable X is induced by a pdf f or a pmf p, then we often subscript the probability function by the name of the random variable, i.e., f_X or p_X.

A *random vector* is simply a finite collection of random variables. For example, if X_i; $i = 0, 1, \cdots, n-1$, are random variables defined on a common sample space, then $\mathbf{X} = (X_0, \cdots, X_{n-1})^t$ is a random vector. The *distribution* $P_\mathbf{X}$ of the random vector is the induced probability measure, that is,

$$P_\mathbf{X}(G) = P(\{\omega : \mathbf{X}(\omega) = (X_0(\omega), \cdots, X_{n-1}(\omega)) \in G\}).$$

As with any probability measure, a distribution can be specified by a cdf $F_\mathbf{X}(\mathbf{x}) = \Pr(X_i \leq x_i;\ i = 0, 1, \cdots, k-1)$ and possibly by a multidimensional pdf $f_\mathbf{X}$ or a pmf $p_\mathbf{X}$. In the continuous case the pdf can be found from the cdf by differentiation:

$$f_\mathbf{X}(\mathbf{x}) = \frac{\partial^k}{\partial x_0 \partial x_1 \cdots \partial x_{k-1}} F_\mathbf{X}(\mathbf{x}).$$

As in the scalar case, we can define a trivial random vector on n-dimensional Euclidean space by $\mathbf{X}(\mathbf{r}) = \mathbf{r}$. If the probability measure is specified by a Gaussian (or other named) cdf, then the random vector is also said to be Gaussian (or other corresponding name).

A pair of events F and G are said to be *independent* if

$$P(F \cap G) = P(F)P(G).$$

This definition has an interpretation in terms of elementary conditional probability. Given two events F and G for which $P(G) > 0$, the conditional probability of F given G is defined by

$$P(F|G) = \frac{P(F \cap G)}{P(G)}.$$

Thus if two events F and G are independent and $P(G) > 0$, then $P(F|G) = P(F)$, the knowledge of G does not influence the probability of F.

A finite collection of events F_i, $i = 0, \cdots n - 1$ is said to be *independent* if for any subset of indices $i_1, i_2, \cdots, i_k$ of $0, 1, \cdots, n - 1$ we have that

$$P(\bigcap_{l=1}^{k} F_{i_l}) = \prod_{l=1}^{k} P(F_{i_l}).$$

Note that a collection of events will be independent only if all intersections of events drawn from the collection have probability equal to the product of the probabilities of the individual events. Thus if a collection of events is independent, so is any subcollection of the same events.

A collection of random variables X_i; $i = 0, \cdots, n - 1$ is said to be independent if for all collections of output events G_i; $i = 0, \cdots, n - 1$, the events $\{\omega : X_i(\omega) \in G_i\}$ are independent. The definition is equivalent to saying that the random variables are independent if and only if

$$\Pr(X_i \in G_i; i = 0, \cdots, n - 1) = \prod_{i=0}^{n-1} P_{X_i}(G_i), \qquad (2.3.1)$$

for all G_i. This is true if and only if the n-dimensional cdf for $\mathbf{X} = (X_0, \cdots, X_{n-1})$ is a product cdf. Thus a random vector has independent components if and only if the appropriate probability functions factor in a manner analogous to (2.3.1), that is, if

$$F_{\mathbf{X}}(\mathbf{x}) = \prod_{i=0}^{n-1} F_{X_i}(x_i),$$

or

$$f_{\mathbf{X}}(\mathbf{x}) = \prod_{i=0}^{n-1} f_{X_i}(x_i),$$

or

$$p_{\mathbf{X}}(\mathbf{x}) = \prod_{i=0}^{n-1} p_{X_i}(x_i).$$

A random vector is said to be independent and identically distributed or *iid* if it has independent components with identical marginals, that is, all of the distributions P_{X_i} (or the corresponding probability functions) are identical.

Gaussian random vectors have several unusual properties that often simplify analysis. The two most important are the following.

- If the coordinates of a Gaussian vector are uncorrelated, e.g., if

$$E(X_i X_j) = E(X_i)E(X_j) \text{ for } i \neq j,$$

  then they are also independent.

- Linear operations on Gaussian vectors yield Gaussian vectors. For example, if $\mathbf{X}$ is a k-dimensional Gaussian random vector with covariance matrix Λ and mean vector $\mathbf{m}$ and $\mathbf{Y} = \mathbf{HX} + \mathbf{b}$ for some $n \times k$ matrix H and n-dimensional vector b, then $\mathbf{Y}$ is an n-dimensional Gaussian random vector with mean vector $\mathbf{Hm} + \mathbf{b}$ and covariance matrix $\mathbf{H\Lambda H}^t$.

2.4   Random Processes

A *random process* is an indexed family of random variables $\{X_t; \ t \in T\}$ or $\{X(t); \ t \in T\}$. The index t corresponds to time. Both forms of notation are used, but X_t is more common for discrete time and $X(t)$ more common for continuous time. If T is continuous, e.g., is $(-\infty, \infty)$ or $[0, \infty)$, then the process is called a *continuous time random process*. If T is discrete, as in the set of all integers or the set of all nonnegative integers, then the process is called a *discrete time random process* or a *random sequence* or a *time series*. When considering the discrete time case, letters such as n and k are commonly used for the index. The condition $t \in T$ is often omitted when it is clear from context and a random process is then denoted simply by $\{X_n\}$ or $\{X(t)\}$.

A random process is usually described via its finite dimensional distributions, that is, by defining the distributions for sample vectors taken at arbitrary sample times. This requires a formula that gives for any fixed number of samples n and any collection of sample times $t_0, t_1, \cdots, t_{n-1}$ the joint cdf

$$F_{X_{t_0}, \cdots, X_{t_{n-1}}}(\mathbf{x}) = P(X_{t_i} \leq x_i; \ i = 0, \cdots, n-1).$$

These cdf's in turn can be specified by joint pdf's or joint pmf's. That the collection of finite-dimensional joint distributions is sufficient to completely define a random process is a fundamental result of probability theory known as the *Kolmogorov extension theorem*. Its practical import is that a random process is completely determined by a rule for computing its distributions for finite collections of samples. We next treat the two most important examples of such specification.

A discrete time process is said to be *independent and identically distributed* or *iid* if there is a cdf F such that all n-dimensional cdf's of the process are product cdf's with marginal cdf F. The name follows since the condition implies that the random variables produced by the process are independent and have identical distributions. Thus all of the pdf's and pmf's thus far considered yield corresponding iid processes. It is straightforward to show that the resulting distributions are consistent and hence the processes well defined.

A process is said to be *Gaussian* if all finite collections of samples are Gaussian random vectors. It is straightforward (but tedious) to show that the consistency conditions are met under the following conditions:

- there exist functions $\{m(t); \ t \in \mathcal{T}\}$, called the *mean function*, and $\{\Lambda(t, s); \ t, s \in \mathcal{T}\}$, called the *covariance function*, where the covariance function is symmetric and positive definite, i.e., $\Lambda(t, s) = \Lambda(s, t)$, for any n and $t_1, \cdots, t_n$

$$\sum_{i=1}^{n} \sum_{j=1}^{n} y_{t_i} \Lambda(t_i, t_j) y_{t_j} > 0;$$

- given n and $t_1, \cdots, t_n$, the vector $(X_{t_1}, \cdots, X_{t_n})$ is Gaussian with $\mathbf{m} = (m(t_1), \cdots, m(t_n))$ and $\mathbf{\Lambda} = \{\Lambda(t_i, t_j); \ i, j = 1, \cdots, n\}$.

A discrete time Gaussian process is iid if and only $m(t)$ is a constant, say m, and $\Lambda(t, s)$ has the form $\sigma^2 \delta_{t-s}$, where δ_k is the Kronecker delta, 1 if $k = 0$ and 0 otherwise.

It is often useful to consider conditional probabilities when specifying a random process. A few remarks will provide sufficient machinery for our purposes. Suppose that we fix n and a collection of sample times $t_1 < t_2 < \cdots < t_n$. Suppose further that the process is described by pmf's and consider the pmf for the random vector $X_{t_0}, X_{t_1}, \cdots, X_{t_{n-1}}$. We can write

$$p_{X_{t_0}, X_{t_1}, \cdots, X_{t_{n-1}}}(x_0, x_1, \cdots, x_{n-1})$$
$$= p_{X_{t_0}}(x_0) \prod_{k=1}^{n-1} \frac{p_{X_{t_0}, X_{t_1}, \cdots, X_{t_k}}(x_0, x_1, \cdots, x_k)}{p_{X_{t_0}, X_{t_1}, \cdots, X_{t_{k-1}}}(x_0, x_1, \cdots, x_{k-1})}.$$

The ratios in the above product are recognizable as elementary conditional probabilities, and hence we define the conditional pmf's

$$p_{X_{t_k}|X_{t_{k-1}}, \cdots, X_{t_0}}(x_k|x_{k-1}, \cdots, x_0)$$
$$= \Pr(X_{t_k} = x_k | X_{t_i} = x_i; \ i = 0, \cdots, k-1)$$
$$= \frac{p_{X_{t_0}, X_{t_1}, \cdots, X_{t_k}}(x_0, x_1, \cdots, x_k)}{p_{X_{t_0}, X_{t_1}, \cdots, X_{t_{k-1}}}(x_0, x_1, \cdots, x_{k-1})}.$$

This can be viewed as a generalization of a product measure and a chain rule for pmf's. By analogy we can write a similar form in the continuous case. Suppose now that the process is described by pdf's. We can write

$$f_{X_{t_0}, X_{t_1}, \cdots, X_{t_{n-1}}}(x_0, x_1, \cdots, x_{n-1})$$
$$= f_{X_{t_0}}(x_0) \prod_{k=1}^{n-1} \frac{f_{X_{t_0}, X_{t_1}, \cdots, X_{t_k}}(x_0, x_1, \cdots, x_k)}{f_{X_{t_0}, X_{t_1}, \cdots, X_{t_{k-1}}}(x_0, x_1, \cdots, x_{k-1})}.$$

Define the conditional pdf's by

$$f_{X_{t_k}|X_{t_{k-1}}, \cdots, X_{t_0}}(x_k|x_{k-1}, \cdots, x_0) = \frac{f_{X_{t_0}, X_{t_1}, \cdots, X_{t_k}}(x_0, x_1, \cdots, x_k)}{f_{X_{t_0}, X_{t_1}, \cdots, X_{t_{k-1}}}(x_0, x_1, \cdots, x_{k-1})}.$$

These are no longer probabilities (they are *densities* of probability), but the conditional probabilities are obtained from the conditional pdf by the following integral:

$$P(X_{t_k} \in G | X_{t_i} = x_i; i = 0, .., k-1)$$
$$= \int_G f_{X_{t_k}|X_{t_{k-1}}, \cdots, X_{t_0}}(x_k|x_{k-1}, \cdots, x_0) \, dx_k.$$

Thus conditional pdf's can be used to compute conditional probabilities.

A simple example of the use of conditional probabilities to describe a process is in the definition of Markov processes. A *Markov process* is one for which for any n and any $t_1 < t_2 < \cdots < t_n$ we have that for all $x_1, \cdots, x_{n-1}$

$$\Pr(X_{t_n} \in G_n | X_{t_{n-1}} = x_{n-1}, \cdots, X_{t_1} = x_1) = \Pr(X_{t_n} \in G_n | X_{t_{n-1}} = x_{n-1});$$

that is, given the recent past, probabilities of future events do not depend on previous outcomes. If the process is described by pdf's this is equivalent to

$$f_{X_{t_n}|X_{t_{n-1}},\cdots,X_{t_0}}(x_n|x_{n-1},\cdots,x_0) = f_{X_{t_n}|X_{t_{n-1}}}(x_n|x_{n-1}).$$

If the process is described by pmf's, it is equivalent to

$$p_{X_{t_n}|X_{t_{n-1}},\cdots,X_{t_0}}(x_n|x_{n-1},\cdots,x_0) = p_{X_{t_n}|X_{t_{n-1}}}(x_n|x_{n-1}).$$

If a process is Gaussian and has a covariance function of the form $\Lambda(t,s) = \sigma^2\rho^{-|t-s|}$, then it is Markov. This process is usually called the *Gauss-Markov process* .

2.5 Expectation

Suppose that X is a random variable described by either a pdf f_X or pmf p_X. The *expectation* of X is defined by the integral

$$EX = \int_{-\infty}^{\infty} x f_X(x)\,dx$$

if described by a pdf, and

$$EX = \sum_x x p_X(x)$$

if described by a pmf.

Given a random variable X defined on a sample space Ω, we can define another random variable, say Y, as a function of X, $Y = g(X)$. Thus, Y is the composition of the two mappings X and g, yielding a new random variable Y defined on the original space by $Y(\omega) = g(X)(\omega) = g(X(\omega))$. In other words, we form the measurement Y on Ω by first taking the measurement $X(\omega)$ and then applying g to the result. We could find the expectation of Y by first finding its cdf F_Y and then taking the appropriate integral with respect to the cdf, but the *fundamental theorem of expectation* provides a shortcut and allows us to use the original pdf or pmf. For example, if the appropriate pdf's are defined.

$$EY = Eg(X) = \int y f_Y(y)\,dy = \int g(x) f_X(x)\,dx.$$

Expectations of a random variable conditioned on an event are frequently used. If $A \in \Omega$ is an event, then

$$E[X|A] = \int x f_{X|A}(x|A)\,dx$$

Let $\{A_i\}$ be a sequence of events that form a partition of the sample space so that

$$\cup_i A_i = \Omega, \ A_i \cap A_j = \emptyset \text{ for } i \neq j,$$

then

$$EX = \sum_i E[X|A_i]P(A_i).$$

Expectations can also be taken with respect to conditional probability functions to form conditional expectations, e.g., given two continuous random variables X and Y

$$E(Y|x) = E(Y|X = x) = \int y f_{Y|X}(y|x)\, dy.$$

It should be clearly understood that $E(Y|x)$ is a deterministic function of x. Then, by replacing x by the random variable X, the expression $E(Y|X)$ defines a new random variable that is a function of the random variable X.

Conditional expectations are useful for computing ordinary expectations using the so-called iterated expectation or nested expectation formula:

$$
\begin{aligned}
E(Y) &= \int y f_Y(y)\, dy = \int y \left(\int f_{X,Y}(x,y)\, dx \right) dy \\
&= \int f_X(x) \left(\int y f_{Y|X}(y|x) dy \right) dx \\
&= \int f_X(x) E(Y|x)\, dx = E(E(Y|X)).
\end{aligned}
$$

Two important examples of expectation of functions of random variables are the *mean*

$$m_X = EX$$

and the *variance*

$$\sigma_X^2 = E[(X - m_X)^2].$$

Expectations of the form $E(X^n)$ are called the nth order *moments* of the random variable. Note that it is usually easier to evaluate the variance and moments using the fundamental theorem of expectation than to derive the new cdf's.

Suppose that $\{X_t; t \in T\}$ is a random process. The *mean function* or *mean* is defined by

$$m_X(t) = EX_t.$$

The *autocorrelation function* or *autocorrelation* of the process is defined by

$$R_X(t,s) = E[X_t X_s].$$

The *autocovariance function* or *covariance* is defined by

$$K_X(t,s) = E[(X_t - m_X(t))(X_s - m_X(s))].$$

The notation $\Lambda_X(t,s)$ is also commonly used to denote covariance functions. Observe that

$$K_X(t,t) = \sigma_{X_t}^2.$$

It is easy to see that

$$K_X(t,s) = R_X(t,s) - m_X(t)m_X(s).$$

A process $\{X_t; t \in T\}$ is said to be *weakly stationary* if

1. $m_X(t) = m_X(t+\tau)$, all t, τ such that $t, t+\tau \in T$, and

2. $R_X(t,s) = R_X(t+\tau, s+\tau)$, all τ such that $t, s, t+\tau, s+\tau \in T$.

A process is weakly stationary if the first two moments—the mean and the correlation—are unchanged by shifting, that is, by changing the time origin. This is equivalent to saying that the mean function is constant, that is, there is a constant m_X such that

$$m_X(t) = m_X,$$

(does not depend on time); and that the correlation depends on the two sample times only through their difference, a fact that is often expressed as

$$R_X(t,s) = R_X(t-s).$$

A function of two variables which depends on the variables only through their difference as above is said to be *Toeplitz* and the theory of weakly stationary processes is intimately connected with the theory of Toeplitz functions[172].

Weak stationarity is a form of *stationarity property* of a random process. Stationarity properties refer to moments or distributions of samples of the process being unchanged by time shifts. We shall later consider other stationarity properties.

If a process $\{X_t; t \in T\}$ is weakly stationary, then we define its *power spectral density* $S_X(f)$ as the Fourier transform of the correlation:

$$S_X(f) = \mathcal{F}(R_X(t)) = \begin{cases} \int R_X(t)e^{-j2\pi ft}\,dt & T \text{ continuous} \\ \sum R_X(k)e^{-j2\pi fk} & T \text{ discrete} \end{cases}$$

2.6   Linear Systems

A *system* is a mapping L of an input time function $x = \{x_t;\ t \in T\}$ into an output time function $y = \{y_t;\ t \in T\} = L(x)$. The system is *linear* if $L(ax_1 + bx_2) = aL(x_1) + bL(x_2)$ for all input functions x_1, x_2 and all constants a, b. An important class of linear systems are those described by convolutions. First suppose that T is continuous, e.g., the real line. Here by convention we denote waveforms by $\{x(t);\ t \in T\}$ or simply by $x(t)$. Suppose there is a function $h_t(v)$ such that

$$y(t) = \int_{s \in T} x(s) h_t(t - s)\, ds.$$

The function $h_t(t - s)$ is called the *impulse response* of the system since it can be considered as the output of the system at time t resulting from a unit impulse or Dirac delta function $x(t) = \delta(t - s)$ applied at time s. A system of this type is called a *linear filter* and the above integral is called a *superposition integral*. If the impulse response depends only on the difference $s - t$ the filter is said to be time-invariant and we denote the impulse response by $h(v) = h_t(v)$. The superposition integral then specializes to a *convolution integral* given by

$$y(t) = \int_{s \in T} x(s) h(t - s)\, ds$$

or equivalently

$$y(t) = \int_{\tau:t-\tau \in T} x(t - \tau) h(\tau)\, d\tau.$$

A linear system is said to be *causal* if for each t, $h_t(\tau) = 0$ whenever $\tau < 0$.

Suppose that the Fourier transforms $X(f) = \mathcal{F}_f(x(t))$ and $H(f) = \mathcal{F}_f(h(t))$ exist, where

$$X(f) = \int_T x(t) e^{-j2\pi ft}\, dt.$$

This will be the case, for example, if x is absolutely integrable, e.g.,

$$\int_T |x(t)|\, dt < \infty.$$

If h is absolutely integrable, it is said to be *stable* since bounded inputs then result in bounded outputs. A basic property of Fourier transforms is that if an input x with Fourier transform $X(f)$ is convolved with an impulse response h with Fourier transform $H(f)$ (called the *transfer function* of the filter), then the output waveform y has Fourier transform $Y(f) = X(f)H(f)$.

In a similar manner we can consider linear filters, convolution, and Fourier transforms for discrete time systems. Here T is discrete, e.g., the set of integers, and the signal is a sequence rather than a waveform. The sequence can arise from an inherently discrete time source such as a digital computer or it might arise from sampling a continuous time process by forming $x_n = x(nT)$, where $x(t)$ is a continuous time process and T is called the *sampling period* and $1/T$ is called the *sampling rate* in number of samples per second.

A discrete time linear filter is described by a sum instead of an integral. For example, a time-invariant linear filter is defined by a sequence $h = \{h_k; \ k \in T\}$ and an input sequence $x = \{x_k; k \in T\}$ produces an output sequence $y = \{y_k; \ k \in T\}$ via the discrete time convolution

$$y_k = \sum_{k:\ k \in T} x_k h_{n-k} = \sum_{k:\ n-k \in T} x_{n-k} h_k.$$

The sequence h_n is called the *pulse response* or *Kronecker delta response* since it is the response to a Kronecker delta δ_n or unit pulse input at time 0. It is also sometimes called an *impulse response* in analogy with the continuous time case. We assume that h_n is real-valued. If only a finite number of the h_n are nonzero, the filter is called an FIR (finite impulse response) filter. Otherwise it is called an IIR (infinite impulse response) filter. Suppose that the Fourier transforms $X(f)$ and $H(f)$ exist, where now

$$X(f) = \sum_{k:\ k \in T} x_k e^{-j 2\pi f k}.$$

When the discrete time signal is obtained by sampling a continuous time signal with a sampling period T, the frequency variable is often scaled and the Fourier transform is defined by

$$X(f) = \sum_{k:\ k \in T} x_k e^{-j 2\pi f k T}.$$

We shall usually incorporate the scaling factor T in the formulas, but it is often taken as $T = 1$ for normalization. Regardless of the scaling, as in the continuous time case, $Y(f) = X(f)H(f)$.

Alternatively, it is common for discrete time systems to describe signals and systems using z transforms obtained by replacing $e^{2j\pi fT}$ by z. In this case the transfer function is defined as

$$G(z) = \sum_{i=-\infty}^{\infty} h_i z^{-i}$$

It is also common to consider the discrete time Fourier transform in terms of the angular frequency $\omega = 2\pi f$, then

$$H(\omega) = G(e^{j\omega T}) = \sum_{k=-\infty}^{\infty} h_k e^{-j\omega kT}.$$

Note that it may also be regarded as a Fourier series for the periodic function $H(\omega)$ with period $2\pi/T$. Usually it suffices to consider only the transfer function defined for the range $(-\pi/T, \pi/T)$.

A filter is said to be *causal* if $h_n = 0$ for $n < 0$. A filter will be called *stable* if its impulse response is absolutely summable, that is if $\sum_{n=-\infty}^{+\infty} |h_n| < \infty$.

If a filter with transfer function $H(\omega)$ satisfies the property that $[H(\omega)]^{-1}$ is a stable transfer function, that is, if

$$[H(\omega)]^{-1} = \sum_{k=-\infty}^{\infty} u_k e^{-jk\omega T}$$

with

$$\sum_{k=-\infty}^{\infty} |u_k| < \infty,$$

then the filter is said to be *invertible*. In the particular case where u_n, as given above, is a causal and stable impulse response, then the filter $H(\omega)$ is said to be *causally invertible*. A *minimum phase* filter (also know as a *minimum delay* filter) is defined as a filter that is causal, stable, and causally invertible. Finally, we define an *all-pass* filter as a filter whose transfer function is constant, independent of frequency ω.

As an example, consider the filters

$$G_1(z) = z^{-L}$$

where $L > 0$,

$$G_2(z) = \frac{z^{-1} - \alpha}{1 - \alpha z^{-1}}$$

and

$$G_3(z) = \frac{1 - \beta z^{-1}}{1 - \alpha z^{-1}},$$

where $0 < \alpha < 1$ and $0 < \beta < 1$. The first filter is clearly stable, causal, and invertible but not causally invertible and hence it is not minimum phase. Note that its inverse is a stable but anticausal filter. The second filter can be expanded into a power series in z (containing terms with nonpositive powers of z) whose coefficients yield a causal impulse response for this filter. This series is convergent in the region $|z| > \alpha$, from which it can be clearly seen to be causal and stable. The inverse transfer function can be expanded into a power series in z consisting of terms with nonnegative powers of z corresponding to an anticausal impulse response. The series converges in $|z| < 1/\alpha$. This corresponds to an anticausal filter so that G_2 is *not* causally invertible. It may also be formally expanded into a power series in z^{-1}, which converges in the region $|z| > 1/\alpha$, but the causal impulse response corresponding to the coefficients in this series is divergent indicating a causal but unstable filter. From either perspective, it is clear that there is no physically meaningful (i.e., stable and causal) filter that inverts the operation performed by the filter G_2. In fact, both G_1 and G_2 are all-pass filters with unit magnitude, as can be readily verified by setting $z = e^{j\omega T}$ and calculating the magnitude of the transfer function. The third filter, G_3, is stable, causal, and causally invertible.

If the input to a linear filter is a random process instead of a deterministic waveform, then the Fourier transform of the input signal will not exist in general. In particular, a sample waveform produced by a random process is not usually absolutely integrable. In this case Fourier techniques are most useful when applied to the moments of the process rather than to its sample waveforms. Suppose that a random process $\{X_t\}$ is weakly stationary and has power spectral density $S_X(f)$. If the process is put into a linear time-invariant filter with transfer function $H(f)$ to produce an output random process $\{Y_t\}$ defined as the convolution of the input sample waveform and the filter, then the output process will have power spectral density

$$S_Y(f) = S_X(f)|H(f)|^2.$$

This relation is true for both continuous and discrete time systems.

2.7 Stationary and Ergodic Properties

We have already seen an example of a stationarity property of a process: A process $\{X_t\}$ is weakly stationary if the mean function and autocorrelation function are not affected by time shifts, that is,

$$EX_t = EX_{t+\tau}$$

and

$$R_X(t, s) = R_X(t + \tau, s + \tau),$$

for all values of t, s, τ for which the above are defined. While the same thing can be stated more concisely as $EX_t = m$ and $R_X(t, s) = R_X(t - s)$, the above form better captures the general idea of stationarity properties. We say that a process is *strictly stationary* or, simply, *stationary* if the probabilities of all events are unchanged by shifting the events. A precise mathematical definition is complicated, but the condition is equivalent to the following: For any dimension n and any choice of n sample times $\mathbf{t} = (t_0, \cdots, t_{n-1})$, the joint cdf for the n samples of the process at these times satisfies

$$F_{X_{t_0}, \cdots, X_{t_{n-1}}}(x_0, \cdots, x_{n-1}) = F_{X_{t_0+\tau}, \cdots, X_{t_{n-1}+\tau}}(x_0, \cdots, x_{n-1})$$

for all real vectors $(x_0, x_1, \cdots, x_{n-1})$ choices of τ and $\mathbf{t}$ for which the cdf's are defined. Strict stationarity is also equivalent to the formulas obtained by replacing the cdf's above by pdf's or pmf's, if appropriate. The key idea is that the probabilistic description of the random process is unchanged by time shifts.

There are more general stationarity properties than the strict stationarity defined above. For example, processes can have nonstationarities due to transients that die out in time or periodicities. In the first case the distributions of samples converge to a stationary distribution as the sample times increase and in the second case a stationary distribution results if the distributions are suitably averaged over the periodic behavior. Making these ideas precise leads to the notions of *asymptotically stationary* and *asymptotically mean stationary* processes, which we mention solely to point out that there is nothing "magic" about the property of being stationary: many of the useful properties of stationary processes hold also for the more general notions. (See, e.g., [155], Chapters 6 and 7.)

There are two uses of the word *ergodic* in describing a random process. The first, that of an *ergodic process*, is a rather abstract mathematical definition, while the second, that of the *ergodic property*, describes a key property of a random process that is important for both theory and application. The first concept is quite technical and would require additional notation and terminology to define precisely. Roughly speaking it states that the process has the property that if a set of signals is unaffected by shifts in time, that is, shifting a signal in the set gives another signal in the set, then that set must have either probability 1 or probability 0. If the process is also stationary (or asymptotically stationary), it can be shown that this technical condition is equivalent to the property that events that are widely separated in time are approximately independent. In other words, ergodicity is a condition on the decay of the memory of a random process.

The concept of an *ergodic property* is more operational. As already discussed, a fundamental goal of probability theory is to relate the long term behavior of random processes to computable quantities, e.g., to relate sample or time averages to expectations. Suppose for example that $\{X_n\}$ is a real-valued discrete time random process. It is of interest to know conditions under which the sample mean

$$\frac{1}{n} \sum_{k=0}^{n-1} X_k$$

converges in some sense as $n \to \infty$. Other limits may also be of interest, such as the sample autocorrelation

$$\frac{1}{n} \sum_{k=0}^{n-1} X_k X_{k+n}$$

or the sample average power

$$\frac{1}{n} \sum_{k=0}^{n-1} X_k^2.$$

We might pass $\{X_n\}$ through a linear filter with pulse response h_k to form a new process $\{Y_n\}$ and then ask if

$$\frac{1}{n} \sum_{k=0}^{n-1} Y_k$$

converges. In general, we could take a series of measurements g_n on the random process (such as the Y_n above) which could in principle depend on the entire past and future and ask if the sample average

$$\frac{1}{n} \sum_{k=0}^{n-1} g_k$$

converges. If the process has continuous time, then we are interested in the time average means obtained by replacing the above sums by integrals, e.g., the limit as $T \to \infty$ of

$$\frac{1}{T} \int_0^T X_t \, dt,$$

and so on.

A random process is said to have the *ergodic property* if *all* such time averages for reasonable measurements on the process converge in some sense. The type of convergence determines the type of ergodic property, e.g., convergence in mean square, in probability, or with probability one. Theorems proving that a given process has the ergodic property are called *ergodic theorems* or *laws of large numbers*. Perhaps the most famous ergodic theorem states that if a discrete time random process is stationary, then it has the ergodic property. If the process is also ergodic, then sample averages converge to expectations, e.g.,

$$\frac{1}{n} \sum_{k=0}^{n-1} X_k \underset{n \to \infty}{\to} E(X_0).$$

If the process is not ergodic, then sample averages still converge, but they will converge in general to a random variable.

It is important to observe that stationarity is sufficient for the convergence of sample averages, but it is not necessary. Similarly, ergodicity alone is not sufficient for the convergence of sample averages, but combined with stationarity it is. It is erroneous to assume that either stationarity or ergodicity are required for a process to possess the ergodic property, that is, to have convergent time averages. For example, processes can have transients or short term behavior which will cause them to be nonstationary, yet sample averages will still converge. Processes may have distinct ergodic modes and hence not be ergodic, e.g., the process may be Gaussian with a known correlation but with a time-independent mean function chosen at random from [0,1] according to an unknown distribution. This process can be viewed as a mixture of ergodic processes, but it is not itself ergodic. It does, however, have the ergodic property and time averages will converge. In fact all stationary processes that are not ergodic can be modeled as a mixture of ergodic sources, that is, as a stochastic process where nature first selects a process at random from a collection of ergodic processes, and then produces the output of that process forever. In other words, if we observe a stationary nonergodic process, then we are actually observing an ergodic process, we just don't know which one. We can determine which one, however, by observing the long term behavior and estimating the probabilities from relative frequencies. This property of stationary processes is called the *ergodic decomposition* and it opens the door to the application of random process tools to nonergodic processes. (See chapters 7 and 8 of [155] for a development of the ergodic decomposition.)

As a final comment on the ergodic property, we note that it can be shown that a sufficient *and necessary* condition for a random process to have the ergodic property is that it be asymptotically mean stationary, that is, stationary in the general sense mentioned previously. Fortunately, nearly all information sources and processes produced by virtually all coding structures encountered in the real world are well modeled by asymptotically mean stationary processes. Furthermore, for any asymptotically mean stationary source there is an equivalent stationary source with the same long term sample averages. For this reason it is often reasonable to assume that processes are stationary in analysis even if they are not, provided that they do satisfy the more general conditions.

2.8 Useful Processes

We have already seen several examples of random processes that are common in the communications literature because they are either good models of real phenomena or because the are tractable enough to yield simple answers which may provide insight into more complicated systems. In particular, we have defined iid processes with arbitrary marginal distributions, Gaussian processes, and Gauss Markov processes. Another general class of random processes is formed by passing memoryless processes through linear filters. The resulting process is named according to the class of linear filters permitted. First suppose that $\{X_n\}$ is an iid process defined over the set of integers n and that h_n is the pulse response of a stable time-invariant filter. Thus the output process $\{Y_n\}$ is represented by the convolution

$$Y_n = \sum_k X_{n-k} h_k. \qquad (2.8.1)$$

A filter of this form, that is, which defines the output by a convolution, is called a *moving average filter*. This term is sometimes reserved for the case where only a finite number of the h_k are nonzero, e.g., for FIR filters. In the case of FIR filters, it is also called a *transversal filter* or an *all-zero filter*. The reason for the latter name is that if the transfer function function $H(f)$ is analytically continued to the complex plane by the substitution $z = e^{j2\pi fT}$, then the resulting complex function contains only zeros and no poles. We denote the resulting complex function by $H(z)$, the *z-transform* of the pulse response. (This is admittedly poor notation because we are using the argument of the function to specify the function, that is, to say whether it is a Fourier transform or a z-transform. The notation is, however, common.)

If an iid process is put into a moving average filter, then the output process is called a *moving average process*. Since the input process is iid, it has a constant spectral density, say σ_X^2. The moving average process therefore must have spectral density $S_Y(f) = \sigma_X^2 |H(f)|^2$.

Instead of an explicit input-output relation of the convolution form of (2.8.1), we could have an implicit expression of the form

$$X_n = \sum_k a_k Y_{n-k}; \qquad (2.8.2)$$

that is, the input is given as the convolution of the output with a sequence $\{a_k\}$ called *regression coefficients*. A linear filter of this form is called an *autoregressive filter*. If $a_k = 0$ for $k < 0$ and $a_0 \neq 0$, then this relation becomes a *linear recursion* since it can be expressed in the form

$$Y_n = \frac{1}{a_0}[X_n - \sum_{k=1}^{\infty} a_k Y_{n-k}].$$

The corresponding transfer function has the form

$$H(f) = \frac{1}{\sum_{k=0}^{\infty} a_k e^{j2\pi fTk}}.$$

If h_k is the pulse response of a stable, causal, and causally invertible transfer function, then a_k is a causal, stable impulse response with $a_0 \neq 0$. If there are only a finite number of nonzero a_k (a_k is the pulse response of an FIR filter), then the analytic continuation of $H(f)$ into the complex plane using the substitution $z = e^{j2\pi fT}$ has only poles and no zeros. For these reasons a finite-order autoregressive filter is known as an *all-pole filter*. Such filters may or may not be stable, depending on the location of the poles. The process resulting from putting an iid process $\{X_n\}$ into an autoregressive filter is called an *autoregressive process*.

For example, suppose we have regression coefficients and define the z-transform

$$A(z) = 1 + \sum_{k=1}^{\infty} a_k z^{-k},$$

then the power spectral density of the autoregressive process formed by passing an iid process $\{X_n\}$ with spectral density σ_X^2 through the autoregressive filter described by $A(z)$ is

$$S_Y(f) = \frac{\sigma_X^2}{|A(e^{j2\pi fT})|^2}.$$

$A(z)$ is often referred to as the *inverse filter* or as a *whitening filter* for the process.

One can also define more general linear models by combining all-zero and all-pole filters, e.g., by a difference equation of the form

$$\sum_k a_k Y_{n-k} = \sum_k h_k X_{n-k}.$$

Such a process can be constructed by passing an iid process through a linear filter with transfer function

$$H(f) = \frac{\sum_k h_k e^{-j2\pi kfT}}{\sum_k a_k e^{-j2\pi kfT}}.$$

Such processes are called *autoregressive-moving average* or *ARMA* processes. If $a_k = 0$ and $h_k = 0$ for $k < 0$ and $a_0 \neq 0$, then this relation becomes a linear recursion in the form

$$Y_n = \frac{1}{a_0} [\sum_{k=0}^{\infty} h_k X_{n-k} - \sum_{k=1}^{\infty} a_k Y_{n-k}].$$

Thus far all the random processes described are mathematical abstractions. What of "real" random processes such as speech or image signals? On one hand, they fit the mathematical abstractions: Continuous speech occupies a bandwidth from a few hundred Hertz to something between 4 kHz and 16 kHz, depending on the desired quality of the reproduced speech. Hence by sampling fast enough, we can obtain a real valued random process $\{X_n\}$ from which the original speech can be perfectly reconstructed. A sequence of images can be considered as a sequence $\{X_n\}$ where each X_n is a finite raster (rectangular matrix) of nonnegative real numbers representing the intensity at each pixel of the image. While we clearly have a physical random process in both cases since we have a sequence of numbers or matrices that are random in the sense of not being known in advance, we do not yet have a mathematical random process since we have not provided the necessary probability distributions or functions.

Here several paths are possible. First, we could estimate distributions by computing long term histograms based on real speech or image data. Mathematically, if the physical processes were asymptotically mean stationary, then these relative frequencies will converge and we can get reasonable estimates. Physically, one could simply try computing the relative frequencies and see if they converge. Practically, however, this is an impossible task. Without assuming some additional structure, one would have to estimate probabilities of arbitrarily large collections of samples. To make matters worse, the model will depend on the real signal used, e.g., the types of images considered and the sex of the speakers.

The task could be simplified by assuming a particular form of distribution, e.g., Gaussian or a first order autoregressive source with Laplacian iid input, and then estimate a small number of defining parameters by looking at data. We could simply assume or approximate the real processes by a common distribution, e.g., we could model real speech as being Gaussian. Such models can, however, be controversial and may or may not yield results relevant for real speech.

Our basic response to the modeling issue is to eschew the use of specific mathematical models for real processes in general, but to assume that such real processes can be treated as mathematical random processes. In other words, we pretend that speech has a corresponding mathematical random process $\{X_n\}$, but we do not know what the distributions are. Thus we can write expectations and probabilities in principle, but we may not be able to evaluate them. We will, however, try to estimate both by looking at long term time averages. We will not try to build complete and explicit models for speech or image data, but we will try to estimate certain attributes of the unknown underlying distributions. This approach will not yield fundamental performance bounds for real speech, but it will yield design techniques for good signal coding systems and a consistent means of comparing different systems designed for real signal sources. Comparisons of theoretical optima with actual systems will have to be confined to those cases where the theoretical optima can be evaluated, e.g., Gaussian and other mathematical random processes.

2.9   Problems

2.1.   (a) Suppose that X is a nonnegative random variable in the sense that $\Pr(X \geq 0) = 1$. Show that for any $\epsilon > 0$

$$\Pr(X \geq \epsilon) \leq \frac{E(X)}{\epsilon}.$$

This is called the *Markov inequality*. *Hint:* Assume that X is described by either a pmf or a pdf and break up the expression for $E(X)$ into two pieces depending on whether the dummy variable is greater than or less than ϵ.

(b) Use the Markov inequality to prove the *Chebychev inequality* that states for a random variable Y with mean m and variance σ^2 the following holds:

$$\Pr(|Y - m| \geq \epsilon) \leq \frac{\sigma^2}{\epsilon^2}.$$

In other words, the probability that a random variable is far from its mean diminishes with the variance of the random variable.

2.2.   Prove the *Cauchy-Schwartz* inequality for random variables which states that if X and Y are two real random variables, then

$$|E(XY)| \leq \sqrt{E(X^2)E(Y^2)}.$$

it Hint: Consider the quantity

$$E[(\frac{X}{\sqrt{E(X^2)}} \pm \frac{Y}{\sqrt{E(Y^2)}})^2].$$

2.3. Given a random variable X, two random variables W and V (considered to be noise) are added to X to form the observations

$$Y = X + W$$

$$Z = X + V.$$

Both W and V are assumed to be independent of X. Suppose that all variables have zero mean and that the variance of X is σ_X^2 and the random variables W and V have equal variances, say σ^2. Lastly, suppose that

$$E(WV) = \rho\sigma^2.$$

(a) What is the largest possible value of $|\rho|$? *Hint:* Use the Cauchy-Schwartz inequality of Problem 2.2.

(b) Find $E(YZ)$.

(c) Two possible estimates for the underlying random variable X based on the observations are

$$\hat{X}_1 = Y$$

and

$$\hat{X}_2 = \frac{Y + Z}{2},$$

that is, one estimate looks at only one observation and the other looks at both. For what values of ρ is $\hat{X}_2$ the better estimate in the sense of minimizing the mean squared error

$$E[(\hat{X}_i - X)^2]?$$

For what value of ρ is this best mean squared error minimized? What is the resulting minimum value?

(d) Suppose now that X, W, V are jointly Gaussian and that ρ is 0. (We still assume that W, V are independent of X.) Find the joint pdf

$$f_{Y,Z}(y, z).$$

(e) Suppose that ρ is arbitrary, but the random variables are Gaussian as in part (d). Are the random variables $Y + Z$ and $Y - Z$ independent?

2.4. Let $\{Y_n\}$ be a stationary first order autoregressive process defined by the difference equation

$$Y_n = \rho Y_{n-1} + X_n,$$

where $\{X_n\}$ is an iid process and $|\rho| < 1$. Suppose that we know that

$$f_{X_n}(y) = \frac{\lambda}{2}e^{-\lambda|y|}, \text{ all } y,$$

a doubly exponential pdf. (This is a popular model for sampled speech, where ρ is typically between .85 and .95, depending on the sampling rate.)

(a) Find $E(X)$, σ_X^2, and the autocorrelation $R_X(k, j)$.

(b) Find the *characteristic function* $M_X(ju) = E(e^{jux})$.

(c) Find the probability $\Pr(Y_n > X_n + 1/2)$ and the conditional pdf $f_{X_n|Y_n}(x|y) = f_{X_n,Y_n}(x, y)/f_{Y_n}(y)$.

(d) What is the power spectral density of $\{Y_n\}$?

(e) Find $f_{Y_n|Y_{n-1}}(y|z)$ and $E(Y_n|Y_{n-1})$.

2.5. A continuous time Gaussian random process $\{X(t); \ t \in (-\infty, \infty)\}$ is described by a mean function $E[X(t)] = 0$ all t and an autocorrelation function $R_X(t, s) = E[X(t)X(s)] = \sigma^2 \rho^{|t-s|}$ for a parameter ρ in $(-1, 1)$. The process is digitized in two steps as follows: First the process is sampled every T seconds to form a new discrete time random process $X_n = X(nT)$. The samples X_n are then put through a binary quantizer defined by

$$q(x) = \begin{cases} +a & \text{if } x \geq 0 \\ -a & \text{if } x < 0 \end{cases}$$

to form a new process $q(X_n)$.

(a) Write an expression for the pdf

$$f_{X_n,X_k}(x, y)$$

(b) Find an expression for the mean squared quantization error defined by
$$\Delta = E[(X_n - q(X_n))^2]$$
What choice of a minimizes Δ?

(c) Suppose now that $\rho = 1$ above which means that
$$E[X_n X_k] = E[X_0^2]$$
for all n and k. Find a random variable Y such that
$$\frac{1}{n}\sum_{i=0}^{n-1} X_i$$

converges in mean square to Y, i.e.,

$$\lim_{n \to \infty} E[(\frac{1}{n} \sum_{i=0}^{n-1} X_i - Y)^2] = 0?$$

Hint: Use the Markov or Chebychev inequality of Problem 2.1 and the above condition on the expectations to suggest a guess for Y, and then prove that the guess works.

2.6. Let $\{X_n\}$ be an iid sequence of random variables with a uniform probability density function on $[0,1]$. Suppose we define an N-level uniform quantizer $q(r)$ on $[0,1]$ as follows: $q(r) = (k + 1/2)/N$ if $k/N \le r < (k+1)/N$. In other words, we divide $[0,1]$ up into N intervals of equal size. If the input falls in a particular bin, the output is the midpoint of the bin. Application of the quantizer to the process $\{X_n\}$ yields the process $\{q(X_n)\}$. Define the quantizer error process $\{\epsilon_n\}$ by $\epsilon_n = q(X_n) - X_n$.

(a) Find $E(\epsilon_n)$, $E(\epsilon_n^2)$, and $E(\epsilon_n q(X_n))$.
    Hint: Break up the integrals into a sum of integrals over the separate bins.

(b) Find the (marginal) cumulative distribution function and the pdf for ϵ_n.

(c) Find the covariance and power spectral density of ϵ_n.

2.7. A k-dimensional real random vector $\mathbf{X}$ is coded as follows: First it is multiplied by a unitary matrix $\mathbf{U}$ to form a new vector $\mathbf{Y} = \mathbf{UX}$. (By unitary it is meant that $\mathbf{U}^* = \mathbf{U}^{-1}$, where $\mathbf{U}^*$ is the conjugate transpose of $\mathbf{U}$). Each component Y_i, $i = 0, 1, \cdots, k-1$ is separately quantized by a quantizer q_i to form a reproduction $\hat{Y}_i = q_i(Y_i)$. Let $\hat{\mathbf{Y}}$ denote the resulting vector. This vector is then used to produce a reproduction $\hat{\mathbf{X}}$ of $\mathbf{X}$ by the formula $\hat{\mathbf{X}} = \mathbf{U}^{-1}\hat{\mathbf{Y}}$. This code is called a *transform code*.

(a) Suppose that we measure the distortion between input and output by the mean squared error

$$(\mathbf{X} - \hat{\mathbf{X}})^*(\mathbf{X} - \hat{\mathbf{X}}) = \sum_{i=0}^{k-1} |X_i - \hat{X}_i|^2,$$

where the asterisk denotes complex conjugate transpose and hence

$$(\mathbf{X} - \hat{\mathbf{X}})^*(\mathbf{X} - \hat{\mathbf{X}}) = \sum_{i=0}^{k-1} |X_i - \hat{X}_i|^2.$$

Show that the distortion is the same in the original and transform domain, that is,

$$(\mathbf{X} - \hat{\mathbf{X}})^*(\mathbf{X} - \hat{\mathbf{X}}) = (\mathbf{Y} - \hat{\mathbf{Y}})^*(\mathbf{Y} - \hat{\mathbf{Y}}).$$

(b) Find a matrix $\mathbf{U}$ such that $\mathbf{Y}$ has uncorrelated components, that is, its covariance matrix is diagonal.

2.8. Consider the coding scheme of Figure 8 in which the encoder is a simple quantizer inside a feedback loop. H and G are causal linear

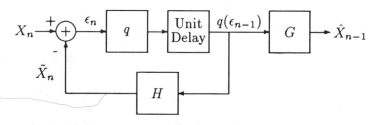

Figure 2.1: Coding Scheme

filters. Show that if the filters are chosen so that

$$\hat{X}_n = \tilde{X}_n + q(\epsilon_n),$$

then

$$|X_n - \hat{X}_n| = |\epsilon_n - q(\epsilon_n)|$$

and hence

$$E(|X_n - \hat{X}_n|^2) = E(|\epsilon_n - q(\epsilon_n)|^2),$$

that is, the overall error is the same as the quantizer error. Suppose that G is fixed. What must H be in order to satisfy the given condition? What constraint is there on G so that H will be causal? (For the purpose of this problem, the "quantizer" can be any nonlinear memoryless mapping.)

2.9. Suppose that X is a random variable described by an exponential pdf

$$f_X(\alpha) = \lambda e^{-\lambda \alpha}; \ \alpha \geq 0.$$

Define a function q which maps real numbers into integers by $q(x) =$ the largest integer less than or equal to x. In other words

$$q(x) = k \text{ if } k \leq x < k + 1.$$

(This function is denoted by $q(x) = \lfloor x \rfloor$.) The function q is a form of quantizer, it rounds all real numbers downward to the nearest integer below the input real number. Define the following two random variables: the quantizer output

$$Y = q(X)$$

and the quantizer error

$$\epsilon = X - q(X).$$

Note: By construction ϵ can only take on values in $[0, 1)$.

(a) Find the pmf $p_Y(k)$ for Y.

(b) Find the expectations $E(X)$ and $E(Y)$.

(c) Find the conditional probability $\Pr(Y = k \mid Y \leq 10)$.

(d) Derive the probability density function for ϵ.

2.10. Write a program to generate pairs of Gaussian random vectors with unit variance and covariance ρ by first generating vectors that are uniformly distributed on the unit circle and using these to generate independent Gaussian variates. Indicate how you map them into correlated pairs. Include your source code with your solutions and a printout of 20 pairs of vectors generated with the program for $\rho = 0.9$. How would you modify your technique if the original random variables are uniform on the unit square instead of on the unit circle.

Chapter 3

Sampling

3.1 Introduction

Physical sources of signals such as speech, images, and all observable electrical waveforms are analog and continuous time in nature. The first step to convert signals to digital form is *sampling*. An analog continuously fluctuating waveform can usually be characterized completely from a knowledge of its amplitude values at a countable set of points in time so that we can in effect "throw away" the rest of the signal. We do not need to observe how it behaves in between any two isolated instances of observation. This is at the same time remarkable and intuitively obvious. It is remarkable that we can discard so much of the waveform and still be able to accurately recover the missing parts. The intuitive idea is that if we sample periodically at regularly spaced instants in time, and the signal does not fluctuate too quickly so that no unexpected wiggles can appear between two consecutive sampling instants, then we can expect to recover the complete waveform by a simple process of interpolation or smoothing, where a smooth curve is drawn that passes through the known amplitude values at the sampling instants.

When we watch a movie, we are actually seeing 24 still pictures flashed on the screen every second. (Actually, each picture is flashed twice.) The movie camera that produced these pictures was actually photographing a scene by taking one still picture every 1/24th of a second. Yet, we have the illusion of seeing continuous motion. In this case, the cinematic process works because our brain is somehow doing the interpolation for us. This is an example of sampling in action in our everyday lives. Another example is watching soap operas on television. It is typically sufficient to tune in to one out of four episodes in one of these ongoing daily serials to keep track of the story and the events in the lives of the characters. Again your brain does the interpolation from the observed episodes by filling in the gaps in the story using implicitly known story rules.

For an electrical waveform, or any other one-dimensional signal, the samples can be carried as amplitudes on a periodic train of narrow pulses. This way of carrying a sampled signal electrically is called *pulse amplitude modulation* (PAM). The key result of sampling is that we have converted a continuous time signal into a discrete time signal and, under conditions to be described, the process will be reversible; that is, we can recover the original waveform from the samples. This is a significant step towards the transition to a digital signal and it is of some interest in its own right. We have achieved a kind of "compression" by replacing a waveform that has an uncountably infinite number of amplitude values by a signal that has a countable number of values or data arriving at a steady rate. Much of the subject of "digital" signal processing in fact deals with such discrete-time, continuous amplitude signals rather than with truly digital signals, i.e., signals that are discrete in time and amplitude.

Sampling provides the gateway to bring physical continuous time signals into the more familiar world of discrete-time signal processing.

3.2   Periodic Sampling

Consider a scalar time function $x(t)$ which has a Fourier transform $X(f)$. We assume there is a finite upper limit on how fast $x(t)$ can wiggle around or vary with time. Specifically, assume that $X(f) = 0$ for $|f| \geq W$. Thus, the signal has a strictly lowpass spectrum with cutoff frequency W Hertz. We require that it be zero at $f = \pm W$ to avoid certain pathological cases. To sample this signal, we periodically observe the amplitude at isolated time instants $t = kT$ for $k = \cdots, -2, -1, 0, 1, 2, \cdots$. The *sampling rate* is $f_s = 1/T$ and T is the *sampling period* or *sampling interval* in seconds.

We consider first the simple idealized case of impulse sampling with a perfect ability to observe isolated amplitude values at the sampling instants kT. The effect of sampling is best modeled as the process of multiplying the original signal $x(t)$ by a *sampling function*, $s(t)$, which is the periodic train of impulses (Dirac delta functions) given by

$$s(t) = T \sum_{k=-\infty}^{\infty} \delta(t - kT). \qquad (3.2.1)$$

where the amplitude scale is normalized to T so that the dc (average) value of $s(t)$ is unity. We will treat the impulse function or Dirac delta in the usual engineering manner as if it were an ordinary function, but caution should be exercised with it, since mathematical rigor requires that it be considered as a generalized function or distribution as treated in the Fourier analysis texts mentioned in Chapter 2.

In the time domain, the effect of this multiplication operation is to generate a new impulse train whose amplitudes are samples of the waveform $x(t)$. Thus,

$$y(t) = x(t)s(t) = T \sum_{k=-\infty}^{\infty} x(t)\delta(t - kT). \qquad (3.2.2)$$

Thus we now have a signal $y(t)$ that contains only the sample values of $x(t)$ and all values in between the sampling instants have been discarded.

The sampling operation is best viewed as a conversion from continuous time to discrete-time and hence from a waveform into a sequence. Additional insight into the implications of sampling, however, comes from viewing sampling as a modulation process. Amplitude modulation in its usual form is the process of multiplying a waveform by a sinusoid. More generally, if we multiply a waveform by a sum of sinusoids we are still doing modulation. The sampling function $s(t)$ is a periodic signal and can be represented as a sum of sine waves by using a formal Fourier series. (We use the caveat "formal" because the series does not exist in an ordinary sense and careful analysis requires using generalized functions to handle the Dirac delta functions.) Thus,

$$s(t) = \sum_{n=-\infty}^{\infty} c_n e^{jn2\pi f_s t}. \qquad (3.2.3)$$

It happens that all the Fourier coefficients are equal to unity. This is easily verified by applying the formula for the Fourier coefficients:

$$c_n = \frac{1}{T} \int_{-T/2}^{T/2} s(t)e^{-jn2\pi f_s t} \, dt = \frac{1}{T} \int_{-T/2}^{T/2} T\delta(t)e^{-jn2\pi f_s t} \, dt = 1 \quad (3.2.4)$$

since $s(t) = T\delta(t)$ within the interval of integration.

Now we can make it more obvious that sampling is modulation by using (3.2.3) to get

$$
\begin{aligned}
y(t) &= \sum_{n=-\infty}^{\infty} x(t)e^{jn2\pi f_s t} \\
&= x(t)[1 + 2\cos 2\pi f_s t + 2\cos 4\pi f_s t + \cdots] \qquad (3.2.5)
\end{aligned}
$$

It is well known that the effect of amplitude modulation is to shift the frequency spectrum of a signal. Each complex sinusoid term in (3.2.3) corresponds to a particular shift of the original signal spectrum. The overall spectrum of the sampled signal $y(t)$ can be seen by summing the effects of each frequency shift. By taking the Fourier transform of (3.2.5), term by term, we get

$$Y(f) = \sum_{n=-\infty}^{\infty} X(f - nf_s). \qquad (3.2.6)$$

This is the key equation for interpreting the effect of the sampling process.

By sketching the spectral magnitude of $Y(f)$ versus frequency, assuming a simple shape for $X(f)$, the critical question is whether or not the "spectral integrity" of the original signal spectrum will be retained without distortion due to overlaps from different shifted versions of the spectrum $X(f)$. Figure 3.1 illustrates the spectral magnitude of $Y(f)$ for three different values of the *oversampling factor*, ρ_s, defined by $\rho_s = f_s/2W$. In Figure 3.1(a),

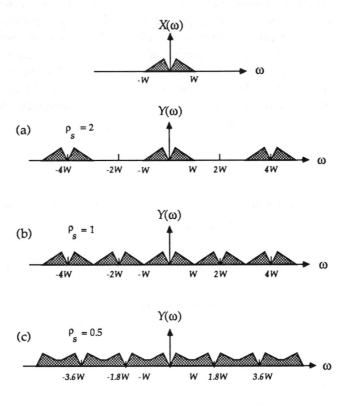

Figure 3.1: Spectra of Sampled Signals

$\rho_s = 2$ and the sampling rate is clearly higher than necessary to preserve the spectral integrity. In Figure 3.1(b), $\rho_s = 1$ and the sampling rate is just barely adequate. Finally, in Figure 3.1(c), where $\rho_s = 0.5$, the spectrum is clearly distorted, due to overlaps of shifted spectra, suggesting that the signal may not be recoverable. Correctly plotting the spectral magnitude of $Y(f)$ in this case would depend on the phase versus frequency characteristics of $X(f)$ as well as on its magnitude since addition of complex numbers is involved here.

The recovery of $x(t)$ from the sampled signal $y(t)$ is most easily considered in the frequency domain. From the spectral plots, it is clear that if the oversampling factor is greater than unity, then an ideal low pass filter $H(f)$ with cutoff w_c Hertz can perfectly recover the original signal as long as the cutoff is chosen so that $W \leq w_c \leq f_s - W$. For $\rho_s > 1$, this condition can always be satisfied. Why? This corresponds to a minimum sampling frequency of $2W$ Hertz, called the *Nyquist frequency* or *Nyquist rate*, which is sufficient to allow (in principle) perfect recoverability of the original signal from its samples. The transfer function of this ideal lowpass filter is given by

$$H(f) = \begin{cases} 1 & \text{if } |f| \leq f_c \\ 0 & \text{if } |f| > f_c \end{cases}, \tag{3.2.7}$$

and its impulse response is

$$h(t) = 2f_c \frac{\sin 2\pi f_c t}{2\pi f_c t} = 2f_c \, \text{sinc}(2f_c t). \tag{3.2.8}$$

In the frequency domain, it is evident that filtering of the signal $y(t)$ by this ideal lowpass filter will produce the output $x(t)$. This discussion, although somewhat casual mathematically, proves the fundamental result:

Theorem 3.2.1 Sampling Theorem: *A signal bandlimited to W Hertz can be exactly reconstructed from its samples when it is periodically sampled at a rate $f_s \geq 2W$.*

We must recognize that the reconstruction filter $H(f)$ as given above is not causal and therefore is not physically realizable. In practice, it can be approximated by a causal lowpass filter with a reasonable "rolloff" interval where the magnitude drops smoothly from unity to zero. To preserve the shape of the original signal spectrum, this rolloff region must be kept outside of the band $|f| \leq W$. Hence, in practice, the oversampling factor must be kept strictly greater than unity to avoid any part of the shifted spectral components from passing through the reconstruction filter. For example, in generating a digital audio signal for storage on compact disc (CD), a sampling rate of 44.1 kHz is used to accommodate an audio signal with bandwidth 20 kHz. This corresponds to an oversampling factor of about

1.1. An even higher sampling rate of 48 kHz (oversampling factor of 1.2) is used for the new consumer digital audio tape (DAT) cassettes. Digital telephony is standardized at a sampling rate of 8 kHz to accommodate a bandwidth of 3.4 kHz for an oversampling ratio of nearly 1.2. The choice of ρ_s is a tradeoff between a desire for the lowest data rate possible and the limit to the complexity of the analog filtering that will be needed: a prefilter (to be discussed later) prior to sampling and the reconstruction filter.

By performing the filtering operation in the time domain, we see that the reconstruction of $x(t)$ is simply a convolution of $h(t)$ with the train of impulses $y(t)$. This immediately simplifies to

$$x(t) \;=\; 2f_cT \sum_{k=-\infty}^{\infty} x(kT)\frac{\sin 2\pi f_c(t-kT)}{2\pi f_c(t-kT)} \tag{3.2.9}$$

$$=\; 2f_cT \sum_{k=-\infty}^{\infty} x(kT)\mathrm{sinc}(2f_c(t-kT)). \tag{3.2.10}$$

This result is primarily of mathematical interest. It does show that the original waveform can be perfectly reconstructed from its samples, but it does not immediately suggest a practical means of accomplishing this. Note that in theory one needs the entire sequence of samples in order to reconstruct the original signal at a single time t.

If, however, we choose the cutoff frequency to be equal to the half-sampling rate, namely, $f_c = f_s/2 = 1/2T$, the reconstruction formula (3.2.10) becomes:

$$x(t) = \sum_{k=-\infty}^{\infty} x(kT)\frac{\sin \pi(t-kT)/T}{\pi(t-kT)/T} = \sum_{k=-\infty}^{\infty} x(kT)\mathrm{sinc}(\frac{t}{T}-k), \tag{3.2.11}$$

which is known as the *sampling expansion*. This important equation explicitly shows how a signal can be recovered from its samples by using the $\mathrm{sinc}(x) = \sin \pi x/\pi x$ functions as interpolation functions. Note that when $t = mT$ for any integer m, all terms in the sum are exactly zero except the mth term which has value $x(mT)$. This interpolation property does not hold for (3.2.10) except for the specific choice of cutoff frequency $f_c = f_s/2$. A similar result holds for the case of random processes. If $\{X(t)\}$ is a random process that is bandlimited in the sense that its power spectral density $S_X(f)$ is identically 0 outside the frequency range $[-W, W]$, then if $f_c > 2W$ the sampling expansion

$$\hat{X}(t) \;=\; 2f_cT \sum_{k=-\infty}^{\infty} X(kT)\frac{\sin 2\pi f_c(t-kT)}{2\pi f_c(t-kT)} \tag{3.2.12}$$

$$= 2f_cT \sum_{k=-\infty}^{\infty} X(kT)\text{sinc}(2f_c(t-kT)) \qquad (3.2.13)$$

gives the original signal in the sense that

$$E[(X(t) - \hat{X}(t))^2] = 0. \qquad (3.2.14)$$

To be precise, the infinite sum must be taken in the quadratic mean sense. (See, for example, Sakrison [280].)

Aliasing

What happens if we violate the condition of the sampling theorem? In other words, suppose the sampling rate is less than twice the maximum frequency component in the spectrum of the signal to be sampled. The answer is easily seen from examining the spectral plots for the case of a modest degree of undersampling, as shown in Figure 3.1(c), and noting from (3.2.6) the spectral overlap. After reconstruction with an ideal lowpass filter with cutoff $f_c = W$, the recovered spectrum in the frequency region close to $f = f_s/2$, will consist of the original spectrum $X(f)$ modified by the addition of $X(f - f_s)$. Thus the low frequency components of the frequency shifted versions of $x(t)$ will be added to the high frequency components close to $f = W$. What has happened is that we recover the sum of the original signal plus an additional undesired waveform whose spectrum overlaps with the high frequency components of the original signal. This undesired component is called *aliasing noise* and the overall effect is referred to as *aliasing* since the noise introduced here is actually a part of the signal itself but with its frequency components shifted to a new frequency. Giving a new name (actually a new frequency) to an old quantity corresponds to the English word *alias*, hence the use of this term. Suppose we choose the cutoff frequency to be $f_c = f_s/2$ instead of the higher cutoff of W in this case of undersampling. What happens? We may avoid aliasing noise (for ρ_s less than unity but fairly close to unity), but what we are doing is completely rejecting the high frequency components of the original signal lying in the interval $f_s/2 \le |f| \le W$. This will typically be more detrimental than aliasing, but any specific conclusion is application dependent.

The idea of aliasing is familiar in the cinematic example. When you watch an old western movie showing the bad guys chasing the stagecoach and the horses panting and sweating as they gallop along at full speed, you may often see the wagon wheels appear to be revolving very slowly. This inconsistency is simply the aliasing of high frequency visual components into a low frequency of motion due to the camera's low sampling rate.

Antialias Filtering

The rate at which data are being produced by the sampling process usually determines the amount of processing or storage that will subsequently be required. Hence, it is desirable to use the lowest possible sampling rate that will satisfy a given application. On the other hand, most signals are not strictly bandlimited. Speech and music for example do not have a well defined cutoff frequency below which significant power density exists and above which no signal power is present. Typically the contribution of the higher frequency signal components gradually diminishes in importance as the frequency increases. So how do we choose a meaningful sampling rate that is not higher than necessary and yet does not violate the sampling theorem?

The answer is that we first decide how much of the original signal spectrum we really need to retain. Then we perform analog lowpass filtering on the analog signal *before* sampling so that we suppress the higher frequency components that we don't really need. This also eliminates out-of-band random noise that might also be present in the original signal and that would otherwise contribute to the aliasing distortion. Consequently, we can now choose a sampling frequency that is not much higher than twice the maximum retained frequency of the signal. Now, we can sample the signal without introducing aliasing noise and still avoid an excessive sampling rate. This analog prefiltering is often called *antialias filtering*.

In digital telephony, the standard antialias filter has a cutoff of 3.4 kHz although the acoustic speech signal contains frequency components extending well beyond this frequency. This cutoff allows the moderate sampling rate of 8 kHz to be used and retains the voice fidelity that was already achieved with analog telephone circuits which already were limited to roughly 3.4 kHz. The antialias filter also has a notch at 60 Hz in North American applications in order to eliminate hum due to electromagnetic pickup in the analog circuitry caused by the presence of the 60 Hz power supply signal.

In summary, analog prefiltering is needed to prevent aliasing of signal and noise components that lie outside of the frequency band that must be preserved and reproduced.

Imaging

Suppose that for sampling we satisfy the requirements of the sampling theorem, with f_s slightly higher than $2W$, but for reconstruction we use a lowpass filter whose cutoff is larger than $f_s/2$ so that it does not remove all frequency components above f_s. This can arise from the use of low-cost, low-order reconstruction filters, with a broader rolloff characteristic extending beyond the half-sampling frequency. From examining the spectrum of the sampled signal using (3.2.6), we can readily see that we will again be reproducing the original signal but added to it is another signal whose spectrum is that part of the shifted input spectrum which is accepted by the reconstruction filter. The result is another form of "noise," but like aliasing, it is signal dependent. The new components are exactly like the original signal but are modulated to a new frequency. (For this reason, it is preferable to avoid the use of the word "noise" altogether.) For a cutoff of $f_c \equiv f_s$, the new spectral component is a mirror image of the original spectrum, shifted into the higher frequency band $f_s/2 \leq |f| \leq f_s$, causing so-called "imaging distortion." In audio signal processing, imaging has a distinctly recognizable effect, sounding "tinny" due to the presence of high frequency components that were not in the original but which are correlated with the original signal.

Just as aliasing is a distortion caused by inadequate analog prefiltering, imaging is a distortion caused by inadequate reconstruction filtering. The two types of analog filtering operations have a complementary role.

3.3 Noise in Sampling

Suppose that a band-limited signal is sampled in such a way that the original signal can be reconstructed from the samples. Prior to reconstruction, however, the samples are corrupted by an additive noise. This might model the inclusion of quantization error, for example. How will this noise in the samples effect the final continuous time reproduction? In this section we use the ideal sampling expansion of the previous section in order to study the average squared reconstruction error. The basic ideas are taken from Sakrison [280], but the proofs are different.

Suppose that a weakly stationary continuous time random process $\{X(t)\}$ is bandlimited to $[-W, W]$ in the sense that its power spectral density $S_X(f)$ is 0 outside this region. Corresponding to (3.2.11), a sampling expansion for $X(t)$ is then

$$X(t) = \sum_{k=-\infty}^{\infty} X(kT) \operatorname{sinc}(2W(t - kT)),$$

where we consider the ideal case with sampling frequency $f_c = 2W$ and sampling period $T = 1/f_c = 1/2W$. Each sample $X_k = X(kT)$ is then corrupted by a discrete time noise term N_k to form a noisy sample $Y_k = X_k + N_k$. We do not assume that the noise N_k are independent or even uncorrelated with the input samples X_k. We do, however, assume that the discrete time noise process is weakly stationary, that is, has constant mean (which for simplicity we assume to be 0) and has an autocorrelation function

$$R_N(k, j) = E(N_k N_j) = R_N(j - k).$$

Both of these assumptions are met, for example, in the case to be considered in Chapters 5 and 6 where the samples $X_k = X(kT)$ are quantized to form a reproduction $\hat{X}_k = q(X_k)$ provided that the input pdf and the quantizer are symmetric about 0. The noisy samples Y_k are then used to reconstruct the final continuous time signal

$$
\begin{aligned}
Y(t) &= \sum_{k=-\infty}^{\infty} Y_k \operatorname{sinc}(2W(t - kT)) \\
&= \sum_{k=-\infty}^{\infty} (X(kT) + N_k) \operatorname{sinc}(2W(t - kT)) \\
&= X(t) + \sum_{k=-\infty}^{\infty} N_k \operatorname{sinc}(2W(t - kT)).
\end{aligned}
$$

An important question is how good an approximation $Y(t)$ will be to $X(t)$. In particular, if the discrete time noise or errors N_k are small on the average, than one would hope that the resulting continuous time reproduction would be good.

To quantify this question we will measure the quality of the overall reproduction (at time t) by the expected squared error, that is, we will evaluate

$$\Delta(t) \equiv E[(X(t) - Y(t))^2].$$

Note that if we define the continuous time process $W(t)$ by

$$W(t) = \sum_{k=-\infty}^{\infty} N_k \operatorname{sinc}(2W(t - kT)),$$

then $X(t) - Y(t) = W(t)$ by construction and hence

$$\Delta(t) = E[W(t)^2].$$

Because of the properties of the sinc function, the samples of the $W(t)$ process are just the noise amplitudes N_k; that is, $W(kT) = N_k$. Thus if we knew that the continuous time process were weakly stationary, we would be done since in that case we would have that

$$
\begin{aligned}
\Delta(t) &= E[W(t)^2] = E[W(0)^2] \\
&= E[N_0^2] = R_N(0). \tag{3.3.1}
\end{aligned}
$$

Thus the overall reproduction average squared error would be independent of time and equal to the average squared sample error. We will demonstrate that this is indeed the case.

The mean of $W(t)$ is easily seen to be 0 and hence is constant. Thus we need only prove that $R_W(t, s)$ depends only on $t - s$. To do this we begin by writing

$$
\begin{aligned}
R_W(t, s) &= E[W(t)W(s)] \\
&= E[\sum_{k=-\infty}^{\infty} N_k \operatorname{sinc}(2W(t - kT)) \sum_{l=-\infty}^{\infty} N_l \operatorname{sinc}(2W(s - lT))] \\
&= \sum_{k=-\infty}^{\infty} \sum_{l=-\infty}^{\infty} \operatorname{sinc}(2W(t - kT)) \operatorname{sinc}(2W(s - lT)) E[N_k N_l] \\
&= \sum_{k=-\infty}^{\infty} \sum_{l=-\infty}^{\infty} \operatorname{sinc}(2W(t - kT)) \operatorname{sinc}(2W(s - lT)) R_N(l - k).
\end{aligned}
$$

We now change variables setting $m = l - k$ to write

$$
R_W(t, s) = \sum_{m=-\infty}^{\infty} R_N(m) \sum_{k=-\infty}^{\infty} \operatorname{sinc}(2W(t - kT)) \operatorname{sinc}(2W(s - (m + k)T)).
$$

$$(3.3.2)$$

Applying the ordinary sampling expansion to the delayed sinc function $\operatorname{sinc}(2W(t + \psi))$ yields (see Problem 3.3)

$$
\operatorname{sinc}(2W(t + \psi)) = \sum_{k=-\infty}^{\infty} \operatorname{sinc}[2W(t - kT)] \operatorname{sinc}[2W(kT + \psi)].
$$

Applying this result to right-most sum of (3.3.2) with $\psi = mT - s$ we obtain

$$
\begin{aligned}
\sum_{k=-\infty}^{\infty} &\operatorname{sinc}(2W(t - kT)) \operatorname{sinc}(2W(s - (m + k)T)) \\
&= \sum_{k=-\infty}^{\infty} \operatorname{sinc}(2W(t - kT)) \operatorname{sinc}(2W((m + k)T - s)) \\
&= \operatorname{sinc}(2W(t + mT - s))
\end{aligned}
$$

and hence (3.3.2) becomes

$$
R_W(t, s) = \sum_{m=-\infty}^{\infty} R_N(m) \operatorname{sinc}(2W(t - s + mT)),
$$

which depends only on the difference $t - s$. Thus $W(t)$ is weakly stationary as claimed and (3.3.1) holds as claimed.

3.4   Practical Sampling Schemes

In practice we do not have impulses at our disposal. The results of the previous section still provide a basis for modeling the sampling process in various realistic situations. One form of sampling called "natural sampling" corresponds to periodically observing a short chunk of a waveform. This is analogous to the soap opera example mentioned earlier. Rather than observe an isolated value, we assume here that we have the freedom to observe the actual waveform for a short time interval τ where $\tau < T$ and this observation is made every T seconds. A second realistic case is where we observe the signal at one instant and we store that value as the amplitude of a rectangular pulse of duration τ thereby generating a piecewise constant waveform that approximates the original. This case corresponds to a *sample-and-hold* system, sometimes called *flat-top sampling*, and is commonly used. In a third type of sampling, which might be called *averaged sampling*, we observe the waveform for a short interval of length τ, but measure the average value of the signal in this interval. This is repeated every T seconds to produce a data sequence. Many other situations can be modeled by combinations of the above. For each case some consideration must be given to determining the appropriate reconstruction method to recover the original waveform.

Natural Sampling

In this case we are selecting "chunks" or segments of the actual signal $x(t)$ observed at periodic intervals, rather than observing the instantaneous value of the signal at isolated, periodically spaced points. As the width of these chunks shrinks, we can expect this case to approach the case of ideal impulse sampling treated above. Intuitively, natural sampling should do as well as ideal sampling since we are retaining more information. In fact, as we depart from ideal reconstruction filtering, it turns out that natural sampling degrades more slowly.

To model natural sampling, we define the periodic sampling function as:

$$s(t) = T \sum_{k=-\infty}^{\infty} q_\tau(t - kT). \tag{3.4.1}$$

The pulse $q_\tau(t)$ as shown in Figure 3.2 is given by

$$q_\tau(t) = \begin{cases} 1/\tau & \text{for } |t| \leq \tau/2 \\ 0 & \text{otherwise.} \end{cases}$$

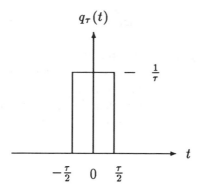

Figure 3.2: Natural Sampling Pulse

Note that this pulse has the Fourier transform

$$Q_\tau(f) = \frac{\sin \pi f \tau}{\pi f \tau} = \text{sinc}(f\tau), \qquad (3.4.2)$$

and that as $\tau \to 0$, the pulse $q_\tau(t)$ "approaches" the delta function $\delta(t)$ (that is, it can be used to define the Dirac δ as a generalized function). We again take the viewpoint that sampling is a modulation operation and consider the Fourier series of $s(t)$ which has the form

$$s(t) = \sum_{k=-\infty}^{\infty} c_k e^{jk2\pi f_s t}$$

where the Fourier coefficients are easily found to be

$$c_k = \frac{1}{T} Q(kf_s)$$

for $k \neq 0$ and $c_0 = 1$. Note that $c_k \to 1/T$ as $\tau \to 0$ so that the results of idealized sampling are approached as the observation width τ approaches zero. Now we multiply $x(t)$ by $s(t)$ to get the sampled signal

$$y(t) = x(t)s(t) = x(t)[1 + 2c_1 \cos 2\pi f_s t + 2c_2 \cos 4\pi f_s t + \cdots] \qquad (3.4.3)$$

and in the frequency domain, we have

$$\begin{aligned}
Y(f) = &X(f) + c_1[X(f - f_s) + X(f + f_s)] \\
&+ c_2[X(f - 2f_s) + X(f + 2f_s)] + \cdots.
\end{aligned} \qquad (3.4.4)$$

From this equation we see the same effect as in the ideal sampling case except that each shifted spectrum has its amplitudes attenuated by a factor

c_k that is less than unity. The unshifted spectrum is reproduced without any attenuation. Hence, the same conditions are required for reconstruction as in ideal sampling and the same issues of aliasing and imaging are applicable here. The only minor difference is that for the same quality of antialias filtering and reconstruction filtering, the distortions due to aliasing and imaging will be less severe since the shifted spectra are somewhat reduced in amplitude due to the nonzero observation window of natural sampling.

Sample-And-Hold

We define an idealized sample-and-hold circuit as a system that maps an input waveform $x(t)$ into the output waveform of the form

$$y(t) = \sum_{k=-\infty}^{\infty} x(kT)g_\tau(t - kT) \tag{3.4.5}$$

as illustrated in Figure 3.3. Typically $\tau = T$ in single channel sampling

Figure 3.3: Sample-and-Hold

systems but other applications may require values of τ much smaller than the sampling period T. This type of sampling is easily modeled by the structure of Figure 3.4 which shows an ideal periodic impulse sampling scheme followed by a filter G whose impulse response is given by

$$g_\tau(t) = \begin{cases} 1 & \text{for } 0 \le t \le \tau \\ 0 & \text{otherwise} \end{cases}$$

and whose transfer function is

$$G_\tau(f) = \tau \frac{\sin(\pi f \tau)}{\pi f \tau} e^{-j\pi f \tau}$$

If we take the Fourier transform, $Y(f)$, of $y(t)$ as given in (3.4.3) we can again deduce the appropriate reconstruction technique for recovering $x(t)$.

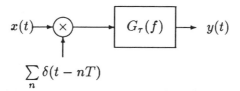

Figure 3.4: Sample-and-Hold as Filtered Impulse Train

We then have

$$Y(f) = G_\tau(f) \sum_{n=-\infty}^{\infty} X(f - nf_s), \qquad (3.4.6)$$

where we have factored out $G_\tau(f)$ from each term in the sum. Note that
(3.4.6) differs from the spectrum in ideal sampling, given in (3.2.6), only
by the factor $G_\tau(f)$. Except for this factor, the spectra of $y(t)$ will be
the same as shown in Figure 3.1. Hence, it is possible to reconstruct the
original waveform from $y(t)$ assuming the sampling theorem is satisfied and
that $\tau < T$. We will again want to use an ideal lowpass filter to remove
the shifted spectral components. But that is not enough. From the above
equation it is clear that the effect of ideal lowpass filtering (with suitable
cutoff frequency) is to produce the waveform $\tilde{x}(t)$ whose spectrum is

$$\tilde{X}(f) = G_\tau(f)X(f)$$

from which it is seen that an additional filtering operation is needed, namely
one that inverts the characteristic frequency response of $G_\tau(f)$ over the
frequency range $|f| < W$ where W is the bandwidth of $x(t)$. Since $\tau < T$,
the first zero crossing of $G(f)$ occurs at $1/\tau > 1/T = f_s > 2W$. Hence,
the transfer function is invertible in the frequency range $|f| \leq W$ without
any difficulty. Figure 3.5 shows the magnitude of the reconstruction filter
which will (apart from a time delay) reproduce $x(t)$ exactly. This modified
lowpass filtering task is referred to as "$\sin x/x$ compensation" and is often
included in digital to analog conversion chips.

Averaging Sampling

Averaging sampling can be described as follows. We observe the sequence
of numbers derived from the original signal $x(t)$ according to

$$u_n = \frac{1}{\tau} \int_{nT-\tau}^{nT} x(t)\, dt.$$

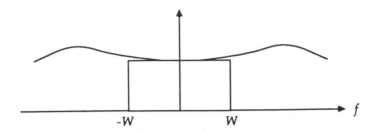

Figure 3.5: Sample-and-Hold Reconstruction Filter

Thus we periodically observe chunks of the original signal of width τ and we take the average value in each chunk to get the data sample u_n. We now determine under what conditions and how we can recover $x(t)$ from the data $\{u_n\}$.

This situation can be modeled as follows. Let $u(t)$ be the output of a linear filter, G, with impulse response $\frac{1}{\tau}g_\tau(t)$ ($g_\tau(t)$ was defined earlier) and whose input is the signal $x(t)$. Thus by convolution,

$$u(t) = \frac{1}{\tau} \int_{-\infty}^{\infty} g(s)x(t-s)\,ds = \frac{1}{\tau} \int_{0}^{\tau} x(t-s)\,ds = \frac{1}{\tau} \int_{t-\tau}^{t} x(\nu)\,d\nu$$

and therefore by sampling $u(t)$ at T second intervals, we have

$$u_n = u(nT).$$

Thus we see that the averaging operation can simply be modeled as a prefiltering operation on $x(t)$. Figure 3.6 shows the overall model of the averaging sampling scheme. When the signal spectrum is nonzero only

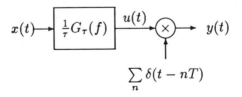

Figure 3.6: Averaging Sampling

in the band $0 \le |f| \le W$, we can perform ideal sampling on $u(t)$ and perfectly reconstruct $u(t)$ and then postfilter with the inverse of $G_\tau(f)$ as in the sample-and-hold case. We thereby recover $x(t)$ perfectly, provided we satisfy the sampling theorem condition that $f_s = 1/T \geq 2W$ where W is the maximum frequency limit of $x(t)$.

This discussion illustrates the almost universal applicability of ideal impulse sampling as a building block model for more practical realistic sampling situations. Similar techniques can be used to model other practical sampling circuitry.

3.5 Sampling Jitter

In analog to digital conversion, the clock that is used to determine the time at which each sample is observed may contain phase jitter. Similarly, in digital to analog conversion, the clock that determines the time at which successive samples are applied to the reconstruction circuit may contain phase jitter. One common source of timing jitter arises in digital communications when the clock is derived by a phase-locked loop or other timing recovery mechanism in a receiver. Jitter refers to a random perturbation of the actual sampling instant around the nominal value. The frequency of the clock is assumed to be fixed in this discussion while the jitter is assumed to be a zero mean random variable that is added to the nominal sampling instants.

Here we consider the effect of sampling jitter introduced in the analog to digital converter. Suppose that each ideal sampling instant nT is perturbed by a random variable θ_n with zero mean and variance σ_θ^2 (independent of n) with $\sigma_\theta^2 \ll T$. We define the *jitter fraction* as $\rho_J = \sigma_\theta/T$. Typically, the rms jitter is below 1% of the sampling period, so that $\rho_J < 0.01$. The actual sampling instants in the presence of jitter are

$$t_n = nT + \theta_n. \tag{3.5.1}$$

In order to study the effect of jitter, it is convenient to assume the input signal $x(t)$ to be sampled is a stationary random process. Let $R_x(\tau)$ and $S_x(f)$ denote respectively the autocorrelation function and spectral density of $x(t)$. As a measure of the effect of jitter, it is useful to determine the mean squared error in the sample amplitude due to jitter. This is given by

$$D_J = E\left([x(nT) - x(nT + \theta_n)]^2\right) \tag{3.5.2}$$

Performing the squaring operation and taking the conditional expectation given θ_n gives

$$D_J = 2E[R_x(0) - R_x(\theta_n)]$$

since $R_x(0) = E[x^2(nT)] = E[x^2(nT + \theta_n)]$ and $R_x(\theta_n) = E[x(nT)x(nT + \theta_n)]$ where we have assumed the jitter is statistically independent of the input signal so that using nested expectations can be used. Now we average over the distribution of θ_n, and for notational convenience suppressing the dependence of θ_n on the index n, yielding

$$
\begin{aligned}
D_J &= 2E \left(\int_{-\infty}^{\infty} S_x(f)\, df - \int_{-\infty}^{\infty} S_x(f) e^{j2\pi f \theta}\, df \right) \\
&= 2E \left(\int_{-\infty}^{\infty} S_x(f)[1 - \cos 2\pi f \theta]\, df \right) \\
&\approx E \left(\int_{-\infty}^{\infty} S_x(f)(2\pi f)^2 \theta^2\, df \right) \\
&= 4\pi^2 \sigma_\theta^2 \int_{-\infty}^{\infty} f^2 S_x(f)\, df,
\end{aligned}
\tag{3.5.3}
$$

where we have used the Fourier transform relationship between the auto-correlation function and the spectral density and the fact that $S_x(f)$ is real and even; also, we have replaced the cosine term by a quadratic approximation to its power series representation. This approximation is accurate in the frequency band containing the input signal under the assumption that the jitter fraction is small.

Now the integral can be recognized as the product of the mean squared bandwidth W_{rms}^2 of $x(t)$ and the variance σ_x^2 of $x(t)$. Hence we obtain the convenient relation,

$$
D_J = 4\pi^2 \sigma_x^2 W_{\text{rms}}^2 \sigma_\theta^2.
\tag{3.5.4}
$$

Let η denote the ratio of rms bandwidth W_{rms} to signal cutoff frequency W, a characteristic of the spectral shape. Then, assuming that $f_s = 2W$, we find that the ratio of the mean squared jitter noise to mean squared signal level is given by

$$
\frac{D_J}{\sigma_x^2} = \pi^2 \eta^2 \frac{\sigma_\theta^2}{T^2}.
$$

Thus we obtain a convenient formula for the signal to jitter noise power ratio in dB, given by

$$
\text{SNR} = 20 \log_{10}(T/\sigma_\theta) + C.
\tag{3.5.5}
$$

where the constant, $C = -20 \log_{10}(\pi\eta)$, is typically in the neighborhood of 0 dB. For example, a jitter fraction of 1% gives an approximate SNR of 40 dB. For a more extensive treatment of sampling jitter, see Liu and Stanley [217].

3.6 Multidimensional Sampling

Can a painting such as the Mona Lisa be exactly determined from knowledge of the amplitude values of each color component at a regularly spaced set of points on the canvas? Sampling is indeed possible, useful, and needed for signals such as images that are functions of a vector variable. By analogy with one-dimensional signals, we use the notation $\mathbf{t}$ to denote the vector variable that in the one-dimensional case is usually the time variable. For an image, the variable is $\mathbf{t} = (t_1, t_2)$ where t_1 and t_2 are the horizontal and vertical coordinates of an image. For a "moving" or time-varying image, the vector would be three-dimensional, using time as the third component. Occasionally there is a need to study multidimensional signals where the argument $\mathbf{t}$ is three-dimensional or even higher. Most frequently, the dimension is two and the application is in sampling of an image for subsequent digitization and digital processing. Hence the variable $\mathbf{t}$ is often called a *space* variable and the description of the signal as a function of spatial position $\mathbf{t}$ is often called the *space domain* corresponding to the use of the *time domain* for one-dimensional signals. Usually the term "multidimensional signal" is used in the literature to refer to signals whose independent variable is a vector whereas "vector signals" refers to vector-valued signals that are functions of the scalar t representing time, such as might arise in multiple channel processing. Occasionally, there is a need to consider "multidimensional vector-valued signals." For example in color images, we need three amplitude values (for three primary color components) at each point in the plane. To avoid confusion in this case, we use the term "vector-valued signals" in place of vector signals.

Our object now is to examine how we can describe a signal $x(\mathbf{t})$ by a countable set of sample values, thereby reducing the data needed to specify (and reconstruct) the signal in order to allow us to operate in the world of sampled data (corresponding to discrete-time signal processing for one-dimensional signals). It is convenient to have a concept similar to *periodic* sampling in the one-dimensional case. But how do we define a periodic signal in the multidimensional case? In two dimensions, this problem has effectively been solved long ago by fabric or carpet designers (very likely without the aid of mathematics) who have designed intricate but periodic patterns to decorate their products. Henceforth, we assume all vectors are two-dimensional. Three and higher dimensional sampling can be readily derived by a straightforward extension of the two-dimensional results. We next consider the concept of periodicity in the plane and then we formulate the sampling and reconstruction of two-dimensional signals using a lattice of sampling points in the plane.

Periodic Signals in the Plane

A signal $x(t)$ is *periodic in two dimensions* if there exist two linearly independent vectors **a** and **b**, such that

$$x(t) = x(t + a) \text{ for all } t$$

and

$$x(t) = x(t + b) \text{ for all } t.$$

The vectors **a** and **b** are called *periodicity vectors*. If we define the *periodicity matrix*, **C**, whose columns are the periodicity vectors, then **C** is a nonsingular matrix, given by

$$\mathbf{C} = \begin{bmatrix} a_1 & b_1 \\ a_2 & b_2 \end{bmatrix}.$$

The periodicity condition can then be simply restated as follows: a signal $x(t)$ is periodic with periodicity matrix **C** if for any integer vector $\mathbf{m} = (m_1, m_2)^t$,

$$x(t + Cm) = x(t)$$

for all **t**. For an alternate presentation of two-dimensional sampling, see [102]. A more general treatment of sampling of multidimensional signals can be found in [254].

Note that certain patterns in the plane could have periodicity along one particular direction without having two-dimensional periodicity. A simple example of this is a pattern of alternating black and white diagonal stripes of constant width. Such signals correspond to a degenerate case of periodicity where the periodicity matrix is singular. Sampling of such signals can be handled using one-dimensional techniques. On the other hand, a black and white checkerboard pattern of squares (infinitely extended in each direction) is an example of a nondegenerate signal with two-dimensional periodicity. For example, the checkerboard with unit length squares parallel to the coordinate axes has the periodicity vectors $(1, 1)^t$ and $(2, 0)^t$.

A periodic pattern in the plane may be described as a tiling or *tessellation* of the plane by congruent tiles each containing the same pattern and each located at a different location by a translation (without any rotation) from the location of a reference tile. In order to cover the plane without overlap and without any gaps between tiles, the only allowable convex tile shapes are squares, rectangles, parallelograms, and hexagons with opposite edges parallel. (A triangle is not acceptable since it requires rotations as well as translations to tile the plane.) If the periodicity vectors are orthogonal, then the tiles are squares or rectangles. Otherwise, it is always possible to find a hexagon and a parallelogram as the basic tile shape either of which can be separately used for tiling the plane to generate the periodic signal.

In the case of one-dimensional signals, a periodic signal with period T is also periodic with period $2T$ but we normally consider the true period, or *fundamental* period to be T. Similarly, in two-dimensional signals, the *fundamental* periodicity matrix $\mathbf{C}$ is that periodicity matrix whose determinant has the smallest magnitude. This corresponds to a tessellation with tiles of smallest area. In fact, we shall see that the determinant of the periodicity matrix is equal to the area of the corresponding tile.

Figure 3.7 illustrates two examples of a two-dimensional periodic signal and shows how each pattern can be tiled with rectangles, parallelograms, and hexagons. Figure 3.7(a) is a rectangle, 3.7(b) is a parallelogram, and 3.7(c) is a hexagon Note that in each case the tiles have the same area.

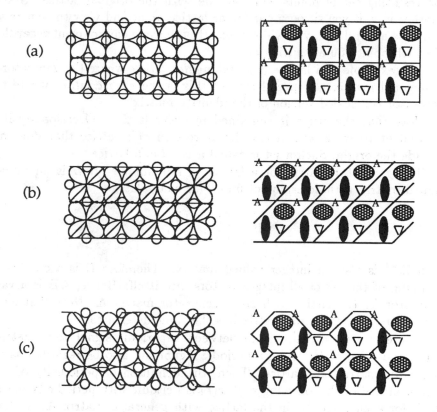

Figure 3.7: Two-Dimensional Periodic Signals in the Space Domain.

Lattices

Intimately tied to periodic sampling is the concept of a lattice. A (two-dimensional) lattice is in essence a set of points with a regular arrangement in the plane. Specifically, we define a *lattice* as a set of points in the plane of the form:

$$\mathcal{L} = \{\mathbf{t} \in \mathbf{R}^2 : \mathbf{t} = \mathbf{A}\mathbf{m} \text{ for all pairs of integers } \mathbf{m} = (m_1, m_2)\}$$

where $\mathbf{A}$ is a nonsingular (2×2) matrix called the *generator matrix* or simply the *generator* of the lattice and $\mathbf{m}$ is an integer vector. A lattice is a very natural set of points to use for sampling a signal.

A key feature of a lattice is the regularity with which the points are distributed in the plane. In particular, if we select any point, $\mathbf{u}$, in the lattice and then shift the entire lattice by subtracting $\mathbf{u}$ from each point, the resulting set of points will coincide with the original lattice. Stated another way, if you sit on a particular lattice point and view the entire set of points around you, you will see exactly the same environment regardless of which point you are sitting on.

One example of a lattice is a *rectangular lattice* where the generator is a diagonal matrix. A special case of this is the *square lattice* where the generator is a scaled version of the identity matrix.

Note that the origin is contained in every lattice. Occasionally it is convenient to work with a translated version of a lattice that does not include the origin. Such a set is called a *coset* of a lattice.

The generator matrix for a lattice is not unique. If B is an integer valued matrix with determinant unity, for example,

$$\mathbf{B} = \begin{bmatrix} 1 & -1 \\ 0 & 1 \end{bmatrix},$$

then $\mathbf{B}^{-1}$ is also an integer valued matrix. Therefore $\mathbf{B}$ is a one-to-one mapping of the set of all integer vectors into itself. Hence, $\mathbf{AB}$ is a valid generator for the lattice with given generator matrix $\mathbf{A}$. Note that every such equivalent generator matrix has the same determinant.

There is a close relationship between a periodic signal and a lattice. The periodicity matrix $\mathbf{A}$ of a periodic signal can also be regarded as the generator matrix of a lattice. Then, the definition of periodicity of $x(\mathbf{t})$ can be stated in the following way. $x(\mathbf{t})$ is periodic with periodicity matrix $\mathbf{A}$, if for each point $\mathbf{t_n}$ in the lattice with generator matrix $\mathbf{A}$, we have $x(\mathbf{t}) = x(\mathbf{t} + \mathbf{t_n})$. From the previous paragraph, it also follows that the periodicity matrix of a given periodic signal is not unique; however, each valid periodicity matrix for the given signal will have the same determinant.

Lattice Transformations on Periodic Signals

It is always possible to perform a warping of the plane, by means of a linear transformation, to map any lattice into a square lattice whose points consist of all pairs of integers (m, n), i.e., the lattice with generator is the identity matrix $\mathbf{I}$. Such a warping maps a signal with arbitrary periodicity into a signal whose periodicity is rectangular. Specifically, if a signal $x(\mathbf{t})$ is periodic with periodicity matrix $\mathbf{C}$, let $\tau = \mathbf{C}^{-1}\mathbf{t}$, we define a transformed signal

$$y(\tau) = x(\mathbf{C}\tau).$$

Now we demonstrate that we have a square lattice, and correspondingly, rectangular periodicity in the τ plane. Let $\mathbf{u}_1 = (1, 0)^t$ and $\mathbf{u}_2 = (0, 1)^t$ so that we have the periodicity matrix $\mathbf{I}_2$, the 2×2 identity matrix. Thus if $x(\mathbf{t})$ is periodic in the τ plane with periodicity matrix $\mathbf{C}$, we have

$$y(\tau + \mathbf{u}_1) = x(\mathbf{C}\tau + \mathbf{C}\mathbf{u}_1) = x(\mathbf{t} + \mathbf{a}) = x(\mathbf{t}) = y(\tau).$$

Similarly,

$$y(\tau + \mathbf{u}_2) = y(\tau).$$

Without vector notation, the periodicity conditions on $y(\tau_1, \tau_2)$ are:

$$y(\tau_1 + 1, \tau_2) = y(\tau_1, \tau_2)$$

and

$$y(\tau_1, \tau_2 + 1) = y(\tau_1, \tau_2)$$

for all values of τ_1 and τ_2, which shows that $y(\tau_1, \tau_2)$ is a periodic function of τ_1 when τ_2 is held fixed and vice versa. This is a specialization of the general definition of periodicity to the case where the periodicity vectors are $\mathbf{u}_1$ and $\mathbf{u}_2$.

Two-Dimensional Fourier Analysis

To proceed further, we assume the reader is familiar with the basic idea of a two-dimensional Fourier transform and with Fourier series for rectangular periodicity in two dimensions. We first review these relations without any derivation. Then we extend the Fourier series to include more general periodic signals in two dimensions.

The two-dimensional Fourier transform $U(\mathbf{f})$ is defined by

$$U(\mathbf{f}) = \int_{-\infty}^{\infty} \int_{-\infty}^{\infty} u(\mathbf{t}) e^{-j 2\pi \mathbf{f}^t \mathbf{t}} dt,$$

and the corresponding inverse relation is given by

$$u(t) = \int_{-\infty}^{\infty} \int_{-\infty}^{\infty} U(\mathbf{f}) e^{j 2\pi \mathbf{f}^t \mathbf{t}}\, d\mathbf{f}.$$

The vector frequency variable is $\mathbf{f} = (f_1, f_2)^t$. Note that for a real-valued function $u(t)$, the magnitude of the Fourier transform is an even function in the sense that $|U(\mathbf{f})| = |U(-\mathbf{f})|$.

Two-dimensional Fourier analysis is widely used in image processing. In this context, the variables f_1 and f_2 are called *spatial frequency* variables when t_1 and t_2 have the physical meaning of space variables, i.e., they measure length rather than time or other physical quantities.

Let $y(\tau)$ denote a signal with unit square periodicity so that it is separately periodic in τ_1 and in τ_2 with period 1 for each variable. Then it can be represented as

$$y(\tau) = \sum_{n_1=-\infty}^{\infty} \sum_{n_2=-\infty}^{\infty} d_{n_1 n_2} e^{j 2\pi (n_1 \tau_1 + n_2 \tau_2)} = \sum_{\mathbf{n}} d_{\mathbf{n}} e^{j 2\pi (\mathbf{n}^t \tau)} \qquad (3.6.1)$$

with the Fourier coefficients give by

$$d_{\mathbf{n}} = \int_{-\frac{1}{2}}^{\frac{1}{2}} \int_{-\frac{1}{2}}^{\frac{1}{2}} y(\tau) e^{-j 2\pi (n_1 \tau_1 + n_2 \tau_2)} d\tau_1\, d\tau_2 = \int_S y(\tau) e^{-j 2\pi (\mathbf{n}^t \tau)}\, d\tau, \quad (3.6.2)$$

where $\mathbf{n}$ is the integer vector (n_1, n_2) and S denotes the unit square centered at the origin:

$$S = \{\tau : |\tau_1| \le \frac{1}{2},\ |\tau_2| \le \frac{1}{2}\}.$$

This is a standard two-dimensional Fourier series which follows directly from the usual one-dimensional Fourier series. It is applicable to any two-dimensional signal that is separately periodic in each of its variables with period 1. We call this representation a *rectangular* Fourier series. It can be derived by first representing $y(\tau)$ with a Fourier series in τ_1, obtaining coefficients that are periodic in τ_2 and then the coefficients are themselves represented by a Fourier series in the variable τ_2.

Suppose now that we are interested in a signal $x(t)$ with a periodicity matrix $\mathbf{C}$. The rectangular Fourier series of (3.6.1) is not directly applicable to the case of two-dimensional periodicity when C is not a diagonal matrix. However, we can derive a more general Fourier series representation for arbitrary periodic signals by introducing a change of variables that maps (i.e., warps) the signal into one with rectangular periodicity.

We apply the rectangular Fourier series defined above to the periodic signal $y(\tau) = x(\mathbf{C}\tau)$ whose periodicity matrix is $\mathbf{I}_2$. By simply substituting $\tau = \mathbf{C}^{-1}\mathbf{t}$ into the Fourier series representation of $y(\tau)$, we get

$$x(\mathbf{t}) = \sum_{\mathbf{n}} d_{\mathbf{n}} e^{j2\pi \mathbf{n}^t \mathbf{C}^{-1} \mathbf{t}} \tag{3.6.3}$$

where

$$d_{\mathbf{n}} = \frac{1}{|\det \mathbf{C}|} \int_P x(\mathbf{t}) e^{-j2\pi \mathbf{n}^t \mathbf{C}^{-1} \mathbf{t}} \, d\mathbf{t}. \tag{3.6.4}$$

The region of integration P is the set

$$P = \{\mathbf{t} : \mathbf{C}^{-1}\mathbf{t} \in S\},$$

which is the inverse image of the unit square $\mathbf{S}$ and is a parallelogram, as shown in Figure 3.8, formed from the two column vectors $\mathbf{a}$ and $\mathbf{b}$ of $\mathbf{C}$. Note that the Jacobian of the transformation $\tau = \mathbf{C}^{-1}\mathbf{t}$ is the magnitude of the determinant of $\mathbf{C}^{-1}$ and is equal to $|\det \mathbf{C}|^{-1}$. Equation (3.6.3) is a general Fourier series representation for any periodic signal $x(\mathbf{t})$ in two dimensions with periodicity matrix $\mathbf{C}$ and (3.6.4) gives the formula for finding the Fourier coefficients for representing the signal. The set P identifies a particular region of the periodic signal $x(\mathbf{t})$ corresponding to a tile that generates the signal by a tessellation of the plane. It can be shown that any alternate set corresponding to a tessellating tile (with the same area) can be used as the region of integration in (3.6.4) to evaluate the integral.

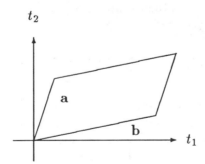

Figure 3.8: Parallelogram in the Space Domain.

Lattice Sampling in Two Dimensions

We follow the approach of Section 3.2 where sampling is treated as a modulation process. It is again convenient to use the Dirac delta function but now it must be generalized to the two-dimensional case. Following the usual defining property of the one-dimensional delta function, we let $\delta(\mathbf{t})$ be that function which satisfies

$$\int_G \delta(\mathbf{t})g(\mathbf{t})\,dt = \begin{cases} g(0) & \text{if } \mathbf{0} = (0,0) \in G \\ 0 & \text{otherwise} \end{cases}$$

for any function $g(\mathbf{t})$ that is continuous at the origin, where G is any open region in $\mathbf{R}^2$. Thus $\delta(\mathbf{t})$ has total volume equal to unity and is zero for any $\mathbf{t} \neq (0,0)$.

We define the two-dimensional sampling function

$$s(\mathbf{t}) = |\det \mathbf{C}| \sum_{\mathbf{m}} \delta(\mathbf{t} - \mathbf{Cm})$$

where the sum is taken over all integer vectors $\mathbf{m}$ and $\mathbf{C}$ is a nonsingular matrix. Observe that $s(\mathbf{t})$ is a periodic signal with periodicity matrix $\mathbf{C}$. Note that $|\det \mathbf{C}|$ is the area of the parallelogram P, so that $s(\mathbf{t})$ has been defined in such a way that it has an average value of unity. The two-dimensional analog of the term "sampling rate" is the *sampling density* defined as the number of signal samples per unit area. Since the area of the parallelogram P is $|\det \mathbf{C}|$, the absolute value of the determinant of $\mathbf{C}$, we see that the sampling density is $|\det \mathbf{C}|^{-1}$ points per unit area. As discussed earlier, the periodicity matrix is not unique; however each equivalent periodicity matrix has the same determinant, so that the sampling density is well-defined.

The sampling process is then simply modeled by taking the product of $x(\mathbf{t})$ and $s(\mathbf{t})$ to form $y(\mathbf{t})$ which is given by

$$y(\mathbf{t}) = |\det \mathbf{C}| \sum_{\mathbf{m}} x(\mathbf{Cm})\delta(\mathbf{t} - \mathbf{Cm}).$$

This clearly consists only of sample values of the original signal $x(\mathbf{t})$.

The sampling function is amenable to Fourier analysis (with the same qualifying remarks about the lack of mathematical rigor as in the one-dimensional case). Applying the Fourier series representation (3.6.3) and (3.6.4) to $s(\mathbf{t})$, we find that $d_{\mathbf{n}} = 1$ for all $\mathbf{n}$. The result is that we have the two-dimensional Fourier series for the sampling function given by

$$s(\mathbf{t}) = \sum_{\mathbf{n}} e^{j 2\pi \mathbf{n}^t \mathbf{C}^{-1} \mathbf{t}}.$$

Now, to determine the conditions needed for recovery of the original signal we take the viewpoint of sampling as modulation, viewing the sampling function as a sum of sinusoids. This gives

$$y(t) = s(t)x(t) = \sum_n x(t)e^{j2\pi n^t C^{-1}t}.$$

As always, modulation simply introduces frequency-shifted copies of the spectrum. Taking the two-dimensional Fourier transform of $y(t)$ above, we get

$$Y(f) = \sum_n X(f - C^{-t}n),$$

where $C^{-t} = (C^{-1})^t$. This is the key equation that allows us to determine the conditions for reconstructing $x(t)$ from its sample values taken on a lattice generated by the matrix C. We see that the spectrum of $x(t)$ is periodically repeated in the frequency plane with the periodicity matrix $D = C^{-t}$, which is the transpose of the inverse of the periodicity matrix in the spatial domain. The two column vectors, g and h of D, are the periodicity vectors associated with the periodic spectrum, $Y(f)$, of the sampled signal. These vectors constitute what is known as the *reciprocal* or *dual* basis to the set a and b, the periodicity vectors corresponding to C defined earlier.

To recover $x(t)$ from the sampled signal $y(t)$, it is now evident that we must avoid spectral overlap of the original spectrum $X(f)$ with the shifted spectral components. A sufficient condition for recoverability is that $X(f)$ have a limited bandwidth in the sense that it must be zero outside a suitably bounded set B containing the origin in the f plane. This set is called the *baseband*. Thus, we assume $X(f) = 0$ for $f \notin B$. Since the spectrum must have even magnitude, then for any f_0 in B, the set B must have the property that $-f_0$ is also in B. Figure 3.9 illustrates examples of different sampled spectra $Y(f)$ where the shaded areas are regions with nonzero spectral content in the frequency plane. If

$$Y(f) = \sum_n X(f - C^{-t}n),$$

for all f and for any $n \neq 0$, we can recover $x(t)$ by applying a lowpass filtering operation to $y(t)$ where the filter transfer function is

$$H(f) = \begin{cases} 1 & \text{if } f \in B \\ 0 & \text{if } f \notin B \end{cases}.$$

Thus, $H(f)Y(f) = X(f)$, and so $x(t)$ is given by the convolution of $y(t)$ and $x(t)$. This gives

$$x(t) = |\det C| \sum_m x(Cm)h(t - Cm)$$

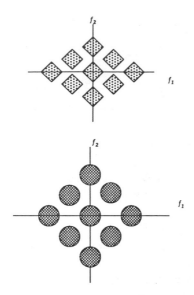

Figure 3.9: Two-Dimensional Examples of Sampled Spectra

where

$$h(\mathbf{t}) = \int\!\!\int_B e^{j2\pi\mathbf{f}^t\mathbf{t}}\, d\mathbf{f}.$$

In particular, if $B = \{f_1, f_2 : |f_1| < F_1 \ |f_2| < F_2\}$ then

$$h(\mathbf{t}) = 4F_1 F_2 \operatorname{sinc}(2F_1 t_1)\operatorname{sinc}(2F_2 t_2)$$

For a given sampling lattice with generator matrix $\mathbf{C}$, the spectrum $Y(\mathbf{f})$ is periodic with periodicity matrix $\mathbf{D}$ and the sampling density is $d_s = |\det\mathbf{C}|^{-1}$, and there is indeed more than one baseband set B with maximal area d_s that will allow perfect recoverability. A necessary but not sufficient condition to allow perfect recovery of the signal is that the spectrum be bandlimited to a region of area not greater than d_s. The spectral limitations imposed on the signal $x(\mathbf{t})$ can take on several forms. Consider in particular the set B in the $\mathbf{f}$ plane defined by

$$B = \{\mathbf{f} : \mathbf{f} = (\frac{\alpha}{2}\mathbf{g} + \frac{\beta}{2}\mathbf{h}) \text{ for all } \alpha \text{ and } \beta \in (-1, 1)\}$$

This set, shown in Figure 3.10, is a parallelogram centered at the origin having area equal to $|\det\mathbf{C}|^{-1}$ and which is a tile that when shifted to every point on the lattice generated by $\mathbf{C}^{-t}$ will tile the plane without overlapping and without leaving any untiled space. This parallelogram in the frequency domain is determined by the reciprocal lattice vectors to the periodicity vectors used to generate the samples. Of course, this is not the only region that can be used to define an acceptable region for the spectrum that will avoid aliasing. In general, it is also possible and preferable to find a hexagonal spectral region using a hexagonal pattern that tiles the frequency plane. Figure 3.11 illustrates alternate baseband regions with the same area and corresponding to the same periodicity vectors in the frequency plane.

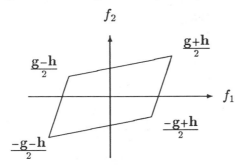

Figure 3.10: Baseband Parallelogram in the Two-Dimensional Frequency Domain.

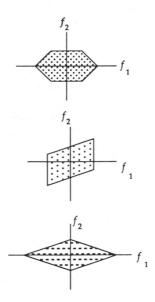

Figure 3.11: Alternate Baseband Regions for the same Sampling Lattice

3.7 Problems

3.1. In this problem we provide a rigorous derivation of the sampling expansion that does not require either the notion of modulation or that of Dirac deltas. Throughout the problem assume that $x(t)$ is a signal with Fourier transform $X(f)$ such that $X(f) = 0$ for $|f| \geq W$.

(a) Show that $X(f)$ can be expressed as a Fourier series on $[-W, W]$, that is, show that

$$X(f) = \sum_{n=-\infty}^{\infty} c_n e^{-2\pi j \frac{f}{2W} n}; f \in [-W, W],$$

where

$$c_n = \frac{1}{2W} \int_{-W}^{W} X(f) e^{j 2\pi \frac{f}{2W} n} \, df; n \in \mathcal{Z}$$

and $\mathcal{Z}$ is the set of all integers.

(b) Use the fact that $X(f) = 0$ outside of $[-W, W]$ to show that

$$c_n = \frac{1}{2W} x\left(\frac{n}{2W}\right)$$

for all n. This proves that

$$X(f) = 1_{[-W,W]}(f) \sum_{n=-\infty}^{\infty} \frac{x\left(\frac{n}{2W}\right)}{2W} e^{-j 2\pi \frac{f}{2W} n},$$

where the indicator function $1_B(f)$ of a set B is defined as 1 if $f \in B$ and 0 otherwise.

(c) Take the inverse Fourier transform of the previous relation and show that

$$x(t) = \sum_{n=-\infty}^{\infty} x\left(\frac{n}{2W}\right) \text{sinc}[2W(t - \frac{n}{2W})].$$

3.2. Suppose that $x(t)$ is a signal bandlimited to $[-W, W]$. Show that for $T < 1/2W$ and any θ the following sampling expansion is valid:

$$x(t) = \sum_{k=-\infty}^{\infty} x(kT - \theta) \text{sinc}[2W(t - kT + \theta)].$$

Hint: Find the sampling expansion for $x(t + \theta)$ and change variables.

3.3. Show that for any fixed ψ the following sampling expansion is valid for the sinc function itself:

$$\text{sinc}(2W(t + \psi)) = \sum_{k=-\infty}^{\infty} \text{sinc}[2W(t - kT)]\,\text{sinc}[2W(kT + \psi)]; \ t \in \mathcal{R}.$$

Hint: Since the sinc function is bandlimited (even with a delay) you can apply the sampling expansion to it.

3.4. Show that the energy of a signal bandlimited to $[-W, W]$ can be expressed as

$$\int_{-\infty}^{\infty} |x(t)|^2 \, dt = \frac{1}{2W} \sum_{n=-\infty}^{\infty} |x(\frac{n}{2W})|^2.$$

3.5. Suppose that $x(t)$ is bandlimited to $[-W, W]$. If all we wish to know is the area

$$A_x = \int_{-\infty}^{\infty} x(t) \, dt,$$

what is the slowest possible rate at which we can sample $x(t)$ and still be able to find A_x? How do we find A_x from the samples?

3.6. A signal bandlimited to $[-W, W]$ is to be sampled with sampling period T where $1/T < 2W$. Specify the bandwidth of the widest rectangular low-pass filter to be used prior to sampling which will eliminate aliasing. If no filter is used prior to sampling, specify the widest rectangular low-pass filter that can be used after sampling to provide an output free from aliasing.

3.7. A signal $x(t)$ is known to have a spectrum that is non-zero only in the range $4W < |f| < 6W$. The signal could be recovered from samples taken at a rate of $12B$ samples per second or more, but one would hope that it could be recovered from samples taken at a much slower rate using a single linear time-invariant interpolation filter. What is the slowest rate for which this is true? Describe the interpolation filter.

3.8. A waveform $x(t)$ with bandwidth limited to F Hz is sampled at a rate $f_s > 2F$. Each sample $x(kT)$ is stored on an MOS capacitor for T sec. and is then dumped. The dumping is imperfect so that a small fraction c, with c much less than unity, of the voltage remains behind and is added to the next sample value applied to the capacitor. The resulting voltage is again stored for T sec. then dumped leaving behind a residual fraction c of the voltage, and so on. The staircase voltage $v(t)$ on the capacitor is applied to an ideal lowpass filter with cutoff F Hz. Determine how the filter output $y(t)$ is related to $x(t)$ in both the time and frequency domains.

3.9. Let $\mathbf{C}$ be a periodicity matrix for a signal that is periodic in two dimensions. Let T denote a shape in the plane of smallest area that is periodically replicated in this signal. Show that $|\det\mathbf{C}|$ is the area of T.

3.10. Consider a general periodic image in the plane whose periodicity vectors are not orthogonal. Show that the image can be characterized by a hexagonal pattern which repeats throughout the image so that the image can be constructed by tiling the plane with infinitely many copies of one standard hexagonal tile. It is sufficient to show the result pictorially with an example, using a clearly drawn picture and an explanation of your reasoning.

3.11. An image plane is periodically sampled with periodicity vectors $(0,1)$ and $(1,2)$. Determine and sketch the bandlimited region of largest area in the frequency plane (f_1, f_2) for which perfect reproduction of the image from its samples is possible.

3.12. Prove the claim that the area of the parallelograms resulting from two-dimensional periodic sampling using the periodicity matrix $\mathbf{C}$ is given by $|\det C|$.

3.13. An image has two-dimensional spectrum with baseband confined to the region where $|f_1|+|f_2| < F$. Find a lattice in the space domain for sampling this image which has the lowest sampling density possible without introducing aliasing.

3.14. Consider the signal $u(y,t)$, the time varying intensity of a fixed vertical cross section of a monochrome video signal for a particular fixed horizontal location. The signal is sampled so that values $y = 2n\delta$ when $t = 2mT$ and $y = (2n+1)\delta$ when $t = (2m+1)T$ for all integers n and m.

    (a) Determine the generating matrix of the lattice of samples.

    (b) Find the region in the two-dimensional frequency plane in which $u(y,t)$ should be bandlimited to avoid any aliasing effects. Note that the frequency plane variables, f_1, f_2, has the meaning that f_1 is a spatial frequency while f_2 is a "temporal" frequency (as in one-dimensional signal analysis). Assume $\delta = 1$ mm and $T = 1/60$ second.

(c) Explain why an ideal rectangular 2D filter (with impulse response of the form $h_1(y)h_2(t)$) cannot effectively reconstruct the signal from its sampled version.

3.15. Derive a general interpolation formula for a two-dimensional signal $x(\mathbf{t})$ that is suitably bandlimited. It should give the signal $x(\mathbf{t})$ as a function of the samples $x(\mathbf{Am})$ for all integer vectors $\mathbf{m}$ where $\mathbf{A}$ is a nonsingular 2×2 matrix. State the result clearly and precisely.

Chapter 4

Linear Prediction

4.1 Introduction

Can we predict the future? Perhaps surprisingly, the answer is a qualified "yes." The key qualification is simply that we cannot predict the future with certainty. In general there is so much built-in structure and inertia in the flow of events that we can do quite a good job estimating the future based on observations of the past and present.

How does this relate to signal compression? We are interested in finding efficient, compact descriptions for random processes. Linear prediction can help to eliminate redundancy so that there is less waste. Consequently, fewer bits are needed to describe each second of a waveform.

Linear prediction is a beautiful and elegant subject. It is intuitively satisfying, it pervades many different disciplines, and is powerful yet fairly simple to understand and use. In fact, linear prediction is just a specialized version of least-squares estimation theory or linear regression. Yet because of its great importance and relevance to signal compression, it warrants a separate treatment in its own right. This chapter is intended to present the most important and most basic results, including the essential theory and design algorithms for putting it to use.

Linear prediction is actually not the ultimate answer in predicting the future. For non-Gaussian processes, linear prediction cannot fully squeeze out all possible information about the past that will help to predict the future. Nonlinear prediction can, but it is much harder to do in practice and we will consider it only briefly in this chapter.

Much of linear prediction consists of variations on a common theme of predicting one vector given another. This topic is both basic and enlightening. Hence it forms the beginning and a common ground for the subsequent results and applications. Section 4.2 derives the elementary results of linear estimation theory needed for the study of linear prediction. Section 4.3 covers the main result of finite memory linear prediction. Section 4.4 presents the concept of forward and backward prediction that is useful for the derivation of the Levinson-Durbin algorithm, a computational technique for linear prediction, the relation between prediction and reflection coefficients, and also the basis of lattice filters for implementing linear prediction, whitening and shaping filters. Section 4.5 presents the important Levinson-Durbin algorithm for computing linear predictor coefficients from autocorrelation values and Section 4.6 describes how to obtain the needed autocorrelation values from empirical data. Sections 4.7, 4.8 and 4.9 present some relevant theoretical topics, including the theory of infinite memory prediction. Finally, Section 4.10 describes how to generate random processes with prescribed spectral densities for computer simulation studies and is based on the theory of infinite memory prediction.

Readers wishing to learn the more practical techniques quickly without the theory can proceed directly to Section 4.4 and continue with Sections 4.5 and 4.6. For a review of results on linear prediction and an extensive bibliography on this subject, see Makhoul [228].

4.2   Elementary Estimation Theory

Prediction is a statistical estimation procedure where one or more random variables are estimated from observations of other random variables. It is called prediction when the variables to be estimated are in some sense associated with the "future" and the observable variables are associated with the "past" (or past and present). The most common use of prediction is to estimate a sample of a stationary random process from observations of several prior samples. Another application is in image compression where a block of pixels is estimated from an observed "prior" block of pixels in a block raster scan of the image. Various other applications arise where prediction of a vector from one or more other vectors is needed. In this section we consider the basic theory needed for a wide variety of applications where prediction is used in signal compression.

Prediction of One Vector Given Another

The simple problem of predicting one vector given another serves to illustrate the basic ideas. It also provides a convenient means of introducing the fundamental results required both in predictive waveform coding for random processes and in block predictive coding used in image coding. Suppose that we observe a random (column, real-valued) vector $\mathbf{X} = (X_1, \cdots, X_N)^t$ and we wish to predict or estimate a second random vector $\mathbf{Y} = (Y_1, \cdots, Y_K)^t$. Note that the dimensions of the two vectors need not be the same. In particular, choosing $K = N = 1$ gives the simplest case of prediction of one scalar random variable from another. Another special case of frequent utility is when $K = 1$ corresponding to the prediction of a random variable from a set of N observed random variables. Note also that in order to do probabilistic analysis we assume that these random vectors are described by a known probability distribution $P_{\mathbf{X}, \mathbf{Y}}$ (which might be estimated based on an empirical distribution).

The prediction $\hat{\mathbf{Y}} = \hat{\mathbf{Y}}(\mathbf{X})$ should be chosen to be the "best" in some sense. Here we will use the most common measure of performance of a predictor: the mean squared error (MSE). For a given predictor $\hat{\mathbf{Y}}$, define the resulting mean squared error by

$$
\begin{aligned}
\epsilon^2(\hat{\mathbf{Y}}) &= E(\|\mathbf{Y} - \hat{\mathbf{Y}}\|^2) \equiv E[(\mathbf{Y} - \hat{\mathbf{Y}})^t(\mathbf{Y} - \hat{\mathbf{Y}})] \\
&= \sum_{i=1}^{K} E[(Y_i - \hat{Y}_i)^2].
\end{aligned}
\tag{4.2.1}
$$

The MSE as a performance criterion can be viewed as a measure of how much the energy of the signal is reduced by removing the predictable information based on the observables from it. Since the goal of a predictor is to remove this predictable information, a better predictor corresponds to a smaller MSE. A predictor will be said to be *optimal* within some class of predictors if it minimizes the mean squared error over all predictors in the given class. The most common classes of predictors are

- linear predictors, and

- unconstrained, possibly nonlinear predictors.

The first class constrains the predictor to be a linear operation on the observed vector. The second class essentially places no constraints on the predictor; it can be an arbitrary function of the observed vector. Such possibly nonlinear predictors are generally capable of better performance because of the lack of restrictions, but linear predictors are usually much simpler to construct. The second class is often referred to as the class of "nonlinear predictors," but a better (if clumsier) name would be "possibly nonlinear predictors" since in some cases the best unconstrained predictor is in fact linear.

We shall have need of two quite different special cases of the above structure. In the first case, the vector $\mathbf{X}$ will consist of N consecutive samples from a stationary random process, say $\mathbf{X} = (X_{n-1}, X_{n-2}, \cdots, X_{n-N})$ and Y will be the next, or "future", sample $Y = X_n$. In this case the goal is to find the best *one-step predictor given the finite past.* In the second example of importance, $\mathbf{Y}$ will be a rectangular subblock of pixels in a sampled image intensity raster and $\mathbf{X}$ will consist of similar subgroups above and to the left of $\mathbf{Y}$. Here the goal is to use portions of an image already coded or processed to predict a new portion of the same image. This vector prediction problem is depicted in Figure 4.1 where subblocks A, B, and C would be used to predict subblock D.

Figure 4.1: Vector Prediction of Image Subblocks

Nonlinear Predictors

First consider the class of nonlinear predictors. The following theorem shows that the best nonlinear predictor of $\mathbf{Y}$ given $\mathbf{X}$ is simply the conditional expectation of $\mathbf{Y}$ given $\mathbf{X}$. Intuitively, our best guess of an unknown vector is its expectation or mean *given* whatever observations that we have.

Theorem 4.2.1 *Given two random vectors* $\mathbf{Y}$ *and* $\mathbf{X}$, *the minimum mean squared error (MMSE) estimate of* $\mathbf{Y}$ *given* $\mathbf{X}$ *is*

$$\hat{\mathbf{Y}}(\mathbf{X}) = E(\mathbf{Y}|\mathbf{X}). \qquad (4.2.2)$$

Proof: Our method of proof will be used repeatedly and is intended to avoid calculus or Lagrange minimizations. The technique works nicely when one knows the answer in advance. We compute the MSE for an arbitrary estimate and use simple inequalities to show that it is bounded below by the MSE for the claimed optimal estimate. The more complicated variational techniques are only necessary when tackling an optimization problem whose solution is not yet known. Hindsight proofs are often far easier than the original developments.

Suppose that $\hat{\mathbf{Y}}$ is the claimed optimal estimate and that $\tilde{\mathbf{Y}}$ is some other estimate. We will show that $\tilde{\mathbf{Y}}$ must yield an MSE no smaller than does $\hat{\mathbf{Y}}$. To see this consider

$$\epsilon^2(\tilde{\mathbf{Y}}) \quad = \quad E(\|\mathbf{Y} - \tilde{\mathbf{Y}}\|^2) = E(\|\mathbf{Y} - \hat{\mathbf{Y}} + \hat{\mathbf{Y}} - \tilde{\mathbf{Y}}\|^2)$$

$$\begin{aligned} &= E(\|\mathbf{Y} - \hat{\mathbf{Y}}\|^2) + E(\|\hat{\mathbf{Y}} - \tilde{\mathbf{Y}}\|^2) + 2E[(\mathbf{Y} - \hat{\mathbf{Y}})^t(\hat{\mathbf{Y}} - \tilde{\mathbf{Y}})] \\ &\geq \epsilon^2(\hat{\mathbf{Y}}) + 2E[(\mathbf{Y} - \hat{\mathbf{Y}})^t(\hat{\mathbf{Y}} - \tilde{\mathbf{Y}})]. \end{aligned}$$

We will prove that the rightmost term is zero and hence that $\epsilon^2(\tilde{\mathbf{Y}}) \geq \epsilon^2(\hat{\mathbf{Y}})$, which will prove the theorem. Recall that $\hat{\mathbf{Y}} = E(\mathbf{Y}|\mathbf{X})$ and hence

$$E[(\mathbf{Y} - \hat{\mathbf{Y}})|\mathbf{X}] = 0.$$

Since $\hat{\mathbf{Y}} - \tilde{\mathbf{Y}}$ is a deterministic function of $\mathbf{X}$,

$$E[(\mathbf{Y} - \hat{\mathbf{Y}})^t(\hat{\mathbf{Y}} - \tilde{\mathbf{Y}})|\mathbf{X}] = 0.$$

Then, by iterated expectation (see, e.g., [164]), we have

$$E(E[(\mathbf{Y} - \hat{\mathbf{Y}})^t(\hat{\mathbf{Y}} - \tilde{\mathbf{Y}})|\mathbf{X}]) = E[(\mathbf{Y} - \hat{\mathbf{Y}})^t(\hat{\mathbf{Y}} - \tilde{\mathbf{Y}})] = 0$$

as claimed, which proves the theorem. □

Unfortunately the conditional expectation is mathematically tractable only in a few very special cases, e.g., the case of jointly Gaussian vectors. In the Gaussian case the conditional expectation given $\mathbf{X}$ is formed by a simple matrix multiplication on $\mathbf{X}$ with possibly a constant vector being added; that is, the optimal estimate has a linear form. (Strictly speaking, this is an *affine* form and not a linear form if a constant vector is added. $y = ax$ is a linear function of x, $y = ax + b$ is affine but not linear unless $b = 0$.)

Constructive methods for implementing nonlinear predictors have received relatively little attention in the literature. The techniques that have been considered range from Volterra series [297], neural networks [307], and interpolative vector quantization [324]. For some general results and bounds on nonlinear prediction of moving-average processes, see [292].

Even when the random vectors are not Gaussian, linear predictors or estimates are important because of their simplicity. Although they are not in general optimal, they play an important role in signal processing. Hence we next turn to the problem of finding the optimal *linear* estimate of one vector given another.

Optimal Linear Prediction

Suppose as before that we are given an N-dimensional vector $\mathbf{X}$ and wish to predict a K-dimensional vector $\mathbf{Y}$. We now restrict ourselves to estimates of the form

$$\hat{\mathbf{Y}} = \mathbf{A}\mathbf{X},$$

where the $K \times N$-dimensional matrix $\mathbf{A}$ can be considered as a matrix of N-dimensional row vectors $\mathbf{a}_k^t$; $k = 1, \cdots, K$:

$$\mathbf{A} = [\mathbf{a}_1, \mathbf{a}_2, \cdots, a_K]^t$$

so that if $\hat{\mathbf{Y}} = (\hat{Y}_1, \cdots, \hat{Y}_K)^t$, then

$$\hat{Y}_k = \mathbf{a}_k^t \mathbf{X}$$

and hence

$$\epsilon^2(\hat{\mathbf{Y}}) = \sum_{k=1}^{N} E[(Y_k - \mathbf{a}_k^t \mathbf{X})^2]. \tag{4.2.3}$$

The goal is to find the matrix $\mathbf{A}$ that minimizes ϵ^2, which can be considered as a function of the predictor $\hat{\mathbf{Y}}$ or of the matrix $\mathbf{A}$ defining the predictor. We shall provide two separate solutions which are almost, but not quite, equivalent. The first is constructive in nature: a specific $\mathbf{A}$ will be given and shown to be optimal. The second development is descriptive: without actually giving the matrix $\mathbf{A}$, we will show that a certain property is necessary and sufficient for the matrix to be optimal. That property is called the *orthogonality principle*, and it states that the optimal matrix is the one that causes the error vector $\mathbf{Y} - \hat{\mathbf{Y}}$ to be orthogonal to (have zero correlation with) the observed vector $\mathbf{X}$. The first development is easier to use because it provides a formula for $\mathbf{A}$ that can be immediately computed in many cases. The second development is less direct and less immediately applicable, but it turns out to be more general: the descriptive property can be used to derive $\mathbf{A}$ even when the first development is not applicable.

The error $\epsilon^2(\mathbf{A})$ is minimized if each term $E[(Y_k - \mathbf{a}_k^t \mathbf{X})^2]$ is minimized over $\mathbf{a}_k$ since there is no interaction among the terms in the sum. We can do no better when minimizing a sum of such positive terms than to minimize each term separately. Thus the fundamental problem is the following simpler one: Given a random vector $\mathbf{X}$ and a random variable (one-dimensional vector) Y, we seek a vector $\mathbf{a}$ that minimizes

$$\epsilon^2(\mathbf{a}) = E[(Y - \mathbf{a}^t \mathbf{X})^2]. \tag{4.2.4}$$

One way to find the optimal $\mathbf{a}$ is to use calculus, setting derivatives of $\epsilon^2(\mathbf{a})$ to zero and verifying that the stationary point so obtained is a global minimum. As previously discussed, variational techniques can be avoided via elementary inequalities if the answer is known. We shall show that the optimal $\mathbf{a}$ is a solution of

$$\mathbf{a}^t \mathbf{R}_X = E(Y\mathbf{X}^t), \tag{4.2.5}$$

so that if the autocorrelation matrix

$$\mathbf{R}_X = E[\mathbf{X}\mathbf{X}^t] = \{R_X(k, i) = E(X_k X_i); \ k, i = 1, \cdots, N\}$$

is invertible, then the optimal $\mathbf{a}$ is given by

$$\mathbf{a}^t = E(Y\mathbf{X}^t)\mathbf{R}_X^{-1}. \tag{4.2.6}$$

To prove this we assume that $\mathbf{a}$ satisfies (4.2.6) and show that for any other vector $\mathbf{b}$

$$\epsilon^2(\mathbf{b}) \ge \epsilon^2(\mathbf{a}). \tag{4.2.7}$$

To do this we write

$$
\begin{aligned}
\epsilon^2(\mathbf{b}) &= E[(Y - \mathbf{b}^t\mathbf{X})^2] = E[(Y - \mathbf{a}^t\mathbf{X} + \mathbf{a}^t\mathbf{X} - \mathbf{b}^t\mathbf{X})^2] \\
&= E[(Y - \mathbf{a}^t\mathbf{X})^2] + 2E[(Y - \mathbf{a}^t\mathbf{X})(\mathbf{a}^t\mathbf{X} - \mathbf{b}^t\mathbf{X})] \\
&\quad + E[(\mathbf{a}^t\mathbf{X} - \mathbf{b}^t\mathbf{X})^2].
\end{aligned}
$$

Of the final terms, the first term is just $\epsilon^2(\mathbf{a})$ and the rightmost term is obviously nonnegative. Thus we have the bound

$$\epsilon^2(\mathbf{b}) \ge \epsilon^2(\mathbf{a}) + 2E[(Y - \mathbf{a}^t\mathbf{X})(\mathbf{a}^t - \mathbf{b}^t)\mathbf{X}]. \tag{4.2.8}$$

The crossproduct term can be written as

$$
\begin{aligned}
2E[(Y - \mathbf{a}^t\mathbf{X})(\mathbf{a}^t - \mathbf{b}^t)\mathbf{X}] &= 2E[(Y - \mathbf{a}^t\mathbf{X})\mathbf{X}^t(\mathbf{a} - \mathbf{b})] \\
&= 2E[(Y - \mathbf{a}^t\mathbf{X})\mathbf{X}^t](\mathbf{a} - \mathbf{b}) \\
&= 2\left(E[Y\mathbf{X}^t] - \mathbf{a}^t E[\mathbf{X}\mathbf{X}^t]\right)(\mathbf{a} - \mathbf{b}) \\
&= 2\left(E[Y\mathbf{X}^t] - \mathbf{a}^t \mathbf{R}_X\right)(\mathbf{a} - \mathbf{b}) \\
&= 0 \tag{4.2.9}
\end{aligned}
$$

invoking (4.2.5). Combining this with (4.2.8) proves (4.2.7) and hence optimality. Note that because of the symmetry of autocorrelation matrices and their inverses, we can rewrite (4.2.6) as

$$\mathbf{a} = \mathbf{R}_X^{-1} E[Y\mathbf{X}]. \tag{4.2.10}$$

Using the above result to perform a termwise minimization of (4.2.3) now yields the following theorem describing the optimal linear vector predictor.

Theorem 4.2.2 *The minimum mean squared error linear predictor of the form* $\hat{\mathbf{Y}} = \mathbf{A}\mathbf{X}$ *is given by any solution* $\mathbf{A}$ *of the equation:*

$$\mathbf{A}\mathbf{R}_X = E(\mathbf{Y}\mathbf{X}^t).$$

If the matrix $\mathbf{R}_X$ is invertible, then $\mathbf{A}$ is uniquely given by

$$\mathbf{A}^t = \mathbf{R}_X^{-1} E[\mathbf{X}\mathbf{Y}^t],$$

that is, the matrix A has rows $\mathbf{a}_k^t$; $k = 1, \cdots, K - 1$, with

$$\mathbf{a}_k = \mathbf{R}_X^{-1} E[Y_k \mathbf{X}].$$

Alternatively,

$$\mathbf{A} = E[\mathbf{Y}\mathbf{X}^t]\mathbf{R}_X^{-1}. \tag{4.2.11}$$

Optimal Affine Prediction

Having found the best linear predictor, it is also of interest to find the best predictor of the form

$$\hat{\mathbf{Y}}(\mathbf{X}) = \mathbf{A}\mathbf{X} + \mathbf{b}, \tag{4.2.12}$$

where now we allow an additional constant term. This is also often called a linear predictor, although as previously noted it is more correctly called an affine predictor because of the extra constant. We explore how such a constant term can be best incorporated. As the result and proof strongly resemble the linear predictor result, we proceed directly to the theorem.

Theorem 4.2.3 *The minimum mean squared predictor of the form $\hat{\mathbf{Y}} = \mathbf{A}\mathbf{X} + \mathbf{b}$ is given by any solution $\mathbf{A}$ of the equation:*

$$\mathbf{K}_X \mathbf{A} = E[(\mathbf{Y} - E(\mathbf{Y}))(\mathbf{X} - E(\mathbf{X}))^t] \tag{4.2.13}$$

where the covariance matrix $\mathbf{K}_X$ is defined by

$$\mathbf{K}_X = E[(\mathbf{X} - E(\mathbf{X}))(\mathbf{X} - E(\mathbf{X}))^t] = \mathbf{R}_{X-E(X)},$$

and

$$\mathbf{b} = E(\mathbf{Y}) - \mathbf{A}E(\mathbf{X}).$$

If $\mathbf{K}_X$ is invertible, then

$$\mathbf{A} = E[(\mathbf{Y} - E(\mathbf{Y}))(\mathbf{X} - E(\mathbf{X}))^t]\mathbf{K}_X^{-1}. \tag{4.2.14}$$

Note that if $\mathbf{X}$ and $\mathbf{Y}$ have zero means, then the result reduces to the previous result; that is, affine predictors offer no advantage over linear predictors for zero mean random vectors. To prove the theorem, let $\mathbf{C}$ be any matrix and $\mathbf{d}$ any vector (both of suitable dimensions) and note that

$$
\begin{aligned}
E(\|\mathbf{Y} &- (\mathbf{C}\mathbf{X} + \mathbf{d})\|^2) \\
=\ & E(\|(\mathbf{Y} - E(\mathbf{Y})) - \mathbf{C}(\mathbf{X} - E(\mathbf{X})) + E(\mathbf{Y}) - \mathbf{C}E(\mathbf{X}) - \mathbf{d}\|^2) \\
=\ & E(\|(\mathbf{Y} - E(\mathbf{Y})) - \mathbf{C}(\mathbf{X} - E(\mathbf{X}))\|^2) \\
& + E(\|E(\mathbf{Y}) - \mathbf{C}E(\mathbf{X}) - \mathbf{d}\|^2) \\
& + 2E[\mathbf{Y} - E(\mathbf{Y}) - \mathbf{C}(\mathbf{X} - E(\mathbf{X}))]^t[E(\mathbf{Y}) - \mathbf{C}E(\mathbf{X}) - \mathbf{d}].
\end{aligned}
$$

From Theorem 4.2.2, the first term is minimized by choosing $\mathbf{C} = \mathbf{A}$, where $\mathbf{A}$ is a solution of (4.2.13); also, the second term is the expectation of the squared norm of a vector that is identically zero if $\mathbf{C} = \mathbf{A}$ and $\mathbf{d} = \mathbf{b}$, and similarly for this choice of $\mathbf{C}$ and $\mathbf{d}$ the third term is zero. Thus

$$E(\|\mathbf{Y} - (\mathbf{C}\mathbf{X} + \mathbf{d})\|^2) \geq E(\|\mathbf{Y} - (\mathbf{A}\mathbf{X} + \mathbf{b})\|^2).$$

$\square$

We often restrict interest to linear predictors by assuming that the various vectors have zero mean. This is not always possible, however. For example, groups of pixels in a sampled image intensity raster can be used to predict other pixel groups, but pixel values are always nonnegative and hence always have nonzero means. Hence in some problems affine predictors may be preferable. Nonetheless, we will often follow the common practice of focusing on the linear case and extending when necessary. In most studies of linear prediction it is assumed that the mean is zero, i.e., that any dc value of the process has been removed. If this assumption is not made, linear prediction theory is still applicable but will generally give inferior performance to the use of affine prediction.

The Orthogonality Principle

Although we have proved the form of the optimal linear predictor of one vector given another, there is another way to describe the result that is often useful for deriving optimal linear predictors in somewhat different situations. To develop this alternative viewpoint we focus on the error vector

$$\mathbf{e} = \mathbf{Y} - \hat{\mathbf{Y}}. \qquad (4.2.15)$$

Rewriting (4.2.15) as $\mathbf{Y} = \hat{\mathbf{Y}} + \mathbf{e}$ points out that the vector $\mathbf{Y}$ can be considered as its prediction plus an error or "noise" term. The goal of an optimal predictor is then to minimize the error energy $\mathbf{e}^t\mathbf{e} = \sum_{n=1}^{N} e_n^2$. If the prediction is linear, then

$$\mathbf{e} = \mathbf{Y} - \mathbf{A}\mathbf{X}.$$

As with the basic development for the linear predictor, we simplify things for the moment and look at the scalar prediction problem of predicting a random variable Y by $\hat{Y} = \mathbf{a}^t\mathbf{X}$ yielding a scalar error of $e = Y - \hat{Y} = Y - \mathbf{a}^t\mathbf{X}$. Since we have seen that the overall mean squared error $E[\mathbf{e}^t\mathbf{e}]$ in the vector case is minimized by separately minimizing each component $E[e_k^2]$, we can later easily extend our results for the scalar case to the vector case.

Suppose that $\mathbf{a}$ is chosen optimally and consider the crosscorrelation between an arbitrary error term and the observable vector:

$$
\begin{aligned}
E[(Y - \hat{Y})\mathbf{X}] &= E[(Y - \mathbf{a}^t\mathbf{X})\mathbf{X}] \\
&= E[Y\mathbf{X}] - E[\mathbf{X}(\mathbf{X}^t\mathbf{a})] \\
&= E[Y\mathbf{X}] - \mathbf{R}_X\mathbf{a} = 0
\end{aligned}
$$

using (4.2.5).

Thus for the optimal predictor, the error satisfies

$$ E[e\mathbf{X}] = \mathbf{0}, $$

or, equivalently,

$$ E[eX_k] = 0; \quad k = 1, \cdots, K - 1. \tag{4.2.16} $$

When two random variables e and X are such that their expected product $E(eX)$ is 0, they are said to be *orthogonal* and we write

$$ e \perp X. $$

We have therefore shown that the optimal linear predictor for a scalar random variable given a vector of observations causes the error to be orthogonal to all of the observables and hence orthogonality of error and observations is a necessary condition for optimality of a linear predictor.

Conversely, suppose that we know a linear predictor $\mathbf{a}$ is such that it renders the prediction error orthogonal to all of the observations. Arguing as we have before, suppose that $\mathbf{b}$ is any other linear predictor vector and observe that

$$
\begin{aligned}
\epsilon^2(\mathbf{b}) &= E[(Y - \mathbf{b}^t\mathbf{X})^2] \\
&= E[(Y - \mathbf{a}^t\mathbf{X} + \mathbf{a}^t\mathbf{X} - \mathbf{b}^t\mathbf{X})^2] \\
&\geq \epsilon^2(\mathbf{a}) + 2E[(Y - \mathbf{a}^t\mathbf{X})(\mathbf{a}^t\mathbf{X} - \mathbf{b}^t\mathbf{X})],
\end{aligned}
$$

where the equality holds if $\mathbf{b} = \mathbf{a}$. Letting $e = Y - \mathbf{a}^t\mathbf{X}$ denote the error resulting from an $\mathbf{a}$ that makes the error orthogonal with the observations, the rightmost term can be rewritten as

$$ 2E[e(\mathbf{a}^t\mathbf{X} - \mathbf{b}^t\mathbf{X})] = 2(\mathbf{a}^t - \mathbf{b}^t)E[e\mathbf{X}] = 0. $$

Thus we have shown that $\epsilon^2(\mathbf{b}) \geq \epsilon^2(\mathbf{a})$ and hence no linear predictor can outperform one yielding an error orthogonal to the observations and hence such orthogonality is sufficient as well as necessary for optimality.

Since the optimal predictor for a vector $\mathbf{Y}$ given $\mathbf{X}$ is given by the componentwise optimal predictions given $\mathbf{X}$, we have thus proved the following alternative to Theorem 4.2.2.

Theorem 4.2.4 *The Orthogonality Principle:*

A linear predictor $\hat{\mathbf{Y}} = \mathbf{A}\mathbf{X}$ is optimal (in the MSE sense) if and only if the resulting errors are orthogonal to the observations, that is, if $\mathbf{e} = \mathbf{Y} - \mathbf{A}\mathbf{X}$, then

$$E[e_n X_k] = 0; \ n = 1, \cdots, N - 1; \ k = 1, \cdots, K - 1.$$

The orthogonality principle is equivalent to the classical projection theorem in vector spaces where it has a simple geometric interpretation [224].

4.3   Finite-Memory Linear Prediction

Perhaps the most important and common application of linear prediction is to the prediction of future values of a stationary random process given its past values. Consider a discrete time stationary real-valued random process $\{X_n\}$ with zero mean, autocorrelation function,

$$r_j = R_X(j) = E(X_n X_{n-j})$$

and variance $\sigma_X^2 = r_0$. Since the process has zero mean, the autocorrelation function coincides with the autocovariance function.

It will later be convenient to describe the process by its *spectral density*, which is given by the Fourier transform

$$S(f) = \sum_{m=-\infty}^{\infty} r_m e^{-j2\pi f m}.$$

If the discrete time process is formed by sampling a continuous time process using a sampling period of T, then f on the right hand side is often scaled to fT. We effectively normalize f by the sampling frequency when the factor T is replaced by unity. The spectral density is by definition a periodic function of the frequency f with period 2π.

In general, there is nonzero correlation between samples so that knowledge of one sample value provides some partial information about another sample. Specifically, observations of the past of the process give some *a priori* knowledge of the present value. A *one-step predictor* is a function that operates on past observations, $X_{n-1}, X_{n-2}, \cdots$, to estimate X_n, the present value. In this case, we predict one time unit ahead. Occasionally there is interest in *s-step prediction* where we wish to predict s time units ahead, that is to predict X_{n+s-1} from observations of $X_{n-1}, X_{n-2}, \cdots$. In this chapter, however, we focus on one-step prediction because it has the most important applications and is most frequently used.

In the language of the previous section, we wish to predict a scalar random variable $Y = X_n$ given an observation vector $\mathbf{X} = (X_{n-1}, X_{n-2}, \cdots)$. We can not immediately apply the results of the previous section because the observation vector is in general an infinite dimensional vector and our proofs were only for finite dimensional vectors.

As before, the most general predictor will be a nonlinear function of all past sample values. For a *linear predictor* the function is linear and has the form of a linear combination of the past values. A *finite-memory predictor with size m* is one which uses only a finite number of previous values, $X_{n-1}, X_{n-2}, \cdots, X_{n-m}$ to predict X_n. A predictor with infinite memory uses the entire past to predict the present value. Clearly if we restrict the predictor to have finite memory, then we can immediately apply our earlier results to find the form of the predictor. The infinite memory results will then hopefully be limiting versions of the finite memory results. For the moment we will emphasize finite memory predictors, but we will return in a later section to the infinite memory case.

As before, we will measure the performance of a predictor by mean squared error. Replacing Y by $\hat{X}_n$ our performance measure is

$$D_m = E[(X_n - \hat{X}_n)^2],$$

where the subscript m denotes the predictor memory size. The performance of a predictor is usually assessed by the *prediction gain ratio* which is defined as

$$G_m = \frac{\sigma_X^2}{D_m}$$

or by the *prediction gain*, which is measured in dB units and given by

$$10 \log_{10} G_m.$$

To summarize, we wish to find a finite memory one-step predictor $\hat{X}_n$ of X_n as a function of the m prior samples, $X_{n-1}, X_{n-2}, \cdots, X_{n-m}$, that minimizes the MSE $E[(X_n - \hat{X}_n)^2]$. In general, this predictor could have the form:

$$\hat{X}_n = f(X_{n-1}, X_{n-2}, \cdots, X_{n-m}), \qquad (4.3.1)$$

where f is a function of m real variables. From Theorem 4.2.1, the optimal estimate is given by

$$\hat{X}_n = E(X_n | X_{n-1}, X_{n-2}, \cdots, X_{n-m}), \qquad (4.3.2)$$

namely, the conditional expectation of X_n given the past observations. In general, this produces a nonlinear estimate which may be intractable and lack a closed form expression. Furthermore, computation of a conditional expectation would require an explicit knowledge of the joint probability distribution of $m + 1$ consecutive samples of the random sequence.

As discussed in the basic vector prediction development, however, in the special case of a Gaussian process the optimal estimate reduces to a linear combination of the past values. As before, we do not need to restrict ourselves to Gaussian processes to avoid dealing with nonlinear estimation. We can restrict our predictor to be linear and then find the best possible linear predictor for the actual process. The resulting performance may be below what a nonlinear predictor would achieve, but it is usually still very effective and avoids analytical and computational difficulties.

We now assume a linear predictor of the form

$$\hat{X}_n = -\sum_{i=1}^{m} a_i X_{n-i} \qquad (4.3.3)$$

where the parameters a_i are called *linear prediction coefficients* and the minus sign is included for convenience in describing the prediction error. The prediction error is defined by

$$e_n \equiv X_n - \hat{X}_n = X_n + \sum_{i=1}^{m} a_i X_{n-i}. \qquad (4.3.4)$$

This is exactly in the form of the previous section except for the minus sign, that is, we can replace Y by X_n, K by m, $\mathbf{X}$ by $(X_{n-1}, \cdots, X_{n-m})$, and $\mathbf{a}$ by $(-a_1, -a_2, \cdots, -a_m)$. Application of Theorem 4.2.2 then yields

$$\mathbf{a} = -\mathbf{R}_m^{-1}\mathbf{v}, \qquad (4.3.5)$$

where here $\mathbf{a} = (a_1, a_2, \cdots, a_m)^t$ is the predictor coefficient vector, $\mathbf{v} = (E(X_n X_{n-1}), \cdots, E(X_n X_{n-m}))^t = (r_1, r_2, \cdots, r_m)^t$, and the matrix $\mathbf{R}_m$ is the mth order autocorrelation matrix of the stationary random sequence X_n with autocorrelation values r_i:

$$\mathbf{R}_m = \begin{bmatrix} r_0 & r_1 & \cdots & r_{m-1} \\ r_1 & r_0 & \cdots & r_{m-2} \\ \vdots & \vdots & \ddots & \vdots \\ r_{m-1} & r_{m-2} & \cdots & r_0 \end{bmatrix}. \qquad (4.3.6)$$

A square matrix that has the property that the elements along each diagonal have the same value is called a *Toeplitz* matrix and the autocorrelation matrix is therefore a Toeplitz matrix. (Much of the theory of stationary processes is based on the theory of Toeplitz matrices and Toeplitz functions. The definitive reference is [172]. See also [152], [149].)

Alternatively (and more generally) we can apply the orthogonality principle (Theorem 4.2.4) to conclude that

$$e_n \perp X_{n-j} \text{ for } j = 1, 2, \cdots, m. \tag{4.3.7}$$

More explicitly, this gives $Ee_n X_{n-j} = 0$, or,

$$E[(\sum_{i=1}^{m} a_i X_{n-i}) X_{n-j}] = \sum_{i=1}^{m} a_i E(X_{n-i} X_{n-j}) = -E(X_n X_{n-j}) \tag{4.3.8}$$

for each $j = 1, 2, \cdots, m$. Simplifying this expression gives the set of so-called *normal equations* (also known as the *finite-memory Wiener-Hopf equation in discrete time* and the *Yule-Walker equations*) for the optimal linear predictor coefficients:

$$\sum_{i=1}^{m} a_i r_{j-i} = -r_j \text{ for } j = 1, 2, \cdots, m. \tag{4.3.9}$$

This system of m equations with m unknowns can be expressed as the matrix equation:

$$\mathbf{R}_m \mathbf{a} = -\mathbf{v}. \tag{4.3.10}$$

If $\mathbf{R}_m$ is nonsingular, the solution will be unique, given by (4.3.5).

It is of interest to compute the prediction gain for a finite memory predictor as a function of m. This is of particular interest for assessing how large a memory, m, of past values is worth using for the prediction. From the prediction error expression (4.3.4), we find that in general the mean squared error, D_m, for any particular choice of $\mathbf{a}$, is given by

$$
\begin{aligned}
D_m &= Ee_n^2 = E[(X_n + \sum_{i=1}^{m} a_i X_{n-i})^2] \\
&= r_0 + 2 \sum_{i=1}^{m} a_i r_i + \sum_{i=1}^{m} \sum_{j=1}^{m} a_i r_{i-j} a_j \\
&= r_0 + 2\mathbf{a}^t \mathbf{v} + \mathbf{a}^t \mathbf{R}_m \mathbf{a} \tag{4.3.11}
\end{aligned}
$$

which shows explicitly the quadratic nature of the mean squared error as a function of the coefficient vector $\mathbf{a}$. An alternative and common form of writing the error D_m which will prove useful is obtained by defining an extended form of the $\mathbf{a}$ vector: Set $a_0 = 1$ and define the $(m+1)$-dimensional vector $\mathbf{c}$ by

$$\mathbf{c} = (a_0, a_1, \cdots, a_m)^t = (1, \mathbf{a})^t.$$

Then rearranging terms in (4.3.11) yields

$$D_m = \mathbf{c}^t \mathbf{R}_{m+1} \mathbf{c} \tag{4.3.12}$$

and the optimization problem can be restated as minimizing the quadratic form of (4.3.12) over all vectors $\mathbf{c}$ subject to the constraint that $c_0 = 1$. Clearly finding the optimal $\mathbf{c}$ for (4.3.12) is equivalent to finding the optimal $\mathbf{a}$ for (4.3.11)

It is interesting to observe from (4.3.12) that the mean squared prediction error D_m cannot be zero unless the autocorrelation matrix $\mathbf{R}_{m+1}$ is singular. A random process whose mth order autocorrelation matrices are nonsingular for each $m > 0$, is said to be a *nondeterministic* process. Such a process can not be perfectly predicted from a finite number of past values.

By setting $\mathbf{a}$ to its optimal value given by (4.3.5), we find the minimum mean squared error for the optimal value of $\mathbf{a}$ is given by:

$$D_m = r_0 - \mathbf{v}^t \mathbf{R}_m^{-1} \mathbf{v}.$$

From (4.3.5), this simplifies to

$$D_m = \sum_{i=0}^{m} a_i r_i \tag{4.3.13}$$

where the a_i are now understood to be the optimal predictor coefficients and we define the coefficient $a_0 \equiv 1$ for notational convenience. In fact, (4.3.13) can be obtained more directly by noting that

$$D_m = E e_n (X_n + \sum_{i=1}^{m} a_i X_{n-i}) = E e_n X_n \tag{4.3.14}$$

since e_n is uncorrelated with X_{n-i} for $i = 1, 2, \cdots m$ by the orthogonality condition. This immediately gives

$$D_m = r_0 + \sum_{i=1}^{m} a_i r_i \tag{4.3.15}$$

which is the same as (4.3.13).

It is often helpful to consider the prediction error as a separate stationary random sequence that is generated by applying the input sequence X_n to a finite-impulse response (FIR) filter with transfer function $A(z)$. From (4.3.4) it follows immediately that

$$A(z) = \sum_{i=0}^{m} a_i z^{-i}. \tag{4.3.16}$$

This filter is called the *prediction error filter* and plays an important role in speech processing.

This completes the basic properties of one-step linear prediction of stationary processes. In principle, we can solve for the optimal predictor for any memory length m and use the above formula to determine the prediction gain as a function of m. There is much more to say, however, that is of value both for insight and practical use.

4.4 Forward and Backward Prediction

We return to the finite memory linear prediction problem from a different perspective. This treatment leads directly to the important computational technique, the Levinson-Durbin algorithm. It also leads to the important lattice filter structures for implementing certain filtering operations associated with linear prediction, adaptive filtering, and spectral shaping.

Consider a discrete time signal where m past values,

$$X_{n-1}, X_{n-2}, \cdots, X_{n-m},$$

are considered as the observable data for linear prediction of the present X_n as already discussed. The prediction is given by

$$\hat{X}_n = -\sum_{k=1}^{m} a_k X_{n-k}, \tag{4.4.1}$$

which we now call *forward prediction* since we are predicting *forward* in time. In contrast, we can also predict *backward* in time, using the very same observables to estimate the signal value, X_{n-m-1}, at one time unit prior to the oldest observable value. This is called *backward prediction* and the predictor is given by

$$\hat{X}_{n-m-1} = -\sum_{k=1}^{m} b_k X_{n-k} \tag{4.4.2}$$

where *future* values (relative to the time index $n - m - 1$) of the data are used for prediction and the coefficients b_k denote the backward prediction coefficients. (One might question the use of the word "prediction" for this backward estimation task, but it is particularly suitable due to the symmetrical relation with forward prediction to be elaborated in a moment.)

The forward prediction error, previously denoted by e_n, will now be denoted by $e_n^+ = X_n - \hat{X}_n$. The orthogonality condition of linear least-squares estimation gives the optimality condition

$$e_n^+ \perp X_{n-i} \quad \text{for } i = 1, 2, \cdots, m \tag{4.4.3}$$

yielding

$$\sum_{k=0}^{m} a_k X_{n-k} \perp X_{n-i} \quad \text{for } i = 1, 2, \cdots, m \tag{4.4.4}$$

where $a_0 \equiv 1$.

Similarly, the backward prediction error will be denoted as e_n^- and is given by $e_n^- = X_{n-m-1} - \hat{X}_{n-m-1}$ so that

$$e_n^- = \sum_{k=1}^{m+1} b_k X_{n-k}, \tag{4.4.5}$$

where $b_{m+1} \equiv 1$. The orthogonality condition of linear least-squares estimation gives the optimality condition

$$e_n^- \perp X_{n-m-1+j} \quad \text{for } j = 1, 2, \cdots, m \tag{4.4.6}$$

yielding

$$\sum_{s=1}^{m+1} b_s X_{n-s} \perp X_{n-j} \quad \text{for } j = 1, 2, \cdots, m. \tag{4.4.7}$$

Thus both forward and backward prediction estimates are linear combinations of the same observable data, but the forward prediction error includes the variable X_n weighted by the coefficient a_0 (which is unity) and the backward prediction error includes the variable X_{n-m-1} weighted by the coefficient b_{m+1} (which is unity). The only difference in form of these two prediction errors is that they each have different prediction coefficients. But do they? Since the process is stationary, (4.4.4) can be expressed as

$$\sum_{k=0}^{m} a_k R_X(k - i) = 0; \quad \text{for } i = 1, 2, \cdots, m \tag{4.4.8}$$

with $a_0 = 1$, and (4.4.7) is equivalent to

$$\sum_{s=1}^{m+1} b_s R_X(s - j) = 0; \quad \text{for } j = 1, 2, \cdots, m \tag{4.4.9}$$

with $b_{m+1} = 1$. These are the same set of equations if we identify

$$b_s = a_{m-s+1} \quad \text{for } s = 1, 2, \cdots, m + 1. \tag{4.4.10}$$

To verify this, introduce the change of variable $l = m + 1 - s$ in (4.4.9) and apply (4.4.10). This yields for $j = 1, 2, \cdots, m$

$$0 = \sum_{s=1}^{m+1} a_{m-s+1} R_X(s - j) = \sum_{l=0}^{m} a_l R_X(j - m + l - 1) \tag{4.4.11}$$

where we have used the symmetry of the autocorrelation function. The last expression becomes (4.4.8) if we replace $m - j + 1$ by i. Thus the backward prediction coefficients are simply the forward coefficients reversed in time.

The benefit of studying backward prediction will soon become evident. Let us assume that we know the optimal predictor for some value m of memory size. We would like to make as much use as possible of the known results, specifically of the known predictor values $a_1, a_2, \cdots, a_m$ to determine the optimal predictor for the increased memory size $m + 1$. Before proceeding, it should be made clear that we now need to solve a completely new problem. The value of a_1 that is optimal for memory m is in general no longer the optimal value of a_1 for a predictor with memory $m + 1$. Reviewing (4.2.6) shows that the entire vector **a** depends on the value of m. But rather than start all over, we can indeed find a quick and easy way to get the best set of predictor coefficients for the case $m + 1$ from a knowledge of the best set for memory size m. This leads to a convenient and general-purpose recursion for rapid computation of optimal predictor coefficients.

Let $\tilde{X}_n$ denote the forward one-step prediction of X_n from the $m + 1$ past observables, $X_{n-1}, X_{n-2}, \cdots, X_{n-m-1}$. The tilde is used to distinguish from the previous predictor $\hat{X}_n$ which used memory size m. Then

$$\tilde{X}_n = - \sum_{k=1}^{m+1} a'_k X_{n-k} \tag{4.4.12}$$

and

$$\tilde{e}_n = X_n - \tilde{X}_n \tag{4.4.13}$$

is the corresponding forward prediction error. Since we are dealing with a new memory size, we distinguish the predictor coefficients $\{a'_i\}$ for $i = 0, 1, 2, \cdots, m+1$ for the $m + 1$ memory size from the $\{a_i\}$ coefficients in the previous case of memory size m. Following the previous convention, we let $a_0' = 1$. From (4.4.12) the forward prediction error is given by

$$\tilde{e}_n = \sum_{k=0}^{m+1} a'_k X_{n-k}. \tag{4.4.14}$$

What makes the $m + 1$ size forward predictor different from the size m case? It is simply that we now have *one extra piece of information* that can contribute somehow to improving our estimate of the current sample X_n. This piece of information is the observable X_{n-m-1} or more accurately, the extra information is *whatever is new about this additional observable that we do not already know*. What we already know is everything that can be deduced from the old observables $X_{n-1}, X_{n-2}, \cdots, X_{n-m}$. The new information in X_{n-m-1} is precisely the information in X_{n-m-1} that is not linearly determined from the old observables. In fact, the *backward prediction error e_n^- for memory size m is exactly that part of X_{n-m-1} that is not linearly determined from the old observables.* Thus it seems reasonable that the best linear prediction of X_n with memory $m+1$ can be represented in the form

$$\tilde{X}_n = \hat{X}_n - K e_n^- . \tag{4.4.15}$$

The first term is the best prediction of X_n if we had only the old observables to work with and the second term is proportional to the new information that comes from the new observable, but stripped of all irrelevant parts that will not help us. The constant K is a proportionality constant that we have not yet determined. Now, however, we have only a single unknown parameter to determine rather than a whole bunch of parameters. An equivalent form of (4.4.15) in terms of the prediction error is

$$\tilde{e}_n = e_n^+ + K e_n^- \tag{4.4.16}$$

which comes from (4.4.15) by subtracting X_n from both sides.

Eq. (4.4.15) implicitly gives $\tilde{X}_n$ as a linear combination of the $m+1$ observables $X_{n-1}, X_{n-2}, \cdots, X_{n-m}, X_{n-m-1}$. To ensure that it is indeed the optimal predictor for memory size $m+1$ we have only to make sure that the orthogonality condition is satisfied. Thus for optimality we need to have

$$\tilde{e}_n \perp X_{n-i} \quad \text{for } i = 1, 2, \cdots, m \tag{4.4.17}$$

and also

$$\tilde{e}_n \perp X_{n-m-1}. \tag{4.4.18}$$

Condition (4.4.17) is satisfied because e_n^+ is orthogonal to the old observables, since it is the optimal size m forward prediction error; also, e_n^- is also orthogonal to the old observables since it is the optimal size m backward predictor.

We now examine the orthogonality requirement (4.4.18) and solve for the value of K that makes it work. Thus we apply (4.4.18) to (4.4.16) to get

$$E(e_n^+ X_{n-m-1}) + K E(e_n^- X_{n-m-1}) = 0 \tag{4.4.19}$$

or

$$K = -\frac{\beta_m}{\alpha_m} \tag{4.4.20}$$

where

$$\beta_m = E e_n^+ X_{n-m-1} = \sum_{k=0}^{m} a_k E X_{n-k} X_{n-m-1} \qquad (4.4.21)$$

so that

$$\beta_m = \sum_{k=0}^{m} a_k r_{m+1-k} \qquad (4.4.22)$$

and

$$\alpha_m = E e_n^- X_{n-m-1} = \sum_{k=1}^{m+1} b_k E(X_{n-k} X_{n-m-1}) = \sum_{j=0}^{m} a_j r_j \qquad (4.4.23)$$

using (4.4.5) and (4.4.10). Thus, from (4.3.15) we recognize that

$$\alpha_m = D_m. \qquad (4.4.24)$$

Finally, we have the formula

$$K = -\frac{\beta_m}{D_m} \qquad (4.4.25)$$

for the parameter K, frequently called a *reflection coefficient*, where β_m is given conveniently by (4.4.22). Thus having determined K, we now have a predictor that uses all $m + 1$ past observables and satisfies the orthogonality condition that the prediction error is orthogonal to each of these observables. Consequently, it must be the optimal $m + 1$ memory forward predictor that we were seeking. Now, to examine the actual predictor coefficients, we substitute (4.4.5) and (4.4.1) into the predictor expression (4.4.15). Equating coefficients gives

$$a_i' = a_i + K b_i \quad \text{for } i = 1, 2, \cdots, m + 1 \qquad (4.4.26)$$

with the understanding that $a_{m+1} = 0$. This gives the $m + 1$ memory forward predictor coefficients from the m memory forward and backward predictor coefficients. Of course the backward coefficients can be eliminated using (4.4.10), so that we get the recursion

$$a_i' = a_i + K a_{m+1-i} \quad \text{for } i = 1, 2, \cdots, m + 1 \qquad (4.4.27)$$

which determines the predictor coefficients a_i' for memory size $m+1$ directly from the values a_i for memory size m.

From the same approach, we can also calculate the mean squared prediction error recursively since

$$D_{m+1} = E(\tilde{e}_n^2) = \sum_{i=0}^{m+1} a_i' r_i$$

$$= \sum_{i=0}^{m} a_i r_i + K \sum_{i=1}^{m+1} a_{m-i+1} r_i = D_m + K\beta_m. \qquad (4.4.28)$$

But since $\beta_m = -K D_m$, we have

$$D_{m+1} = D_m(1 - K^2), \qquad (4.4.29)$$

a convenient and explicit recursion for the mean squared error.

Some important observations can be made from (4.4.29). First, note that if $|K| > 1$, $D_{m+1} < 0$, which is impossible. Furthermore, if $|K| = 1$, $D_{m+1} = 0$ which is not possible for a nondeterministic process, i.e., one whose autocorrelation matrices $\mathbf{R}_m$ are nonsingular. Thus, *for a nondeterministic process, the reflection coefficients always have magnitude strictly less than unity*. Note also that $K = 0$ implies there is no performance improvement in going from an mth order to $(m+1)$th order optimal predictor. In this case, the optimal mth order and $(m+1)$th order predictors coincide, as can be seen by setting $K = 0$ in (4.4.15).

The previous results provide a recursion for the predictor coefficients and for the mean squared prediction error for a transition from memory m to $m + 1$. This recursion is usually called the Levinson-Durbin Recursion. N. Levinson [211] first noticed how to exploit the symmetry of the Toeplitz matrix for a simplified solution of the discrete time finite memory version of the Wiener-Hopf equation for optimal estimation similar to (4.2.11) but with arbitrary right-hand side $-\mathbf{v}$. James Durbin [105] improved the algorithm so that the computation fully exploits the particular form of $\mathbf{v}$ in the case of one-step prediction. An extensive discussion of the algorithm and a comparison with other similar algorithms may be found in [234]. In the next section we use these results to give an explicit algorithmic solution to the linear prediction problem that is of great importance in speech processing and seismology.

The Levinson-Durbin Recursion provides the basis for an important family of *lattice* filter or *ladder* filter structures, which can be used to implement the prediction error filter $A(z)$ or its inverse $A^{-1}(z)$. Define the *backward prediction error filter*

$$B(z) = \sum_{i=1}^{m+1} b_i z^{-i}. \qquad (4.4.30)$$

Then from (4.4.10) we have

$$B(z) = A(z^{-1}) z^{-m-1} \qquad (4.4.31)$$

so that the backward prediction error filter is readily determined from the forward prediction error filter $A(z)$.

We shall now show how the forward and backward prediction error polynomials of order $m + 1$, denoted respectively by $A'(z)$ and $B'(z)$, can be derived directly from the corresponding pair, $A(z)$ and $B(z)$ of order m. From (4.4.26) we immediately have

$$A'(z) = A(z) + KB(z). \tag{4.4.32}$$

Also, replacing m by $m + 1$ in (4.4.31) we have

$$B'(z) = z^{-m-2}A'(z^{-1}), \tag{4.4.33}$$

which when applied to (4.4.32) gives

$$B'(z) = z^{-m-2}A(z^{-1}) + Kz^{-m-2}B(z^{-1}). \tag{4.4.34}$$

Applying (4.4.31), then gives the polynomial recursion

$$B'(z) = Kz^{-1}A(z) + z^{-1}B(z). \tag{4.4.35}$$

The pair of recursions (4.4.32) and (4.4.35) have several interesting applications. In particular, it suggests an implementation with the lattice (or ladder or two-multiplier ladder) filter section shown in Figure 4.2. Repeated application of the recursion leads to a cascade of these lattice sections, each with a different reflection coefficient as the characterizing parameter. These filter structures can be used to generate the prediction error process e_k for an Mth order predictor from the input signal process X_k or conversely, to generate X_k as output from the input e_k. For a more detailed description of lattice filter structures in linear prediction, see [234].

4.5   The Levinson-Durbin Algorithm

From the Levinson-Durbin recursion, it is a simple step to formulate a complete algorithm to design an optimal forward predictor of memory size m by starting with memory size zero and recursively updating the memory size. The key equations we have obtained in the previous section are (4.4.22), (4.4.23), (4.4.25) and (4.4.27) which provide a path to finding the reflection coefficient K and thereby the new linear prediction coefficients a'_i for memory size $m + 1$ from memory size m according to:

$$a'_i = a_i + Ka_{m+1-i} \tag{4.5.1}$$

and the new mean squared prediction error D_{m+1} from the old value D_m according to

$$D_{m+1} = D_m(1 - K^2). \tag{4.5.2}$$

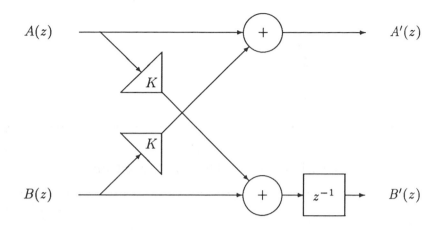

Figure 4.2: Lattice Filter Section Corresponding to the Levinson-Durbin Recursion

We begin with a zero-step predictor, that is, a predictor which has no observations at all of the past. This estimator is simply a constant since it is not dependent on any observations of the past. Under our zero mean assumption, the optimal estimate is therefore the mean value of X_n which is zero. Hence the 0th order prediction error is $e_n = a_0 X_n$ where $a_0 = 1$, and the mean squared prediction error, D_0, is then simply equal to r_0, the variance of X_n. From (4.4.22), we also see that is $\beta_0 = a_0 r_1 = r_1$. This provides the necessary initialization for the recursion.

To avoid ambiguity at this stage, we slightly modify our notation by defining K_{m+1} as the value of the reflection coefficient K used in the transition from memory size m to size $m + 1$ and we let $a_i^{(m)}$ denote the ith prediction coefficient for an optimal predictor with memory size m and for convenience we define $a_{m+1}^{(m)} = 0$.

The Levinson-Durbin Algorithm of order M is summarized in Table 4.1.

The equations in Table 4.1 provide a convenient and complete algorithm for designing linear predictors and for determining the performance as a function of memory size m. The prediction gain as a function of memory size is easily computed from the sequence of mean squared error values found in this algorithm.

In the Levinson-Durbin algorithm, both the predictor coefficients and the reflection coefficients are simultaneously determined from the autocorrelation values. In fact, the set of Mth order prediction coefficients are

Table 4.1: **Levinson-Durbin Algorithm**

Initialization: Let
$$a_0^{(0)} = 1 \text{ and } D_0 = r_0.$$

Recursion: For $m = 0, 1, \cdots, M - 1$, do:

$$\beta_m = \sum_{k=0}^{m} a_k^{(m)} r_{m+1-k} \qquad (4.5.3)$$

$$K_{m+1} = -\frac{\beta_m}{D_m} \qquad (4.5.4)$$

$$a_0^{(m+1)} = 1 \qquad (4.5.5)$$

$$a_k^{(m+1)} = a_k^{(m)} + K_{m+1} a_{m+1-k}^{(m)}$$
$$\text{for } k = 1, 2, \cdots, m + 1 \qquad (4.5.6)$$

$$D_{m+1} = (1 - K_{m+1}^2) D_m. \qquad (4.5.7)$$

completely determined from the set of M reflection coefficients and vice versa. This follows from the pair of recursions (4.4.32) and (4.4.35) which show that the $m + 1$th order predictor coefficients are determined from the $m + 1$th reflection coefficients and the mth order predictor coefficients. By iterating from $m = 0$, the Mth order predictor coefficients can be computed from the M reflection coefficients. More simply, we can use (4.4.27) to obtain an algorithm for finding the Mth order predictor coefficients from the M reflection coefficients.

Step-Up Procedure

(Predictor Coefficients from Reflection Coefficients)

Given: $K_1, K_2, \cdots, K_M$. Let $a_0^{(0)} = 1$.

For $m = 1, 2, \cdots, M$:
$$1 \to a_0^{(m)}$$
$$0 \to a_{m+1}^{(m)}$$
for $i = 1, 2, \cdots, m$:
$$a_i^{(m-1)} + K_m a_{m-i}^{(m-1)} \to a_i^{(m)}$$

To obtain the reflection coefficients from a known linear prediction error polynomial, we need to reverse the order of the Levinson-Durbin recursion and proceed from an mth order prediction error polynomial to an $m - 1$st order polynomial extracting K_m in the process. The pair of recursions (4.4.32) and (4.4.35) can be inverted by solving for $A_m(z)$ and $B_m(z)$ in terms of $A_{m+1}(z)$ and $B_{m+1}(z)$ and applying (4.4.33). This leads to

$$A_m(z) = \frac{1}{1 - K_{m+1}^2}[A_{m+1}(z) - K_{m+1}z^{-m-1}A_{m+1}(z^{-1})].$$

By matching the corresponding coefficients of the polynomials on the two sides of this equation, we obtain the needed recursion. We first see that the coefficient of z_{m+1} in the right hand side must be zero so that $K_m = a_m^{(m)}$. Thus we obtain the following algorithm.

Step-Down Procedure

(Reflection Coefficients from Predictor Coefficients)

Given: Mth order predictor coefficients, $a_i^{(M)}$ for $i = 0, 1, 2, \cdots, M$.
Let

$$K_M = a_M^{(M)}.$$

For $m = M - 1, M - 2, \cdots, 1$,
    for $i = 0, 1, \cdots, m$,

$$\frac{a_i^{(m+1)} - K_{m+1} a_{m+1-i}^{(m+1)}}{1 - K_{m+1}^2} \rightarrow a_i^{(m)}$$

$$a_m^{(m)} \rightarrow K_m$$

Note that the Step-Down procedure requires that the reflection coefficients have magnitude less than unity.

4.6 Linear Predictor Design from Empirical Data

Suppose that we do not have a mathematical description of the statistics of a stationary random process $\{X_n\}$, but we have a finite set of consecutive observations of X_n for $n \in \mathcal{M} \equiv \{0, 1, 2, \cdots, M\}$. We wish to find an mth order 1-step linear predictor that in some sense will be optimal or nearly optimal for the statistics of the process $\{X_n\}$ which are not known to us. We assume that $m \ll M$, that is, the number of observations is much greater than the memory size of the desired predictor.

We could of course simply form empirical estimates of the autocorrelation values r_i from our observations and apply them to the normal equation $(\mathbf{R}_m \mathbf{a} = -\mathbf{v})$ and solve for the desired coefficients. Instead, a more direct approach is to find the predictor that would do the best possible job in performing predictions on the actual data available. This idea leads to two alternative formulations which for historical reasons are called the *autocorrelation method* and the *covariance method* of linear predictor design. It can be pointed out that the general philosophy of optimizing a system directly on a set of data instead of indirectly on a model constructed based on the data will also be followed later when designing quantizers.

It is convenient to define the values

$$s(n) = w(n)X_n \tag{4.6.1}$$

for *all* n, where we have introduced an observation *window* function $w(n)$ to distinguish the part of the original process X_n that we can see from the part that is not observable and to allow some variable weighting values between zero and one. We assume the window function always satisfies

$w(n) > 0$ for all $n \in \mathcal{M}$ and is zero outside of this interval. In the simplest case, we could use a rectangular (or *boxcar*) window,

$$w(n) = \begin{cases} 1 & \text{for } n \in \mathcal{M} \\ 0 & \text{otherwise} \end{cases} \tag{4.6.2}$$

so that $s(n)$ is simply a truncated version of the original random process. As long as we satisfy the condition that the window $w(n)$ is nonzero only for $n \in \mathcal{M}$, we are using only the available observations of the random process. The window $w(n)$ is defined for all n but is always chosen to be zero outside of the finite interval $\mathcal{M}$ of actual observation times. For some applications, a better window will still have a finite duration "opening" but it will drop gradually rather than suddenly as the end points are approached to reduce the artifact of high frequency ripples introduced by the spectrum of the window. These issues are not relevant to our development here. The rationale of choosing windows is widely discussed in the digital signal processing literature.

We begin with the objective of minimizing the short-term average squared prediction error defined by

$$D = \sum_{n \in \mathcal{N}} e^2(n) = \sum_{n \in \mathcal{N}} [s(n) + \sum_{l=1}^{m} a_l s(n - l)]^2 \tag{4.6.3}$$

where $\mathcal{N}$ is a finite interval to be specified later. By differentiating D with respect to a_i, or by applying the orthogonality condition $e(n) \perp s(n - i)$, we get the result that

$$\sum_{n \in \mathcal{N}} [s(n) + \sum_{l=1}^{m} a_l s(n - l)] s(n - i) = 0 \text{ for } i = 1, 2, \cdots, m. \tag{4.6.4}$$

This expression simply says that the error $e(n)$ is orthogonal to the observables $s(n - i)$ where instead of a statistical average, we perform a time-average over the observable set of time instants $\mathcal{N}$. Simplifying (4.6.4) gives the normal equations:

$$\sum_{l=1}^{m} a_l \phi(i, l) = -\phi(i, 0) \text{ for } i = 1, 2, \cdots, m \tag{4.6.5}$$

where

$$\phi(i, l) = \sum_{n \in \mathcal{N}} s(n - i)s(n - l); 1 \leq i, l \leq m. \tag{4.6.6}$$

We next specialize this general result in two different ways to obtain two important computational techniques.

Autocorrelation Method

Our "observable" process $s(n)$ is defined for all $n \in \mathcal{N}$, where here $\mathcal{N}$ is the set of all integers. In the case of a rectangular window, the observable sequence that we use to calculate autocorrelations is simply the truncated version of X_n and it looks like this:

$$\cdots 0000 X_0 X_1 \cdots X_{M-1} X_M 000 \cdots.$$

More generally, the observable values of X_n will be weighted by the window values. The squared prediction error that we are trying to minimize is formed by summing the squared errors in predicting each value of $s(n)$ from the previous m values for all $n \in \mathcal{N}$. Note that $e^2(n) = 0$ for $n > M + m$ and for $n < 0$ and that the error is going to be very large in predicting the first few and last few values of the process $s(n)$, i.e., for $n = 0, 1, 2, \cdots, m-1$ and $n = M+1, M+2, \cdots M+m$. A tapered window will help reduce the effect of these meaningless components on the quantity to be minimized. Of course, for $m \ll M$, the end effects will be relatively small and meaningful results will still be achieved.

Applying the definition of $\phi(i, l)$ to our observable process we see that

$$\phi(i,l) = \sum_{-\infty}^{\infty} s(t)s(t + i - l) = R(i - l) \tag{4.6.7}$$

where $R(j)$ is the time average autocorrelation of $s(n)$ and is given explicitly in terms of the original process by

$$R(j) = \sum_{t=0}^{M-j} X_t X_{t+j} \tag{4.6.8}$$

for $0 \leq j \leq M$ and $R(j) = 0$ for $j > M$ and $R(j) = R(-j)$ for $j < 0$.

The result is that we have the normal equations

$$\sum_{j=1}^{m} a_j R(i - j) = -R(i); \quad i = 1, 2, \cdots, m \tag{4.6.9}$$

which are exactly equivalent to the case treated in the previous sections. The autocorrelation function $R(j)$ defined by (4.6.7) is a valid autocorrelation function in the sense that its Fourier transform is a nonnegative function. This follows by noting from (4.6.7), that $R(j)$ is the pulse autocorrelation function of the finite duration time function $s(t)$. From $R(j)$ we obtain an autocorrelation matrix for each order m that is a "healthy"

Toeplitz matrix. Hence, the Levinson-Durbin algorithm can be applied to calculate the optimal predictor and the mean squared prediction error.

The nice feature of the autocorrelation method is the fact that the solution is easily computed. Of course the objective function to be minimized seems a bit artificial since we are not really interested in minimizing the prediction error for the extreme end regions of the observation interval. However, the artifacts introduced by this objective are reduced by windowing and if the observation interval M is sufficiently large these artifacts have a negligible effect on the resulting predictor. Finally, a bonus of this method is that we always end up with a minimum delay prediction error filter $A(z)$. This property, derived in the following section, becomes very important in LPC synthesis, where we need a stable inverse filter, $A^{-1}(z)$.

Covariance Method

In this method we avoid the end effects of the autocorrelation function and we make sure our objective function only contains terms that fully make sense. Thus we define the squared error objective function as

$$D = \sum_{n=m}^{M} e^2(n) \text{ for } n \in \mathcal{N} = \{m, m+1, m+2, \cdots, M\}. \qquad (4.6.10)$$

Now the prediction error is considered only when past input values and current values are actual data observables (no zeros are inserted here). Then we have

$$\phi(i,j) = \sum_{n=m}^{M} s(n-i)s(n-j) \qquad (4.6.11)$$

defined for $1 \leq i \leq m$ and $0 \leq j \leq m$, where for the values of n used in the summation, $s(n) = X_n$. So that we now have to solve

$$\sum_{j=1}^{m} a_j \phi(i,l) = -\phi(i,0). \qquad (4.6.12)$$

This can be written in matrix form as

$$\mathbf{\Phi a} = -\psi. \qquad (4.6.13)$$

The matrix $\mathbf{\Phi}$ is symmetric, nonnegative definite, and in most cases positive definite. (Recall that a matrix $\mathbf{\Phi}$ is nonnegative definite or positive semidefinite if for any nonzero vector $\mathbf{u}$, $\mathbf{u}^t \mathbf{\Phi u} \geq 0$. A nonnegative definite matrix $\mathbf{\Phi}$ is positive definite or strictly positive definite if $\mathbf{u}^t \mathbf{\Phi u} = 0$ if and only if $\mathbf{u}$ is an all-zero vector.) In fact, it can be shown that $\mathbf{\Phi}$ is always positive definite if the autocorrelation matrices $\mathbf{R}_m$ are positive definite. (See Problem 4.7.) In Section 4.8 it is shown that the autocorrelation matrices of a nondegenerate stationary random process are always positive definite. Unfortunately, $\boldsymbol{\Phi}$ is not a Toeplitz matrix. This piece of bad news means that we cannot use the Levinson-Durbin algorithm to solve (4.6.13). But all is not lost. There is a numerically stable method for solving this equation (when $\boldsymbol{\Phi}$ is nonsingular) that is better than trying to invert a matrix. This method is the well-known Cholesky decomposition or Cholesky factorization which is extensively treated in numerical linear algebra books. In brief, the Cholesky decomposition allows $\boldsymbol{\Phi}$ to be factored into the form $\boldsymbol{\Phi} = \mathbf{LU}$, where $\mathbf{L}$ is a lower triangular matrix with unit values on the main diagonal and $\mathbf{U}$ is an upper triangular matrix. Then (4.6.13) can be solved by first solving the equation

$$\mathbf{Lq} = -\psi \qquad (4.6.14)$$

for the vector $\mathbf{q}$ and then solving

$$\mathbf{Ua} = \mathbf{q}$$

for $\mathbf{a}$. Unfortunately, the prediction error polynomial $A(z)$ obtained from the covariance method is not in general a minimum delay prediction error filter. A modified covariance method, sometimes called the *stabilized covariance* method, has been proposed which has the desired property of producing a minimum delay prediction error polynomial [20] [100].

4.7 Minimum Delay Property

The prediction error filter $A(z)$ for any finite memory optimal linear one-step predictor has the very important property that it is a minimum-delay filter. In other words, all the zeros of the polynomial $A(z)$ lie inside the unit circle. This property can be proven in several ways. One way makes use of the theory of orthogonal polynomials on the unit circle [172]. Another way is based on Rouché's Theorem in complex variable theory; see for example [177]. In this section, we give a simple proof of this result that does not depend on techniques outside of the scope of this book.

Suppose $A(z)$ is a prediction error filter that contains a zero outside the unit circle. We shall show that $A(z)$ cannot be the optimal prediction error filter by showing that the filter can be modified in a way that reduces the mean squared prediction error. Specifically, we replace any zero outside the unit circle with a new zero that is inside the unit circle. As usual, $A(z)$ has constant term unity ($a_0 = 1$), that is, it is a *monic* polynomial of degree m where m is the memory size of the predictor.

First suppose $A(\alpha) = 0$ where alpha is a real-valued zero that is outside the unit circle, i.e., $|\alpha| > 1$. Let

$$Q(z) = \frac{A(z)(1 - \alpha^{-1} z^{-1})}{1 - \alpha z^{-1}}, \tag{4.7.1}$$

so that $Q(z)$ now has the zero α^{-1} instead of α and $Q(z)$ is still a monic polynomial of the same degree as $A(z)$. Next we define the stable, allpass filter

$$H(z) = \frac{1 - \alpha z^{-1}}{1 - \alpha^{-1} z^{-1}} \tag{4.7.2}$$

so that we have

$$A(z) = Q(z)H(z). \tag{4.7.3}$$

(See Chapter 2 for a review of the basic properties of linear filters including minimum delay or minimum phase filters and allpass filters in discrete time.) Let X_n denote the stationary random process for which $A(z)$ is the given prediction error filter. Suppose X_n is applied to the filter $Q(z)$ producing the output u_n which is then fed into $H(z)$ to produce the output v_n as shown in Figure 4.3. Hence v_n is the prediction error associated with

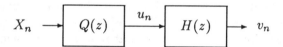

Figure 4.3: Prediction Error Filter Configuration

the overall prediction error filter $A(z)$. Now the mean squared prediction error is given by

$$\sigma_v^2 = \int_{-1/2}^{1/2} S_u(f) |H(e^{j 2\pi f})|^2 \, df, \tag{4.7.4}$$

where $S_u(f)$ denotes the spectral density of u_n. By noting the allpass character of $H(z)$, we have

$$|H(z)|_{z=e^{j2\pi f}} = |\alpha z^{-1} \cdot \frac{1 - \alpha^{-1} z}{1 - \alpha^{-1} z^{-1}}|_{z=e^{j2\pi f}} = |\alpha| > 1. \tag{4.7.5}$$

Thus it follows that

$$\sigma_v^2 = |\alpha| \sigma_u^2 > \sigma_u^2. \tag{4.7.6}$$

This shows that $Q(z)$ produces lower mean squared prediction error than $A(z)$. Since $Q(z)$ is monic and of the same order (memory size), $A(z)$ is not the optimal prediction error filter of memory size m.

Now suppose $A(z)$ has a complex-valued zero, β, that lies outside of the unit circle. Since $A(z)$ is a prediction filter, it has a real-valued impulse response so that any complex-valued zeros must occur in complex conjugate pairs. So β^* must also be a zero of $A(z)$ where the asterisk denotes complex conjugation. We now modify $A(z)$ by replacing this pair of zeros by their reciprocal values to get a new polynomial $Q'(z)$.

Let

$$Q'(z) = \frac{A(z)(1 - \beta^{-*}z^{-1})(1 - \beta^{-1}z^{-1})}{(1 - \beta z^{-1})(1 - \beta^* z^{-1})}, \qquad (4.7.7)$$

where $\beta^{-*} = (\beta^{-1})^*$. Note that $Q'(z)$ is again a polynomial of the same degree as $A(z)$ and is also monic. By defining the second order allpass filter:

$$H(z) = \frac{(1 - \beta z^{-1})(1 - \beta z^{-*})}{(1 - \beta^{-*}z^{-1})(1 - \beta^{-1}z^{-1})} \qquad (4.7.8)$$

and noting that on the unit circle $|z| = 1$, we have $|H(z)| = |\beta|^2$, and it then follows from (4.7.4) that

$$\sigma_v^2 = |\beta|^2 \sigma_u^2 > \sigma_u^2$$

where the configuration of Figure 4.3 again applies if $H(z)$ and $Q(z)$ are replaced by $H'(z)$ and $Q'(z)$, respectively. By the same reasoning as in the case of a real-valued zero, we conclude that the mean squared prediction error for the prediction error filter $Q'(z)$ of memory size m will be smaller than that of $A(z)$. Hence we conclude that if $A(z)$ is an optimal prediction error filter it can not have zeros outside of the unit circle and is therefore a minimum delay, causal, and causally invertible filter.

The inverse filter, $A^{-1}(z)$, corresponding to an optimal prediction error filter is an all-pole filter which plays an important role in the predictive coding systems to be discussed in the following chapter and has important applications in speech coding, synthesis, and recognition.

An important consequence of the minimum delay character of the prediction error filter is that it allows a very useful characterization of the random process X_n. Specifically, for any monic mth order polynomial $A(z)$ with real-valued coefficients and with all zeros inside the unit circle there exists a stationary random process X_n for which $A(z)$ is the optimal mth order prediction error polynomial. The process X_n can be defined as the output of the stable filter $A^{-1}(z)$ when white Gaussian noise with zero mean and nonzero variance is applied as the input. (In Section 4.10, we shall describe how this result can be used to simulate random processes on a computer.) Correspondingly, the inverse filter $A^{-1}(z)$ will map X_n into the original white process. With this construction, the process will be nondeterministic since (a) the m stage prediction error will have nonzero variance, and (b) increasing the prediction order beyond m will not reduce this error variance any further since the output of $A(z)$ to input X_n is already a white process thereby indicating that $A(z)$ is also the optimal infinite memory prediction error filter.

In Section 4.4, we have seen that for any nondeterministic stationary random process the reflection coefficients have magnitude less than unity. Since every monic real polynomial $A(z)$ of order m with roots inside the unit circle represents a valid stationary random process, the associated reflection coefficients will therefore have magnitude less than unity. The Step-Down procedure, described in Section 4.5 can be applied to obtain these reflection coefficients from the polynomial $A(z)$.

Conversely, every set of m real numbers with magnitude less than unity can be used to define a set of reflection coefficients which can be mapped into an mth order linear prediction error polynomial using the Step-Up procedure, also described in Section 4.5. It can also be shown that this polynomial will always be minimum delay. Thus we conclude that *a necessary and sufficient condition for a real monic polynomial to have all its roots inside the unit circle is that the Step-Down procedure generates a set of reflection coefficients that are less than unity in magnitude.* This interesting spin-off of linear prediction theory leads to an important stability testing procedure. In fact, the Step-Down procedure, coincides (with slight variations) with established algorithms for stability testing of polynomials.

As a final note, we observe that the finite memory assumption was not required for proving the minimum delay property of the optimal prediction error filter. If $A(z)$ is a causal infinite memory transfer function with leading term unity, it may still be regarded as a prediction error filter corresponding to an infinite memory linear predictor. The minimum delay property remains applicable and we shall make use of this feature in discussing the infinite memory case of linear prediction later.

4.8 Predictability and Determinism

In this brief section, we introduce some properties of random processes related to linear predictability. This fills in a gap in finite memory prediction and provides a point of transition to infinite memory linear prediction.

In Section 4.2 we defined a stationary random sequence as *nondeterministic* if the mth order autocorrelation matrix is nonsingular for each $m > 0$. This condition is equivalent to the statement that for each $m \geq 0$,

$$\sum_{i=0}^{m} b_i X_{n-i} = 0 \text{ with probability 1} \tag{4.8.1}$$

implies that

$$b_i = 0 \text{ for } i = 0, 1, 2, \cdots, m. \tag{4.8.2}$$

In other words, the random variables $X_n, X_{n-1}, \cdots, X_{n-m}$ are linearly independent with probability one.

Conversely, if X_n is *deterministic* (not nondeterministic) then for some $m > 0$ and for some choice of the coefficients a_i

$$X_n = \sum_{i=1}^{n} a_i X_{n-i} \tag{4.8.3}$$

with nonzero probability. In this case, X_n is *exactly* predictable from the past m observations with nonzero probability. By repeated application of this formula for successive values of n, we can see that the entire future of the random sequence is determined from any particular set of m contiguous past observations.

Theorem 4.8.1 *A stationary sequence X_n is nondeterministic if and only if every FIR filter operating on X_n produces an output sequence with nonzero variance.*

Proof: Since

$$Y_n = \sum_{s=0}^{m} h_s X_{n-s}, \tag{4.8.4}$$

we have

$$E(Y_n^2) > 0 \text{ is equivalent to } E(|\sum_{s=0}^{m} h_s X_{n-s}|^2) > 0. \tag{4.8.5}$$

$\square$

The autocorrelation matrix $\mathbf{R}_m$ that arises in the normal equation (4.3.10) is central to the study of linear prediction theory. As noted earlier, $\mathbf{R}_m$ is a Toeplitz matrix which endows it with certain important consequences. In particular, it is nonnegative definite.

This result clarifies the condition under which a finite memory linear predictor is *unique*. The normal equation (4.3.10) gives a unique solution when the autocorrelation matrix $\mathbf{R}_m$ is nonsingular. From the above discussion, it is clear that when the process to be predicted is nondeterministic this will indeed be the case. Suppose, on the other hand, that the process is deterministic. Then for some value of m, the autocorrelation matrix $\mathbf{R}_m$ is singular. Since this implies a linear dependence among m consecutive random variates X_n, there will also be a linear dependence among m' consecutive variates X_n if $m' > m$, so that all higher order autocorrelation matrices will also be singular. Consequently, for $m' \geq m$, the rank of the autocorrelation matrix will be less than its order and there will be more than one set of predictor coefficients that will yield the optimal predictor for any order $m' \geq m$. This does not exclude the possibility that for memory size $m' < m$ the covariance matrix is nonsingular.

Theorem 4.8.2 *The covariance matrix* $\mathbf{R}_m$ *of a stationary sequence* X_n *is nonnegative definite; if* X_n *is nondeterministic, then* $\mathbf{R}_m$ *is a positive definite matrix.*

Proof: Since

$$0 \leq E(|\sum_{i=0}^{m-1} u_i X_{n-i}|^2) = \sum_{i,j} u_i u_j E(X_{n-i} X_{n-j}) \qquad (4.8.6)$$

$$= \sum_{i=0}^{m-1} \sum_{j=0}^{m-1} u_i r_{i-j} u_j = \mathbf{u}^t \mathbf{R}_m \mathbf{u}, \qquad (4.8.7)$$

which immediately proves the nonnegative definite property. Furthermore, the above equations show that the quadratic form $\mathbf{u}^t \mathbf{R}_m \mathbf{u}$ is simply the mean squared value of the output of an FIR filter whose input is X_n and whose impulse response is u_i, so that if X_n is nondeterministic, then $\mathbf{R}_m$ must be positive definite. $\square$

A random sequence is said to be *purely nondeterministic* if it is nondeterministic and if

$$\lim_{m \to \infty} \min_{\{a_i\}} E(|\sum_{i=0}^{m} a_i X_{n-i}|^2) > 0 \qquad (4.8.8)$$

where $\{a_i\}$ indicates a sequence of coefficients, $a_0, a_1, a_2, \cdots, a_m$ with $a_0 = 1$. If a random sequence is not purely nondeterministic, it can be exactly predicted from the infinite past. For a purely nondeterministic process, nearly perfect prediction (with arbitrarily small mean squared error) can not be achieved no matter how large the memory m.

4.9 Infinite Memory Linear Prediction

Until this point, we have focused on the finite memory case of linear prediction where the m most recent observables are used to predict the current value X_n of the process. Of course, we are free to make m have an extraordinarily large value and it will be reasonable to assume that the resulting prediction gain can be made arbitrarily close to the ultimate value achievable as m goes to infinity. But this is not a practical technique since the computational complexity and the required random-access memory in the prediction filter also grow with m. Theoretically it is not satisfying since we do not get a handle on the ultimate performance in prediction gain and we are forced to resort to large numbers of iterations of the Levinson-Durbin recursion to even estimate what the prediction gain approaches as the memory goes to infinity. In fact, there is an alternative way, both for practical implementation (in hardware or software) and for analytical understanding. The practical solution is to use infinite-impulse response (IIR) filters with a finite number of poles and zeros in the complex z plane to perform the linear prediction operation. The analytical solution to the infinite memory prediction problem is based on spectral factorization and leads to an elegant and explicit formula for the optimal prediction gain achievable with linear prediction in terms of the spectral density of the random sequence.

The Wiener-Hopf Equation

Following the notation adopted in Section 4.3, we can state the infinite memory prediction problem as follows. We wish to find a causal and stable linear filter $P(z)$ with impulse response $\{-a_i\}$ for $i = 1, 2, 3, \cdots$ that minimizes the mean squared value of the prediction error:

$$e_n = X_n + \sum_{i=1}^{\infty} a_i X_{n-i}. \tag{4.9.1}$$

The predictor is given as

$$\hat{X}_n = -\sum_{i=1}^{\infty} a_i X_{n-i} \tag{4.9.2}$$

and has transfer function

$$P(z) = -\sum_{i=1}^{\infty} a_i z^{-i}. \tag{4.9.3}$$

The prediction error filter has transfer function

$$A(z) = \sum_{i=0}^{\infty} a_i z^{-i} \tag{4.9.4}$$

where $a_0 = 1$. By the orthogonality principle,

$$e_n \perp X_{n-j} \text{ for } j \geq 1 \tag{4.9.5}$$

or

$$E(e_n X_{n-j}) = 0 \text{ for } j \geq 1 \qquad (4.9.6)$$

which gives

$$E(\sum_{i=0}^{\infty} a_i X_{n-i} X_{n-j}) = 0$$

so that

$$\sum_{i=1}^{\infty} a_i r_{j-i} = -r_j \text{ for } j = 1, 2, 3, \cdots \qquad (4.9.7)$$

which is the discrete time infinite memory Wiener-Hopf equation.

In principle, this equation can be solved for the optimal predictor filter $P(z)$ thereby obtaining the prediction error filter $A(z)$. Alternatively, the orthogonality condition (4.9.6) can be used directly to derive properties and even a solution to the optimal prediction problem.

Whitening Property of the Prediction Error Filter

An important feature of the infinite memory case that is not present in finite memory prediction is the *whitening* character of the optimal prediction error filter. This critical property of the optimal prediction error filter can be deduced directly from the orthogonality condition.

Theorem 4.9.1 *The prediction error sequence e_n produced by filtering X_n through the optimal prediction error filter is an uncorrelated (white) sequence.*

Proof: From the orthogonality condition, e_n is uncorrelated with all past values X_{n-j} for $j \geq 1$ of the input sequence X_n. But e_{n-i} is fully determined by a linear combination of past input values X_{n-i-j} for $j > 0$. Hence, e_n is uncorrelated with e_{n-i} for $i > 0$. Since e_n is stationary, it is therefore white and its spectral density is given by $S_e(f) = \sigma_p^2$, where σ_p^2 denotes the mean squared prediction error for infinite memory. □

At this point we have some useful information about the optimal prediction error filter $A(z)$. We know that:

(a) it has leading term unity: $a_0 = 1$;

(b) it is causal and has a real-valued impulse response;

(c) it must be minimum delay;

(d) it must be a whitening filter.

The third property follows from Section 4.7 where, as noted, the minimum delay property did not assume finite memory. Observe that the first three properties depend only on the filter itself, while the fourth property depends on the input random process. Let us define an *admissible* filter as one which satisfies the first three properties. Now we raise the following question: Is an admissible filter that whitens a process necessarily the optimal prediction error filter for that process? Think about it before reading on.

The answer to the above rhetorical question is provided by the following theorem.

Theorem 4.9.2 *Let $A(z)$ be an admissible filter that whitens the stationary random sequence X_n. Assume that $S_X(f)$ is nonzero for almost all f. Then the predictor $P(z) = 1 - A(z)$ is the least mean square predictor for the sequence X_n.*

Proof: Since $A(z)$ whitens X_n, from basic linear systems, the spectral density of the whitened process is given by the spectral density, $S_X(f)$, of X_n multiplied by the magnitude squared of the transfer function:

$$S_X(f)|A(e^{j2\pi f})|^2 = \sigma^2 \qquad (4.9.8)$$

where σ^2 is the variance of the response of filter $A(z)$ to input X_n. This equation can be *uniquely* solved for an admissible $A(z)$ by the process of spectral factorization. Since the optimal prediction error filter is admissible and, by Theorem 4.9.1, is a whitening filter for X_n, it is the unique solution for $A(z)$ and $\sigma^2 = \sigma_p{}^2$. $\qquad\qquad\square$

The proof assumes you are familiar with spectral factorization. We now fill in this missing link by examining the problem of finding $A(z)$ from a given spectral density so that (4.9.8) is satisfied.

The Spectral Factorization Problem

To find $A(z)$ we need to find a causal and stable filter with impulse response having leading term of unity ($a_0 = 1$) that solves the Wiener-Hopf equation (4.9.7) and satisfies (4.9.8). We assume the spectral density is nonzero so that it is invertible. Thus, we need to find a minimum delay transfer function $A(z)$ satisfying the requirement that $A(z)A(z^{-1})|_{z=e^{j2\pi f}}$ is a prescribed function of f. This is a frequently recurring problem in signal processing and is known as the *spectral factorization problem*.

Factorization of a Rational Spectral Density Function

Suppose that the input process X_n has a *rational* spectral density function, that is, it can be expressed as the ratio of two polynomials in the variable

$e^{j2\pi f}$. Using (4.9.8), this condition can be written as

$$\frac{1}{\sigma_p^2} \frac{R(z)}{Q(z)}\Big|_{z=e^{j2\pi f}} = \frac{1}{S_X(f)} \qquad (4.9.9)$$

where R and Q are each finite order polynomials in z (positive powers of z apart from a factor of the form z^{-L}). Then we have

$$A(z)A(z^{-1}) = \frac{R(z)}{Q(z)} \qquad (4.9.10)$$

and we must find $A(z)$ such that it is a ratio of polynomials in z^{-1} according to

$$A(z) = \frac{R_1(z)}{Q_1(z)} \qquad (4.9.11)$$

where $R_1(z)$ and $Q_1(z)$ each have all their zeros inside the unit circle so that $A(z)$ will be stable and minimum delay. Furthermore, we require $a_0 = 1$ so that the constant term in each of R_1 and Q_1 can be required to be unity.

This factorization task is in principle very simple now. Since $R(z)$ and $Q(z)$ have real-valued coefficients, all complex valued zeros must occur in conjugate pairs. Furthermore, since replacing z by z^{-1} in (4.9.10) leaves the expression unaltered, all zeros of R and Q (not on the unit circle) must also occur in reciprocal pairs: for each root, α, with modulus not unity there must be a corresponding root α^{-1}. This implies that if we factor R and Q, we can select all factors of R and Q of the form $(1 - \alpha z^{-1})$ with $|\alpha| < 1$ for inclusion in R_1 and Q_1 respectively. With a suitable constant scale factor, this leads to a solution for $A(z)$ having the requisite properties. The actual prediction filter, $P(z)$, is then given as $P(z) = 1 - A(z)$. For a rational spectral density, this is indeed practical and can be performed with the help of a root-finding algorithm. In fact, any spectral density can be approximated by a rational function using established approximation techniques, so that any practical application where spectral approximation is needed can be based on a rational spectral model.

Spectral Density Factorization: General Case

We now present a general formulation of the factorization problem without requiring rational spectra. This will also lead us to a general formula for the least mean squared prediction error based on linear prediction from the infinite past. The derivation gives some additional insight into the factorization problem and linear prediction.

We begin with the spectral density, $S_X(f)$. For a discrete time sequence, as previously defined, $S_X(f)$ is a symmetric, positive, real-valued periodic

function of f with period 1. Hence by taking the logarithm of the spectral density we obtain a new periodic function of f with the same period
and which is also real-valued and symmetric in f. Using a Fourier series
expansion, we obtain

$$\log S_X(f) = \sum_{-\infty}^{\infty} \beta_k e^{-j2\pi fk} \qquad (4.9.12)$$

with the Fourier coefficients

$$\beta_k = \int_{-1/2}^{1/2} \log S_X(f) e^{j2\pi fk}\, df. \qquad (4.9.13)$$

Since $S_X(f)$ is real and symmetric, the coefficients β_k are also real and
symmetric in k. Note that the dc average of the log spectrum is

$$\beta_0 = \int_{-1/2}^{1/2} \log S_X(f)\, df \qquad (4.9.14)$$

and e^{β_0} is the *geometric mean* of the spectral density $S_X(f)$.

Now the factorization is simply performed as follows: Define

$$L(f) = \frac{1}{2}\beta_0 + \sum_{k=1}^{\infty} \beta_k e^{-j2\pi fk} \qquad (4.9.15)$$

so that

$$\log S_X(f) = L(f) + L(-f). \qquad (4.9.16)$$

Note that $L(f)$ can be viewed as the transfer function of a causal linear
filter with real-valued impulse response $\{\beta_0/2, \beta_1, \beta_2, \cdots\}$. We then define
$H(f) = e^{L(f)}$ which is also periodic with period 1. Expanding $H(f)$ in a
Fourier series gives

$$H(f) = \sum_{k=0}^{\infty} h_k e^{-j2\pi fk} \qquad (4.9.17)$$

which does not contain any terms with positive powers of $e^{j2\pi f}$. Why not?
Since e^x can be expanded into a power series with only positive powers
of x, by substituting $L(f)$ for x, we get only positive powers of the terms
contained in $L(f)$. Thus we have obtained the factorization:

$$H(f)H^*(f) = S_X(f). \qquad (4.9.18)$$

Now we observe that by the Parseval equality

$$\sum_{k=0}^{\infty} h_k^2 = \int_{-1/2}^{1/2} |H(f)|^2\, df = \int_{-1/2}^{1/2} S_X(f)\, df = \sigma_X^2 < \infty \qquad (4.9.19)$$

so it follows that the filter $H(f)$ has a finite energy impulse response and is causal. This property which we call *energy-stable* is equivalent to stability for a rational transfer function and is sometimes used as an alternative definition of stability. Furthermore, we can show that $H(f)$ is causally invertible. To see this, simply define $H^-(f) \equiv e^{-L(f)}$ which by the same argument as above can be shown to be an energy-stable and causal filter. But $H^-(f) = [H(f)]^{-1}$. Thus we have shown that $H(f)$ or its equivalent expression as a function of z,

$$G(z) = \sum_{i=0}^{\infty} h_i z^{-i} \tag{4.9.20}$$

is a real, energy-stable, causal, and causally invertible filter that solves the factorization problem.

We began with a specific factorization problem in mind, namely, to find the optimal prediction error filter $A(z)$. Referring to (4.9.8), we see that the solution $H(f)$, or $G(z)$, obtained above immediately yields the desired prediction error filter:

$$A(z) = \frac{\sigma_p}{G(z)}. \tag{4.9.21}$$

From the constraint $a_0 = 1$, it follows that $h_0 = \sigma_p$. This leads to a very useful formula for the mean squared prediction error for optimal infinite past linear prediction.

Infinite Past Prediction Error Formula

We now derive the formula that explicitly gives the mean squared prediction error for the optimal linear predictor that operates on the infinite past of a stationary random process X_n. The formula provides a very useful bound on the performance in the finite memory case and is conveniently expressed directly in terms of the spectral density of the process X_n.

Theorem 4.9.3 *The minimum mean squared prediction error incurred in estimating a stationary process X_n with spectral density $S_X(f)$ from the infinite past $\{X_{n-i}\}$ for $i = 1, 2, \cdots$, is given by*

$$\sigma_p^2 = \exp\left(\int_{-1/2}^{1/2} \log S_X(f)\, df\right). \tag{4.9.22}$$

Proof: From the general factorization result presented earlier, we found that $H(f) = e^{L(f)}$. By expanding e^x into a power series in x, where x is the expression for $L(f)$ as given in (4.9.1), we get

$$
\begin{aligned}
H(f) &= e^{\frac{1}{2}\beta_0 + \sum_{k=1}^{\infty} \beta_k e^{-j2\pi f k}} \\
&= e^{\frac{1}{2}\beta_0} \left[\sum_{n=0}^{\infty} \frac{1}{n!} \left(\sum_{k=1}^{\infty} \beta_k e^{-j2\pi f k} \right)^n \right] \\
&= e^{\frac{1}{2}\beta_0} \left[1 + \gamma_1 e^{-j2\pi f} + \gamma_2 e^{-2j2\pi f} + \cdots \right].
\end{aligned}
\tag{4.9.23}
$$

From this it is apparent that

$$
h_0 = e^{\frac{1}{2}\beta_0}
\tag{4.9.24}
$$

so that

$$
h_0^2 = e_0^\beta = \exp\left(\int_{-1/2}^{1/2} \log S_X(f)\, df \right)
\tag{4.9.25}
$$

where we have substituted (4.9.14) for β_0. But we have seen that $h_0 = \sigma_p$, and therefore

$$
\sigma_p^2 = \exp\left(\int_{-1/2}^{1/2} \log S_X(f)\, df \right).
\tag{4.9.26}
$$

The ultimate performance capability of linear prediction is given by the prediction gain for the infinite memory case. Specifically this is

$$
G_p^\infty = \frac{\sigma_X^2}{\sigma_p^2},
\tag{4.9.27}
$$

but we have seen that σ_p^2 is simply the geometric mean of the spectral density $S_X(f)$ while the variance σ_X^2 of the process X_n is the arithmetic mean of the spectral density. Thus this performance measure is the ratio of arithmetic to geometric means of the spectral density:

$$
G_p^\infty = \frac{\int_{-1/2}^{1/2} S_X(f)\, df}{e^{\int_{-1/2}^{1/2} \log S_X(f)\, df}}.
\tag{4.9.28}
$$

The reciprocal γ_x of the infinite memory prediction gain is called the *spectral flatness measure*. The closer the process is to white noise, the flatter is the spectrum and the closer is the ratio to unity. Thus the degree of predictability of a random process is indicated by the degree to which its spectrum is "curvy," not flat, or nonwhite.

4.10 Simulation of Random Processes

The results on spectral factorization provide a useful spinoff: a convenient method for generating a discrete time random process with a prescribed spectral density. This technique is of considerable value in computer simulation studies.

Suppose we are given a spectral density $S(f)$. Assume it is rational (otherwise it can readily be approximated by Fourier series analysis as a polynomial in $e^{j2\pi f}$ scaled by a factor of the form e^{-jfM}). Apply the spectral factorization technique to obtain a causal minimum delay transfer function $G(z)$ as in Section 4.9. Then simply apply independent zero mean unit variance random variables, Z_n, (say with a Gaussian pdf) to the filter $G(z)$ to produce an output sequence with the desired spectral density. In particular, if $G(z)$ has the form

$$G(z) = \frac{\sum_{i=0}^{M} b_i z^{-i}}{\sum_{i=0}^{N} c_i z^{-i}} \tag{4.10.1}$$

where, for convenience, we assume that $c_0 = 1$, then the output process Y_n is given by the recursion

$$Y_n = -\sum_{i=1}^{N} c_i Y_{n-i} + \sum_{i=0}^{M} b_i Z_{n-i}. \tag{4.10.2}$$

Since the filter is stable, any choice of initial conditions for N consecutive values of Y_n will have only a transient effect. An asymptotically stationary random process will be generated in this way. If $G(z)$ is an *all-pole filter*, then $M = 0$ and the process generated is known as an *autoregressive* process. If $G(z)$ is an *all-zero filter*, then $N = 0$ (i.e., the first sum in the expression above is absent) and the process generated is called a *moving-average* process. In general, when both M and N exceed zero, the process is called an *autoregressive-moving average* (ARMA) process.

4.11 Problems

4.1. Find expressions for the MSE resulting from the optimal (MMSE) predictor for a random vector $\mathbf{Y}$ given a random vector $\mathbf{X}$.

4.2. Assume that $\mathbf{Y}$ and $\mathbf{X}$ are zero mean random vectors. Show that if $\hat{\mathbf{Y}} = \mathbf{AX}$ is the optimal (MMSE) linear predictor of $\mathbf{Y}$ given $\mathbf{X}$, $\mathbf{e} = \mathbf{Y} - \hat{\mathbf{Y}}$, and

$$\sigma_e^2 = E(\|\mathbf{Y} - \hat{\mathbf{Y}}\|^2) = E[(\mathbf{Y} - \hat{\mathbf{Y}})^t(\mathbf{Y} - \hat{\mathbf{Y}})],$$

then
$$\sigma_e^2 = E(\|\mathbf{Y}\|^2) - E(\|\hat{\mathbf{Y}}\|^2) = \sigma_Y^2 - \sigma_{\hat{Y}}^2.$$
Thus the variance of the optimal prediction is smaller than that of the true value.

4.3. Suppose that X does not have zero mean. Do the results of the previous problem remain true for the optimal linear predictor? Derive and compare the expressions for the mean squared error resulting from the optimal linear predictor and the optimal affine predictor. Prove that the latter is strictly smaller than the former.

4.4. It was stated several times that the optimal predictor of one vector given another is in fact linear if the two vectors are jointly Gaussian. Prove this.

4.5. Use the results of Section 4.2 to find the optimal linear M-step predictor, that is, the best linear predictor of X_{n+M-1} given
$$X_{n-1}, \cdots, X_{n-m}.$$

4.6. Suppose that X_n is a Wiener process, e.g., it has the form $X_0 = 0$ and
$$X_n = X_{n-1} + W_n, \ n = 1, 2, \cdots,$$
where the W_n is a zero mean i.i.d. Gaussian sequence with variance σ^2. What is the optimal prediction of X_n given $X_1, \cdots, X_{n-1}$? What is the optimal linear estimate?

4.7. Suppose the stationary random process $\{X_n\}$ has positive definite autocorrelation matrices $\mathbf{R}_m$ for all orders m. Let $s(n)$ denote the windowed samples obtained from X_n according to (4.6.1). Show that a linear combination of these samples $Y = \sum_{j=m}^{N} a_j s(n - j)$ has nonzero second moment EY^2 unless the coefficients a_j are all equal to zero. Use this result to show that the covariance matrix Φ whose elements are defined in (4.6.11) is positive definite.

4.8. Find the optimal predictor coefficients of order 2 ($m = 2$) as a function of the autocorrelation values r_0, r_1, and r_2 for a stationary random process by (a) solving the matrix equation directly and (b) applying the Levinson-Durbin recursion.

4.9. A stationary Gaussian random process with zero mean has spectral density given by $S(\omega) = 17 - 15 \cos 2\omega$.

(a) Find the autocorrelation function of this process.

(b) Show that the process can be modeled as a moving-average process and specify a moving-average equation that generates a process with the above spectral density while starting with a white Gaussian process.

4.10. A unit variance random process $\{X_n\}$ has zero mean and autocorrelation function given by

$$r_k = (0.9)^{|k|}.$$

(a) Show that this process can be modeled as a first-order autoregressive process.

(b) Calculate and plot the prediction gain in dB versus the memory m for the optimal finite memory prediction for $m = 1, 2,$ and 3.

(c) Predict the value of X_{17} when it is observed that $X_{16} = 2.4$ and $X_{15} = -1.9$.

4.11. A segment of a "voiced" speech signal, $s(n)$, for $n = 1, 2, \cdots, N$ can be modeled approximately as a periodic waveform (whose period corresponds to the *pitch* of the speaker) but successive periods may decay or grow exponentially. The segment is long enough to contain several periods. Suppose we wish to find an *optimal* pitch extractor from a speech signal in the sense that it should minimize the average squared prediction error, given by

$$D(\beta, P) = \sum_{n=1}^{N} [s(n) - \beta s(n - P)]^2$$

where β is a scaling factor for the predictor and P is the candidate pitch period. Thus the pitch predictor has transfer function $H(z) = \beta z^{-P}$.

(a) Find a convenient formula for the optimal β to minimize D as a function of the predictor lag value P.

(b) Substitute the above value of β into the formula for D and simplify it to obtain a convenient and compact expression. The optimal predictor lag P^* (which will usually be the "true" pitch) is then found by finding the value of D for all possible lags P in the desired range. (Typically, for 8 kHz sampled speech, the values of interest are 21 to 148.)

4.12. Find the optimal linear predictor coefficients for a 10th order predictor for the moving-average process $X_n = W_n + 3W_{n-1}$ where W_n is zero mean unit variance and white. Also plot the prediction gain in dB versus the memory size achieved for optimal prediction of this process with memory sizes ranging from 0 to 10 samples.

4.13. A linear prediction error polynomial of third order is given as

$$A(z) = 1 + 0.5z^{-1} + 0.25z^{-2} + 0.5z^{-3}.$$

Find the corresponding set of reflection coefficients. Does this polynomial have all of its roots inside the unit circle? Explain.

4.14. Let X_n denote the samples of a stationary random process. Suppose we have already found the optimal linear *one-step* prediction of X_n of order m given by

$$\hat{X}_n^{(1)} = -\sum_{i=1}^{m} c_i X_{n-i}$$

and the optimal linear *two-step* prediction of X_{n+1} of order m given by

$$\hat{X}^{(2)} = -\sum_{i=1}^{m} d_i X_{n-i}.$$

In each of these two predictors, the same observable past data

$$X_{n-1}, X_{n-2}, \cdots, X_{n-m}$$

is used and the prediction error is orthogonal to this set of m past observables.

Now we wish to find the optimal linear one-step prediction of X_{n+1} of order $m+1$ in terms of the $m+1$ past observables

$$X_n, X_{n-1}, X_{n-2}, \cdots X_{n-m}$$

given by

$$\hat{X}_{n+1} = \sum_{i=0}^{m} f_i X_{n-i}.$$

(a) Find a simple formulation for solving this problem, making suitable use of the orthogonality condition, so that, if carried out to completion, it would allow the coefficients f_i to be computed from the known coefficients c_i and d_i and from knowledge of the autocorrelation values $r_i = E\{X_n X_{n-i}\}$.

(b) Find an explicit formula for f_i for $i = 0, 1, 2, \cdots, m$ in terms of d_i, f_i, and r_i for $i = 0, 1, 2, \cdots, m$. (For convenience, define $d_0 = 0$ and $d_{-1} = 1$.)

4.15. A stationary Gaussian process with zero mean has spectral density given by

$$S(\omega) = 17 - 15 \cos 2\omega.$$

(a) Find the transfer function of a causal minimum delay filter that maps X_k into a white Gaussian process with unit variance.

(b) Find the optimal infinite memory linear prediction filter, $G(z)$, that predicts X_k from previous values $X_{k-1}, X_{k-2}, X_{k-3}, \cdots$.

(c) Describe the algorithmic procedure that you would use to generate a Gaussian random process with this spectral density on the computer.

(d) What is the mean square prediction error for the predictor in (b)?

4.16. A random process $\{X_k\}$ is whitened by the filter

$$B(z) = \frac{1}{6 - 5z^{-1} + z^{-2}}$$

Is $B(z)$ causally invertible? What is the optimal infinite memory linear predictor transfer function and what is the mean squared prediction error?

Part II

Scalar Coding

Chapter 5

Scalar Quantization I: Structure and Performance

5.1 Introduction

Quantization is the heart of analog-to-digital conversion. In its simplest form, a quantizer observes a single number and selects the nearest approximating value from a predetermined finite set of allowed numerical values. Ordinarily the input value is *analog*; that is, it may take on any value from a continuous range of possible amplitudes; furthermore, the output is *digital*, being uniquely specified by an integer in the set $\{1, 2, 3, \cdots, N\}$, where N is the size of the set of output values. As a simple example, consider a digital thermometer that can measure temperature in degrees Centigrade and for temperatures in the range of $-.49$ to 99.49 it can display the value rounded to the nearest integer. Clearly, the input value is analog and the output, a two digit number from a set of size 100, is digital.

More precisely, we define an N-point scalar (one-dimensional) quantizer Q as a mapping $Q : \mathcal{R} \to \mathcal{C}$ where $\mathcal{R}$ is the real line and

$$\mathcal{C} \equiv \{y_1, y_2, y_3, \cdots, y_N\} \subset \mathcal{R} \tag{5.1.1}$$

is the output set or *codebook* with size $|\mathcal{C}| = N$. The output values, y_i, are sometimes referred to as *output levels*, *output points*, or *reproduction values*.

In all cases of practical interest, N is finite so that a finite number of binary digits is sufficient to specify the output value. Later, we shall see that it is of theoretical convenience to consider the case where the output

set is countably infinite. For a quantizer with a finite output set, we assume
that the indexing of output values is chosen so that

$$y_1 < y_2 < \cdots < y_N. \tag{5.1.2}$$

For the digital thermometer, the input set, $\mathcal{R}$, is the set of all possible
temperatures to which the thermometer can ever be exposed. The output
set is $\{0, 1, 2, \cdots, 99\}$, consisting of the values that the thermometer can
display. The mapping Q is determined by the physical mechanism of the
sensor and the digital electronics in this device. If it is well designed, we
would expect that for any actual temperature in the range of 0 to 99 degrees,
the output reading will give the true value rounded to the nearest integer.

We define the *resolution* or *code rate*, r, of a scalar quantizer as $r = \log_2 N$ which measures the number of bits needed to uniquely specify the
quantized value. The resolution indicates the accuracy with which the
original analog amplitude is described. Specifically, if r is an integer, one
could assign to each y_i a unique binary r-tuple (by coding the values y_i into
binary vectors in an invertible manner). In this case a *fixed rate* code is
being used to specify the quantized value. Alternatively, different quantized
values can be specified with binary words of differing numbers of digits.
In this case, a *variable rate* code is being used and the *average rate* is the
expected number of digits needed to specify the quantized value. Chapters 9
and 17 present the principal concepts and methods for variable rate coding.

In the case of waveform communication, a sequence of samples is usually
digitized one sample at a time by quantization and a fixed number of binary
digits (bits) is used to specify each quantized sample value. We define the
transmission rate, R, of the quantizer as the number of bits transmitted
per sample to describe the waveform so that $R \equiv \log_2 N$ and the code rate
and transmission rate are numerically equal. We point out the difference
between the two ideas because the code rate is inherently a property of
the code while the transmission rate is a property of the communication
channel. Both coding rate and transmission rate are, however, commonly
abbreviated to the single word *rate* without a qualifying adjective. In a
sense, resolution is a more basic quantity than transmission rate since it
measures the precision offered by a quantizer without regard to the context
in which the quantizer is used. Occasionally the natural logarithm is used
instead of the logarithm with base 2 to define the rate in which case the
units are nats per sample.

If the sampling period is T, the transmission rate normalized to bits per
second is R/T bits per second, a quantity sometimes called simply the *bit
rate*.

Associated with every N point quantizer is a *partition* of the real line
$\mathcal{R}$ into N *cells* or *atoms* R_i, for $i = 1, 2, \cdots, N$. The ith cell is given by

$R_i = \{x \in \mathcal{R} : Q(x) = y_i\} \equiv Q^{-1}(y_i)$, the inverse image of y_i under Q. It follows from this definition that $\bigcup_i R_i = \mathcal{R}$ and $R_i \bigcap R_j = \emptyset$ for $i \neq j$. A cell that is unbounded is called an *overload* cell. Each bounded cell is called a *granular* cell. Together all of the overload (granular) cells are called the *overload region (granular region)*.

Since a quantizer Q is completely described by its output levels $\{y_i; \ i = 1, 2, \cdots, N\}$ and the corresponding partition cells $\{R_i; \ i = 1, 2, \cdots, N\}$, we sometimes write $Q = \{y_i, R_i; \ i = 1, 2, \cdots, N\}$ to make this description explicit.

A quantizer is defined to be *regular* if

a) each cell R_i is an interval of the form (x_{i-1}, x_i) together with one or both of the end points, and

b) $y_i \in (x_{i-1}, x_i)$.

The values x_i are often called *boundary points, decision points, decision levels*, or *endpoints*. A regular quantizer has the characteristic property that if two input values, say a and b with $a < b$, are quantized with the same output value, w, any other input that lies between a and b will also be quantized with the same output w. This property corresponds to our intuitive notion of quantization as a way of approximating a real number with a finite precision representation. As an arbitrary means of assigning endpoints so as to provide a complete description, we shall assume that for regular quantizers

$$R_i = (x_{i-1}, x_i] = \{r : \ x_{i-1} < r \leq x_i\}; \ i = 0, 1, \cdots, N - 1,$$

where the notation $(a, b]$ means the interval (a, b) together with the right endpoint b. The same definition will hold for $i = N$ if x_N is finite. If x_N is infinite, then $R_N = (x_{N-1}, x_N)$.

For a regular quantizer, the interval endpoints are ordered according to:

$$x_0 < y_1 < x_1 < y_2 < x_2 < \cdots < y_N < x_N \qquad (5.1.3)$$

as shown in Figure 5.1. Note that the set of endpoints and output values fully and uniquely specifies the quantizer. Of frequent interest is the special case of a *symmetric* quantizer, defined by the condition $Q(x) = -Q(-x)$ (except possibly on the cell boundaries). In addition, a quantizer is usually designed to be *monotonic*, defined by the condition that $Q(x)$ is a monotone nondecreasing function of x. It is easy to see that a regular quantizer must be monotonic.

A typical graph of the function $Q(x)$ for a symmetric regular quantizer shown in Figure 5.2 illustrates the inevitable staircase character of a quantizer. The input-output characteristic consists of N *treads*, or horizontal

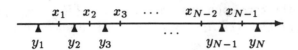

Figure 5.1: One-dimensional quantizer interval endpoints and levels indicated on a horizontal line

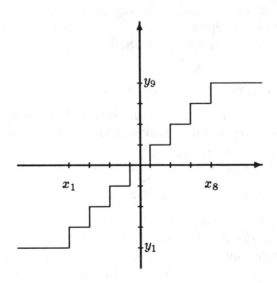

Figure 5.2: Graph of staircase character of a quantizer

steps, with a *riser* or vertical segment indicating the jump discontinuities between two successive steps occurring at each boundary value. The quantizer shown is known as a *mid-tread* quantizer since the origin is in the middle of a tread. In a *mid-riser* quantizer, the origin is a boundary point that is equidistant from the two adjacent output points.

Usually $x_0 = -\infty$ and $x_N = +\infty$ and the cells R_0 and R_N are overload cells. The *range*, B, of a quantizer is defined as the total length of the granular cells, so that for an unbounded regular quantizer, $B = x_{N-1} - x_1$. Occasionally the input values may be restricted to lie on a subset of the real line. In particular, if the input values lie in a bounded region, then x_0 and x_N may be finite since there is no need to define the quantizer mapping for input values that can never occur. In this case, the quantizer does not contain any overload cells and the range is given by $B = x_N - x_0$.

In most (but not all) applications, a quantizer is regular. A notable exception is the class of modulo-PCM quantizers [257]. As an example of a modulo-PCM quantizer, consider time measured in hours as the input to an operation that performs modulo 24 reduction and then rounding to the nearest integer, so that the input 36.7 is quantized to 13. If no modular operation were performed and $y_N = 23$, then 36.7 would be quantized to 23, the nearest reproduction level to the overloading input value. The modulo-PCM operation could equivalently be modeled as a nonlinear and noninvertible mapping followed by a regular quantizer.

Every quantizer can be viewed as the combined effect of two successive operations (mappings), an *encoder*, $\mathcal{E}$, and a *decoder*, $\mathcal{D}$. The encoder is a mapping $\mathcal{E} : \mathcal{R} \rightarrow \mathcal{I}$ where $\mathcal{I} = \{1, 2, 3, \cdots, N\}$, and the decoder is the mapping $\mathcal{D} : \mathcal{I} \rightarrow \mathcal{C}$. Thus if $Q(x) = y_i$ then $\mathcal{E}(x) = i$ and $\mathcal{D}(i) = y_i$. With these definitions we have $Q(x) = \mathcal{D}(\mathcal{E}(x))$. Occasionally in the engineering literature, a quantizer is assumed to generate both an index $\mathcal{E}(x)$ and an output value, y, and the decoder is sometimes referred to as an "inverse quantizer."

In the context of a waveform communication system, the encoder transmits the index i of the selected level, y_i, chosen to represent an input sample, and not the value y_i itself. Thus, if the rate R is an integer, one could assign to each y_i a unique binary R-tuple (one *codes* the y_i into binary vectors in an invertible manner). This binary R-tuple can then be transmitted or stored and then the decoder reconstructs the corresponding reproduction value. Note that the decoder can be implemented by a table-lookup procedure, where the table or *codebook* contains the output set which can be stored with extremely high precision without affecting the transmission rate R. A sequence of reproduction values obtained in this manner provides an approximation to the original sequence of samples and hence, using a suitable interpolating filter, an approximation to the original waveform can be reconstructed.

5.2   Structure of a Quantizer

Unlike linear systems, which are thoroughly understood by signal processing engineers, quantization is intrinsically a nonlinear operation and is not
as clearly understood. To obtain insight and intuition, it is usually helpful
to decompose a complex system into simpler components, each of which
has a more elementary function. If these components perform standard,
or "canonical," operations, a large variety of complex systems can be represented by different structural interconnections of the same elementary
components. If the complex system contains many parameters whose values specify the system, the decomposition is useful if each primitive element
is fully determined by only one or a few of the parameter values. A building block is called *primitive* if it cannot be further decomposed into simpler
components that are both convenient and meaningful. A quantizer may indeed be modeled as such an interconnection of simple building blocks. The
model is also relevant to the implementation (in hardware or software) of a
quantizer. In this section, we introduce a primary and secondary structural
description of a scalar quantizer in terms of simple building blocks.

Primary Structure of a Quantizer

A quantizer is uniquely determined by its partition and output set. The
encoder is specified by the partition of the input set and the decoder is
specified by the set of output values. The job of the encoder is simply to
make a statement of the form: "The current input value lies in cell 27 of the
partition." This operation is conveniently modeled by a *selector function*
defined by $S_i(x) = 1$ if $x \in R_i$ and 0 if $x \notin R_i$. Thus for example, the
selector function S_{49} simply gives a 'yes' or 'no' answer to the question: *Is
the input in cell 49?* Mathematically a selector function is just the *indicator
function* of the partition cell. Thus, if we define the indicator function 1_R
of a set R by

$$1_R(x) = \begin{cases} 1 & \text{if } x \in R \\ 0 & \text{otherwise} \end{cases}$$

then

$$S_i(x) = 1_{R_i}(x).$$

The decoder's job on receiving the index 27 is to look up the 27th entry
in a table and announce the reproduction value corresponding to this index.
We implicitly assume that a unique numbering of the N partition cells and
a corresponding numbering of the output points with the set of integers $\mathcal{I}$

has been defined. The operation of quantization can be expressed in a very useful form in terms of the selector functions:

$$Q(x) = \sum_{i=1}^{N} y_i S_i(x). \tag{5.2.1}$$

Note that for any given input value, x, only one term of the sum is nonzero. This representation of the overall quantization operation is depicted in Figure 5.3 and will be referred to as the *first canonical form* or the *primary structural decomposition* of a quantizer. Note that each S box is a memory-

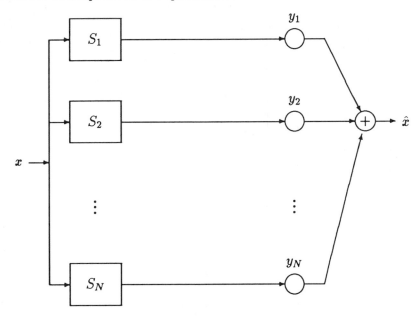

Figure 5.3: Primary structure of a quantizer.

less nonlinear operation on the input value; also, the multipliers, indicated by circles, have weight values given by the corresponding output point for that index value.

To separate explicitly the encoder and decoder functions of a quantizer, we modify the primary structural description. Define the address generator operation A as the mapping $A : \mathbf{B}^N \rightarrow \mathcal{I}$ and its inverse, the address decoder $A^{-1} : \mathcal{I} \rightarrow \mathbf{B}^N$, where $\mathbf{B}^N$ denotes the set of N binary N-tuples $\mathbf{b} = (b_1, b_2, \cdots, b_N)$ where $b_i \in \{0, 1\}$ and only one component is unity (the vector has a *Hamming weight* of one in coding parlance), that is,

$$\mathbf{b} = (S_1(x), S_2(x), \cdots, S_N(x)).$$

Then $A(\mathbf{b}) = i$ if $b_i = 1$ and $b_j = 0$ for $j \neq i$. With these definitions the encoder operation can be written as:

$$\mathcal{E}(x) = A(S_1(x), S_2(x), \cdots, S_N(x)) \tag{5.2.2}$$

and the decoder operation is given by:

$$\mathcal{D}(i) = \sum_{k=1}^{N} y_k A^{-1}(i)_k, \tag{5.2.3}$$

where $A^{-1}(i) = (A^{-1}(i)_1, \cdots, A^{-1}(i)_N) \in \mathbf{B}^N$. The corresponding structural diagrams for the coder and decoder are shown in Figure 5.4. This

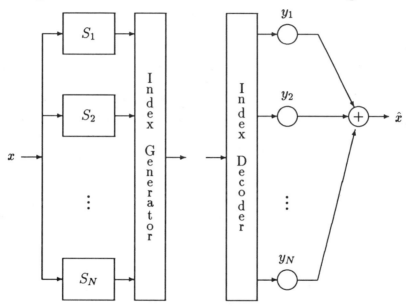

Figure 5.4: Coder/decoder primary structural model.

representation emphasizes that the decoder can be considered as being "linear" in the sense that it is a simple summation, while the encoder has no such simple form and can in general be quite complicated unless the selector functions happen to have additional simplifying structure. (The operation of course is not, however, genuinely linear in the technical sense. Quantization is inherently a nonlinear operation.)

Secondary Structure of a Quantizer

The selector operation, although a relatively simple building block, is not primitive since it can be further decomposed into elementary comparators.

The simplest nontrivial quantizer is the *binary threshold element*, or *comparator*, which partitions the real line into two cells, making a distinction between the left half and right half with respect to a particular threshold value, or boundary point. A comparator operating on a real number x may be viewed as a quantizer described by the indicator function $B_v(x)$, where v is the threshold value, $N = 2$, and the output set is $\{0, 1\}$:

$$B_v(x) = 1_{(-\infty, v]}(x) = \begin{cases} 1 & \text{if } x \leq v \\ 0 & \text{if } x > v \end{cases} \tag{5.2.4}$$

For completeness, define $B_\infty(x) = 1$. For convenience, the values $\{0, 1\}$ assumed by $B_v(x)$ will be taken to be either in the field of real numbers or in the field of binary numbers (modulo 2) depending on the context. The comparator, $B_v(x)$, is shown in Figure 5.5. It is also convenient to define the

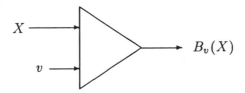

Figure 5.5: Comparator: The binary threshold element, the most primitive quantizer element.

complementary binary threshold element, $\overline{B}_v(x) = 1 - B_v(x) = 1_{(v, \infty)}(x)$.

The selector functions of any N-point coder can be implemented with binary threshold elements and combinatorial logic as long as each cell of the partition is the union of a finite number of intervals. In particular, for a regular quantizer, each cell is a single interval characterized by its two endpoints, x_{i-1} and x_i, and it suffices to note that each selector function can be written as

$$S_i(x) = B_{x_i}(x) - B_{x_{i-1}}(x). \tag{5.2.5}$$

$S_i(x)$ will be one if and only if x is less than or equal to x_i but is not less than or equal to x_{i-1}. This operation can also be written in other ways. Again considering the variables as real valued, we can write

$$S_i(x) = B_{x_i}(x)\overline{B}_{x_{i-1}}(x).$$

If we consider the variables as Boolean binary variables and let AND denote the logical "and" operation (a AND b is 1 if both a and b are 1 and 0 otherwise), then

$$S_i(x) = B_{x_i}(x) \text{ AND } \overline{B}_{x_{i-1}}(x).$$

Thus the selector functions can be formed from the binary threshold devices either using real addition or multiplication or using a logical AND device.

Combining (5.2.1) and (5.2.5) and noting that $B_{-\infty}(x) \equiv 0$ shows that at most $N - 1$ comparators are needed to implement an N-point regular quantizer. To see this, observe that

$$
\begin{aligned}
Q(x) &= \sum_{i=1}^{N} y_i [B_{x_i}(x) - B_{x_{i-1}}(x)] \\
&= \sum_{i=1}^{N-1} (y_i - y_{i+1}) B_{x_i}(x) + y_N B_{x_N} - B_{x_0} y_1).
\end{aligned}
$$

For a cell that is the union of a finite number of intervals, a similar expression involving a larger number of comparator functions can readily be formed.

For a bounded regular quantizer, each of the N selector functions corresponding to a finite quantization cell can be replaced with a combination of two comparators as shown in Figure 5.6. Thus by applying the decomposition shown in Figure 5.6 to the structures of Figure 5.4, we obtain the *secondary* structure for a quantizer and for an encoder. For the case of an unbounded quantizer, there are $N - 2$ internal intervals and associated selector operations each requiring two comparators. In addition, one comparator is required for each of the two unbounded cells, $(-\infty, x_1]$ and $(x_{N-1}, +\infty)$. This gives a total of $2N - 2$ comparators. Of course, there is obvious redundancy in such a structure since $N - 1$ duplicate comparators are being used and could be eliminated. The simpler and efficient coder structure, requiring only $N - 1$ comparators, is shown in Figure 5.7 and is the actual structure of a so-called "flash" analog-to-digital converter.

5.3 Measuring Quantizer Performance

The purpose of quantization is to provide a limited-precision description of a previously *unknown* input value. It is only because the input is not known in advance that it is necessary to quantize. Thus the input must be modeled as a random variable, having some specific statistical character, usually specified by its probability density function (pdf). Consequently, the error introduced in quantizing this value will also be random. To conveniently assess the performance of a particular quantizer, we need a single number that indicates the overall quality degradation or *distortion* incurred over the lifetime of its use with the particular statistically specified input. In addition, some kind of "overall" measure of performance, usually based on

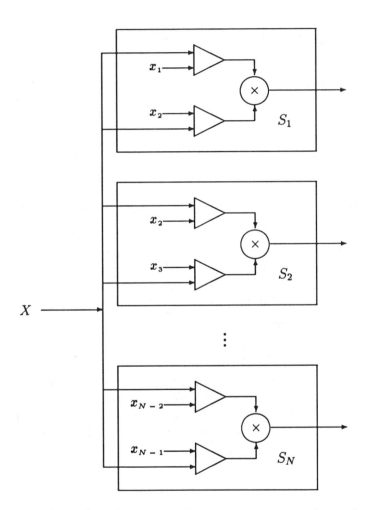

Figure 5.6: Secondary structure: Two comparators replace selector box.

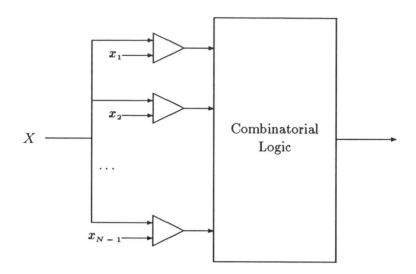

Figure 5.7: Efficient coder structure: The "Flash" Coder

statistical averaging, is required that takes into account the input pdf as well as the specific quantizer characteristic.

The most common measure of the distortion between two numbers is that already encountered in the development of optimal linear prediction: the squared error defined by

$$d(x, \hat{x}) = |x - \hat{x}|^2.$$

When applied to quantizers, x is the input sample and $\hat{x} = Q(x)$ is the reproduced output value of the quantizer (or decoder). Another measure of interest is the *absolute error* $|x - \hat{x}|$. Both are included as special cases of a slightly more general distortion measure often used in studies of scalar quantizers, the mth power of the magnitude error:

$$d_m(x, \hat{x}) = |x - \hat{x}|^m. \tag{5.3.1}$$

The special cases of $m = 2$ and 1 are by far the most popular.

For an overall measure of performance that considers the "lifetime" performance of the quantizer we can use either a *worst-case* value or a *statistical average* of some suitable measure of the distortion. For a bounded input, the *worst-case* absolute error, often more simply called the *maximum error*, is the maximum of $d_1(x, \hat{x})$ taken over all possible inputs x. (The worst-case squared error is simply the square of the worst-case absolute error.) Note that the maximum error depends only on the *support* of the pdf, that is the set of input values for which the pdf is nonzero. For an unbounded input with a finite quantizer (the number of levels N is finite), the worst-case error is infinite and hence is not a meaningful performance measure.

The statistical average of the distortion is usually a more informative and meaningful performance measure. In general, it can be written as

$$D = Ed(X, Q(X)) = \int_{-\infty}^{\infty} d(x, Q(x)) f_X(x)\, dx. \tag{5.3.2}$$

where $f_X(x)$ is the pdf of X. We shall focus most of our attention on the statistical average of the squared error distortion measure and refer to this as the *average distortion*. It is also commonly called the *mean squared error* (*MSE*). For a given input random variable, X, and quantizer, $Q = \{y_i, R_i;\ i = 1, 2, \cdots, N\}$, the average distortion is given by

$$D = E[(X - Q(X))^2] = \sum_{i=1}^{N} \int_{R_i} (x - y_i)^2 f_X(x)\, dx. \tag{5.3.3}$$

For the case of a regular quantizer the average distortion can be expressed as

$$D = \sum_{i=1}^{N} \int_{x_{i-1}}^{x_i} (x - y_i)^2 f_X(x)\, dx. \tag{5.3.4}$$

In waveform coding a signal consists of a sequence of inputs, $\{X_n\}$, successively applied to the quantizer. If the signal is stationary, then the average distortion at any time, n, is given by

$$
\begin{aligned}
D &= E(|X_n - Q(X_n)|^m) \\
&= E(|X - Q(X)|^m) \\
&= \int_{-\infty}^{\infty} |x - Q(x)|^m f_X(x)\, dx,
\end{aligned} \tag{5.3.5}
$$

where $f_X(x) = f_{X_n}(x)$ is the probability density function of X_n, which does not depend on n because of the assumed stationarity. The average squared error distortion is the most common measure of the performance of a data compression or quantization system and we shall usually evaluate performance with this measure.

The squared error distortion is simple to use and often has an intuitive motivation (e.g., it measures the average power in an error signal), but it is also well known that it does not correspond well to subjective assessments of quality for aural or visual signals destined for human perception. In this chapter, however, we shall focus primarily on the squared error measure for its simplicity. Because of its analytical tractability, it allows quantitative results that are practical, insightful, and provide a first step toward handling more difficult perceptually-oriented performance objectives.

For bounded input waveforms, the worst-case distortion, or maximum error, is another useful indicator of performance particularly when the actual probability distribution is not known or may change from one situation to another. The maximum error is also a useful criterion when the number of levels, N, is infinite (but countable) and distributed over the entire real line so that there is no overload region. Although not practicable, infinite level quantizers are theoretically useful both as a model for quantizers with a very large number of levels and for finding performance bounds. A key feature of a worst-case measure is that the value depends only on the range of values that may occur but not on the actual distribution.

The statistically averaged distortion at one time instant will correspond to the time average of the distortion if the signal is a stationary ergodic process as considered in Chapter 2. Specifically, if the average distortion has value D and the input sequence is stationary and ergodic, then the ergodic theorem implies that with probability one

$$\lim_{n \to \infty} \frac{1}{n} \sum_{i=1}^{n} d(X_i, Q(X_i)) = D,$$

that is, the limiting time average distortion is also given by D. In this way the more practically meaningful time average of performance can be studied mathematically by the use of statistical averages when the first-order probability distribution of the process is known or can be adequately modeled.

Granular and Overload Noise

Given a signal $\{X_n\}$ and a scalar quantizer Q, the inevitable error $\epsilon_n \equiv Q(X_n) - X_n$ that arises in the quantization of an analog signal is often regarded as "noise" introduced by the quantizer. It is analogous to the thermal noise that is added to a physical signal which thereby limits the accuracy in observing the signal amplitude. Specifically, *granular* noise is that component of the quantization error that is due to the granular character of the quantizer for an input that lies within the bounded cells of the quantizer. The *overload* noise is that quantization error that is introduced when the input lies in an overload region of the partition, that is, in any unbounded cell. Generally, granular noise is relatively small in amplitude and occurs to varying degrees with the quantization of each input sample while overload noise can have very large amplitudes but for a well-designed quantizer will occur very rarely. It is important to emphasize that overload noise depends very strongly on the signal amplitude.

Signal-to-Noise Ratio

The performance of a quantizer is often specified in terms of a signal-to-noise ratio (SNR) or signal-to-quantization-noise ratio (sometimes denoted SQNR or SQR) defined by normalizing the signal power by the quantization error power and taking a scaled logarithm:

$$\text{SNR} = 10 \log_{10} \frac{E(X^2)}{D} \qquad (5.3.6)$$

measured in units of decibels (dB). Since the size of the quantization errors is most meaningful *relative* to the power of the input signal, the SNR is a more important user-oriented measure of performance. It is usually assumed that the input process has zero mean and hence that the signal power is the same as the variance σ_X^2 of the signal samples. We shall later see that no generality is lost by this assumption since the theory and design can be modified in a straightforward manner to handle nonzero means. Should a signal have nonzero mean, however, the SNR is often (but not always) defined in the literature using the signal variance rather than the average power in the ratio in order not to count a dc level as useful signal power.

Although widely used, in many applications the SNR is an inadequate measure of performance. In particular, in speech and audio applications, varying the spectral distribution of the noise while keeping the SNR constant can make an enormous difference in the *perceived* quality due to the frequency-dependent sensitivity and the complex frequency-dependent masking effects of the human auditory system. Other spectrally weighted measures are preferable although less convenient to use, such as the *articulation index* [195] which is based on partitioning the spectrum into small frequency subbands and averaging the SNR for each subband.

High Resolution Quantization

In most applications of scalar quantization, the number of levels, N, is chosen to be very large so that the quantized output will be a very close approximation to the original input. We shall use the term *high resolution* (or *high rate*) to refer to the case of quantization when the average distortion is much less than the input variance, i.e., $D << \sigma_X^2 = E[(X-\overline{X})^2]$, where $\overline{X}$

is the mean value of the input X and is usually assumed to be zero. An SNR value of 10 dB is a borderline case between low and high resolution. Many of the most useful analytical results on quantization require approximate methods of analysis that apply only to the case of high resolution. We shall distinguish between theoretically exact results applicable for all quantizers and approximate results that apply only for high resolution. Frequently, high resolution results are referred to as "fine quantization" approximations or as "asymptotic" results because they become increasingly accurate as the resolution approaches infinity. Usually, increasing the number of output levels is equivalent to decreasing the average distortion. For a given input signal, however, a quantizer with $N = 100$ can be inferior to a quantizer with $N = 10$ if the levels are poorly located in the former case. This problem is easily avoided by intelligent design.

Additive Noise Model of Quantization

The quantizer error sequence resulting from applying a quantizer Q to an input signal $\{X_n\}$ has been defined by

$$\epsilon_n = Q(X_n) - X_n.$$

The choice of sign (that is, the reason for not defining the error as $X_n - Q(X_n)$) is to permit us to write

$$Q(X_n) = X_n + \epsilon_n; \tag{5.3.7}$$

that is, the output of the quantizer can be written as the input signal plus the "quantizer error" or "quantizer noise" ϵ_n as depicted in Figure 5.8. This

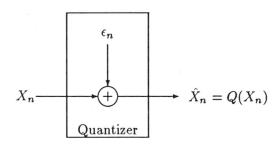

Figure 5.8: Additive noise model of a quantizer

representation is often called the "additive noise model" of a quantizer and allows us to view the quantization process as the addition of a noise term to the original signal. Note that we have not (yet) made any assumptions or approximations; the additive noise model is always valid in the sense that (5.3.7) always holds. The model is often carried farther, however, by making specific assumptions as to the statistical nature of the quantization noise so that the model can be handled analytically as a classical "signal plus noise" system in communications or signal processing theory. We shall see that these assumptions and approximations do make sense in certain situations, but it should be kept in mind that they are only approximations and indeed they may not hold in many applications. The most common approximations encountered in the literature are the following:

Common Quantizer Noise Approximations

1. The signal $\{X_n\}$ and the noise $\{\epsilon_n\}$ are uncorrelated, that is,

$$E(X_n \epsilon_k) = E(X_n)E(\epsilon_k)$$

   for all n and k.

2. The noise process $\{\epsilon_n\}$ has a uniform marginal probability distribution.

3. The noise process $\{\epsilon_n\}$ is white, that is, the quantizer noise is an uncorrelated sequence of random variables.

The first assumption is sometimes strengthened to claim that the signal and noise are independent random processes, an assumption that is clearly inaccurate since the noise is a *deterministic function of the input*. It is conceivable, however, that the two can be uncorrelated. Something like independence, however, holds in high resolution quantization. For example, the conditional mean and variance of the input given the quantization error are each approximately constant for all possible values of the quantization error. (See Problem 5.11.) In other words, given a particular value of quantization error, the possible values of the input that could have caused this error cover a range of values similar to the range of the input pdf. Nevertheless, the conditional distribution of the input given a particular value of the quantization error has a drastically different character than the (unconditioned) input pdf.

If the input signal is itself stationary and memoryless (iid), then the quantization noise will be white. Thus one would expect the noise to be approximately white if the input has little memory. The assumption is often made, however, even for signals with a great deal of memory. We shall see that the assumption is still valid *in certain circumstances*.

As an example pointing out that the assumptions or approximations cannot be universally valid, consider the simple cases of a constant (dc) or sinusoidal input to a uniform quantizer with no overload. If the input is constant, so is the quantization noise. Hence the quantization noise will be neither uniform nor white. If the input is a randomly selected constant according to a uniform distribution (that is, if we randomly select a dc input and then apply it to the quantizer for all time), then the quantization noise may have a uniform marginal distribution, but it will not be white. If the input is a sinusoid, one can use standard techniques for analyzing the behavior of memoryless nonlinear systems to show that the quantization noise will consist of an infinite number of harmonics, their amplitudes weighted by Bessel functions. (See, e.g., [72], [73], [158], and Chapter 6 of [159].) Even if the sinusoidal frequency is high and hence the input is very "active," the noise spectrum is neither continuous nor white; it consists of spikes whose amplitude and frequencies depend on the input.

In spite of their shortcomings, the quantizer noise approximations are extremely popular because they linearize the analysis of data compression systems such as DPCM by eliminating the need to deal with the nonlinear character of a quantizer. For high resolution applications, the model is indeed very valuable for obtaining tractable results and useful design formulas. It is important, however, to remain cognizant of the approximate nature of this assumption and to exercise caution before accepting the validity of formulas or conclusions resulting from the use of the additive noise model. Caveat emptor! (Let the user beware.)

Loading Factor

The performance of a quantizer is ultimately dependent on all the partition boundary values and on all of the output points as well as on the input statistics. Nevertheless, a single parameter that has an important influence on quantizer performance is the *loading factor*, γ, which measures the size of the highest decision level, x_{N-1}, relative to the rms (root mean squared) value σ of the input signal. Specifically, the loading factor is given by

$$\gamma = \frac{V}{\sigma} \qquad (5.3.8)$$

where V is the peak signal magnitude that can be quantized without incurring an excessive overload error. This value is usually taken to be x_{N-1} or y_N and for the usual case of a symmetric quantizer, $V = B/2$ and $B = x_{N-1} - x_1$. For high resolution quantization, either choice gives almost the same result. Observe that we are effectively considering zero mean signals since we take the maximum signal magnitude to be half the range (effectively assuming that zero signal is in the middle of the range) and we consider the square root of the variance and not the average power.

The *loading fraction*, β, is the reciprocal of the loading factor, i.e., $\beta = 1/\gamma$. A reasonable choice of loading factor is typically in the range of 2–4.

The loading factor mediates the trade-off between granular and overload distortion incurred in quantization. It is therefore an essential parameter that indicates how well the input is matched to the quantizer. In practical use of A/D converters, adjustment of the input gain level is essential to obtain a reasonable loading factor for effective digitization of an analog signal. As we shall see later, the SNR performance depends critically on the loading factor.

5.4 The Uniform Quantizer

The most common of all scalar quantizers is the uniform quantizer, sometimes known also as a "linear" quantizer because its staircase input-output response lies along a straight line (with unit slope). All quantizations are inherently nonlinear operations in the proper sense of the term. The familiar operations of truncation or rounding in approximating real numbers are examples of uniform quantization. The digital thermometer, discussed earlier is another example. Most general-purpose analog-to-digital (A/D) converters also use uniform quantization.

A *uniform quantizer* is a regular quantizer in which (a) the boundary points are equally spaced and (b) the output levels for granular cells are the midpoints of the quantization interval. The condition (a) implies that with step size Δ, $y_i - y_{i-1} = \Delta$ for $i = 2, 3, \cdots, N$ and (b) implies that $y_i = (x_{i-1} + x_i)/2$ for $i = 2, 3, \cdots, N - 1$. Thus, the partition of the granular region consists of intervals of length Δ and for unbounded inputs, the quantizer has overload cells $(-\infty, y_1]$ and $(y_N, +\infty)$ with $y_1 = x_1 - \Delta/2$ and $y_N = x_{N-1} + \Delta/2$. Note also that $y_i = x_{i-1} + \frac{\Delta}{2}$.

Average Distortion

Consider first the case of a uniform quantizer where the input is bounded with values lying in the range (a, b) with size $B = b - a$. Then $a = x_0$ and $b = x_N$ and the range is divided into N equal quantization cells, each of size $\Delta = B/N$. The resulting quantizer has N output levels, or treads, with step size Δ corresponding to a staircase input-output characteristic with each step of equal size. The risers also have size Δ.

It is not difficult to show that regardless of the shape of the input pdf, if the range of the input has size B then the maximum possible error is $B/2N = \Delta/2$ and the uniform quantizer minimizes the maximum error. This gives the uniform quantizer a useful robustness so that it maintains reasonably good performance for a wide variety of input signals. Partly for this reason and partly because of its simplicity, the uniform quantizer is widely used in A/D conversion.

When the input pdf is uniformly distributed over the granular region, it is readily seen that the quantizer error $\epsilon = Q(X) - X$ has a uniform pdf on $[-\Delta/2, \Delta/2]$, where $\Delta = B/N$ is the cell width. The average distortion given the cell is then simply the variance of a random variable that is uniformly distributed on an interval of width Δ, that is, $\Delta^2/12$. Thus the mean of the quantization error is given by

$$E(\epsilon) = 0 \qquad (5.4.1)$$

and the overall distortion is given by

$$D = E(\epsilon^2) = \frac{\Delta^2}{12}. \qquad (5.4.2)$$

The uniform distribution assumption for quantization noise holds exactly in this case. It can also be shown that the first of the common quantizer noise approximations is invalid in this case since

$$E(\epsilon X) = -\frac{\Delta^2}{12} \qquad (5.4.3)$$

and hence the signal and quantizer error are *not* uncorrelated. Equation (5.4.2) is an important formula that will recur frequently in the study of quantizer performance. For the mean absolute error measure, the average cell distortion is given by the first absolute moment, $\Delta/4$. Later, we shall see that the uniform quantizer is the optimal quantizer for minimizing both the average distortion (mean squared error) and the mean absolute error when the input is uniformly distributed.

In practice, most input signals of interest are unbounded. In this case, there are two overload cells $(-\infty, x_1]$ and $(x_{N-1}, +\infty)$ and $L = N - 2$ granular cells. The most significant consequence of allowing unbounded inputs is that the maximum error is now infinite. However, most inputs of interest have rapidly decreasing tail probabilities for their pdf's so that for suitable quantizer designs the overload region will have very low probability of containing an input sample. Hence, performance results for uniform quantization with bounded inputs are approximately valid when the input is unbounded as long as the loading fraction is sufficiently small. Nevertheless, it is important to recognize that in general, the average distortion is the combined effect of the granular and overload quantization errors.

We shall see that even when the input pdf is not uniform, if N is large, the quantizer cell width Δ small, the input pdf is sufficiently smooth, and the values of N and Δ are chosen so that overload effects are kept small, then (5.4.2) is approximately true for a uniform quantizer. For this reason, (5.4.2) is one of the most commonly occurring formulas in the analysis of A/D conversion and in various other quantization applications.

Lattice Quantizer

In order to isolate the effect of granular noise without requiring that the input signal be bounded, we define the (one-dimensional) *lattice quantizer* as a uniform quantizer with an infinite number of levels where the entire real line is partitioned into a (countable) set of equal size intervals of length Δ. Again, the midpoint of each interval is the output produced by the quantizer to an input occurring in that interval. In this case, the maximum error is again given by $\Delta/2$ as in the bounded input case since there is no overload effect possible. The finite uniform quantizer can be viewed as a truncated version of the lattice quantizer where L contiguous granular cells are retained and each remaining output cell is merged into either a left overload cell or a right overload cell.

The lattice quantizer can be viewed as an approximate model of a finite uniform quantizer with the same step size, since both give the same results for input values in the granular region of the finite quantizer. For any input value outside of this interval, the lattice quantizer will underestimate the actual overload error produced by the finite quantizer.

High Resolution Uniform Quantization

In most cases, a uniform quantizer is designed to operate with high resolution so that N is large, the probability of overload is very small, and the step size is much smaller than the rms signal level. As a result, some convenient approximations can be made to obtain general performance results that are independent of the specific input pdf.

If we neglect overload noise, the average distortion can be calculated using the lattice quantizer model. Assume the input pdf is sufficiently smooth so that the variation in amplitude of the pdf over any interval of width Δ is very small. Then we can approximate the pdf as having a constant amplitude within a given cell of the partition so that the conditional pdf of the error given the input is in the cell of interest is a uniform pdf over the interval $(-\Delta/2, +\Delta/2)$. Thus the conditional error has the variance $\Delta^2/12$ and since this value is independent of the specific cell in question, we find that the unconditional mean squared error or average distortion of the lattice quantizer is given approximately by (5.4.2). This calculation is a special case of the general high-resolution distortion approximation derived in Section 5.6. Although it is not easy to obtain quantitative bounds on

the inaccuracy of this approximate formula, an upper bound on the average
distortion of the lattice quantizer is given simply by the square of the maximum error, that is, $\Delta^2/4$. This upper bound is attained when the input
pdf has all of its probability mass concentrated exactly at boundary values
of the partition cells.

The SNR for a uniform quantizer with step size Δ is calculated easily in
the high resolution case. Using the fact that $\Delta = B/N = 2V/N = 2\sigma\gamma/N$,
the average granular distortion $D_{\text{gran}} = \Delta^2/12$ can be rewritten as

$$D_{\text{gran}} = \frac{1}{3}\sigma^2\gamma^2 N^{-2} = \frac{1}{3}\sigma^2\gamma^2 2^{-2r} \qquad (5.4.4)$$

where the resolution $r = \log_2 N$ and the subscript "gran" has been added
to emphasize that the overload noise has not been taken into account in the
above calculations. This result shows that the distortion goes to 0 exponentially as $r \to \infty$ and confirms the intuitive expectation that the distortion
must approach zero quite rapidly as the resolution in bits increases. Assuming that the signal has zero mean (or, equivalently, measuring signal
power by the variance) the SNR is given by the convenient formula

$$\text{SNR} = 10\log_{10}(3\beta^2 2^{2r})$$

using the loading fraction, $\beta = 1/\gamma$. This simplifies to the formula

$$\text{SNR} = 6.02r + C_1 \qquad (5.4.5)$$

where the constant $C_1 = 10\log 3\beta^2$ is ordinarily negative for reasonable
loading factors. For example, when $\gamma = 4$, the case of "four-sigma loading"
occasionally quoted in the literature, $C_1 = -7.27$ dB. Note that the SNR
increases with β (or with σ_X) until overloading occurs. Although these results assume high resolution, it turns out that they are often fairly accurate
in cases of moderately low resolution.

Equation (5.4.5) gives the well known rule of thumb: *the SNR increases
6 dB for each additional bit used to quantize an input sample.* This formula
suggests that the performance will continue to improve without bound as
the loading fraction, β, is increased. The derivation of this result, however,
was based on the assumption that the overload noise can be neglected. As
the loading factor decreases to values below 2 or 3, the contribution of
overload noise is no longer negligible. A more careful derivation of SNR
must begin with the formula for the total average distortion:

$$D = D_{\text{gran}} + D_{\text{ol}} \qquad (5.4.6)$$

where D_{ol} denotes the overload distortion, which is a function of the loading fraction. Specifically,

$$D_{ol} = \int_{-\infty}^{x_1} (x - y_1)^2 f_X(x)\,dx + \int_{x_{N-1}}^{+\infty} (x - y_N)^2 f_X(x)\,dx. \qquad (5.4.7)$$

This equation comes directly from (5.3.4). It is not difficult to show that for a symmetric quantizer D_{ol} is a function only of the loading fraction β, that is, for a fixed loading value β it does not vary with σ_X. Using this formula we can compute the SNR from (5.4.6), taking into account both granular and overload noise. It is of particular value to understand how the SNR depends on the loading factor when we include overload effects. Figure 5.9 shows the dependence of SNR for a uniform quantizer on the loading fraction which is proportional to the input signal level for a fixed quantizer. For a fixed rate r, the SNR increases steadily with a slope of

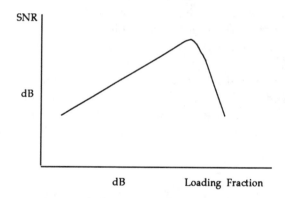

Figure 5.9: SNR vs. loading for uniform quantizers

unity on the log-log scale till the overload noise becomes significant and then it decreases fairly rapidly. The rate of decrease due to overload noise depends on the tail characteristics of the particular input pdf. This curve can also be viewed as showing SNR versus input signal power level for a particular quantizer with a fixed overload point. Hence the curve shows how the performance varies as the input signal power changes.

Note that for each additional bit of resolution, the SNR curve rises vertically by 6 dB. These curves show that a uniform quantizer has a relatively modest range of input power levels for which the SNR remains close to its peak value. In other words, the performance of a uniform quantizer is quite sensitive to the input signal level. As a result, very high resolutions are needed to obtain satisfactory performance for signals whose power levels are not accurately known in advance or vary with time.

Signals such as speech or music are nonstationary and exhibit a short term power level that varies widely over time. The *dynamic range* of a signal is defined as the ratio, usually specified in dB units, of the maximum to minimum short-term power levels. The voice of one speaker can typically have a dynamic range of 30 dB and music can have a dynamic range of 60 dB or even higher. The number of bits of resolution required to accommodate such signals depends on the minimum SNR specification. To design such a quantizer in principle requires knowledge of the overload noise as well as the granular noise and hence requires specific knowledge of the "tails" of the input signal distribution. This can be evaluated analytically for certain distributions such as Gaussian and Laplacian and simulated or measured in others. In practice, it is usually sufficient to know at what signal level overloading begins to occur and then to control the signal levels so that overloading virtually never occurs. In order to accommodate the dynamic range objective for a particular application, a minimum SNR can be specified that must be maintained over a desired dynamic range. This implies that a sufficiently high rate is needed to achieve the desired SNR for the lowest power level signals of interest. For example, for telephone voice communication 12 bits are typically required so that the resulting SNR will exceed 25 dB at signal levels as low as 30 dB below the overload point. High-fidelity audio signals, particularly classical music, have higher SNR requirements and much wider dynamic range requirements so that 16 bits of amplitude resolution has become standard in compact discs and digital audio tapes.

5.5 Nonuniform Quantization and Companding

There are two major advantages to using nonuniform spacing of quantization levels. First, it is possible to significantly increase the dynamic range that can be accommodated for a given number of bits of resolution by using a suitably chosen nonuniform quantizer. Second, it is possible to design a quantizer tailored to the specific input statistics so that considerably superior SNR is attained for a given resolution and given input pdf when the levels are allowed to be nonuniformly spaced. In this section, we introduce a general way of describing nonuniform scalar quantizers. Then we focus on the most important and general purpose class of nonuniform quantizers that provide robust performance for signals with a wide dynamic range such as voice or music. Finally, we consider the implications of high resolution for a general nonuniform quantizer (which includes the uniform quantizer as a special case). Later, in Chapter 6, while discussing optimal quantization we shall return to the subject of nonlinear mappings as a tool for designing optimal quantizers.

Compandor Modeling of Quantizers

A general model for any nonuniform quantizer with a finite number of levels is shown in Figure 5.10. The input x is first transformed with a memoryless

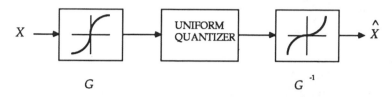

Figure 5.10: Compandor model of uniform quantization

monotonic nonlinearity G to produce an output $y = G(x)$, then it is quantized with a uniform quantizer producing $\hat{y}$ and finally it is transformed with the inverse nonlinearity G^{-1}. The final (nonlinearly) quantized output is $\hat{x} = G^{-1}(\hat{y})$. The first nonlinearity G is called a *compressor* and the inverse nonlinearity G^{-1} is called an *expandor*. This terminology stems from the case of usual interest where G has a small slope for large amplitude inputs and therefore compresses (reduces the spread) of large amplitude values. The expandor reverses this process and expands the large amplitudes. This structure has actually been used for digital transmission of speech with simple diode resistor circuits to implement the nonlinearities for the μ-law compander described in the next section with $\mu = 100$ [131], but it is today primarily of historical and conceptual interest. The term *compandor* comes from a combination of the words compressor and expandor and refers to the overall structure of a nonuniform quantizer consisting of a compressor, a uniform quantizer, and an expandor in cascade.

The generality of the compandor model is established by the following result:

Generality of the Compandor Model: There exists a compandor model for any finite regular quantizer.

Proof: Given the boundary points $\{x_1, x_2, \cdots, x_{N-1}\}$ and output points $\{y_1, y_2, \cdots, y_N\}$, we construct a compandor model as follows: Define the following two sets, jointly comprising $2N - 1$ points in the plane:

$$\mathbf{P} = \{(y_i, i\Delta + \kappa), \text{ for } i = 1, 2, \cdots, N\},$$

$$\mathbf{Q} = \{(x_i, i\Delta + \Delta/2 + \kappa) \text{ for } i = 1, 2, \cdots, N - 1\}.$$

The collection of points in **P** and **Q** can be viewed graphically, each axis being labeled in multiples of Δ (with a fixed offset in each axis). Connecting consecutive interleaved boundary and reproduction points by straight line segments yields a continuous monotone increasing curve $G(x)$ defined over the range (y_1, y_N) with

$$G(x_i) = i\Delta + \Delta/2 + \kappa$$

and

$$G(y_i) = i\Delta + \kappa.$$

Suitable selection of the value of κ can be used to satisfy the usual and convenient condition that $G(0) = 0$.   □

Figure 5.11 gives an illustration of the construction stated above. It is

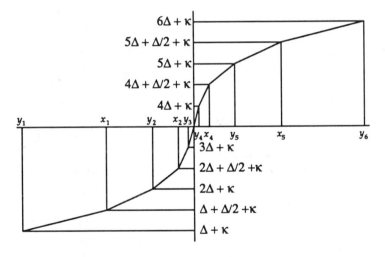

Figure 5.11: An illustration of how the compander may be constructed ($N = 6$).

intuitively clear that this piecewise linear curve can be replaced by a smooth curve which retains monotonicity, passes through the $2N-1$ points, and has a continuously varying slope. This degree of smoothness can be achieved using spline approximation methods with third order polynomial segments [129]. Thus, without formal proof, we assume that any nonuniform quantizer can be modeled by a compander model with a smooth compressor nonlinearity.

For high resolution quantization, the compressor slope has a special significance. For a nonuniform quantizer, the step size, $\Delta_i \equiv x_i - x_{i-1}$, varies from one cell to another of the partition. We can approximate the derivative by the difference

$$G'(y_i) \approx \frac{G(x_i) - G(x_{i-1})}{x_i - x_{i-1}} = \frac{\Delta}{\Delta_i}, \tag{5.5.1}$$

where we assume that the compressor curve is continuously differentiable and that Δ_i becomes very small asymptotically as N gets large. We get the condition

$$G'(y_i) = \frac{\Delta}{\Delta_i} \tag{5.5.2}$$

which shows that the compressor slope determines the local step size of the quantizer.

Logarithmic Compandors

The most important family of compandors is the logarithmic compandors, where the compressor characteristic G is symmetric and is designed to approximate the logarithmic curve for positive values. The motivation for this is that for low signal power levels, most samples are small in magnitude and therefore small step sizes are desirable for an adequate SNR. For high signal levels, a large fraction of the samples have very large magnitude and a larger step size can be used to maintain a given SNR objective. This suggests that the ratio y_i/Δ_i should be kept constant rather than keeping Δ_i constant as in the uniform case. Hence, (5.5.2) implies that $G'(x)$ is proportional to the reciprocal of x, implying a logarithmic curve for G. Of course, a logarithmic curve would be unbounded at the origin, corresponding to the ideal, but unrealizable, objective of having the step size approach zero as x approaches 0, so that some modification of the curve is needed.

Two modified logarithmic compressor curves have become widely used as a design guideline for nonuniform quantization of speech in digital telephony. The μ-law characteristic is given by

$$G_\mu(x) = V \frac{\ln(1 + \mu|x|/V)}{\ln(1 + \mu)} \mathrm{sgn}(x); \ |x| \le V \tag{5.5.3}$$

and the A-law characteristic is given by

$$G_A(x) = \begin{cases} \frac{A|x|}{1 + \ln A} \mathrm{sgn}(x) & \text{for } 0 \le |x| \le V/A \\ \frac{V[1 + \ln(A|x|/V)]}{1 + \ln A} \mathrm{sgn}(x) & \text{for } V/A \le |x| \le V. \end{cases} \tag{5.5.4}$$

The parameters μ and A control the degree of compression, determining the ratio of smallest to largest step sizes. The amount of compression is conveniently measured by the *companding advantage* which is the slope of the compressor curve at the origin. Using (5.5.2), we note that this quantity is the ratio of the step size for the uniform quantizer to the smallest step size for the compressor. As the companding advantage increases from its minimum value of unity (corresponding to $\mu = 0$ or $A = 1$), the degree of compression increases and the step size for small levels decreases. With increased compression, the dynamic range capability improves, i.e., the SNR for low signal levels increases, however, the SNR for high signal levels gradually decreases. As usual, there is a trade-off involved and a reasonable compromise must be made. The prevailing values used in practice are $\mu = 255$ and $A = 87.6$.

Piecewise Uniform Quantization

Although the smooth and differentiable compressor characteristics are convenient for mathematical manipulations, there are problems of accurately implementing analog nonlinearities. Today's technology allows accurate implementation of uniform quantizers or of piecewise linear compressor characteristics that can approximate the smooth curves of the logarithmic or other compressors. A *piecewise uniform quantizer* is a quantizer whose range consists of several segments, each of which contains several quantization cells and output points corresponding to a uniform quantizer. Within each segment the quantizer appears to be uniform with a particular step size. Different segments, however, may have different step sizes. Thus if (a, b) is a segment of a piecewise uniform quantizer, then a and b are endpoints of quantization intervals, there are $n > 2$ steps in this segment with each step having size $\Delta = (b - a)/n$, and the output points are the midpoints of these intervals. Each segment may have a different value of step size Δ and a different number of steps, n. Further details of this quantization scheme are considered in Problem 5.15.

In the current North American standard for digital telephony, (CCITT G.711) a symmetric piecewise uniform quantizer with 8 bits of resolution is used with 8 positive segments, increasing in length by a factor of 2 for each successive segment in order of increasing amplitude, and with $n = 16$ steps on each segment. This characteristic was chosen by forming a piecewise linear approximation to the μ-law compressor characteristic for $\mu = 255$. The prevailing international standard for telephony is based on a piecewise uniform approximation to the A-law characteristic with $A = 87.6$ in digital telephony. The two companding laws give similar performance. Digital audio for commercial 8 mm video cassettes also uses a version of logarithmic companding.

The average distortion of a piecewise uniform quantizer can be calculated by a slight generalization of the result for a uniform quantizer under the assumption that the input pdf is uniform. Let s_i denote the step size of each quantization cell in the ith segment and M the number of segments. A slight extension of the argument leading to (5.4.2) yields a formula for the average distortion:

$$D = \frac{1}{12} \sum_{i=1}^{M} q_i s_i^2 \qquad (5.5.5)$$

where q_i is the probability that the input lies in segment i. It will be seen later that (5.5.5) is also applicable to the case of a smooth nonuniform input pdf. This formula has a variety of applications. In particular, the segmented versions of μ and A law characteristics fit within the above category of piecewise uniform quantizers and can be analyzed either by applying the continuous companding curves that are approximated by the segmented version, or by applying the above formula.

5.6 High Resolution: General Case

We now consider the important case of high resolution regular quantizers, where N is very large and the quantization cell widths are very small so that mathematically tractable performance results can be derived. This approach to performance analysis, sometimes known as the *asymptotic quantization* approach, yields useful approximate results that become increasingly accurate as the resolution, R, or codebook size, $N = 2^R$, increases. The key assumptions for high resolution analysis are that (a) N is large, (b) the maximum step size is small compared to the range of the granular region, and (c) the input pdf is reasonably smooth.

Nonuniform regular quantizers produce an average distortion given by

$$D = \sum_{i=1}^{N} \int_{x_{i-1}}^{x_i} (x - y_i)^2 f_X(x)\, dx \qquad (5.6.1)$$

where $x_0 = -\infty$ and $x_N = +\infty$ for an unbounded pdf. This formula is an exact result and, in principle, can be used directly for calculating the performance of a given nonuniform quantizer when the pdf is known. However, for large N, it is not convenient for computational use and does not offer any insight for design or analysis. A far more convenient expression for average distortion can be derived from the fundamental theorem of calculus to approximate the above sum by an integral. The basic idea is that for a smooth input pdf and very high resolution quantization, in any local interval of amplitude values, the quantizer's behavior is very close to that of a uniform quantizer with a uniform pdf input. To obtain the average distortion of the quantizer, we can add up the contributions of each local cell while allowing the step size and the pdf to vary with the amplitude location variable x. This leads to an integral approximation for the distortion, first derived by W. R. Bennett in 1948 [30] and later rediscovered as a valuable tool for theoretical studies of quantization.

Suppose that with probability nearly one the random variable takes on values in a finite interval (a, b). This corresponds to the case of a small overload probability. The quantizer partition divides the finite range (a, b) into N disjoint quantization cells. If the cells are small enough (assuming N is large) and the pdf for X is smooth enough, the pdf is roughly constant over individual cells, that is,

$$f_X(x) \approx f_i; \ x \in R_i \qquad (5.6.2)$$

and from the fundamental theorem of calculus

$$P_i = \Pr(X \in R_i) = \int_{x_{i-1}}^{x_i} f_X(x) \, dx \approx (x_i - x_{i-1})f_i$$

and hence

$$f_i = \frac{\Pr(X \in R_i)}{x_i - x_{i-1}} \equiv \frac{P_i}{\Delta_i}. \qquad (5.6.3)$$

Thus we can approximate the average distortion due to granular noise by

$$D = \sum_{i=1}^{N} P_i \int_{x_{i-1}}^{x_i} \frac{(x - y_i)^2}{\Delta_i} \, dx. \qquad (5.6.4)$$

The interval $(x_{i-1}, x_i]$ is very small and the input pdf is nearly constant over the interval. For reasons discussed in Section 6.2, the reproduction level y_i for the cell R_i can be chosen to be the midpoint of the cell. Therefore the integral is just the variance of a uniformly distributed random variable on R_i which is $\Delta_i^2/12$ and hence the approximation becomes

$$D \approx \frac{1}{12} \sum_{i=1}^{N} P_i \Delta_i^2. \qquad (5.6.5)$$

This result has several applications and is a stepping stone to the derivation of the distortion integral. In particular, for the special case of a piecewise uniform quantizer, each quantization cell in the jth segment has step size s_j and q_j is the sum of the probabilities P_i for those cells i which lie in the jth segment. Hence (5.6.5) immediately yields the average distortion formula (5.5.5) mentioned earlier. We next use (5.6.5) to derive the convenient integral formula for the average distortion of a quantizer in the high resolution case.

In the uniform quantizer case with cell size $\Delta_i = \Delta$ for all i, this result reduces to (5.4.2).

The Distortion Integral

Suppose we have a family of nonuniform quantizers each with the same relative concentration of output levels but with a successively increasing number of levels, N. As N gets large let $N(x)dx$ denote the number of quantization levels that lie between x and $x + dx$ and assume that as $N \to \infty$ we have a limiting density of reproduction levels

$$\lambda(x) = \lim_{N \to \infty} \frac{N(x)}{N}, \tag{5.6.6}$$

which we call the *point density function* for the sequence of quantizers. Thus for sufficiently large N in any interval Δx around x there will be approximately $N\lambda(x)\Delta x$ quantization levels. This has two implications: First, it implies that $\lambda(x)\Delta x$ is the fraction of quantization levels in $(x, x + \Delta x)$ and hence integrating $\lambda(x)$ over the entire range must result in 1. Second, if there are approximately $N\lambda(x)\Delta x$ uniformly spaced levels in a cell Δx in length, the spacing between the levels must be

$$
\begin{aligned}
\Delta_i \quad &\equiv \quad \text{width of cell } i \\
&= \quad \frac{\text{length of interval}}{\text{number of levels in the interval}} \\
&= \quad \frac{\Delta x}{N\lambda(x)\Delta x} \approx \frac{1}{N\lambda(y_i)}.
\end{aligned}
\tag{5.6.7}
$$

From (5.6.5), our approximation to the distortion becomes

$$D \approx \frac{1}{12} \sum_{i=1}^{N} P_i (N\lambda(y_i))^{-2}. \tag{5.6.8}$$

Recall that $P_i \approx f_X(x)\Delta_i$ and $f_X(x) \approx f_X(y_i)$ for any $x \in R_i$ and hence

$$D \approx \frac{1}{12} \sum_{i=1}^{N} f_X(y_i)\Delta_i (N\lambda(y_i))^{-2}$$

which we can approximate by the integral

$$D \approx \frac{1}{12} \frac{1}{N^2} \int_{x_1}^{x_{N-1}} f_X(y)\lambda(y)^{-2} \, dy. \tag{5.6.9}$$

This formula, due to Bennett [30], specifies the average distortion in the high resolution case as a function of the point density function $\lambda(x)$, the input pdf $f_X(x)$, and the number of levels, N. The distortion integral becomes increasingly accurate asymptotically as the number of levels, N, approaches infinity. If the probability of overload can be neglected, the integral can be simplified by increasing the range of integration to the infinite interval $(-\infty, +\infty)$, often leading to a more computationally tractable formula.

It is important to understand that the validity of (5.6.9) depends on the high resolution assumption which means that not only is N very large, but also that the maximum step size, $\max_i |\Delta_i|$ is sufficiently small. For a specific point density function, increasing N directly implies that Δ_{max} decreases inversely with N so that the high resolution condition is indeed being satisfied as N increases.

This may seem to be a valid result in the context of a family of quantizers with a given point density function, but often we are interested in one particular quantizer with a specified partition and output set having a fixed size N. How do we apply the distortion formula to this case? The answer is to establish first that we have reasonably high resolution. A rule of thumb is that Δ_{max} is at least an order of magnitude smaller than the quantizer's granular range B. (A somewhat better rule of thumb will also depend on the input pdf.)

Secondly, we must find a point density function $\lambda(x)$ that corresponds approximately to the specific manner in which the output levels are distributed over the range. This is essentially the same task as discussed earlier of finding a compandor model for a given quantizer. In fact, it is easy to recognize that the compressor slope function, $G'(x)$ is indeed a point density function apart from a scale factor. Indeed, it was this approach that led Bennett to the distortion integral. Specifically, combining the facts $G'(y_i) = \Delta/\Delta_i$ and $\Delta_i = 1/N\lambda(y_i)$, and $\Delta = B/N$ yields

$$\lambda(x) = \frac{1}{B}G'(x). \qquad (5.6.10)$$

Thus for a specific nonuniform quantizer, the distortion integral can be evaluated by finding a convenient analytical approximation for the compressor curve. Of course, when a straight line approximation to the compressor curve is used, the distortion integral will reduce to a summation. For numerical analysis an integral would ordinarily be approximated by a sum. For analytical studies, however, a mathematical model for the compressor curve or the point density curve is far more convenient.

As an example of the use of the distortion integral, suppose that $\lambda(x)$ is assumed to be constant over the range (a, b) of the random variable. This corresponds to a uniform quantizer as in the previous section. Since $\lambda(x)$ must integrate to 1 over the range (a, b), we must have

$$\lambda(x) = \frac{1}{b - a}$$

and hence

$$D \approx \frac{1}{12} \frac{1}{N^2} B^2$$

where the range value $B = |b - a|$. This is sometimes expressed by defining the cell width $\Delta = (b - a)/N$ and writing

$$D \approx \frac{\Delta^2}{12}, \tag{5.6.11}$$

which is exactly (5.4.2). The tradeoff between distortion and resolution (or rate) as measured by $r = \log_2 N$ bits per sample can be made explicit by writing

$$D = \frac{|b - a|^2 2^{-2r}}{12},$$

and confirms the result (5.4.4) previously obtained for uniform quantization.

A particularly useful application of the distortion integral is in the evaluation of companding on the dynamic range of a quantizer. To determine the SNR for a particular compressor characteristic, it is necessary to compute the granular noise for the particular companding characteristic of interest. This can be conveniently done with Bennett's distortion integral. Computation of the overload distortion is the same as previously discussed in connection with uniform quantizers. The details of this computation are left as an exercise for the reader.

In Figure 5.12, we plot SNR versus loading for a μ-law quantizer with $N = 256$ and compare it with the previously illustrated curve (Figure 5.9) for a uniform quantizer with the same number of levels. The vastly improved dynamic range is evident.

Additive Noise Model of Quantization

A few observations regarding this result are in order. As with the uniform quantizer and the uniformly distributed input of Section 5.4, one can interpret the above result to mean that the quantization error $\epsilon = Q(X) - X$ has an approximately uniform distribution over a cell of width Δ_i when the signal is encoded into the ith cell. The conditional average distortion for each cell is approximately $\Delta_i^2/12$. Intuitively, if N is very large, the cells are very small, and hence if f_X is sufficiently smooth it is approximately constant over the cell. Thus the conditional expected squared error given the cell is approximately the variance of a uniformly distributed random variable on the cell.

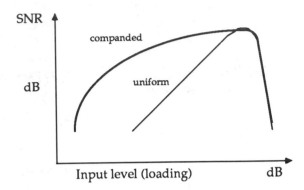

Figure 5.12: SNR vs. loading comparing companded and uniform quantization

The quantization noise is, however, strongly dependent on and, in general, correlated with the input signal, and hence cannot be precisely modeled as signal-independent additive noise. To emphasize the point even further, recall that the quantization noise is in fact a deterministic function of the input signal since it is defined by $\epsilon_n = Q(X_n) - X_n$. If the input sequence $\{X_n\}$ is itself memoryless, then so is the quantization noise sequence $\{\epsilon_n\}$. Nonetheless, it is commonly assumed for more general input signals in the analysis of uniform quantizers that the action of the quantizer can be approximated by the additive noise model where the noise is memoryless, signal-independent, and uniformly distributed. We repeat that this approximation must be used with care since, as we have seen,

1. the quantization noise is not independent or (in general) even uncorrelated with the signal,

2. the quantization noise is uniformly distributed only if the input random variable has a uniform distribution or it has a smooth pdf and there are a large number N of small quantization cells, and

3. the quantization noise is independent, identically distributed (iid) only if the input sequence is.

In spite of all of these observations, the additive noise model does indeed play a useful role in the approximate analysis of coding systems. In later chapters we shall illustrate how it can be used and discuss the limitations to the validity of the results that are based on this model. A more detailed discussion of these approximations may be found in [158] and Chapter 6 of [159].

Spectrum of Quantization Noise

In the high resolution case, we can show that under reasonable regularity conditions the quantization error sequence associated with the quantization of a stationary but correlated sequence of input samples is approximately white. The key assumption is that the input process has a joint pdf $f_{UV}^{(k)}(u, v)$ for two samples spaced k time units apart $(k \neq 0)$, with the property that it is a smooth function of u and v and that the joint pdf is approximately constant over any rectangular region of the plane with length and width each less than $\Delta_{\max}$, the maximum cell size of the quantizer. This assumption means in essence that the input sequence must have a spectral density that (depending on the value of $\Delta_{\max}$) is approximately flat. A slowly varying random input sequence whose spectrum has a low-pass character with cutoff well below the half-sampling frequency will not produce approximately white quantization noise. Roughly speaking, a very slowly varying sequence may often have values spaced k time units apart that differ by less than $\Delta_{\max}$ and hence will be quantized by the same quantization level and may have highly correlated quantization error values. In this case, the quantization noise would not have a wideband or white character.

An extreme example is an input sequence that is constant for all time. In this case the joint pdf will have all of its probability concentrated on the line $u = v$. Similarly, for an extremely slowly varying input, the joint pdf will have most of its probability mass on or very near to the line $u = v$ and will drop very steeply for points moving away from this line. Both these cases would clearly violate the stated assumption.

The correlation value, r_k, of the quantizing error is given by

$$
\begin{aligned}
r_k &= E[(U - Q(U))(V - Q(V))] \\
&= \int_{-\infty}^{\infty} \int_{-\infty}^{\infty} (u - Q(u))(v - Q(v)) f_{UV}(u, v)\, du\, dv \quad (5.6.12)
\end{aligned}
$$

so that we have

$$
r_k = \sum_i \sum_j \int_{R_i} \int_{R_j} (u - y_i)(v - y_j) f_{UV}(u, v)\, du\, dv. \quad (5.6.13)
$$

Now using the assumption that the pdf is approximately constant in the rectangle $\{R_i \times R_j\}$ we get

$$
r_k \approx \sum_i \sum_j f_{UV}(y_i, y_j) \int_{R_i} (u - y_i)\, du \int_{R_j} (v - y_j)\, dv \quad (5.6.14)
$$

and finally assuming that y_i is approximately the midpoint of the interval R_i, we see that each integral above has value zero, so that $r_k \approx 0$ for $k \neq 0$. Therefore, under reasonable sampling rates and high resolution quantization, the quantization noise is approximately white. This result is approximate and is derived in the "time domain." Therefore considerable care must be taken before assuming that the spectral density is truly white, since the Fourier spectrum can be sensitive to relatively subtle changes in the time domain. To be more specific, we note that a direct bound on the degree of "nonwhiteness" of the spectrum is given by

$$|S(\omega) - r_0| \leq 2 \sum_{i=1}^{\infty} |r_i|$$

which shows that it is not sufficient for each r_i to be quite small in order to achieve a flat spectrum.

5.7 Problems

5.1. Compute and plot SNR in dB versus loading fraction (reciprocal of loading factor) in dB ranging from -40 to +5 dB for a 7 bit quantizer when the signal input is Laplacian (two-sided exponential pdf) for both a uniform quantizer and a μ-law quantizer. In each case specify the dynamic range achievable in dB for a 25 dB SNR performance requirement and include overload effects. Plot both curves on one sheet of linear/linear graph paper. Show all formulas used.

5.2. A uniform B bit quantizer has integral nonlinearity of M LSB's, that is, the maximum deviation of the actual digital reconstruction from that given by an ideal uniform quantizer over the full input range is M times the uniform quantizer step size. Assume this is due to second order nonlinearity of the form

$$y = x + cx^2,$$

followed by an ideal uniform quantizer.

(a) Derive a formula relating M, B, and c.

(b) If a 12 bit A/D converter has 1 LSB of integral nonlinearity and an overload point of 5 volts, what is the second harmonic distortion level in dB when a 1 volt peak sine wave is quantized? Assume that no aliasing occurs.

5.3. A uniform quantizer with $\Delta = 1$ and $N = \infty$ (infinite number of intervals, no overload) maps any real number to the nearest integer. If the input random variable has pdf $p_X(x)$, find an explicit and exact expression for the pdf of the quantization error.

5.4. What SNR in dB is achieved for a sine wave of 100 Hz which is sampled at 8 KHz and quantized with a 12 bit symmetric uniform quantizer whose overload is equal to the peak amplitude of the sine wave? How will the result be changed as the frequency of the sine wave is increased?

5.5. The mean square error of a quantizer with $N = 128$ levels, overload point $V = 5$, and compressor slope of 100 at the origin is estimated by Bennett's integral. The quantizer was designed for a given normalized probability density and for the nominal input power level of $\sigma^2 = 4.0$. Explain why the integral becomes invalid as the input level becomes small. What is the correct asymptotic formula for the SNR as σ approaches zero? Consider separately two cases, first when σ is in the neighborhood of 0.003, and second when σ is less than 0.0005. Assume a smooth and well behaved compressor curve.

5.6. A one-dimensional optimal (in the mean squared error sense) quantizer has 256 output levels. For a Gaussian random variable input is it possible to achieve an SNR of 49 dB? Justify your answer.

5.7. Prove (5.4.1)–(5.4.3) for the example described there of a uniformly distributed input.

5.8. Suppose that Q is a uniform quantizer with an even number M of levels separated by Δ and that

$$\epsilon = \epsilon(x) = Q(x) - x$$

is the resulting quantization error. Prove that the quantization error can be expressed as a function of the input as

$$\epsilon(x) = \Delta(\frac{1}{2} - < \frac{x}{\Delta} >), \qquad (5.7.1)$$

where $< r >$ denotes the fractional part of r or $r \bmod 1$, that is, $< r >= r - \lfloor r \rfloor$ where $\lfloor r \rfloor$ is the largest integer less than or equal to r. Provide a labeled sketch of this quantizer error function.

5.9. Dithering means adding random noise to the signal to be quantized. This is sometimes done in practice for two reasons. The first is that a small amount of random noise can ease the subjective perception of blockiness of a quantizer in speech and images. The second reason is that it can modify the statistical relationship between the quantization noise and the signal, as we explore in this problem. A uniform quantizer with step size Δ has the dithered input $X + Z$, where Z is a uniform random variable over $(-\Delta/2, +\Delta/2)$. The random variable X has an arbitrary continuous distribution and Z is independent of X. Two forms of decoder are used: *Case 1:* $\hat{X} = Q(X + Z)$, *Case 2:* (Subtractive dither) $\hat{X} = Q(X + Z) - Z$. In Case 2 the dither signal is removed from the quantizer output. In practice this would be done by having a common pseudo-random noise sequence at encoder and decoder. Assume high resolution quantization.

(a) Find the overall MSE in both Cases 1 and 2.

(b) Neglecting overload effects, show that the coding error $\hat{X} - X$ is statistically independent of X in Case 2. Is the same true in Case 1?

Note: Part (b) is fairly hard.

5.10. This problem shows that sometimes dithering can actually improve the quality of a digitized signal. This problem is, however, completely independent of the previous dithering problem. Suppose that Q is a uniform quantizer with M levels each spaced Δ apart. Suppose that $\{X_n\}$ is a random process (the input signal) and that $\{W_n\}$ is another random process (the dither signal) and that the two processes are independent and that W_n is an iid sequence uniformly distributed on $[-\Delta/2, \Delta/2]$. Further assume that the range of X_n is such that $X_n + W_n$ does not overload the quantizer, i.e., the quantizer error satisfies

$$|Q(X_n + W_n) - (X_n + W_n)| \leq \frac{\Delta}{2}.$$

The key idea in this problem is that X_n varies very slowly so that we can quantize it several times while it has roughly the same value. We assume in particular that $X_n = X$ is a constant (approximately) and that we can take N quantized values. If we simply quantize the constant X N times, we accrue no benefits, the decoder just gets the value $Q(X)$ N times and gains nothing from the extra samples. Suppose instead that the encoder quantizes $X + W_n$ to form $Q(X + W_n)$ for $n = 0, 1, \cdots, N - 1$ and sends these to the decoder. The decoder then forms an estimate

$$\hat{X} = \frac{1}{N} \sum_{n=0}^{N-1} Q(X + W_n).$$

(a) For a fixed X, what is the expected value of a single received reproduction $Q(X + W_n)$?

(b) Show that if N is large $\hat{X}$ can provide a much better reproduction than the decoder could get without dithering. (*Hint:* A correct answer to the previous part and the ergodic theorem provide the answer.)

5.11. Show that for large resolution, the conditional mean and variance of the input given the quantization error are each approximately constant for all possible values of the quantization error. Hence the input and quantization noise have a form of second order independence.

5.12. A 2.3 kHz sine wave with 3 volt peak is to be sampled at 10 kHz and digitized so that the reconstructed signal has SNR of at least 30 dB. Determine the number of quantizer levels needed for a uniform quantizer to achieve this objective and calculate the total ROM storage space in bits needed to store one cycle of the sine wave. Explain. Will storing these samples with sufficient accuracy (i.e., as many bits per sample as needed) provide sufficient data to synthesize a pure sine wave using this ROM and a D/A converter that is clocked at 10 kHz? Explain.

5.13. A random variable, X, has standard deviation $\sigma = a/2$ and a smooth symmetric probability density function limited to the region $-2a \le X \le 2a$ and $\Pr[|X| \le a] = 0.3$. It is to be quantized with 256 levels with 192 equally spaced levels in the region $|X| \le a$ and 64 equally spaced levels in the remaining region of interest. Calculate (approximately) the SNR (signal-to-quantizing noise ratio in dB).

5.14. Find a formula for the SNR of a μ-law quantizer (compressor function $G_\mu(x)$ as given in the text) as a function of μ and β where the input pdf has zero mean and the ratio $\beta = \frac{(E|X|)}{\sigma_x}$ where σ_x is the standard deviation of X.

5.15. The standard quantizer for voice digitization in North American telephone systems has output levels given by the formula

$$y(P, S, B) = (1 - 2P)[(2B + 33)(2^S) - 33]$$

where the parameters can take on the values $P = 0, 1$, $S = 0, 1, \cdots, 7$, and $B = 0, 1, \cdots, 15$.

(a) How many output levels are there? How many finite decision boundaries are there? What is largest positive output value?

(b) Assuming that each boundary value is equidistant between two adjacent output levels, determine and specify all of the boundary values that separate partition regions. (It is not necessary to explicitly tabulate all the numerical values if you can find a formula that determines these values.)

(c) Find and specify clearly a simple algorithm to determine the parameters (P, S, B) that identify the output level for any given real number x to be quantized.

(d) If $x = +132$ find the values of the three parameters that specify the quantization level.

5.16. We are given an input signal X that we wish to quantize. It has a pdf given by

$$f_X(x) = \begin{cases} A - x, & 0 \le x \le 1 \\ A + x, & -1 \le x \le 0 \\ 0, & \text{otherwise.} \end{cases}$$

We are given a 2-level quantizer for this input signal with output levels $y_1 = -\frac{1}{2}$ and $y_2 = \frac{1}{2}$.

(a) Find the constant A.

(b) Find the pdf $f_\epsilon(\alpha)$ of the quantizer error $\epsilon_X = Q(X) - X$.

(c) Find the distortion $D = E(\epsilon_X^2)$.

(d) Now we redesign the quantizer and put the output levels at $y_1 = -\frac{1}{3}$ and $y_2 = \frac{1}{3}$. Find the new distortion $D = E(\epsilon_X^2)$. Compare this to the result you got in part c.

(e) Now we are given a new input signal Z that we wish to quantize, again with the first two-level quantizer $y_1 = -\frac{1}{2}$ and $y_2 = \frac{1}{2}$. The signal has a pdf $f_Z(z)$ given by

$$f_Z(z) = \begin{cases} 1 + z, & -1 \le z \le 0 \\ z, & 0 \le z \le 1 \\ 0, & \text{otherwise.} \end{cases}$$

Find the pdf on the new quantizer error $\epsilon_Z = Q(Z) - Z$ and compare with the result you found in part b.

Chapter 6

Scalar Quantization II: Optimality and Design

6.1 Introduction

In the previous chapter, we focused on the characterization of quantizers and the assessment of the performance of a given quantizer. The second and more important issue from the engineer's perspective is the design and implementation of quantizers to meet performance objectives. As always in engineering, there are conflicting objectives and compromise is needed. To provide the tools to find the best trade-off, it is first necessary to understand what is the best that can be achieved under the given constraints. We first focus on the question of quantizer optimality. Specifically, given the constraint that the number of levels, N, is fixed, we examine the conditions that the optimal quantizer must satisfy in order to minimize the average distortion for a particular input pdf. Having established the two key necessary conditions for optimality, we then examine how they can be used to obtain design algorithms of practical use for particular situations. We then go on to discuss implementation issues.

6.2 Conditions for Optimality

The principal goal of quantizer design is to select the reproduction levels and the partition regions or cells so as to provide the minimum possible average distortion for a fixed number of levels N or, equivalently, a fixed resolution r. In general, this problem does not have any explicit, closed-form solution. Effective algorithms are, however, readily available.

More explicitly, for the average (mean square) distortion measure, the goal, as usually stated, is to find the output points y_i and partition cells R_i that minimize

$$D = \sum_{i=1}^{N} \int_{R_i} (x - y_i)^2 f_X(x) \, dx, \tag{6.2.1}$$

where $f_X(x)$ is the pdf of the random variable X.

Before proceeding to explore solutions to this problem it is helpful to briefly review some basic minimizations that arise in the study of random variables. (For a basic review of estimation theory, see Section 4.2.) We first list the properties and then sketch the proofs.

1. The unique value of a that minimizes the mean squared error

$$E[(Y - a)^2]$$

   is

$$a = EY,$$

   the mean value of Y.

2. Given an event G, the unique value of a that minimizes the conditional mean squared error

$$E[(Y - a)^2 | Y \in G]$$

   is

$$a = E(Y | Y \in G),$$

   the conditional mean of Y given that the event G occurred.

3. The function $g(u)$ that minimizes $E[(Y - g(U))^2]$ for a pair of random variables $\{Y, U\}$ is $g(u) = E[Y | U = u]$, the conditional mean of Y given that U has the value u. Note that this is a direct generalization of the first observation.

4. The function $g(\mathbf{u}) \equiv g(u_1, u_2, \cdots, u_n)$ that minimizes $E[(Y - g(\mathbf{U}))^2]$, where $\mathbf{U} = (U_1, U_2, \cdots, U_n)$, is again the conditional expectation, $g(\mathbf{u}) = E[Y | \mathbf{U} = \mathbf{u}]$.

The first property is a special case of Theorem 4.2.1. It is also easily proved using calculus or from the following inequality (with $\overline{Y} = E(Y)$):

$$
\begin{aligned}
E[(Y - a)^2] &= E[(Y - \overline{Y} + \overline{Y} - a)^2] \\
&= E[(Y - \overline{Y})^2] + (\overline{Y} - a)^2 \\
&\geq E[(Y - \overline{Y})^2].
\end{aligned}
$$

The second property is simply the first applied to conditional expectation given an event. The third and fourth properties follow from the first applied to conditional expectation given a random variable or vector.

The problem of finding an optimal quantizer is not only important but it is also an intriguing one. The lack of any straightforward solution to this problem is a result of the difficulty in dealing with the highly nonlinear nature of quantization. Nevertheless, there are two critically important conditions that are necessary for optimality and these conditions are simple to derive and understand. They provide the gateway to all the available algorithmic solutions. These conditions follow from the decomposition of a quantizer into an encoder and a decoder.

From the structural decomposition of a quantizer, we noted that the encoder is the nonlinear operation whereas the decoder can be represented as a linear operation in a certain sense. By assuming that one part is constrained, it becomes easy to specify a condition for optimality of the second part. Specifically, the encoder part of an optimal quantizer must be optimal for the given decoder while the decoder must be optimal for the given encoder. The two conditions are therefore *necessary* for a quantizer to be optimal. We now focus on these two conditions separately. We shall find a sufficient condition for the optimality of the encoder when the decoder is fixed. We shall also derive a necessary and sufficient condition for the optimality of the decoder with respect to the squared error distortion measure when the encoder is given. A generalization to other distortion measures will also be sought. We begin with the encoder as the corresponding condition is simpler, more intuitive (at least on first reading), and independent of the probability distribution of the original signal.

The Optimal Encoder for a Given Decoder

Consider the task of finding the optimal encoder for a given decoder. Equivalently, we wish to find the best partition for a given codebook. Before formally stating and proving the condition, we note that the condition is so obvious as to seem almost trivial: If the overall goal is to minimize distortion, then no encoder can be better than that encoder which maps input values into the output reproduction level having the minimum distortion with respect to the given input. In other words, the best encoder for a given codebook satisfies the *nearest neighbor condition* which requires that the ith region of the partition should consist of all input values closer to y_i than to any other output level. This result holds as easily for general measures of distortion as for the simple squared error distortion. We now make a formal statement of this property.

Nearest Neighbor Condition

For a given set of output levels, C, the partition cells satisfy

$$R_i \subset \{x : d(x, y_i) \leq d(x, y_j); \text{ all } j \neq i\}; \qquad (6.2.2)$$

that is,

$$Q(x) = y_i \text{ only if } d(x, y_i) \leq d(x, y_j) \text{ all } j \neq i. \qquad (6.2.3)$$

Thus, given the decoder, the encoder is a minimum distortion or nearest neighbor mapping, and hence

$$d(x, Q(x)) = \min_{y_i \in C} d(x, y_i). \qquad (6.2.4)$$

We now prove that the nearest neighbor condition is indeed sufficient for optimal encoding with a given decoder.

Proof: For a given codebook, C, $Q(x)$ takes on values in C, so that

$$D = \int d(x, Q(x)) f_X(x) \, dx \geq \int [\min_{y_i \in C} d(x, y_i)] f_X(x) dx$$

and this lower bound is indeed attained when $Q(x)$ performs the nearest neighbor mapping with the given codebook C. $\square$

Note that the nearest neighbor condition does not uniquely assign boundary points to a specific region when an input value x happens to be equidistant from two output levels. In order to have a well-defined partition of the real line, each point must be uniquely assigned to a particular cell. In the common case of a continuously distributed signal, the assignment of boundary points is arbitrary since the average distortion will not be affected by a finite collection of isolated values of x with zero probability. The simplest resolution of such ambiguity is to arbitrarily assign the boundary value to be a member of the cell to its left, so that $R_i = \{x : x_{i-1} < x \leq x_i\}$.

For both the squared error and the absolute error distortion measures (defined in Chapter 5), the nearest neighbor rule implies that for a given input, x, the output y_i is chosen to minimize $|x - y_i|$. In other words, if x lies between two output levels, y_{i-1} and y_i, the rule is to choose the closer of the two levels. This is accomplished by choosing x_{i-1} as the *midpoint* between two adjacent output levels,

$$x_{i-1} = (y_{i-1} + y_i)/2.$$

The nearest neighbor rule fully defines a partition of the input range once a convention is adopted to handle an input point that happens to lie on a boundary between two cells.

It is possible, although uncommon, that for a given decoder the nearest neighbor partition can have one or more cells that have zero probability. In this case the partition is said to be *degenerate*. For example, for an exponentially distributed random variable (where negative values have zero probability), suppose the given codebook is $\{-1, 1\}$. Then the nearest neighbor cell for the output value -1 is the set of all negative numbers which has zero probability. Clearly, a degenerate nearest neighbor partition will not arise for continuously distributed inputs when the output points each lie in an interval with positive probability. There are situations, however, where degenerate partitions can arise and some care may be needed to avoid this pathology.

The Optimal Decoder for a Given Encoder

We now examine the second necessary condition for optimality which is obtained by fixing an encoder (partition) and optimizing the decoder (codebook). The *centroid condition* is found to be both necessary and sufficient for this optimization provided the squared error distortion measure is used. The centroid condition is simply the condition that the optimal output level, y_i, for the ith cell of the partition is the *centroid*, or *center of mass*, of that part of the input pdf that lies in the region R_i. This property is a variation of the well known result in physics that a moment of inertia of an object around a point (analogous to an average squared error about a reproduction) is smallest when the point is the centroid of the object.

Centroid Condition

Given a nondegenerate partition $\{R_i\}$, the unique optimal codebook for a random variable X with respect to the mean squared error is given by

$$y_i = E[X | X \in R_i]. \tag{6.2.5}$$

Three distinct proofs are presented below. Each gives a somewhat different insight and it is worthwhile examining each proof.

First Proof: In general, the average distortion of a scalar quantizer can be written as:

$$D = \sum_{i=1}^{N} \int_{R_i} (x - y_i)^2 f_X(x) \, dx. \tag{6.2.6}$$

First, we note that given a partition, the minimization of D with respect to y_j involves only one term of the sum. Second, we recognize that this term can be expressed as a conditional expectation:

$$\int_{R_j} (x - y_j)^2 f_X(x) \, dx = P_j \int_{-\infty}^{\infty} (x - .y_j)^2 f_{X|R_j}(x) \, dx$$

$$= P_j E((X - y_j)^2 | X \in R_j),$$

where $f_{X|R_j}(x)$ is the conditional pdf of X given that it is in the jth partition cell, R_j, and P_j is the probability that X lies in R_j. It is immediately obvious from the second minimization property of expectation that the value of y_j that minimizes this integral is the centroid $E(X|X \in R_j)$ of the conditional pdf $f_{X|R_j}(x)$. More explicitly,

$$y_i = \int_{R_i} x f_{X|R_i}(x)\, dx = \frac{\int_{R_i} x f_X(x)\, dx}{\int_{R_i} f_X(x)\, dx}. \tag{6.2.7}$$

□

The second proof is based on the linear representation of the decoder and explicitly shows that the decoder's task can be viewed as a linear estimation problem. Specifically, the decoder has available a set of observable random variables, $\{S_i = S_i(X), \ i = 1, 2, \cdots, N\}$ which provide partial information about the input X. Each observable or selector function S_i is a binary random variable that tells only whether or not the input lies in the ith region. Collectively, these variables simply specify in which region of the partition the input lies.

Second Proof: Using Eq. (5.2.1) we can describe the quantized output value, $Y = Q(X)$, as

$$Y \equiv Q(X) = \sum_{j=1}^{N} y_j S_j. \tag{6.2.8}$$

From this representation of the output, it may be seen that the task of finding the optimal values of each y_j to minimize $E[(Q(X) - X)^2]$ is therefore a problem of optimal linear estimation where the observable random variables are S_i. By the orthogonality principle (Theorem 4.2.4), optimality is achieved when the quantization error $X - Y$ is orthogonal to each of the observable variables, S_i. Hence, for each i, $E[(X - Y)S_i] = 0$, so that

$$E(XS_i) = \sum_{j=1}^{N} y_j E(S_j S_i). \tag{6.2.9}$$

But $E(S_j S_i)$ is equal to zero for $j \neq i$ (since only one selector function can have a nonzero output for any given input) and is equal to $\Pr(S_i = 1) = P_i$ when $j = i$. Also

$$E(XS_i) = E[X|S_i = 1]P_i \tag{6.2.10}$$

then (6.2.9) reduces to

$$y_i = \frac{E(XS_i)}{P_i} = E[X|S_i = 1] \tag{6.2.11}$$

from which the desired result follows. □

The orthogonality condition used in this proof states that the error is uncorrelated with each observable, S_i. Hence, from (6.2.8) it is obvious that the error is also uncorrelated with the output of the quantizer! This is an important and interesting result of the centroid condition. Finally, we give the last proof which also uses the primary decomposition of the decoder described in Chapter 5.

Third Proof: For a given partition, the quantizer output can be expressed as $Y = f(\mathbf{S})$ where $\mathbf{S} = (S_1, S_2, \cdots, S_N)$ and the function f can be arbitrarily chosen by the designer. To minimize $D = E[(X - f(\mathbf{S}))^2]$, the optimal function is immediately recognized from the fourth minimization property of expectation as the conditional mean

$$f(\mathbf{s}) = E[X|\mathbf{S} = \mathbf{s}],$$

where $\mathbf{s} = (s_1, s_2, \cdots, s_N)$. Now, recognizing that only one component of the binary vector s can be nonzero, there are only N distinct values that can be taken on by s. In particular, suppose $\mathbf{s} = \mathbf{u}$ where $u_i = 1$ and $u_j = 0$ for $j \neq i$. Then,

$$f(\mathbf{u}) = y_i = E[X|S_i = 1],$$

which once again proves the necessity of the centroid condition for optimal quantization to minimize the average distortion. □

Note that although the optimal estimation performed by the decoder could be a nonlinear function of the observables, S_i, because of the restricted character of these binary observables, the optimal estimator is always a linear combination of these observables. An immediate consequence of the centroid condition in the case of a uniform input pdf is that the best output value is simply the midpoint of the cell.

The nearest neighbor condition was easily proved for general distortion measures. The centroid condition, however, was proved only for the special case of squared error distortion. We shall later see a natural extension of this condition to quite general distortion measures.

The two necessary conditions for optimality were first reported by Lukaszewicz and Steinhaus in 1955 [225], and were independently observed by Lloyd in 1957 [218] and later by Max [235]. Each study applied these conditions to the design of quantizers. Traditional proofs involve the use of calculus and differentiate the expression (5.6.1) for D with respect to the boundary points and reproduction levels. This leads one to the same conditions. The calculus approach was also developed by Lloyd [218] and subsequently by Max [235] and the quantizer that satisfies these conditions is called the Lloyd-Max quantizer. The proofs provided here show that calculus is not necessary and hence differentiability need not be assumed.

This will provide a simple generalization to vector quantization in a later chapter.

Several authors have computed tables of optimal quantizers for Gaussian and other important pdf's arising in signal compression applications. In particular, Max [235] gives output and boundary levels for N ranging from 2 to 36 in the Gaussian case. Paez and Glisson [249] give similar data for the Laplacian and Gamma distributions. With today's computational resources and with the simplicity of a quantizer design algorithm to be presented below, it is a simple matter to generate extensive and accurate tables for various distributions; these references serve as a convenient check for validating the data.

Implications of Optimal Quantization

The theorem immediately provides some interesting properties of quantizers satisfying one of the necessary conditions for optimality.

Lemma 6.2.1 *A quantizer is regular if it satisfies the necessary conditions for optimality with a distortion measure $d(x, y)$ that is a monotone increasing function of the error magnitude $|x - y|$.*

Proof: If the distortion measure increases monotonically with the absolute error, we can observe the *convexity of the nearest neighbor cell:* if x_a and x_b are both closer to y_α than to any other output point, any point lying between these two points must also be closer to y_α than to any other output point. Consequently, each cell must consist of a single interval. Furthermore, the centroid of a pdf over an interval always lies within the interval, hence from the centroid condition, $y_i \in R_i$. Thus, the conditions of regularity are fulfilled. □

Lemma 6.2.2 *Suppose that $Q(X)$ is a quantizer whose codebook satisfies the centroid condition. Then*

$$E(Q(X)) = E(X), \qquad (6.2.12)$$

the mean of the quantizer output is the same as the mean of the input. Further,

$$E[Q(X)(Q(X) - X))] = 0, \qquad (6.2.13)$$

that is, the quantizer output is uncorrelated with the quantizer error. Finally,

$$E[(X - Q(X))^2] = E[X^2] - E[Q(X)^2] = \sigma_X^2 - \sigma_{Q(X)}^2, \qquad (6.2.14)$$

i.e., the expected squared quantizer error is the difference of the second moments (or the variances) of the signal and the quantized output.

Alternatively, if we define the quantizer error $\epsilon = Q(X) - X$, these properties may be restated as

$$E(\epsilon) = 0, \tag{6.2.15}$$
$$E(Q(X)\epsilon) = 0, \tag{6.2.16}$$
$$E(\epsilon^2) = \sigma_\epsilon^2 = \sigma_X^2 - \sigma_{Q(X)}^2. \tag{6.2.17}$$

Before going into the proof of this lemma, we note that as a consequence of (6.2.14) or (6.2.17) and the fact that second moments or variances are nonnegative:

$$E(Q(X)^2) \leq E(X^2), \tag{6.2.18}$$
$$\sigma_{Q(X)}^2 \leq \sigma_X^2, \tag{6.2.19}$$

that is, the second moment or variance of the quantized output is smaller than that of the input.

Proof: A quantizer whose codebook is optimal for the given partition must satisfy the orthogonality condition given in the second proof of the centroid condition:

$$E[(X - Q(X))S_i] = 0 \text{ for each } i. \tag{6.2.20}$$

By summing over i, we get

$$E[(X - Q(X))(\sum_{i=1}^{N} S_i)] = EX - E(Q(X)) = 0$$

since $\sum_{i=1}^{N} S_i = 1$. This proves (6.2.12).

By multiplying (6.2.20) by y_i and then summing over i, we get

$$E[(X - Q(X))Q(X)] = 0 \tag{6.2.21}$$

which proves (6.2.13).

From Eq. (6.2.21), we have $E(XQ(X)) = E(Q(X)^2)$. Hence,

$$E[(X - Q(X))^2] = E[X^2] - 2E[XQ(X)] + E[Q(X)^2]$$
$$= E[X^2] - E[Q(X)^2], \tag{6.2.22}$$

which proves the first equality in (6.2.14). Since X and $Q(X)$ have the same mean value, the expression for the variances follows by subtracting the square of the mean of X from this result. □

While the lemma states that the quantization error (the quantization "noise") is uncorrelated with the output, it is important to realize that the quantization error is a deterministic function of the *input* and Lemma 6.2.2 in fact shows (from (6.2.14)) that the error $Q(X) - X$ is *always* positively correlated with the input for an optimal quantizer. In particular,

$$E[(X - Q(X))(X - EX)] = E[(X - Q(X))X] = E[(X - Q(X))^2]$$

and hence a nonzero quantization mean squared error implies a nonzero correlation between the error and the input for an optimal quantizer. This important observation means that the additive noise model assumptions discussed in Chapter 5 can never be exactly correct for a quantizer with an optimal codebook, even if the weaker uncorrelated assumption is used instead of the stronger independence assumption! Furthermore, the independence assumption of the additive noise model implies that the output will have larger variance than the input which again contradicts the actual situation for an optimal quantizer. For high resolution, however, the average distortion (variance of the quantization noise) is much smaller than the input variance, so that from Lemma 6.2.2, the difference between input and output variances of an optimal quantizer will be very small.

Simple Examples

We develop two specific simple special cases to exemplify the basic properties of optimal quantizers: a binary quantizer and a uniform quantizer. Suppose that the pdf for X is symmetric about the origin, that is, that $f_X(x) = f_X(-x)$ for all x. Suppose that we wish to design a binary quantizer (also called a *hard limiter*) with only two output levels. Since two of the three boundary points are determined by the maximum and minimum values of X (which may be infinite), we need only select one boundary point. Intuitively a reasonable value is 0, that is, the quantizer will produce one level for a nonnegative input and another for a negative input. Given this boundary point, if one reproduction level is c, the other must be $-c$ so that the boundary point is at the midpoint of the two reproduction levels and satisfies the first necessary condition. The second necessary condition will be satisfied if $c = E[X|X > 0]$. Thus, for example, if the pdf is uniform over a range $[-a, a]$, then a quantizer meeting the necessary conditions for optimality is

$$Q(x) = \begin{cases} c & \text{if } x > 0 \\ -c & \text{if } x \leq 0 \end{cases} \qquad (6.2.23)$$

where $c = a/2$. If the input density is Gaussian with 0 mean and variance σ_X^2, then $c = \sqrt{2/\pi}\ \sigma_X$. Thus the binary quantizer of (6.2.23) meets necessary conditions for optimality.

The next example focuses on the uniform quantizer introduced in Section 5.4. Fix a quantization interval (a, b) and define the range $B = b - a$. Suppose that the input random variable has a uniform pdf in this interval. Given N, divide the quantization interval into N equal cells, each of size $\Delta = B/N$ and let the reproduction levels be the midpoints of the cells. Then it is easy to see that the quantizer satisfies the nearest neighbor condition since the cell boundaries are the midpoints of the reproduction levels. Furthermore, the quantizer satisfies the centroid condition since the reproduction levels are the conditional means of the cells since the distribution is uniform. Thus the uniform quantizer satisfies the necessary conditions for optimality for a uniform distribution over the same interval. The average distortion given the cell is simply $\Delta^2/12$, the variance of a random variable that is uniformly distributed on an interval of width Δ. Thus the overall distortion is given exactly by $\Delta^2/12$.

The Generalized Centroid Condition

The nearest neighbor condition as stated above is directly applicable to other distortion measures $d(x, y)$ such as the absolute error as well as the squared error measure. On the other hand, some modification of the centroid condition is needed to make it applicable to other distortion measures. We define the *generalized centroid* cent(R) of a random variable X in a cell R with respect to a particular distortion measure $d(x, y)$ as the value of y that minimizes $E(d(X, y)|X \in R)$. This is sometimes denoted as

$$\text{cent}(R) = \min_y{}^{-1} E(d(X, y)|X \in R),$$

where the inverse minimum means that the left hand side is equal to the value of y that minimizes $E(d(X, y)|X \in R)$, provided that such a value exists. If the minimizing value is not unique, then some means of specifying which value is to be taken as the cenroid is required. It is convenient to simply refer to the generalized centroid simply as the "centroid" when the specific type of centroid is clear from the context. With this definition the generalization of the centroid condition is immediate.

Generalized Centroid Condition

Given a partition $\{R_i\}$, the optimal codebook for a random variable X with respect to a distortion measure $d(x, y)$ is given by

$$y_i = \text{cent}(R_i). \tag{6.2.24}$$

Proof: Using conditional expectations

$$Ed(X, Q(X)) = \sum_i P_i E(d(X, y_i)|X \in R_i)$$

$$\geq \sum_i P_i \min_y E(d(X,y)|X \in R_i)$$

and the inequality is an equality if the y_i are the centroids.   □

An important technical detail in the above is that the centroids for a particular distribution and distortion measure must exist, that is, such a minimum actually exists. This is the case for most distributions and distortion measures encountered in practice. For the squared error distortion, it is clear that the generalized centroid reduces to the usual meaning as the center of mass of a cell with respect to the pdf of X. It should be noted that for some measures, the generalized centroid is not necessarily unique and there can be infinitely many values that satisfy the definition of the centroid.

It is instructive to consider the centroids for a different distortion measure and hence we now focus on the absolute error distortion measure. First we make the basic observations:

1. A value of b that minimizes $E(|X - b|)$ is $b = \mathcal{M}(X)$ where $\mathcal{M}(X)$ is the *median* of the random variable X. The median is that value b for which $\Pr(X < b) = \Pr(X > b)$ thus, b splits the real line into two equiprobable subsets. (See Problem 6.13.) Note that b is not always unique. (For example, if the pdf is bimodal and a median, b, lies internal to an interval with zero probability, then any slight shift of the value of b will also be a valid median.)

2. A function $g(x)$ that minimizes $E(|Y - g(X)|)$ is $g(x) = \mathcal{M}[Y|X = x]$ where $\mathcal{M}[Y|X = x]$ is the *conditional median* of Y given $X = x$.

For the absolute error, the second minimization observation above immediately leads to

$$y_i = \mathcal{M}[X|X \in R_i]. \tag{6.2.25}$$

Thus, the generalized centroid for each region of the partition is the conditional median of the pdf in that region [225].

Sufficiency Conditions

In general the two optimality conditions (the optimality of the encoder with a given codebook and vice versa) are not sufficient to guarantee the optimality of the overall quantizer. Lloyd [218] provides counterexamples. In some special cases, however, the conditions can be shown to be sufficient. (See Fleischer [120], Trushkin [311] and Kieffer [200] for general conditions on the pdf for quantizers meeting the necessary conditions to be unique and hence optimal.) An example of such a sufficient condition is that $\log f_X(x)$

be a concave function of its argument, that is, have a strictly negative second derivative over its entire range. In particular, the Gaussian pdf satisfies this condition and therefore has a unique optimal quantizer for each value of N.

6.3 High Resolution Optimal Companding

The distortion integral derived in Chapter 5 can be used as the starting point for optimizing the point density function and thereby determining the optimal compressor function to minimize granular quantizing noise for the high resolution case when the loading points are given. Recall the distortion integral,

$$D = \frac{1}{12N^2} \int_{-V_-}^{V_+} \frac{f_X(x)}{\lambda^2(x)}\, dx \qquad (6.3.1)$$

where $\lambda(x) \equiv G'(x)/B$ is the quantizer point density function, G is the compressor characteristic and B is the quantizer range. Hölder's inequality states that for any positive a and b with

$$\frac{1}{a} + \frac{1}{b} = 1$$

the following inequality holds:

$$\left(\int u(x)v(x)\, dx \right) \le \left(\int u(x)^a\, dx \right)^{1/a} \left(\int v(x)^b\, dx \right)^{1/b},$$

with equality if $u(x)^a$ is proportional to $v(x)^b$. In the definition for D choose

$$u(x) = \left(\frac{f_X(x)}{\lambda^2(x)} \right)^{1/3},$$
$$v(x) = \lambda(x)^{2/3},$$

and $a = 3$ and $b = 3/2$. Then, with these values, Hölder's inequality yields

$$\int \left\{ \frac{f_X(x)}{\lambda^2(x)} \right\}^{\frac{1}{3}} [\lambda(x)]^{\frac{2}{3}}\, dx \le \left[\int \frac{f_x(x)}{\lambda^2(x)}\, dx \right]^{\frac{1}{3}} \left[\int \lambda(x)\, dx \right]^{\frac{2}{3}}$$

and since $\int \lambda(x)\, dx = 1$,

$$\int f_X(x)^{1/3}\, dx \le \left(\int \frac{f_X(x)}{\lambda^2(x)}\, dx \right)^{\frac{1}{3}}$$

which combined with the definition of D gives the bound:

$$D \geq \frac{1}{12} \frac{1}{N^2} \left(\int_{-V_-}^{V_+} f_X(x)^{1/3} dx \right)^3 \tag{6.3.2}$$

with equality if the point density satisfies

$$\lambda(x) = \frac{f_X(x)^{1/3}}{\int f_X(y)^{1/3} dy}. \tag{6.3.3}$$

It is often convenient to describe the input random variable X by its *normalized pdf* $\tilde{f}_X(y) = \sigma_X f_X(\sigma_X y)$ which has unit variance. Then the asymptotic expression for the average distortion with the optimal point density function is obtained from the right hand side of the inequality (6.3.2):

$$D_{\mathrm{opt}} = \frac{\sigma_X^2}{12 N^2} \left(\int_{-\gamma_-}^{\gamma_+} \tilde{f}_X(y)^{1/3} dy \right)^3, \tag{6.3.4}$$

where we have used the change of variable $y = x/\sigma_X$ and $\gamma_- = V_-/\sigma_X$ and $\gamma_+ = V_+/\sigma_X$. Usually a symmetric quantizer is assumed and $\gamma_- = \gamma_+ = \gamma$, the usual loading factor.

The asymptotic expressions can in principle provide insight as to the optimal selection of quantizer levels. Given the overload points, $-V_-$ and V_+, the compressor characteristic $G(x)$ for the quantizer is fully determined from $\lambda(x)$ according to:

$$G(x) = B \int_{-V_-}^{x} \lambda(y) dy - V_-. \tag{6.3.5}$$

In order to actually use the above results for quantizer design, there remains the task of finding a quantizer with a finite number of levels satisfying the necessary conditions and having output levels and cells that approximately fit this compressor curve. We cannot expect to obtain an exact optimal solution for finite N using a point density function that is only *asymptotically* optimal. It should be noted, however, that (6.3.5) does not provide a complete design by itself since the optimal point density function is based on specified overload points. The loading factor must also be chosen to optimize the trade-off between overload and granular quantizing noise. The result is at least of conceptual and theoretical interest and it could be used as a starting point for developing a design algorithm that attempts to find an optimal loading factor as well as optimal quantization levels.

Partial Distortion Theorem

The asymptotically optimal point density function shows that the optimal distribution of reproduction levels is not the same as the probability distribution of the input. The distortion D can be written as

$$D = \sum_{i=1}^{N} D_i \tag{6.3.6}$$

where D_i is the ith term in the sum (6.2.1) and is called the *partial distortion*. It is the contribution of the ith quantization cell to the average distortion of the quantizer. Note that D_i can be expressed as the product:

$$D_i = E[(X - Q(X))^2 | X \in R_i] P_i \tag{6.3.7}$$

so that D_i/P_i is the conditional distortion given that X lies in the ith cell. An interesting result of the asymptotic theory is the following.

Theorem 6.3.1 *For quantization with the asymptotically optimal point density function, the partial distortions D_i are asymptotically constant with value D/N as N approaches infinity.*

Proof: From (5.6.5), D_i can be expressed as

$$D_i = \frac{1}{12} \Delta_i^2 P_i$$

Since $P_i = \Delta_i f_X(y_i)$ and the compressor slope satisfies $G'(y_i) \approx \Delta/\Delta_i$, we find that D_i reduces to:

$$D_i \approx \frac{\Delta^3}{12} \frac{f_X(y_i)}{G'^3(y_i)}$$

but from (6.3.3), we see that $G'^3(y_i)$ is directly proportional to $f_X(y_i)$. Hence, the partial distortion is asymptotically constant, e.g., independent of i. □

6.4 Quantizer Design Algorithms

The necessary conditions for optimality provide the basis for the most widely used design algorithms. In this section, we present the main idea of these algorithms and focus in detail on an algorithm due to Lloyd which has become of particular importance because it is also applicable to vector quantizer design.

The Lloyd Quantizer Design Algorithm

We have seen that when N is large, nonuniform quantizers can provide a better match to an input probability density function and yield smaller average distortion. This does not, however, provide a means of actually finding the best quantizer for a fixed and possibly not large N. The necessary conditions for optimality suggest an iterative means of numerically designing at least a good quantizer. The basic design algorithm was first described by Lloyd in an unpublished 1957 report at Bell Laboratories [218] that much later appeared in published form [219]. The algorithm is sometimes called the *Lloyd I* algorithm to distinguish it from a second algorithm also introduced by Lloyd in the same report. The Lloyd I algorithm has been used widely for the design of quantizers and vector quantizers. Essentially the same algorithm is a commonly used procedure in the design of pattern recognition systems.

Before discussing the design algorithm, one more definition is needed. In dealing with a sequence of codebooks, we need to have a measure of quality for any given codebook, $\mathcal{C}$. We define the *average distortion of a codebook* with respect to a particular distortion measure $d(x, y)$ and a statistically specified input X, as the average distortion $D = Ed(X, Q(X))$ of that quantizer $Q(x)$ which has the given codebook $\mathcal{C}$ and a nearest neighbor partition with respect to that codebook. Note that this quantizer need not be optimal although it has the optimal partition for the given codebook. The basic algorithm is quite simple and we formalize it below.

The key to this iterative algorithm is a mapping that converts a given codebook into a new and improved codebook. The design algorithm begins with an initial codebook and repeats this mapping until a suitable stopping criterion is satisfied. The codebook improvement mapping is the basic iteration of the algorithm.

The Lloyd Iteration for Codebook Improvement

(a) Given a codebook, $\mathcal{C}_m = \{y_i\}$, find the optimal partition into quantization cells, that is, use the nearest neighbor condition to form the nearest neighbor cells:

$$R_i = \{x : d(x, y_i) \leq d(x, y_j); \text{ all } j \neq i\}.$$

(b) Using the centroid condition, find $\mathcal{C}_{m+1}$, the optimal reproduction alphabet (codebook) for the cells just found.

The basic form of the Lloyd iteration assumes that the input pdf is known in order to compute the centroids. Figure 6.1 illustrates the Lloyd iteration and indicates the dependence of the mapping on the specified pdf.

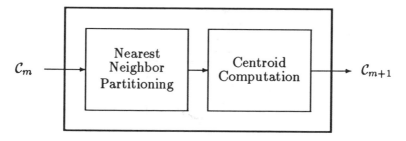

Figure 6.1: Flow chart of the Lloyd iteration for codebook improvement

Later we consider the case where an analytical description of the input pdf is not mathematically tractable or simply not available, but a sample distribution based on empirical observations is used instead to generate the improved codebook. From the necessary conditions for optimality each application of the iteration must reduce or leave unchanged the average distortion. The actual algorithm of Lloyd can now be stated very concisely as in Table 6.1.

Table 6.1: **The Lloyd Algorithm**

Step 1. Begin with an initial codebook C_1. Set $m = 1$.

Step 2. Given the codebook, C_m, perform the Lloyd Iteration to generate the improved codebook C_{m+1}.

Step 3. Compute the average distortion for C_{m+1}. If it has changed by a small enough amount since the last iteration, stop. Otherwise set $m + 1 \to m$ and go to Step 2.

The initial codebook may have the lattice pattern of a uniform quantizer or it may be any reasonable codebook that has N points which roughly represent the range of values taken on by the input. A variety of initialization methods will be considered when the vector extension of the Lloyd algorithm is considered later. The flow chart for the Lloyd algorithm is shown in Figure 6.2. Although various stopping criteria can be used, a particular version that is common and effective is to test if the fractional drop in distortion, $(D_m - D_{m+1})/D_m$, is below or above a suitable threshold. With a threshold value of zero, the algorithm will necessarily produce a sequence of codebooks with monotone nonincreasing values of average distortion. If the algorithm converges to a codebook in the sense that further iterations

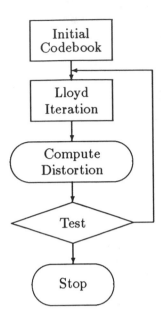

Figure 6.2: Flow chart of the Lloyd algorithm for quantizer design

no longer produce any changes in the set of reproduction values, then the resulting codebook must simultaneously satisfy both the necessary conditions for optimality. In the language of optimization theory, the algorithm converges to a *fixed point* or *stationary point* with respect to the iteration operation.

We shall discuss below how the algorithm can be run on a sample distribution based on training data if a probabilistic model for the input random variable is not tractable or is not available. The basic operation of the algorithm as stated above and shown in Figure 6.2 remains unchanged. Only the actual form of the Lloyd iteration is modified for this case.

Design Based on Empirical Data

In practice, it is often difficult to find convenient analytical models for the pdf of the signals to be quantized. Furthermore, even when the pdf is known, the integration to compute centroids may have no closed form solution. Thus, some kind of numerical integration is needed to find the centroids. One form of numerical integration is the Monte Carlo method where the integrand is evaluated a set of randomly generated numbers as arguments and then an average is computed. This procedure can be made computationally efficient using the technique of importance sampling [173].

In the case where the pdf is unknown, it can be estimated from empirical data, and then numerical integration techniques can be used. This introduces two stages of approximation, first in estimating the pdf and second in numerical integration using the estimated distribution. A more direct approach for the case of an unknown pdf, is to use the empirical observations directly as the random numbers in a Monte Carlo integration. It is this approach that has become the standard technique for quantizer design in recent years, although it is not generally viewed as being a Monte Carlo technique. Specifically, the Lloyd algorithm is applied to the design of a quantizer for the *empirical distribution*, that is the distribution of the set of observations.

It should be noted that there are many ways to generate independent random variables with a known distribution by simple operations on uniform random variables [207]. Uniform random variables can be generated with utilities readily available on most computer systems.

Suppose we have a set of observations (or samples) of the signal to be quantized, called the *training set*, $T = \{v_1, v_2, \cdots, v_M\}$, where M is the size of the training set and v_i are the *training points*. We defined the *training ratio* as $\beta \equiv M/N$, where N is the size of the codebook. This important parameter indicates how effectively the training set describe the true pdf of the source and how well the Lloyd algorithm can yield a codebook that is nearly optimal for the true pdf. The training set can be used to statistically define a random variable U by assigning the probability mass function (pmf) to have mass $1/M$ for each value of v_i in T. Although a pdf does not exist for a discrete distribution (unless we use delta functions), the cumulative distribution function is well defined for U. Specifically, the cdf of the random variable U is given by

$$F^{(M)}(x) = \frac{1}{M} \sum_{i=1}^{M} 1_{(-\infty, x]}(v_i) = \frac{1}{M} \sum_{i=1}^{M} u(x - v_i),$$

where $u(x)$ denotes the unit step function which is zero for negative x and one for nonnegative x. The distribution formed in this way is called a *sample distribution* or an *empirical distribution*. If the v_i are a stationary and ergodic sequence, then application of the ergodic theorem implies that with probability one the limiting time average as $M \to \infty$ of the indicator functions is just the expectation

$$E(1_{(-\infty, x]}(X)) = F_X(x),$$

the true cdf for the random variable X.

It is useful to make a philosophical comment on this approach. One possible approach to designing a quantizer for a training set would be to first construct a probabilistic model for the training set and then to run the Lloyd algorithm on the resulting distribution. One could, for example, assume the data was Gaussian and estimate the underlying mean and variance. The problem with this approach is the necessity of assuming a model and hence of a particular structure to the data. Estimation of the sample average distortion directly avoids such assumptions and estimates the quantity actually needed for the optimization.

Quantization of Discrete Random Variables

Until now, we have focused attention on quantization of continuously distributed random variables which can be described by a pdf. Now, we must consider the quantization of a discrete valued random variable where the number of values M that it takes on exceeds the number of levels N of the quantizer to be designed. (The training ratio always exceeds unity.) Note that M can be infinity (as in the case of a Poisson distribution). Re-examination of the derivation of the centroid and nearest neighbor conditions leads to the conclusion that they remain valid for a discrete input random variable. There are, however, some differences in the use of these conditions. Rather than partition the real line into cells, it is more directly relevant to partition the set $\mathcal{X}$ of input values with positive probability into N cells $\{R_i;\ i = 1, \cdots, N\}$, each cell being a subset of $\mathcal{X}$. In the case of an empirical distribution formed from a training set $\mathcal{T}$, $\mathcal{X} = \mathcal{T}$. Thus, the cells are now countable sets and are disjoint subsets of $\mathcal{X}$. The only difference is that the boundary points of a cell must be well-defined in the case where an input value with positive probability coincides with a cell boundary point.

Optimality Conditions for Discrete Inputs

For a discrete input, the selector functions $S_i(x) = 1_{R_i}(x)$ (defined in Chapter 5) are now defined for a given partition $\{R_i;\ i = 1, \cdots, N\}$ of the training set $\mathcal{T}$. Hence, the centroid condition for a cell R_i is given by

$$y_j = E[X|X \in R_j] = \frac{E(XS_j)}{ES_j}.$$

Thus, we have for a sample distribution formed from a training sequence $\{x_i\}$ that

$$y_j = \frac{\frac{1}{M}\sum_{i=1}^{M} x_i S_j(x_i)}{\frac{1}{M}\sum_{i=1}^{M} S_j(x_i)}$$

for $j = 1, 2, \cdots, N$. The index i counts training points and index j counts partition regions.

The nearest neighbor condition for the discrete input case becomes

$$R_j = \{u \in \mathcal{T} : |u - y_j| \leq |u - y_i| \text{ all } i\}.$$

However, this definition should be modified with a tie-breaking condition to handle the case where a training point is equidistant from two or more output points y_i. The simplest way to do this is to assign the training point to the region R_j that satisfied the nearest neighbor condition with the lowest index value j.

For the discrete case, the average distortion can be expressed as:

$$D = \frac{1}{M} \sum_{i=1}^{M} \sum_{j=1}^{N} (v_i - y_j)^2 S_j(v_i)$$

where the input X is confined to the training set $\mathcal{T} = \{v_i\}$ and the codebook, $\mathcal{C}$, is $\{y_j\}$.

There is a third necessary condition for quantizer optimality that arises only in the case of discrete random variables. Suppose that we have a quantizer that satisfies the two necessary conditions for optimality and it has two adjacent output levels $y_1 < y_2$ so that the corresponding decision level separating cells 1 and 2 is given by $x = (y_1 + y_2)/2$ and x is a member of cell 1. Now suppose that this value of x happens to be a point where the input pmf has positive mass. We modify the quantizer by reassigning the point x from cell 1 to cell 2. Clearly this does not alter the mean square distortion achieved with this quantizer. However, the centroid of cell 1 is now shifted so that y_1 is no longer optimal and the distortion can be reduced by replacing y_1 with the new centroid. Therefore the quantizer is not optimal even though it satisfies the two necessary conditions. We therefore have a third necessary condition for optimality: *the input random variable must have zero probability of occurring at a boundary between nearest neighbor cells.*

The Lloyd iteration can now be directly applied to the discrete input distribution defined from the training set $\mathcal{T}$ to obtain a locally optimal quantizer for this distribution. At each iteration, a test for probability mass on the boundary could be performed and if found, the codebook should be modified by reassigning such a point to the adjoining partition region. This test is not, however, usually peformed as part of the algorithm because the probability of such ties is small (zero for truly analog data). It is a good idea to perform it at when the final codebook is chosen since ties are possible when the algorithm is digitally implemented and it is easy to verify that this additional necessary condition is met.

The general conclusion here is that if we design an optimal N point quantizer for a finite training ratio, it is reasonable to expect that for M

sufficiently large, it will be very nearly optimal for the true distribution of the input random variable. More explicit statements of this result may be found in [167], [279].

Figure 6.3 provides an example of the iterative algorithm. The squared

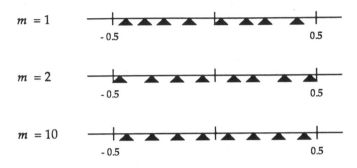

Figure 6.3: Example of Lloyd algorithm applied to a training set for a uniform random input on $[-1/2, 1/2]$

error distortion and a three level quantizer are used for simplicity. The source is a uniform random variable on $[-1/2, 1/2]$ and the training set is obtained from a sequence of samples produced by a random number generator.

The Lloyd II Algorithm

The Lloyd algorithm described above was actually the first of two algorithms proposed by Lloyd. The second, or Lloyd II algorithm, is also based on the necessary conditions for optimality but does not use what we have called the Lloyd iteration. Instead, several passes through the set of quantizer parameters $\{y_1, x_1, y_2, x_3, ..., y_N\}$ are performed. In each pass, the output points and decision boundaries are iterated one at a time while proceeding from left to right on the real number line. This algorithm was later rediscovered by Max [235], and has been widely used for scalar quantizer design. It does not generalize to vector quantization. The algorithm is as in Table 6.2.

The values of ϵ and α are design parameters.

6.5   Implementation

In Chapter 5 we discussed the structural decomposition of a quantizer and its relevance for implementation of quantizers. We now return to the topic of implementation.

Table 6.2: **Lloyd II Algorithm**

Step 1. Pick an initial value of $\{x_j;\ j\ =\ 1, 2, \cdots, N-1\}$ and $\{y_j;\ j = 1, 2, \cdots, N\}$. Set $k = 1$, $x_0 = -\infty$.

Step 2. Find y_k such that it is the centroid of the interval (x_{k-1}, x_k).

Step 3. Find x_k so that it is the midpoint of (y_k, y_{k+1}).

Step 4. If $k = N-1$ go to Step 5, otherwise set $k + 1 \rightarrow k$ and go to Step 2.

Step 5. Compute c, the centroid of the interval $(x_{N-1}, +\infty)$. If $|y_N - c| < \epsilon$, stop. Otherwise, go to Step 6.

Step 6. Let $y_N - \alpha(y_N - c) \rightarrow y_N$ and set $k = 1$. Go to Step 2.

A scalar quantizer that operates on analog inputs must be implemented with analog circuit technology with such elements as comparators and resistors. Digital technology is generally more convenient for most signal manipulations, however, and it is of interest to consider the implementation of a quantizer for inputs that are already digitized but must be requantized with a codebook of smaller size (usually much smaller) than the size of the set of input values. Usually, a uniform quantizer is most commonly implemented as an analog integrated circuit. The number of bits of resolution can be made relatively large in this way. As a second stage of quantization a lower resolution nonuniform quantizer can then be implemented entirely with digital circuitry. For optimal codebooks obtained via the Lloyd algorithm, the reproduction values are irregularly spaced and it is far more convenient to store these values in a digital memory, than to attempt to implement them with analog resistors.

A brute force digital implementation of a scalar quantizer would involve computing the distortion between the input sample and each of the reproduction levels and then testing for the smallest distortion to select the desired word. This is called a *full search quantizer* and involves the computation and minimization of

$$(x - y_i)^2$$

or, equivalently,

$$|x - y_i|.$$

Alternatively, one can eliminate the x^2 term common to all the computations with the squared error and find and minimize

$$y_i^2 - 2xy_i,$$

which involves finding $N = 2^R$ products and then scaling each by a stored bias. Thus a full search scalar quantizer can be constructed either as a sequence of multiplications or a group of parallel multiplications followed by a comparator. The complexity grows exponentially with the rate.

An alternative architecture is best illustrated for an integer value of the resolution, r. Suppose that the quantizer has reproduction levels $y_1 < y_2 < \cdots < y_N$. If r is an integer, then $N = 2^r$ is even. Consider an encoder which works as follows: First the input is compared to the midpoint between the two middle levels, $(y_{N/2} + y_{\frac{N}{2}+1})/2$. If it is larger than this threshold, then the minimum distortion symbol cannot be one of the y_i for $i \leq N/2$ and we have eliminated half the candidates. Suppose that this is the case. Compare the input to the midpoint between the two middle remaining levels. Again the result of the comparison implies that the level must be in either the larger half or the smaller half of the remaining levels. The appropriate collection is selected. The algorithm continues in this way until it makes its final comparison from a collection of only two symbols to obtain the final reproduction. The operation of the encoder is depicted for $N = 8$ in Figure 6.4. The encoder algorithm is depicted in a tree structure. Each node is labeled with the inequality which the input must satisfy to reach that node. The branches of the tree are labeled with the binary symbols produced by the encoder if that branch is taken.

The algorithm is called *successive approximation* because each time it narrows its choices to a smaller number of smaller distortion possibilities. As demonstrated by Figure 6.4, the algorithm is an example of a *tree structured* algorithm. In this case the tree is binary because at each stage one of two choices is made depending on whether the input is larger or smaller than a threshold. Unlike the full search implementation which requires 2^r multiplications, this algorithm requires no multiplications and r comparisons. Since comparisons are no more complicated (and usually far simpler) than multiplications, the successive approximation architecture is significantly superior to the full search implementation (at least for a software or sequential implementation) since its complexity grows only linearly with rate instead of exponentially. The successive approximation quantizer has the further attribute that it is an embedded code, high rate codes are formed from low rate codes by simple concatenation of bits resulting from coding

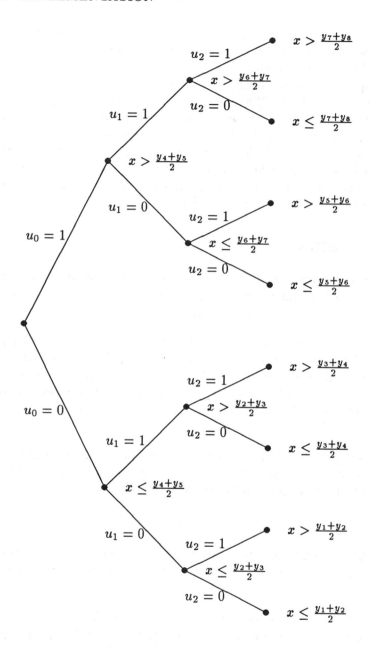

Figure 6.4: Successive Approximation Quantizer

deeper into the tree. In other words, one can use only the first few layers of a tree if one is constrained in rate, one need not completely change the code as would be the case with a flash quantizer.

Another approach to quantizer implementation was introduced recently by Tank and Hopfield [305]. This analog circuit is based on the Hopfield neural network for a content addressable memory which has also been shown to have effective computational capabilities. The method involves the use of r computing units, each of which performs a weighted sum of its analog inputs followed by a memoryless nonlinearity which has a sigmoid shape. The outputs of the computing units are fed into a matrix of resistive weights which generate the inputs to the sigmoid nonlinearity.

6.6 Problems

6.1. Construct an example of a pdf that is symmetric, continuously distributed (no delta functions) and for which the optimal (mean square error) quantizer for $N = 3$ is not symmetric.

6.2. A bounded random variable, X, lies between $-a$ and $+b$, with pdf $f_X(x)$. Determine the value of c that minimizes the l_∞ norm of $x - c$, that is it minimizes $\max_x |x - c|$ for x in the range of values taken on by X. Suggest any ideas you may have for finding the necessary conditions for optimality of a quantizer that minimizes $\max |X - Q(X)|$ with probability one.

6.3. A *random quantizer* Q is defined by a partition R_i of the real line, a set of output points y_j (for $j = 1, 2, \cdots, N$) with y_i contained in R_i, and a set of conditional probabilities $q(j|i)$ such that

$$q(j|i) = P[Y = y_j | X \in R_i].$$

For a given partition, and a given set of output points determine the optimal set of conditional probabilities to minimize mean square quantization error.

6.4. Design an optimal 4 point quantizer ($N = 4$) for the one sided exponential density function with mean value unity using the mean square error distortion measure. Use either a deterministic computation or the Lloyd I algorithm. Aim for at least 2 decimal places of accuracy. Avoid infinite loops; your computer budget is finite!

6.5. Describe explicitly the algorithm you would use to design the same quantizer as specified in Problem 6.4 using the Lloyd II algorithm for the particular density function given. You should indicate how you would satisfy the centroid condition where it is needed. You do not need to actually program and carry out this design.

6.6. Find an explicit formula for the SNR of an optimal quantizer for a Gaussian random variable input when the number of output points, N, is very large (high resolution case). Assume that the overload point is much larger than the standard deviation of the input and that overload noise can be neglected..

6.7. Determine the optimal step size for minimum mse for a 7 bit uniform quantizer with a Laplacian input signal with zero mean and standard deviation of 1 volt. Show how you get your answer!

6.8. A quantizer with $N = 3$ is to be designed to minimize mean squared error using the generalized Lloyd algorithm with the training set.

$$T = \{1.0, 2.0, 3.0001, 4.0, 8.0, 9.0, 12.0\}.$$

(a) Find the optimal codebook, starting with the initial codebook $\mathcal{C} = \{1.0, 5.0, 9.0\}$. Show all intermediate codebooks obtained in the process of computing your answer. (Final answer within one decimal place of precision is adequate.) (A computer is not needed.)

(b) Do the same problem with a different initial codebook.

(c) Replace the training point 3.0001 by 3.0 and repeat part (a).

6.9. A quantizer has 10 output levels given by:

$$\{-0.1996, -0.6099, -1.058, -1.591, -2.34,$$

$$0.1996, 0.6099, 1.058, 1.591, 2.34\}$$

Generate 100 zero mean, unit variance, independent, Gaussian random variables, x_i, and compute the average squared distortion in encoding this set of data using a nearest neighbor encoding rule:

$$D_{\text{av}} = \frac{1}{100} \sum_{i=1}^{100} d(x_i, Q(x_i))$$

6.10. A two-stage quantizer has input X, first stage output $\hat{X}$, second stage input $U = X - \hat{X}$, and second stage output $\hat{U}$. The overall system output is $\tilde{X} = \hat{X} + \hat{U}$. Assume b_1 and b_2 bits for stages 1 and 2, respectively. Assume that asymptotic theory applies for each stage, that the optimum λ is used (as in (6.3.3)), that the input is a Gaussian random variable with unit variance and that the error U after the first stage is approximately Gaussian. Derive an approximate formula for the overall system mean square error.

6.11. Define a distortion measure

$$d(x, y) = \frac{(x - y)^2}{x^2}.$$

Describe in detail the nearest neighbor and centroid conditions for a quantizer for a random variable X and this distortion measure. Are the conditions different?

6.12. Suppose that X is a random variable which can take on only a finite number of values and $d(x, y)$ is the Hamming distortion defined as 1 if $x \neq y$ and 0 if $x = y$. Describe the centroid of a set R with respect to this distortion measure.

6.13. Prove that a value of b that minimizes $E(|X - b|)$ is $b = \mathcal{M}(X)$ where $\mathcal{M}(X)$ is the *median* of the random variable X. Thus, b splits the real line into two equiprobable subsets.

6.14. Suppose that a random variable X has the two-sided exponential pdf

$$f_X(x) = \frac{\lambda}{2} e^{-\lambda |x|}.$$

A three level quantizer q for X has the form

$$q(x) = \begin{cases} +b & x > a \\ 0 & -a \leq x \leq +a \\ -b & x < -a \end{cases}$$

(a) Find an expression for b as a function of a so that the centroid condition is met.

(b) For what value of a will the quantizer using b chosen as above satisfy both the Lloyd conditions for optimality? What is the resulting mean squared error?

(c) Specialize your answer to the case $\lambda = 1$. (You should find that $a = (\sqrt{3} - 1)/2$.)

6.15. A nonuniform quantizer with 1024 levels covers a range from 0 to the positive overload point $V = 80$, so that the range $B = V = 80$. The quantizer is to be optimized for quantizing a nonnegative random variable Y with an exponential pdf, $f(x) = 0.1e^{-x/10}$ for $x > 0$.

(a) Find the optimal compressor $G(x)$, defined for $x > 0$.

(b) Find the approximate mean squared distortion achieved with this quantizer.

6.16. A random variable X is optimally quantized to minimize mean squared error. The quantizer has three output levels, -2, 1, and 3 which are selected with probabilities 0.5, 0.3, and 0.2 respectively. The mean squared error is 0.2. Find

(a) the mean of X,

(b) the correlation $E[XQ(X)]$ between input and output, and

(c) the variance of X.

6.17. A clever farmer would like to classify her pumpkins into three categories small (1), medium (2), and large (3) and she will classify a pumpkin as category 1 if its weight is less than or equal to x_1, as category 3 if its weight is greater than x_2, and as category 2 in all other cases, where $x_1 < x_2$. She will then load up separate trucks for each category and deliver truckloads of pumpkins to different supermarkets. She will charge the supermarkets a fixed price for pumpkins in each category regardless of their individual weight. She needs to find three standard weight sizes, y_1, y_2, and y_3 (one for each of the three categories) as the average weight of pumpkins in the respective categories. She wishes to minimize, D, the average squared error between the weight of a pumpkin (selected at random from her pumpkin fields) and the standard weight of that category in which that pumpkin belongs. The only information available about the distribution of the weight of her pumpkin crop is the following: twelve pumpkins picked at random from her fields have the following weights (in pounds):

1, 3, 4, 6, 8, 10, 13, 14, 16, 20, 24, 30

Find a solution that satisfies the necessary conditions for optimality and compute the value of D based on this data.

6.18. A digital voltmeter can display voltages in the range of -9.99 to +9.99. Suppose we can model the input voltage as a random variable with a smooth probability density function confined to this range of voltages. The voltmeter can be designed in two ways, (a) rounding, where the input voltage is rounded, i.e., mapped to the nearest value that can be displayed, or (b) truncation, where the exact decimal value of the voltage is modified by dropping the digits beyond the second decimal place (e.g, $3.289 \rightarrow 3.28$, $-4.3137 \rightarrow -4.31$, $5.3289 \rightarrow 5.32$).

(a) Find the mean squared error in the displayed voltage for each case. Explain any assumptions.

(b) Sketch the input output characteristics for each voltmeter in the range of voltages $|V| < 0.04$. (displayed voltage versus input voltage).

Chapter 7

Predictive Quantization

7.1 Introduction

This book is about data compression, an operation which includes the removal of redundancy from a signal to allow a more compact representation. Clearly redundancy is at the heart of the subject. Having covered the main ideas of linear prediction, it should be evident that the predictability of a signal is intimately related to the idea of redundancy. The better we can predict the future of a process from the past, the more redundancy the signal contains and the less new information is contributed by each successive observation of the process. We now wish to exploit this redundancy as an adjunct to scalar quantization. We will view a quantizer as an essential building block of a coding system, but now the quantizer will be regarded as only one component of a system that can incorporate other processing elements to facilitate the efficient digitization of a signal.

As a simple but trivial illustration of redundancy removal as an adjunct to quantization, suppose that we have a white stationary random process with zero mean added to a sampled sine wave with a large amplitude (relative to the rms level of the white process). The amplitude distribution of the resulting signal suggests that a very wide range of quantization levels will be required compared to what would have been needed if the sine wave were absent. By predicting the sine wave from past samples of the observed signal, we could quite accurately estimate it and subtract it from all future observations of the composite signal. Then a much lower power level signal can be quantized with the same resolution but with a much smaller amplitude range (hence a lower bit rate) and the sine wave can later be added

back to the quantized signal in order to reconstruct a good approximation
to the original signal. Data compression has been achieved by using prediction to remove the predictable part and then quantizing the remaining
unpredictable part. The overall error between reconstructed and original
signals is the same as the error in quantizing the unpredictable component
signal, so the same overall SNR is achieved with much fewer bits.

In general, any partial knowledge of the past of a process to be quantized
gives some *a priori* information about the current value of the process.
Brute force application of scalar quantization ignores this redundancy. We
now consider how this redundancy can be efficiently exploited.

7.2   Difference Quantization

Before focusing on the use of prediction for quantization, it is convenient
to establish the following simple but important result. Suppose a sequence
U_n is subtracted from the input process X_n, the difference $e_n = X_n - U_n$ is
quantized and then the same sequence U_n is added to the quantized value
$\hat{e}_n = Q(e_n)$ as shown in Figure 7.1 to form a signal $\hat{X}_n = Q(e_n) + U_n$. We

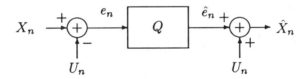

Figure 7.1: Difference Quantizer

call this *difference quantization*. Since

$$e_n = X_n - U_n$$

$$\hat{e}_n = \hat{X}_n - U_n,$$

clearly

$$E[(X_n - \hat{X}_n)^2] = E[(e_n - \hat{e}_n)^2] \qquad (7.2.1)$$

or, in other words, the overall mean squared error in reproducing X_n with
$\hat{X}_n$ is equal to the mean squared quantization error incurred in quantizing
e_n. Trivial as this may appear, it is a very important result. It shows that
we have the freedom to subtract any signal we wish from X_n if we add it
back to the quantized signal $\hat{e}_n$. We can gain some advantage in doing this
if U_n is chosen wisely.

If in particular U_n is a constant, this scheme reduces to the removal
of a constant or dc term from the input prior to quantization. The most

obvious choice would be to subtract the mean value of X_n before quantizing and add it back after quantizing. This can be seen as equivalent to shifting the decision boundaries and the output levels of the quantizer by the mean value.

For a particular fixed quantizer, removing the mean before and after quantization can improve performance for some input distributions. For example, a quantizer might be optimized for a zero mean unit variance Gaussian signal, but the actual signal is a mean m unit variance Gaussian random variable. In this case removing the mean can help. If one is free to design the quantizer for the given signal, however, the removal of a constant mean provides no advantage since the dynamic range of the input signal to the quantizer is not altered and the quantizer can be optimized for either the original signal or the zero mean signal. Suppose, however, that the mean of X_n varies with time. Then by setting $U_n = EX_n$ what we are doing in difference quantization can be regarded as shifting the location of the set of quantizer levels and decision boundaries with time so that the quantizer is centered at the current mean value of the process. In effect, the input X_n "sees" a quantizer that is sliding back and forth tracking the changing center of gravity of the input pdf. Intuitively, this makes sense since the limited number of quantization levels are being located in that region where they can do the most good instead of having to be spread out over a wider range of values when difference quantization is not used.

Let us go one step further. Suppose that instead of having a non-stationary mean value, the process has zero mean but we use difference quantization by letting U_n be a *prediction* of X_n based on some information about the past of X_n. A prediction $\tilde{X}_n$ of X_n can be thought of as the conditional mean of X_n given some information about the past of X_n. The variance of the difference signal e_n will be less than the variance of X_n so that a reduced range of amplitudes can be quantized, but the number of quantizer levels can be the same as would have been needed for quantizing X_n directly. This gives an immediate advantage: The effective resolution is increased without increasing the number of bits needed. Alternatively, the same SNR can be achieved with a reduced number of bits by reducing the number of quantization levels and keeping the mean square quantizing noise the same as would be used for direct quantization of X_n. We have so far not addressed the question of how the decoder can have an identical copy of the signal U_n. This will soon become clear.

Let X_n be a zero mean stationary random process with known second order statistics which contain redundancy so that the spectral flatness measure is less than unity (the process is not white). Then, the expected value of the current observation given some past values is nonzero:

$$E[X_n | X_{n-1}, X_{n-2}, \cdots, X_{n-m}] > 0 \qquad (7.2.2)$$

and this provides some partial information about the current value. This of course is the optimal nonlinear predictor for memory size m. To avoid nonlinear predictors, we can replace this with the optimal linear predictor $\tilde{X}_n$ for this memory size.

Suppose we use $U_n = \tilde{X}_n$ in Figure 7.1, where an optimal linear predictor of memory size m is assumed and $\tilde{X}_n$ is a linear combination of the past inputs $X_{n-1}, X_{n-2}, \cdots, X_{n-m}$. This can easily be generated at the encoder location. In order to have an identical copy at the decoder location, however, we would have to quantize U_n and transmit it to the receiver. This would defeat the objective of data compression. Instead, we avoid this problem by the technique of *closed-loop prediction* as explained next.

7.3   Closed-Loop Predictive Quantization

If the coding system is working reasonably well, the receiver will successfully reproduce a good approximation $\tilde{X}_n$ of X_n so that we could use the *previously·reproduced values* of $\tilde{X}_n$ to predict X_n. Thus we can form an estimate of X_n as a "prediction" from the previously reproduced values according to

$$\tilde{X}_n = f(\hat{X}_{n-1}, \hat{X}_{n-2}, \cdots, \hat{X}_{n-m}), \qquad (7.3.1)$$

where the function f must be suitably chosen so that we are making a good estimate of X_n from the m past values of the reproduced process $\hat{X}_n$. In practice what is usually chosen is

$$\tilde{X}_n = - \sum_{i=1}^{m} a_i \hat{X}_{n-i}, \qquad (7.3.2)$$

where the a_i values are the optimal predictor coefficients that would be used in predicting X_n from its m past values. We take this expression as the definition of the quantity $\tilde{X}_n$, but we note that this deviates from the definition of a linear predictor in Chapter 4, that is, we are using the linear prediction coefficients that are optimal for the *true* past of the signal in order to predict from the *coded approximation* to the past. This suboptimal selection is done for two reasons: First, it is simple to find the coefficients given the true past, while it is usually impossible to find the optimal coefficients given the coded past. Second, if the reproduction is reasonably good, then the coefficients based on the true past should not be too far from the optimal coefficients based on the coded past. If one wishes to find the truly optimal coefficients for the coded past, it can be done approximately with conjugate gradient numerical techniques (see, e.g., [58]) or, in a few rare cases, it can be found analytically by Hermite polynomial expansions (see, e.g., [18]).

Predicting X_n from the *quantized* past, allows us to generate $U_n = \tilde{X}_n$ in both the transmitter (encoding system or encoder) and the receiver (decoding system or decoder) as shown in Figure 7.2 without requiring that any "side information" be transmitted between coder and decoder. A system of this general form is called a *predictive quantizer*. The box

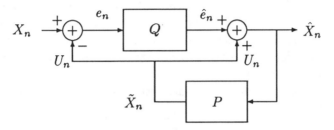

Figure 7.2: Predictive Quantizer

labeled P represents the operation described by (7.3.1), but we shall focus specifically on the case when it is given by (7.3.2), corresponding to a linear predictor. The operation P corresponds to the transfer function $P(z) = 1 - A(z)$ defined in Chapter 4.

A remarkable thing now happens. By redrawing Figure 7.2, making no change in the structure or signal flow, we get Figure 7.3, which has the familiar appearance of what is widely known as DPCM, or differential PCM (differential pulse code modulation). Also, by allowing the quantizer Q to

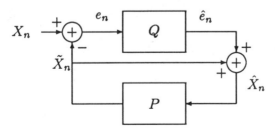

Figure 7.3: DPCM

generate both an index i_n identifying a particular level y_{i_n} in the quantizer codebook $\mathcal{Y}$ and the reconstructed approximation $\hat{e}_n = y_{i_n}$, we can separate the transmitter and receiver sections as shown in Figure 7.4. The inverse quantization operator is defined by $Q^{-1}(i) = y_i$, where y_i is the codeword indexed by i.

Thus the concept of closed-loop predictive quantization which we described from elementary reasoning coincides with the classical signal compression scheme known as DPCM. Note that the DPCM receiver generates the reconstructed output signal by applying a difference signal to a linear filter. The difference signal is of smaller variance than the original input; it is also approximately white when the predictor memory is large enough for the predictor performance to be fairly close to that of an infinite memory filter. This source/filter structure, where the source is roughly white and the filter recreates the spectral shape of the original signal, is a recurring paradigm in the field of speech coding. In fact, it is the basis of a simplified but useful model of the speech production mechanism in humans. The synthesis filter can be viewed as a model of the human vocal tract and the difference signal as a model of the acoustic excitation signal produced at the glottis.

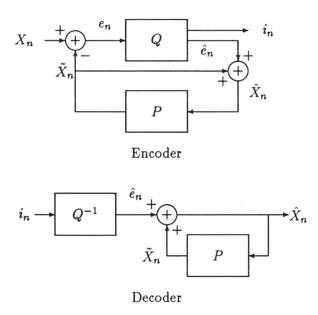

Figure 7.4: DPCM Encoder/Decoder

In examining this structure, note that the operation P always has one unit of delay so that the current output depends on prior inputs and not on the current input. This is necessary for any discrete-time signal flow around a closed-loop.

Note that the operation $B = (1 - P)^{-1}$ must be performed in the receiver. If we use the optimal predictor of memory size m for the operation P, this is simply the inverse of the optimal prediction error filter which is indeed stable since $A(z)$ is minimum delay.

The basic rule of difference quantization still applies to the structure of Figure 7.2. Now we give it the fancier name:

Theorem 7.3.1 Fundamental Theorem of Predictive Quantization:
The overall (input-output) mean squared reproduction error in predictive encoding is equal to the mean squared error in quantizing the difference signal presented to the quantizer.

The difference signal e_n presented to the quantizer can be called a *closed-loop prediction error signal.* This is in contrast to the *open-loop prediction error* that would be produced by applying X_n directly to the prediction error filter $A(z) = 1 - P(z)$. The mean squared prediction error, σ_e^2, will generally be somewhat larger than in the open-loop case for the same memory size.

We define the *closed-loop prediction gain* ratio as the ratio

$$G_{\text{clp}} = \frac{\sigma_X^2}{\sigma_e^2}, \qquad (7.3.3)$$

where σ_X^2 is the variance of X_n. This quantity is a measure of the effectiveness of the predictor in the closed-loop environment of DPCM. If $\hat{X}_n$ is indeed very close to X_n, it is reasonable to expect that the closed-loop prediction gain will be close to the *open loop prediction gain* which is the usual prediction gain as defined in Chapter 4 for the process X_n and the same memory size m. However, some caution is needed here. If two different input signals to a filter (in our case the predictor $P(z)$) are close to one another, one cannot in general conclude that the outputs will be close to one another. Furthermore, if the quantizer has a low-resolution, the reconstructed output cannot be very close to $\hat{X}_n$ and so the optimal closed-loop predictor can be quite different from the optimal open-loop predictor.

The *coding gain* of the quantizer is defined as the ratio

$$G_{\text{Q}} = \frac{\sigma_e^2}{E(e_n - \hat{e}_n)^2}, \qquad (7.3.4)$$

which is simply the signal-to-noise power ratio achieved by the quantizer in coding the difference signal.

With these definitions we can see that the overall signal-to-noise power ratio of the predictive quantization system can be expressed as

$$\frac{\sigma_X^2}{E(X_n - \hat{X}_n)^2} = \frac{\sigma_X^2}{E(e_n - \hat{e}_n)^2} = \frac{\sigma_X^2}{\sigma_e^2}\frac{\sigma_e^2}{E(e_n - \hat{e}_n)^2} = G_{\text{clp}}G_{\text{Q}}, \quad (7.3.5)$$

where the first equality follows from the fundamental theorem of predictive quantization. This result shows that the overall system performance is the product of the closed-loop prediction gain and the coding gain of the quantizer. For a fixed number of bits, the coding gain of the quantizer designed to match its input signal will be roughly the same with or without the predictor loop. This assumes that an appropriate loading factor is chosen for each case and that the pdf's of X_n and of e_n are fairly similar in shape. Consequently, the advantage of predictive quantization is to increase the signal-to-noise power ratio by the prediction gain G_{clp}. Expressed in dB units, the result becomes

$$\text{SNR}_{\text{sys}} = 10 \log G_{\text{clp}} + \text{SNR}_{\text{Q}} \qquad (7.3.6)$$

where SNR_{sys} is the overall system SNR. This is the most important rule used in the study of DPCM systems. Although it is an *exact* result, it can only be evaluated approximately. First of all, the closed-loop prediction gain cannot be determined exactly so that we must use the open-loop prediction gain to estimate it. Second, the SNR of the quantizer is dependent on the input pdf, and subtracting $\tilde{X}_n$ from X_n produces a new signal e_n whose statistics are different from those of X_n. Even if we know the SNR that a quantizer would achieve when operating on X_n directly, we cannot guarantee that the same quantizer (after scaling the range to keep the loading factor unchanged) will produce the same SNR when operating on the difference signal e_n.

If we assume that the closed-loop prediction gain is approximately equal to the open-loop prediction gain for the process X_n and that the coding gain of the quantizer is primarily determined by the number of bits of quantization, then we can roughly compare the performance of predictive quantization of X_n with direct quantization of X_n without prediction (or with a null predictor, $P = 0$). The result above shows that predictive quantization increases the SNR performance by (approximately) the prediction gain in dB. This result assumes that the quantizer resolution is high enough to justify equating the open- and closed-loop prediction gains. The rigorous analysis of predictive quantization is very difficult and it is not amenable to explicit closed form solutions. Nevertheless, the basic approximate result is a useful engineering approximation and is commonly used in practice.

One of the interesting benefits of DPCM in comparison with direct sample by sample quantization or PCM is the greatly reduced sensitivity of DPCM to channel errors. In the DPCM receiver, the received bit stream is decoded to generate the quantized difference values. Then, the quantized difference samples are filtered through the inverse of the prediction error filter. As long as the inverse filter is stable (as it must be for an optimal predictor design), the effect of a transmission error is to cause an error in one quantized difference sample. Since the difference samples have a small variance compared to the reconstructed speech sample, an error in one sample after being filtered through the inverse filter contributes a very small error in many of the output samples and has a very small effect on the time-averaged distortion. Thus, in DPCM the effect of a transmission error is "smeared out" in time. In contrast, a bit error in PCM can cause a very large amplitude error in a single sample of the reconstructed signal. In speech signals, the perceptual effect of an isolated bit error is relatively negligible in DPCM whereas it often produces an audible click in the case of PCM.

Performance Bound for Predictive Quantization

As the memory size of the predictor increases, we can expect the closed-loop prediction gain to improve, approaching a limiting performance as the memory approaches infinity. We now show that the closed-loop prediction gain for any memory size will not exceed the optimal, infinite memory, open loop prediction gain G_p^∞, given by (3.4.6). More precisely:

Theorem 7.3.2 *For a zero mean Gaussian input process to a predictive quantization system with a given quantizer, the closed-loop prediction gain for any predictor is bounded by the infinite-memory open-loop prediction gain G_p^∞, that is*

$$G_{clp} \leq G_p^\infty. \tag{7.3.7}$$

Proof: The closed loop prediction $\tilde{X}_n$ is a function of the previous samples $\hat{X}_{n-1}, \hat{X}_{n-2}, \cdots, \hat{X}_{n-m}$, where m may be finite or infinite. However, $\hat{X}_{n-j}$ is a function of

$$X_{n-j}, X_{n-j-1}, X_{n-j-2}, \cdots.$$

Therefore, $\tilde{X}_n = g(X_{n-j}, X_{n-j-1}, X_{n-j-2}, \cdots)$; that is $\tilde{X}_n$ is some function g of the past input samples. Now, since the process is zero mean Gaussian, the predictor with the highest prediction gain (or lowest mean squared prediction error) is a *linear* combination of the past inputs. Hence the highest prediction gain achievable with any linear or nonlinear function of the past is G_{P}^{∞}. □

Although the Gaussian assumption is needed for a rigorous proof, it is reasonable to expect that the result will remain valid for non-Gaussian inputs. In practice, for moderately high resolution quantizers, i.e., for $N > 3$, the open-loop and closed-loop predictions for a given memory size are nearly the same. Hence the bound (7.3.7) can be closely approached in practice for high resolution quantization. If we combine Theorem 7.3.2 with the system SNR formula (7.3.6), we find that

$$\mathrm{SNR}_{\mathrm{sys}} \leq 10 \log G_{\mathrm{P}}^{\infty} + \mathrm{SNR}_{\mathrm{Q}}. \tag{7.3.8}$$

This, however, is not a rigorous result since any change in the predictor implies that the statistics of e_n will change and therefore the quantizer SNR may be altered as a result. The right hand side is nevertheless usually considered to be a valid and reasonably accurate upper bound on the attainable performance of predictive quantization (DPCM). See for example [192].

Pole-Zero Prediction

A useful improvement in DPCM is the use of pole-zero prediction. Figure 7.5 shows the predictive quantization structure modified by the use of two predictors P, and P^*. Each takes a linear combination of past values from its input. The new predictor, P^*, operates directly on the quantized difference samples, while P as before generates its prediction from the preceding reconstructed output values. By adding the outputs of the two filters an overall prediction (or estimate) $\hat{X}_n$ of X_n is obtained. Note that the corresponding decoder structure, also shown in Figure 7.5 has a pole-zero filter, where P^* contributes zeros and P poles to the filter.

Adaptive Techniques in DPCM

Several important enhancements to DPCM have been developed to handle the nonstationary character of speech signals. In particular, there are two main types of adaptation that are routinely used in Adaptive DPCM (ADPCM). These are *gain-adaptive quantization* (usually called more simply *adaptive quantization* and *adaptive prediction*). These methods help to circumvent performance degradations that would result when the coding system is not matched to the short-term statistical character of the input signal. In effect, they maintain near-optimal performance at each local segment when a non-adaptive DPCM coder would do poorly because it is

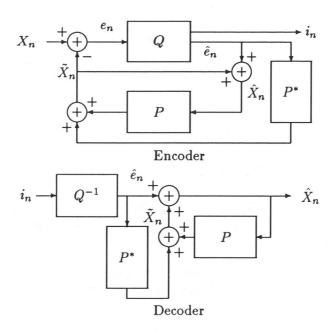

Figure 7.5: DPCM with Pole-Zero Prediction

optimized for the long term statistics of the signal rather than for the local statistical character.

Gain-adaptation compensates for slowly varying input signal power by dynamically scaling the quantizer output and decision levels according to signal level estimates generated from the previously quantized data. The effect is to approximately maintain a constant and near-optimal loading factor over time. (In Chapter 16 gain adaptation techniques are presented for the more general case of vector quantization.)

Adaptive prediction compensates for slow variations in the autocorrelation function of the input process or, equivalently, for the time variations in the input spectral density. In adaptive prediction, the predictor coefficients are dynamically adjusted with small increments at each discrete time instant based on observations of the past reconstructed signal. The adaptation algorithm that is generally used is conceptually based on the well-known LMS algorithm for adaptive filtering. (See for example [177].) An extensive and comprehensive treatment of adaptive prediction is given in [146].

The most widely known version of ADPCM as an international standard for speech coding at 32 kb/s that has been adopted by the CCITT and is known as Recommendation G.721. In this algorithm, both the quantizer amplitude scale and the coefficients of a pole-zero predictor are adapted. The robustness to transmission errors of DPCM is retained in this and other adaptive versions. For a detailed discussion of adaptive techniques in DPCM as well as a discussion of the effects of transmission errors, see [195].

7.4   Delta Modulation

Delta Modulation (DM or ΔM) is a special and extreme case of predictive quantization that arises if we constrain the quantizer to have only two levels and a simple first order predictor ($m = 1$) and if we assume that the original analog signal is sampled at a rate f_s that is much higher than the minimum Nyquist rate. These conditions fit together because

(a) oversampling offers the possibility of compensating for the fact that the quantizer resolution is inadequate since sampling much faster than the Nyquist rate yields a slowly varying discrete time signal, and

(b) the simple predictor is adequate since successive samples are highly correlated.

In fact, delta modulation preceded DPCM historically and is perhaps the simplest of all possible ways to do analog to digital conversion. For an extensive presentation of delta modulation and its many variations see [300].

The basic structure of delta modulation is shown in Figure 7.6. To emphasize the fact that oversampling is used, we show the original analog signal $x(t)$, the low pass antialias filter with cutoff W Hertz, and the sampler operating with an oversampling factor $\rho_s >> 1$. The sampling period is $T = 1/f_s$ and the samples so produced are denoted by $X_n = x(nT)$. We assume a symmetric pdf for the difference signal e_n so that a symmetric one-bit quantizer Q is reasonable. The optimal predictor coefficient $-a_1 = \alpha$ will be very close to unity since the unit lag autocorrelation coefficient r_1 of X_n approaches unity as the oversampling factor becomes large. A digital lowpass filter $H(z)$ with cutoff W is included in the recovery process of an "improved" reproduction sequence $\check{X}_n$ and its use will be explained later. This structure is only a model of delta modulation and does not correspond to the actual, largely analog, circuit that is usually used for its implementation.

By directly specializing the DPCM results to DM, or directly by examining the structure of Figure 7.6, we can obtain an explicit recursive formula for the input output relationship of a delta modulator (ignoring for now

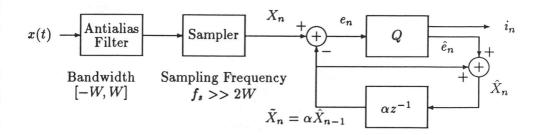

Encoder

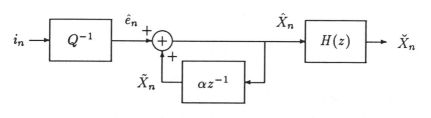

Decoder

Figure 7.6: Delta Modulation

the use of the filter $H(z)$). The reproduction sequence $\hat{X}_n$ is given by

$$\hat{X}_n = \alpha\hat{X}_{n-1} + \hat{e}_n \qquad (7.4.1)$$

and from the quantization operation we have

$$\hat{e}_n = \Delta\operatorname{sgn}(X_n - \alpha\hat{X}_{n-1}) \qquad (7.4.2)$$

where $\operatorname{sgn}(x)$ is the signum function (equal to $+1$ if its argument is nonnnegative and -1 if its argument is negative). Combining these two equations, we obtain the explicit recursion:

$$\hat{X}_n = \alpha\hat{X}_{n-1} + \Delta\operatorname{sgn}(X_n - \alpha\hat{X}_{n-1}). \qquad (7.4.3)$$

Equation (7.4.3) in principle fully characterizes the operation of the delta modulator. The current output $\hat{X}_n$ is formed by taking the previous output $\hat{X}_{n-1}$ scaled by α and incrementing it by either $+\Delta$ or $-\Delta$ according to whether or not the actual input sample X_n is greater or less than the predicted value $\alpha\tilde{X}_{n-1}$ based on the past.

The bit-rate in DM is simply equal to the sampling rate since one bit is transmitted at each sampling instant. Thus, increasing the sampling rate increases the bit rate, but as you might expect, the SNR performance also increases with the sampling rate.

Leaky Integration

If the predictor coefficient is $\alpha < 1$, then DM operates in the mode of *leaky integration*. An integrator based on a real operational amplifier virtually always has some leakage, although the amount of leakage can be quite small. The recursion (7.4.3) shows that there is a "forgetting" behavior where the output at the present time makes less and less use of older past values of the reproduced process $\hat{X}_n$. In fact repeatedly applying the recursion gives

$$\hat{X}_n = \Delta \sum_{i=0}^{\infty} \alpha^i s_{n-i} \qquad (7.4.4)$$

where $s_n = \text{sgn}(X_n - \alpha \hat{X}_{n-1})$.

Ideal Integration

We noted that the optimal predictor coefficient $-a_1 = \alpha$ will be very close to unity. If we set $\alpha = 1$, the implementation is further simplified. In the case where $\alpha = 1$, called *ideal integration*, the scheme is simply accumulating a sum of all past quantizer outputs (adding or subtracting at each sampling instant the fixed step size Δ). Equation (7.4.4) simplifies to

$$\hat{X}_n = \Delta \sum_{i=0}^{\infty} s_{n-i}. \qquad (7.4.5)$$

Thus, the overall feedback path from quantizer output from quantizer input is simply an accumulator. However, caution here is needed because the inverse prediction error filter $(1 - z^{-1})^{-1}$ is not stable. For example, a dc input to the filter will produce an output that grows without bound.

SNR Performance of Delta Modulation

In order to determine the SNR of delta modulation, we need to define the concept of slope overload. The *slope overload* of a delta modulator is the overloading of the one bit quantizer by difference values e_n with excessively large magnitudes.

Figure 7.7 illustrates the original and reproduction waveforms in delta modulation where a sample and hold operation is assumed after the output samples $\hat{X}_n$ are obtained. When the slope of the input waveform exceeds the maximum, Δf_s, that can be tracked, overloading of the quantizer occurs. During low slope regions when the input is varying slowly, *granular noise* occurs due to the alternating step increments that must occur. Clearly, a small value of Δ is preferable for low slopes and a large value for large

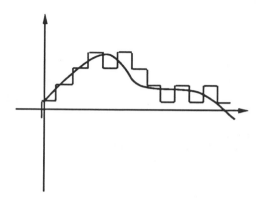

Figure 7.7: Delta Modulation: Original and Reproduction Signals

slope segments. A measure of the degree of compromise between these two conflicting objectives is given by the *slope loading factor*

$$\gamma = \frac{\Delta}{\sigma_e}. \tag{7.4.6}$$

The SNR of delta modulation will be a function of γ for a given input process and oversampling factor. The optimal choice of γ is a complicated function of the statistics of the input process and there is no exact solution. Typical values that are likely to be reasonable are in the neighborhood of unity. What is clear is that the SNR will decrease as γ becomes either too large or too small.

Note that the fundamental theorem of predictive quantization is applicable in the case of DM. In this case, it states that the overall mean squared error between input and output is equal to the mean squared error in quantizing the difference signal e_n with a one bit quantizer with output levels $+\Delta, -\Delta$. To find the overall average distortion, we therefore need to find the average distortion for the binary quantizer. In spite of the deceptively simple form of the delta modulator, this is in fact a difficult task. To derive a formula for the mean squared error of delta modulation, the assumption is often made that the quantization error is a white noise process. This is in fact an extremely common assumption in the literature. Unfortunately, however, the conditions required for the assumption to be valid do not hold. There is not an asymptotically large number of levels (only two), the levels are not close to each other, and the joint probability distribution of the quantizer input is not smooth because of the digital term fed back. Perhaps surprisingly, it can be shown that even though the conditions for the white noise approximation discussed in Chapter 5 are violated (in the extreme), at least the average distortion approximation remains valid when the over-sampling ratio is high and there is no overload. The quantization noise is not, however, at all white: it has a discrete and nonuniform spectrum that is strongly dependent on the specific input. The interested reader is referred to the classical work by Iwersen [191] or the treatment of uniform quantizer noise in PCM and oversampled feedback quantizers in [158] and Chapter 6 of [159].

In spite of the incorrectness of the white noise assumption, it turns out that a useful and reasonable approximation for the mean squared quantization error can found. Specifically,

$$E[(e_n - \hat{e}_n)^2] \approx \frac{\Delta^2}{3} \tag{7.4.7}$$

An intuitive argument justifying the approximate validity of (7.4.7) can be made as follows. Assume ideal integration and suppose for simplicity that $X_0 = 0$. Then, from (7.4.3), the possible values that $\hat{X}_n$ can take on are $\pm\Delta$ for $n = 1$; $0, \pm 2\Delta$ for $n = 2$; $\pm\Delta, \pm 3\Delta$ for $n = 3$; and so on. As n gets large, the values taken on approach the lattice with spacing 2Δ between successive points. Also, note that at any instant, the output value will always lie at one of two levels spaced 2Δ apart with the input X_n lying between these two values. Thus, we see that, neglecting overload, the input samples "see" a *virtual* infinite uniform quantizer with step size 2Δ for which case the mean squared error is given by $(2\Delta)^2/12 = \Delta^2/3$ using the basic result for a uniform quantizer as given in Chapter 6.

The variance of the input e_n to the quantizer is given by

$$\sigma_e^2 = E(e_n^2) \approx E[(X_n - X_{n-1})^2]$$

where we have used $\tilde{X}_{n-1} \approx X_{n-1}$ and $\alpha \approx 1$. This variance can be evaluated approximately by essentially the same analysis used in Section 3.5 on sampling jitter. Thus,

$$\sigma_e^2 = E[[x(nT) - x(nT - T)]^2].$$

Let $R_X(\tau)$ denote the autocorrelation function and $S_X(f)$ denote the spectra density of the continuous time input signal $x(t)$ to the delta modulation system. Then

$$\begin{aligned}
\sigma_e^2 &= 2[R_X(0) - R_X(T)] \\
&= 2\int_{-\infty}^{\infty} S_X(f)[1 - \cos(2\pi fT)]\, df
\end{aligned}$$

$$\approx 2 \int_{-\infty}^{\infty} S_X(f) \frac{1}{2}(2\pi fT)^2 \, df$$

$$= (2\pi T)^2 \int_{-\infty}^{\infty} f^2 S_X(f) \, df,$$

where the autocorrelation values $R_X(0)$ and $R_X(T)$ are each expressed as inverse Fourier transforms of the spectral density, the exponential term $e^{2\pi fT}$ is replaced by its real part, $\cos(2\pi fT)$, since $S_X(f)$ is even and the imaginary part of the integral is zero. We have replaced the cosine term by a quadratic approximation to its power series representation for the frequency band containing the input signal. Mathematically, the derivation above is very similar to the analysis of sampling jitter treated in Chapter 5.

Now, the last integral can be recognized as the product of the squared rms bandwidth f_{rms}^2 of $X(t)$ and the variance σ_X^2 of $X(t)$. Hence, we obtain the convenient relation,

$$\sigma_e^2 = \sigma_X^2 \frac{(2\pi)^2 f_{\text{rms}}^2}{f_s^2}.$$

Let $\eta = f_c/f_{\text{rms}}$, the *spectral shape factor* of the input signal $x(t)$, measuring the ratio of cutoff frequency f_c to the rms bandwidth f_{rms}. This factor depends on the particular spectral density function of the input. Typically, the value of η is in range of 2–4. Setting $f_s = 2f_c\rho_s$, we find that the mean squared prediction error σ_e^2

$$\sigma_e^2 = (2\pi)^2 \sigma_X^2 \frac{f_c^2}{\eta^2 f_s^2} = (2\pi)^2 \frac{\sigma_X^2}{4\eta^2 \rho_s^2}$$

where $\rho_s = f_s/2f_c$ is the oversampling factor and σ_X^2 is the variance of the input.

From (7.4.6), for a given choice of slope loading factor γ, the square of the step size Δ can be expressed as

$$\Delta^2 = \frac{(2\pi)^2 \gamma^2 \sigma_X^2}{4\eta^2 \rho_s^2}.$$

Now, we obtain an expression for the system signal-to-noise ratio from the above results:

$$\text{SNR} = \frac{\sigma_X^2}{E(X_n - \hat{X}_n)^2}$$

$$= \frac{\sigma_X^2}{E(e_n - \hat{e}_n)^2}$$

$$= \frac{\sigma_X^2}{\Delta^2/3} = \kappa \rho_s^2, \tag{7.4.8}$$

where the second equality follows from the fundamental theorem of predictive quantization, the third equality comes from (7.4.7), and the constant κ is given by

$$\kappa = \frac{3\eta^2}{\pi^2\gamma^2}.$$

This is the final result for "unsmoothed" delta modulation.

To easily and substantially improve performance, the reproduced high rate signal $\hat{X}_n$ can be digitally lowpass filtered to remove out of band quantization noise (implicitly assuming that this quantization noise is uncorrelated with the desired signal). For the purpose of estimating the performance enhancement from this "smoothing" operation, the quantization noise can be roughly approximated as white over the bandwidth from 0 to $f_s/2$ and the signal is bandlimited with cutoff $f_c = f_s/2\rho_s$, the effect of an ideal lowpass filter with cutoff f_c is to reduce the white noise power by a fraction $1/\rho_s$. Consequently, the signal-to-noise power ratio after smoothing, denoted $\mathrm{SNR}_{\mathrm{dm}}$, is given by

$$\mathrm{SNR}_{\mathrm{dm}} = \frac{\sigma_X^2}{E(X_n - \hat{X}_n)^2} = \kappa\rho_s^3. \tag{7.4.9}$$

Thus, a factor of 2 increase in sampling rate increases the SNR in dB by $10\log_{10}(2^3) \approx 9$, so that the SNR grows at the rate of 9 dB per octave increase in the sampling rate.

As with DPCM, several adaptation techniques have been developed which have been found to be of great practical value to improve performance of delta modulation. Specifically, most of these techniques are centered on methods for varying the step size Δ as a function of previously transmitted bits. The intent is to maintain a constant slope loading factor over time in spite of the changes in the short-term energy of the slope.

While these methods do not significantly affect the SNR achievable, they allow the coder to track a wide range of signal power levels (and consequently a wide range of signal slopes). This is particularly important for speech or audio signals which have a wide dynamic range. See [195] for a description of such schemes.

Delta modulation with leaky integration and its adaptive variations also have the important feature of being highly robust to the effect of transmission errors. As with DPCM, this is explained by the fact that the effect of a particular bit error is "smeared" over time.

7.5 Problems

7.1. A discrete-time stationary process, $\{X_k\}$, has zero mean and correlation values $r_0 = 4$, $r_1 = 2$, $r_2 = 1$, and $r_k = 0$ for $k > 2$. A first order predictor is used in a DPCM coder for the input process X_k. The quantizer design assumes approximately uniform distribution of prediction error. The loading factor is $\gamma = 4$ and the number of levels is 128. What is the approximate overall SNR of the DPCM system assuming negligible overloading? Explain!

7.2. A third order predictor is used in a DPCM coder. The predictor was designed to be optimal for the input process X_k that is to be coded by the DPCM coder. The (open-loop) prediction gain is 75/16. The quantizer design assumes approximately uniform distribution of the prediction error and 64 output levels are used. An overload factor of 3 is chosen for which the overload noise is negligible. What is the approximate overall SNR of the DPCM system assuming negligible overloading of the quantizer? Explain! What is the bit rate in bits/sample? If the predictor is replaced with an optimal 6th order predictor the prediction gain increases by 2 dB. What is the SNR improvement achieved by simultaneously switching to an optimal 6th order predictor and a 256 level quantizer? Explain your reasoning.

7.3. A zero mean unit variance 4 kHz bandwidth input process is coded at 32 kbits/sec with a delta modulator with step size $\Delta = 0.01$. Estimate the SNR achieved. What sampling rate is needed to increase the SNR by 5.3 dB?

7.4. Describe or specify the set of values taken on by the output, $\hat{X}_n$ for $n \geq 0$, of a delta modulator with leaky integration and leak factor α for the cases (a) $\alpha < 1/2$, (b) $\alpha = 1/2$, and (c) $\alpha > 1/2$. Assume that $\hat{X}_0 = 0$.

7.5. Suppose a quantizer Q is optimal for a random variable X. Describe the optimal quantizer for the random variable $X - E(X)$.

7.6. If you assume that the quantization noise in a delta modulator ($\alpha = 1$) is indeed independent of the quantizer input and white (in spite of all the warnings given), what is the power spectral density of the overall error signal $X_n - \hat{X}_n$?

If the output $\hat{X}_n$ is then passed through an ideal low pass filter $H(z)$ with cutoff frequency W to form a signal $\check{X}_n$, what is the overall average error power $E[(X_n - \check{X}_n)^2]$?

7.7. Suppose that before Delta modulation encoding the input signal is passed through a discrete time integrator as shown in Figure 7.8 and that the decoder is replaced by the low pass filter $H(z)$ of the previous problem.

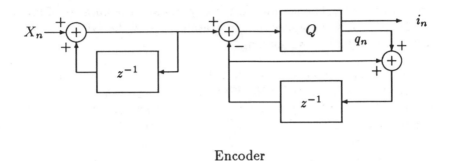

Encoder

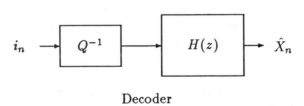

Decoder

Figure 7.8: Modified Delta Modulator

This modified architecture is called a *Sigma Delta modulator* (or *Delta Sigma modulator*) [88] [188] [46] [47] [48] [50] [49] [154] [156] [163]. Show that the encoder can be redrawn as in Figure 7.9. (This is actually a slight variation on the usual model which puts the delay in the feedforward path of the integrator instead of the feedback.)

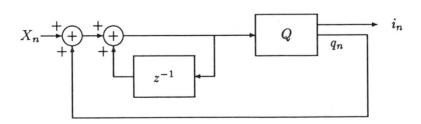

Figure 7.9: Sigma Delta Modulator

As in the previous problem, assume that the quantization noise is white and signal independent and derive an expression for the power spectral density of the overal error signal $X_n - \hat{X}_n$ and of the average squared error $E[(X_n - \hat{X}_n)^2]$ as a function of the low pass cutoff frequency W. Compare your answer with that of the previous problem. Which architecture appears to give the better performance? How accurate do you think the quantizer noise approximations are in this case?

7.8. Consider the difference quantizer setup of Figure 7.1. Suppose that the quantizer is uniform with M levels for M even. Let Δ be the quantizer cell width, the length between the reproduction levels. Suppose that the input process is stationary with a marginal pdf $p_X(x)$ and that U_n is an iid random process that is completely independent of the input process $\{X_n\}$. Let $M_U(ju)$ denote the characteristic function of U_n. Assume that the pdf's of X_n and U_n are such that the quantizer input $e_n = X_n + U_n$ does not overload the quantizer. Find an expression for the joint characteristic function of the quantizer error $\epsilon = q(e_n) - e_n$ and the input signal X_n:

$$M_{\epsilon,X}(ju, jv) = E(e^{ju\epsilon_n + jvX_n}).$$

Show that if the process U_n is such that

$$M_U(j\frac{\pi l}{\Delta}) = 0 \text{ for } l \neq 0,$$

then

$$M_{\epsilon,X}(ju, jv) = M_\epsilon(ju)M_X(jv)$$

and hence the input and the quantizer error are independent. (This is an alternative treatment of *dithering* a quantizer to that presented in the exercises in Chapter 5.)

Chapter 8

Bit Allocation and Transform Coding

8.1 Introduction

Often a signal coding system contains several different quantizers, each of which has the task of encoding a different parameter that is needed to characterize the signal. Each such parameter may have a different physical meaning and may require a different relative accuracy of reproduction to achieve a desired overall quality of the reconstructed signal. The number of bits available to collectively describe this set of parameters is inevitably limited. Consequently, a major concern of the designer of a coding system is *bit allocation*, the task of distributing a given quota of bits to the various quantizers to optimize the overall coder performance. In a broader sense, this is an interdisciplinary problem which arises in everyday life: how do we efficiently distribute a limited or scarce resource to achieve the maximum benefit?

In traditional waveform coding methods, each sample is quantized with the same number of bits since each sample has the same degree of *a priori* importance of its effect on the overall signal reproduction. On the other hand, in some coding systems distinct parameters or features of varying importance are individually quantized and it becomes important to intelligently allocate bits among them. An example is *transform coding*, a system in which a waveform is efficiently coded by a two-step process. First blocks of consecutive samples are formed and a linear transformation of each block is taken to obtain a set of parameters, the so-called transform coefficients. These parameters are then individually quantized in a traditional transform coding system. In contrast with direct waveform coding, these coefficients are not of equal significance and there is good reason to allocate bits in an unequal way to quantize these values. It is worth pointing out that transform coding can be considered as a special case of vector quantization where an entire input vector or block is examined as one entity in order to code the input signal. The actual quantization, however, is performed on scalars, the individual transformed variables. Hence transform coding is naturally included in the treatment of scalar quantizers. A general treatment of vector quantization begins in Chapter 10 and in Chapter 12 the use of vector quantization in combination with transform coding is discussed.

In this chapter, we first examine the general problem of allocating bits to a set of quantizers and present some techniques for doing this in an efficient manner. The results will be broadly applicable to a variety of coding systems. We then describe the technique of transform coding, where bit allocation is an essential component. Transform coding is currently one of the most popular approaches to signal compression. It has been used for speech coding and is currently the basis of the most widely used coding methods for audio, image, and video signals.

8.2   The Problem of Bit Allocation

Suppose that we have a set of k random variables, $X_1, X_2, \cdots, X_k$, each with zero mean value and with variances $EX_i^2 = \sigma_i^2$, for $i = 1, 2, \cdots, k$. Assuming the pdf of each X_i is known and a particular distortion measure is selected, we can design an optimal quantizer Q_i for X_i, based on a particular distortion measure, for any choice of N_i quantizer levels, where N_i is a nonnegative integer. It is often desirable that N_i be an integer power of 2 so that the resolution $b_i = \log_2 N_i$ is an integral number of bits. In fact, this is not really necessary in practice. It is sufficient that the product of the number of quantization levels for each quantizer be a power of two. For example, suppose $k = 2$ so that we must send an integral number of bits to specify each possible pair of quantization levels N_1 and N_2. Suppose the two quantizers have $N_1 = 3$ and $N_2 = 10$. Then, for a pair of parameters, we do not need to send 2 bits for the first variable and 4 bits for the second. Instead, we can send only 5 bits, since $N_1 N_2 = 30 < 32$, by coding two parameters jointly. (This is an elementary form of vector quantization of the parameters since we are coding the parameters jointly as pairs instead of individually as scalars.)

We shall henceforth assume the squared error distortion measure is used. This completes the specification of the optimal quantizer Q_i for any particular value of N_i (or b_i). Let $W_i(b)$ denote the mean squared error incurred in optimally quantizing X_i with b bits of resolution. Thus, treating each random variable in isolation, we have, at least in principle, a complete solution for the performance that can be achieved in quantizing that random variable. The function $W_i(b)$ tells us how we can reduce average distortion by increasing the resolution, or in other words, we have a "price-performance" trade-off available. The *price* is the number of bits we allocate and the *performance* is determined by the mean distortion that results. Now we can proceed to precisely formulate the bit allocation problem.

In most cases of a multi-parameter quantization task, the overall performance of the coding system can simply be determined from the sum of the mean distortions. Thus, we can define the *overall distortion*, D, as a function of the bit allocation vector, $\mathbf{b} = (b_1, b_2, \cdots, b_k)$ according to

$$D = D(\mathbf{b}) = \sum_{i=1}^{k} W_i(b_i). \qquad (8.2.1)$$

The bit allocation problem is to determine the optimal values of $b_1, b_2, \cdots, b_k$ subject to a given fixed quota, B, of available bits:

The Bit Allocation Problem: Find b_i for $i = 1, 2, \cdots, k$ to minimize $D(\mathbf{b}) = \sum_{i=1}^{k} W_i(b_i)$ subject to the constraint that $\sum_{i=1}^{k} b_i \leq B$.

At this point there are two basic approaches for finding an optimal or near-optimal bit allocation. The first is to attempt to find an algorithm that *exactly* minimizes (8.2.1) over all choices of b. If we require $\mathbf{b}$ to have integer components, one such algorithm is obvious in theory but usually impractical: exhaustion. There is only a finite number of vectors $\mathbf{b}$ with integer components summing to B or less. One could in theory find $D(\mathbf{b})$ for all of these and choose the $\mathbf{b}$ yielding the minimum. Unfortunately this algorithm has a second serious drawback in addition to its computational absurdity: the optimal distortion functions $W_i(b_i)$ are rarely known exactly. Nonetheless we shall later see that this general approach does have its merits if we at least know the values of $W_i(b)$ for good (if not optimal) quantizers at a variety of rates b. There exist efficient computer programs for finding optimal allocations in this case and one such algorithm is given by a special case of a decision tree pruning algorithm to be encountered in Chapter 17. Another closely related algorithm, given in [294], will be discussed briefly in Chapter 16.

In the second approach to minimizing (8.2.1), which we shall adopt for the remainder of this chapter, we invoke the high resolution quantization approximations in order to write explicit expressions for the optimal distortions $W_i(b_i)$. The resulting expression for $D(\mathbf{b})$ can then be minimized with familiar Lagrange techniques of calculus or common inequalities.

If we assume that the overload distortion is negligible and that the high resolution approximation applies, then from (6.3.2), we have

$$W_i(b_i) \approx h_i \sigma_i^2 2^{-2b_i},$$ (8.2.2)

where the constant h_i is determined by the pdf $f_i(x)$ of the normalized random variable X_i/σ_i, then

$$h_i = \frac{1}{12} \left\{ \int_{-\infty}^{\infty} [f_i(x)]^{1/3} \, dx \right\}^3.$$ (8.2.3)

For example, if the pdf is zero mean Gaussian (as is often assumed to be the case), then the constant is given by

$$\begin{aligned} h_g &\equiv \frac{1}{12} \left\{ \int_{-\infty}^{\infty} \left[\frac{e^{\frac{-x^2}{2}}}{\sqrt{2\pi}} \right]^{1/3} dx \right\}^3 \\ &= \frac{\sqrt{3}\pi}{2}. \end{aligned}$$ (8.2.4)

The use of the high resolution approximation is certainly not correct for small values of b_i such as 1 or 2 bits, but it often turns out to be a reasonable approximation even in the medium to low resolution cases.

For notational simplicity, we first treat the case where each X_i has the same normalized pdf, that is, where the X_i/σ_i are identically distributed. In this case $h_i = h$ is a constant independent of i. The results can be easily modified to allow for random variables with different normalized pdf's and also for the case of a *weighted overall distortion*, D_w, given by

$$D_w = \sum_{i=1}^{k} g_i W_i(b_i)$$ (8.2.5)

for some set of nonnegative weights g_i. This permits the distortion in different parameters to be assigned varying importance levels in the overall optimization. We consider first the simplifying assumption that the allocations b_i are real numbers (without the restriction to be nonnegative integers).

8.3   Optimal Bit Allocation Results

The classical solution (see Huang and Schultheiss [186]) to the bit allocation problem using the high resolution quantization approximations neglects the practical requirement that $b_i \geq 0$, and obtains an "optimal" solution that could allow negative values of b_i. The solution is then modified by an *ad hoc* procedure that eliminates negative bit allocations and adjusts the initial solution to obtain integer values of b_i. For simplicity, we first make the additional assumption that the random variables have identical normalized pdf's and the overall distortion is unweighted.

Identically Distributed Normalized Random Variables

We shall show that the optimal bit assignment using the high rate approximations with no nonnegativity or integer constraints on the bit allocations is given by the following formula when the normalized random variables are identically distributed and the component distortions are not weighted. Let

$$b_i = \overline{b} + \frac{1}{2} \log_2 \frac{\sigma_i^2}{\rho^2} \tag{8.3.1}$$

where

$$\overline{b} = \frac{B}{k}$$

is the average number of bits per parameter, k is the number of parameters, and

$$\rho^2 = (\prod_{i=1}^{k} \sigma_i^2)^{\frac{1}{k}}$$

is the geometric mean of the variances of the random variables.

The minimum overall distortion attained with this solution is given by

$$D = kh\rho^2 2^{-2\overline{b}}$$

and each quantizer incurs the same average distortion, that is, $W_i(b_i) = h\rho^2 2^{-2\overline{b}}$. In other words, the overall distortion is the same as that which would be achieved if each random variable had variance ρ^2 and each quantizer were allocated $\overline{b}$ bits. Note that the optimal bit allocation is independent of the pdf of the normalized random variables.

Equation (8.3.1) shows that the allocation to quantizer i exceeds the average allocation $\overline{b}$ if the variance of X_i is greater than the geometric average ρ^2 of the variances. Conversely, if σ_i^2 is less than ρ^2, the bit allocation is less than the average allocation and, in fact, if σ_i^2 is sufficiently small, the allocation can even be negative. This simple and elegant result provides a very useful guideline for efficient bit allocation and is widely used in the design of signal compression schemes. Although it assumes identical pdf's of the normalized input random variables, this condition can easily be relaxed as will soon be seen. It does have some serious shortcomings that must be circumvented in practice. Specifically, the solution does not require integer values of resolution so that the number of quantization levels, $N_i = 2^{b_i}$, is not necessarily an integer; furthermore, it does not exclude the possibility of negative bit allocations.

In practice, this bit allocation rule is modified by replacing negative resolutions by zero. Also, if an integer valued allocation is needed, then each noninteger allocation b_i is adjusted to the nearest integer. These modifications can lead to a violation of the allocation quota, B, so that some incremental adjustment is needed to achieve an allocation satisfying the quota. The final integer valued selection can be made heuristically. Alternatively, a local optimization of a few candidate allocations that are close to the initial solution obtained from (8.3.1) can be performed by simply computing the overall distortion for each candidate and selecting the minimum.

Proof of the Bit Allocation Solution

The optimal allocation result (8.3.1) can be derived with Lagrangian optimization techniques, but we instead present a simpler proof in the spirit of the proofs of Chapter 4 based on an elementary inequality to prove a known answer. We wish to prove that if $N_i = 2^{b_i}$; $i = 1, 2, \cdots, k$, and

$$\sum_{i=1}^{k} \log N_i = B,$$

then

$$\sum_{i=1}^{k} h\sigma_i^2 2^{-2b_i} \geq kh\rho^2 2^{-2\bar{b}}$$

and that equality is achieved by the b_i of (8.3.1). This is equivalent to the inequality

$$\frac{1}{k} \sum_{i=1}^{k} \sigma_i^2 2^{-2b_i} \geq \rho^2 2^{-2\bar{b}} = \left(\prod_{i=1}^{k} \sigma_i^2\right)^{\frac{1}{k}} 2^{-2\bar{b}}. \qquad (8.3.2)$$

The arithmetic/geometric mean inequality states that for any positive numbers a_i, $i = 1, 2, \cdots, k$,

$$\frac{1}{k} \sum_{i=1}^{k} a_i \geq \left(\prod_{j=1}^{k} a_j\right)^{\frac{1}{k}} \qquad (8.3.3)$$

with equality if the a_i are all equal. Apply this inequality to (8.3.2) with $a_i = \sigma_i^2 2^{-2b_i}$ to obtain

$$\frac{1}{k} \sum_{i=1}^{k} \sigma_i^2 2^{-2b_i} \geq \left(\prod_{i=1}^{k} \sigma_i^2 2^{-2b_i}\right)^{\frac{1}{k}} = \left(\prod_{i=1}^{k} \sigma_i^2\right)^{\frac{1}{k}} 2^{-\frac{2}{k} \sum_{i=1}^{k} b_i}$$

$$= (\prod_{i=1}^{k} \sigma_i^2)^{\frac{1}{k}} 2^{-2\bar{b}},$$

which is (8.3.2). Equality holds if for each i, $\sigma_i^2 2^{-2b_i} = C$ a constant independent of i. From (8.3.2), where the inequality becomes an equality, we see that C is given by

$$C = \rho^2 2^{-2\bar{b}}$$

so that

$$\sigma_i^2 2^{-2b_i} = \rho^2 2^{-2\bar{b}},$$

which implies (8.3.1). □

The result in a nutshell, simply says that *the number of quantization levels N_i of the ith quantizer should be proportional to the standard deviation of the random variable that it quantizes.* To the extent that this rule cannot be exactly achieved with positive integer values for each N_i, it becomes necessary to modify the bit allocation to a "nearby" achievable allocation. The classical treatment of bit allocation does not concern itself with the meaning of "nearby." Any simple heuristic procedure, however, can be used to perform this modification.

Nonidentically Distributed Random Variables

We now consider the more general case where the random variables X_i are nonidentically distributed while retaining the high resolution assumption and the overall distortion measure is weighted as given by (8.2.5). Then the distinctive distortion constants h_i, are determined from the normalized pdf of the corresponding random variables X_i. The optimization problem is changed (very slightly) to the task of minimizing:

$$\sum_{i=1}^{k} h_i \sigma_i^2 2^{-2b_i}$$

subject to the same bit quota B. It is easily seen that this quantity is the same as the previous objective function except that what was before the variance σ_i^2 is now the scaled variance factor $h_i \sigma_i^2$ so that the solution is now as follows:

$$b_i = \bar{b} + \frac{1}{2} \log_2 \frac{\sigma_i^2}{\rho^2} + \frac{1}{2} \log_2 \frac{h_i}{H} \tag{8.3.4}$$

where $\bar{b}$ and ρ are as before and H is the geometric mean of the coefficients h_i. Note that in the case where the distortion constants h_i are constant, independent of i, the allocation rule reduces to (8.3.1).

The minimum overall distortion attained with this solution is given by

$$D = kH\rho^2 2^{-2\bar{b}} \tag{8.3.5}$$

and again, each quantizer incurs the same average distortion.

Weighted Overall Distortion Measure

As an additional generalization, we now consider the case where the bit allocation is to be optimized for a weighted overall distortion measure as given in (8.2.5) where the weights g_i are specified. We continue to allow non-identically distributed normalized pdf's as described above. This is again a very simple modification of the optimization problem and the solution is given by

$$b_i = \bar{b} + \frac{1}{2}\log_2\frac{\sigma_i^2}{\rho^2} + \frac{1}{2}\log_2\frac{h_i}{H} + \frac{1}{2}\log_2\frac{g_i}{G} \tag{8.3.6}$$

where G is the geometric mean of the weight values g_i.

The minimum overall distortion attained with this solution is given by

$$D = kHG\rho^2 2^{-2\bar{b}}. \tag{8.3.7}$$

In this case, the individual quantizers each contribute an average distortion inversely proportional to the corresponding weight value g_i. Thus, a quantizer whose weight value is large compared to the geometric mean G of the weights has been assigned a relatively important role in achieving the overall quality objective. For the optimal allocation, this quantizer will therefore have a relatively small average distortion.

Nonnegativity Constraint

One improvement to the bit allocation problem is to add the constraint that the resolution of each quantizer must be nonnegative (or the number of levels N_i must be at least unity). This requirement would eliminate the occurrence of negative bit allocation without requiring the solution to have integer values of resolution. This approach was taken by Segall [288] who used some basic techniques of functional analysis to obtain a solution to this problem. He also generalized the problem to apply to any convex distortion function $W_i(b)$ that is strictly convex with a continuous first derivative $W_i'(b)$ such that $W_i'(\infty) = 0$. This condition implies that the quantizer distortion decreases smoothly as the number of bits increases and that the rate of decrease of distortion drops monotonically to zero as the number of bits increases without bound.

Assume that each X_i has identical normalized pdf. Let $w(b) = W_i(b)/\sigma_i^2$, which is independent of i, be the mean squared error incurred in optimally quantizing X_i/σ_i^2. Let the inverse of $w(b)$ be $J(w)$. Since the distortion function is monotonically decreasing with increasing resolution b, the inverse exists and is well defined.

We state, without proof, Segall's solution to the bit allocation problem:

Segall's Solution for Nonnegative Bit Allocations

The optimal allocation is given by

$$b_i = \begin{cases} b_i^* = J\left(\frac{\theta^*}{\sigma_i^2} w'(0)\right), & \text{if } 0 < \theta^* < \sigma_i^2 \\ 0, & \text{if } \theta^* \geq \sigma_i^2 \end{cases}$$

where θ^* is the unique root of the equation

$$S(\theta) = \sum_{i: \sigma_i^2 \geq \theta^*} J\left(\frac{\theta}{\sigma_i^2} w'(0)\right) = B,$$

where B is the total quota of bits.

8.4 Integer Constrained Allocation Techniques

While Segall's solution takes a step in the right direction, it still does not address the concern that any realizable solution must have an integer number of levels, and often it is desirable to implement a solution with an integer number of bits for each quantizer. In the economics literature there have been several treatments of a resource allocation problem that is closely analogous, and in some ways mathematically equivalent, to the bit allocation problem, including the constraint that the number of bits must be a nonnegative integer. One solution is due to Fox [127]. In this approach, the distortion functions are allowed to be distinct and each is assumed to be convex and strictly monotone decreasing with the resolution.

In the following, we present a simple and intuitively appealing algorithm for finding a bit allocation for nonnegative integer resolutions. The algorithm is not optimal, but it allocates bits incrementally, one bit at a time, in a way that yields good assignments. The basic idea is that in each of B iterations, one bit is allocated where it will do the most good at this point. In other words, at each step the overall distortion is minimized *given the current partial allocation*. It is *greedy* in the sense that it is short-sighted in preferring to get the maximum immediate reduction in distortion without regard to the global effect of its choice.

Greedy Integer-Constrained Allocation Algorithm:

Let $W_i(b)$, called the *distortion function*, denote the average distortion of the ith quantizer for a bit allocation of b bits. Let $b_i(m)$ denote the total number of bits allocated to the ith quantizer after iteration m, i.e., after m bits have been distributed among the quantizers. Now we define the *demand* $s_i(m)$ associated with the ith quantizer after the mth iteration of the allocation algorithm according to:

$$s_i(m) = W_i(b_i(m)),$$

that is, the demand $s_i(m)$ after the mth bit has been assigned is simply the average distortion of the ith quantizer for its current allocation of bits. The general greedy algorithm proceeds as in Table 8.1.

Table 8.1: **Greedy Bit Allocation Algorithm**

Step 0. Initialize the bit allocation to zero, so that $b_i(0) = 0$ for each i and $m = 0$. Set $s_i(0) = W_i(0)$ as the initial values of demand.

Step 1. Find the index j with the maximum demand.

Step 2. Set $b_j(m+1) = b_j(m) + 1$, set $b_i(m+1) = b_i(m)$ for each $i \neq j$, and set $s_i(m+1) = W_i(b_i(m+1))$.

Step 3. If $m < B - 1$, increment m by 1 and go to Step 1. Otherwise stop.

This algorithm carries out a very simple and intuitive idea. Simply give away bits to the most needy quantizer, one bit at a time until you run out of bits to give. Measure the degree of neediness of each quantizer by the average distortion it will yield if it were to operate with its current bit assignment. Note that no assumptions have been made on the characteristics of the quantizers.

In the special case where the high resolution assumption holds, the idealized distortion function (8.2.2), yields a demand that is proportional the variance of the ith quantizer times the factor 2^{-2b_i}. Then, the greedy algorithm simplifies as follows: Use the standard deviations as the initial demands and each time a bit is assigned to a quantizer reduce its demand by a factor of 2. This scheme was proposed by Ramstad [262]. (A variation of this scheme that is sometimes used is to reduce the demand by a factor θ whenever a bit is allocated, where θ is empirically chosen in the range of 1.4 to 2.3.) The true distortion function does not follow the formula (8.2.2) very closely when the resolution is very low, but this simple method works fairly well in practice. Also it satisfies the constraint of a nonnegative integer bit allocation.

The more general version of the algorithm presented in Table 8.1 requires explicit knowledge of the distortion function for each quantizer. This is easily obtained as part of the quantizer design process, where a separate quantizer is designed for each bit allocation. The distortion (averaged over the training set) for each bit allocation over the range of interest is stored in a table. The table then represents the needed distortion functions $W_i(j)$ for each i and j of interest.

Two alternative and closely related algorithms that lead to optimal or virtually optimal allocations for integer constraint and unrestricted distortion functions will be described in Chapters 16 and 17 in the context of vector quantizers. The above greedy algorithm is also applicable to vector as well as scalar quantizers.

8.5 Transform Coding

Suppose that we have a block of consecutive samples of a stationary random process, and we wish to efficiently code this block with a specified number of bits. Let $\mathbf{X}$ denote the sample vector

$$\mathbf{X} = (X_1, X_2, \cdots, X_k)^t.$$

These samples will typically have substantial correlation and separate quantization of each would be an inefficient way to encode them when the quota of bits is relatively modest. Note also that due to stationarity these samples have the same marginal pdf and of course the same variance.

The idea of transform coding is that by performing a suitable linear transformation on the input vector, $\mathbf{X}$, we can obtain a new vector, $\mathbf{Y}$, also with k components, often called *transform coefficients* or simply, *coefficients*, with the feature that these coefficients are much less correlated than the original samples. In addition, the information may be much more "compact" in the sense of being concentrated in only a few of the transform coefficients. Having in this sense removed redundancy, we hope as a result to be able to quantize these components more efficiently. We emphasize that there is no general "theorem" that states that uncorrelated quantities can be more efficiently quantized than can correlated variables. Even if the variables were independent (e.g., if they are uncorrelated and Gaussian), a basic result of Shannon's rate-distortion theory [290][134][32][159] is that one can always code vectors more efficiently than scalars. As a matter of practice and experience, however, the more uncorrelated or independent a set of variables, the more efficient scalar coding becomes. In other words, as the elements of a vector become more independent, there is less to be gained by the more complicated vector quantization algorithms to be considered later rather than the simple scalar quantization algorithms. Furthermore, when the resolution is high, the difference in achievable compression with and without memory can be quantified by the Bennett approximations [221]. In spite of this theoretical shortcoming, transform coding has long proved to be a simple and effective way of obtaining good compression. The idea was introduced by Kramer and Mathews in 1956 [209], but the first detailed treatment including a high resolution analysis of performance and bit rate allocation was done by Huang [185] and Huang and Schultheiss [186].

The reconstructed approximation to the original vector, $\hat{\mathbf{Y}}$, is then obtained by performing the corresponding inverse transform on the quantized transformed vector $\hat{\mathbf{X}}$. Figure 8.1 illustrates the structure of transform coding where $\mathbf{T}$ is a $k \times k$ invertible matrix that performs the linear transformation. Note that a distinct quantizer is needed for each transform coefficient since the transform coefficients may in general have different pdf's. This implies that a different bit allocation may be needed for each transform coefficient and a bit allocation strategy will be required for assigning a given quota of B bits for the entire vector $\mathbf{X}$.

We have argued that intuition suggests that a transform which decorrelates the input samples will result in a more efficient coding. There are two other reasons which also support the general approach, one intuitive and one subjective. A transform tends to mix the information of the input samples together, to interleave the information so that each of the transformed samples contains information about the other samples. One can view this as analogous to error control (or channel) coding where one mashes up the information among input symbols so that the decoder can use the increased redundancy to reconstruct missing or noisy information from the available received signals. This is just another way of saying that a transform coder is in fact operating as a simple vector quantizer: it takes advantage of the redundancy in the input vector to better code an entire vector, it just does it in a simple and *ad hoc* fashion that permits the actual digitization to be done by scalar quantizers.

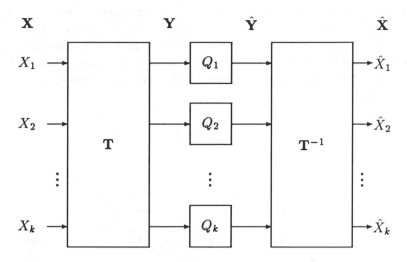

Figure 8.1: Transform Coding

The subjective reason supporting transform coders is that some biological systems, such as the eye and the ear, seem to operate in a transform domain and perceive superior quality if a good job is done of coding in that domain. In particular, transforms based on the Fourier transform (or some variation thereof such as the discrete cosine transform) are popular for both speech and image coding applications. It has long been known that for a given bit rate, signals coded with transform coding will sound or look significantly better than when scalar quantization is performed directly on the input vector of signal samples. This can be explained quantitatively by models of the perceptual masking effects of the human auditory and visual systems. Roughly speaking, an adequate signal-to-noise ratio in each frequency band is more important perceptually than an adequate overall signal-to-noise ratio where quantization noise is distributed uniformly across the frequency spectrum. Transform coding with suitable bit allocation offers a natural way to attain this objective.

The most convenient distortion measure to use for assessing performance of this coding scheme is the *overall squared error distortion*, D_{tc} given by

$$D_{tc} = \sum_{i=1}^{k} E[|X_i - \hat{X}_i|^2] = E[||\mathbf{X} - \hat{\mathbf{X}}||^2]. \qquad (8.5.1)$$

It is important to note that the actual quantization errors incurred by the coefficient quantizers will be transformed by $\mathbf{T}^{-1}$ so that the overall distortion in general depends on the choice of transform matrix. Thus, not every invertible matrix will be of use for transform coding. Even if the matrix does indeed perform some degree of decorrelation, the potential benefit can be lost if the inverse transform significantly magnifies the quantizer errors.

In the simple case of $k = 2$, we can visualize the joint pdf of the input vector, $\mathbf{X}$, distributed over a plane, with contours of constant height as shown in Figure 8.2.

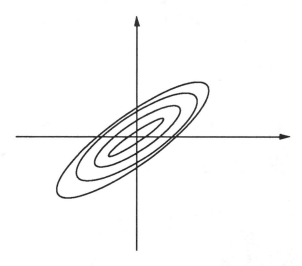

Figure 8.2: Two-Dimensional Example

It is apparent in this case that a change in the coordinate system, as shown, can eliminate the correlation between the two random variables, X_1 and X_2. Such a change is simply a *rotation* and is an *orthogonal* transformation since the new coordinate system has orthogonal coordinates. This is equivalent to the requirement that the distance between any two points in the original coordinate system remains unchanged when the two points are transformed to the new coordinate system.

In general, for an arbitrary vector dimension k, we can define an *orthogonal transformation* as a linear operation on a point $\mathbf{x}$ into a transformed point, $\mathbf{y}$ given by the matrix operation

$$\mathbf{y} = \mathbf{Tx}$$

where the real-valued $k \times k$ matrix $\mathbf{T}$ satisfies the orthogonality condition

$$\mathbf{T}^t = \mathbf{T}^{-1}, \tag{8.5.2}$$

and (as usual) the superscript t denotes transpose. It is useful to point out that the development extends from real k-dimensional vectors to complex

k-dimensional vectors with only minor modifications. In particular, if we are considering complex k-dimensional vectors and a complex $k \times k$ matrix $\mathbf{T}$, then the orthogonality condition of (8.5.2) is replaced by

$$\mathbf{T}^* = \mathbf{T}^{-1}, \tag{8.5.3}$$

where $\mathbf{T}^*$ denotes the conjugate transpose of T. While we focus on the real case, the extension of much of the development to complex vectors is straightforward (and is considered in the exercises). In the complex case (8.5.1) remains valid where now

$$||\mathbf{X} - \hat{\mathbf{X}}||^2 = (\mathbf{X} - \hat{\mathbf{X}})^*(\mathbf{X} - \hat{\mathbf{X}}).$$

An immediate property of this definition is that distances are preserved by the transformation, that is if $\mathbf{y}_1 = \mathbf{T}\mathbf{x}_1$ and $\mathbf{y}_2 = \mathbf{T}\mathbf{x}_2$, then

$$||\mathbf{y}_1 - \mathbf{y}_2|| = ||\mathbf{x}_1 - \mathbf{x}_2||.$$

To prove this, let $\mathbf{x} = \mathbf{x}_2 - \mathbf{x}_1$ and $\mathbf{y} = \mathbf{y}_2 - \mathbf{y}_1$, so that $\mathbf{y} = \mathbf{T}\mathbf{x}$. Then,

$$||\mathbf{y}||^2 = ||\mathbf{x}^*\mathbf{T}^*\mathbf{T}\mathbf{x}||^2 = ||\mathbf{x}||^2.$$

In particular, the length of a vector is preserved by an orthogonal transformation.

We have therefore identified a very valuable property possessed by an orthogonal matrix that makes it suitable for transform coding. Specifically, the distance preserving property implies that the sum of the squared quantization errors for the k quantizers will be equal to the overall distortion, that is:

$$D_{\text{tc}} = E[||\mathbf{Y} - \hat{\mathbf{Y}}||^2] = E[||\mathbf{X} - \hat{\mathbf{X}}||^2].$$

Note that this property is analogous to the fundamental theorem of DPCM discussed in Chapter 7 which states that the mean squared quantizer noise is equal to the overall mean squared reconstruction error. Furthermore, we recognize that minimization of this overall distortion for a given quota of bits and for the transform coding structure of Figure 8.1 implies optimal quantization of the set of transform coefficients with the optimal bit allocation for the given quota, B, of bits.

A second property of an orthogonal transform is that it does not alter the determinant of the autocorrelation matrix of the input vector. Since

$$\mathbf{R}_Y = E[\mathbf{Y}\mathbf{Y}^t] = \mathbf{T}E[\mathbf{X}\mathbf{X}^t]\mathbf{T}^t = \mathbf{T}\mathbf{R}_X\mathbf{T}^t,$$

and $|\det \mathbf{T}| = 1$, then, $\det \mathbf{R}_X = \det \mathbf{R}_Y$, where $\det \mathbf{A}$ denotes the determinant of a matrix $\mathbf{A}$. This result will be of value later.

Since there is an infinite variety of orthogonal matrices to choose from, all of which prevent magnification of quantizing noise, a natural approach is to find a matrix $\mathbf{T}$ which eliminates or nearly eliminates the correlation between the transform coefficients. We shall see that the effect of eliminating correlation is to reduce the product of the variances of the vector components, which in turn will reduce the average distortion attainable for the given bit quota when the optimal bit allocation under the high rate approximation of (8.3.1) is used.

8.6   Karhunen-Loeve Transform

In general, the components of $\mathbf{X}$ are correlated with one another. However, it is indeed possible to select a matrix $\mathbf{T}$, for a given pdf describing $\mathbf{X}$, that will make $\mathbf{Y} = \mathbf{TX}$ have pairwise uncorrelated components. This choice of transform matrix that makes the linear transformation have this desirable property is called the *discrete time Karhunen-Loeve transform* or the *Hotelling transform* and is defined below.

Let $\mathbf{R}_X = E[\mathbf{XX}^t]$, denote the autocorrelation matrix of the input (column) vector $\mathbf{X}$. (If the vectors are complex this becomes $\mathbf{R}_X = E[\mathbf{XX}^*]$.) Let $\mathbf{u}_i$ denote the eigenvectors of $\mathbf{R}_X$ (normalized to unit norm) and λ_i the corresponding eigenvalues. Since any autocorrelation matrix is symmetric and nonnegative definite, there are k orthogonal eigenvectors and the corresponding eigenvalues are real and nonnegative. Without loss of generality we assume the indexing is such that

$$\lambda_1 \geq \lambda_2 \geq \cdots \geq \lambda_k \geq 0.$$

The Karhunen-Loeve transform matrix is then defined as $\mathbf{T} = \mathbf{U}^t$, where

$$\mathbf{U} = [\mathbf{u}_1 \mathbf{u}_2 \cdots \mathbf{u}_k],$$

that is, the columns of $\mathbf{U}$ are the eigenvectors of $\mathbf{R}_X$.

Then the autocorrelation matrix of $\mathbf{Y}$ is given by

$$
\begin{aligned}
\mathbf{R}_Y &= E[\mathbf{YY}^t] = E[\mathbf{U}^t\mathbf{XX}^t\mathbf{U}] \\
&= \mathbf{U}^t\mathbf{R}_X\mathbf{U} = \begin{bmatrix} \lambda_1 & 0 & \cdots & 0 \\ 0 & \lambda_2 & \cdots & 0 \\ \vdots & \vdots & \ddots & \vdots \\ 0 & \cdots & 0 & \lambda_k \end{bmatrix}
\end{aligned}
$$

Thus we see that the Karhunen-Loeve transform does indeed decorrelate the input vector. It also follows that the variances of the transform coefficients are the eigenvalues of the autocorrelation matrix R_X.

Suppose we apply optimal quantization to each transform coefficient. For optimal bit allocation, the mean square distortion resulting from transform coding with the Karhunen-Loeve transform is given by

$$D_{tc} = kH(\prod_{i=1}^{k} \lambda_i)^{\frac{1}{k}} 2^{-2\bar{b}}$$

where H is the geometric mean of the quantization coefficients h_i for the normalized pdf's Y_i. It is important to note that H depends on the transform matrix $\mathbf{T}$ and on the joint pdf of the input vector $\mathbf{X}$. In the special case where $\mathbf{X}$ is Gaussian, each Y_i is also Gaussian so that $H = h_g$, the quantization coefficient for the Gaussian pdf obtained by applying (8.2.3) for the Gaussian case and given by (8.2.4). We have used the fact that the variance of the coefficient Y_i is λ_i. Since the determinant of a matrix is the product of its eigenvalues, we have

$$D_{tc} = kH(\det \mathbf{R}_X)^{\frac{1}{k}} 2^{-2\bar{b}}. \tag{8.6.1}$$

We have not yet shown that the Karhunen-Loeve transform is indeed the best possible transform for minimizing the overall distortion D_{tc} for a given bit allocation. This will be proved under the assumptions that the input variables are Gaussian, that optimal mean squared error quantization of the transform coefficients is performed, and that the optimal allocation result (8.3.3) (based on the high resolution assumption) holds.

Before proceeding, we note the following bound on the determinant of an autocorrelation matrix that will be useful later.

Covariance Determinant Bound

For any autocorrelation (covariance) matrix $\mathbf{R}_v$ of a real-valued random vector $\mathbf{V}$ whose components have zero mean and variances $E[V_i^2] = \sigma_i^2$, we have

$$\det \mathbf{R}_v \leq \prod_{i=1}^{k} \sigma_i^2.$$

Furthermore, equality holds if and only if $\mathbf{R}_v$ is a diagonal matrix.

As an illustration of this result, consider the case where $k = 2$. The determinant is then given by $\sigma_1^2 \sigma_2^2 (1 - \rho)$ where ρ is the correlation coefficient. This clearly satisfies the bound. In this case, and in general, it can be shown that the more positively correlated the vector components are, the smaller is the determinant. It should also be noted that the determinant of an autocorrelation matrix is always nonnegative.

Proof: Let $Z_i = V_i/\sigma_i$ be the normalized variables, components of the vector $\mathbf{Z}$. Then the covariance matrix $\mathbf{R}_Z$ of $\mathbf{Z}$ has unit valued diagonal elements and it is simply related to the covariance matrix $\mathbf{R}_V$ according to

$$\mathbf{R}_V = \mathbf{S}\mathbf{R}_Z\mathbf{S}$$

where $\mathbf{S}$ is the diagonal $k \times k$ matrix whose ith diagonal element is σ_i. Hence, the determinant of $\mathbf{R}_V$ is related to that of $\mathbf{R}_Z$ by

$$\det\mathbf{R}_V = (\prod_{i=1}^{k} \sigma_i^2)\det\mathbf{R}_Z.$$

It is a property of matrices that the trace of a matrix equals the sum of its eigenvalues. Since $\mathbf{R}_Z$ has unit diagonals, its trace is k and therefore we have from the arithmetic/geometric mean inequality of (8.3.3) that

$$\text{Tr}(\mathbf{R}_Z) = k = \sum_{i=1}^{k} \mu_i \geq k(\prod_{k=1}^{k} \mu_i)^{1/k}$$

where Tr denotes the trace and μ_i, $i = 1, \cdots, k$ denote the eigenvalues of $\mathbf{R}_Z$. Equality holds if and only if μ_i is a constant, independent of i, Noting that the product of the eigenvalues of a matrix is equal to the determinant, we then get

$$\det\mathbf{R}_Z = \prod_{i=1}^{k} \mu_i = \left[\frac{1}{k}\sum_{i=1}^{k}\mu_i\right]^k \leq 1.$$

Equality holds if and only if the μ_i are equal to some constant value μ. In this case, $\mathbf{R}_Z = \mathbf{U}\mathbf{\Lambda}\mathbf{U}^t = \mu\mathbf{U}\mathbf{U}^t = \mu\mathbf{I}$, where $\mathbf{\Lambda}$ is the diagonal matrix of eigenvalues of $\mathbf{R}_Z$, $\mathbf{U}$ is the orthogonal matrix that diagonalizes $\mathbf{R}_Z$, and $\mathbf{I}$ is the identity matrix. $\qquad\square$

With this result, we can now show the optimality of the Karhunen-Loeve transform.

High Resolution Optimality of the Karhunen-Loeve Transform

Assume that the input random variables are Gaussian, that each quantizer has a distortion function that is adequately modeled with the high resolution approximation (and overload distortion is neglected) and that we can allocate arbitrary real values to the resolution of each quantizer. Then, the KL transform achieves the lowest overall distortion D_{tc} of any orthogonal transform.

Proof: Letting σ_i^2 be the variance of the ith transform coefficient Y_i for an arbitrary orthogonal transformation on the input vector $\mathbf{X}$, we have

$$
\begin{aligned}
D_{\text{tc}} &= E[\|\mathbf{Y} - \hat{\mathbf{Y}}\|^2 = kH_Y 2^{-2\bar{b}}(\prod_{i=1}^{k}\sigma_i^2)^{1/k} \\
&\geq kH_Y 2^{-2\bar{b}}(\det\mathbf{R}_Y)^{1/k} = kh_g 2^{-2\bar{b}}(\det\mathbf{R}_X)^{1/k}
\end{aligned}
$$

where H_Y denotes the geometric mean of the quantization coefficients for the random variables Y_i and h_g is the quantization coefficient for the Gaussian case. The upper equality follows from the result of optimal bit allocation and the high resolution formula for the quantizer distortion function and the inequality follows from the covariance determinant bound. The last equality follows from the Gaussian assumption and the fact that the determinant of the autocorrelation matrix is invariant to an orthogonal transform of a random vector as given in (8.6.1). The inequality above becomes an equality in the limiting case where $\mathbf{R}_Y$ is a diagonal matrix and only in this case. This lower bound on overall distortion is indeed achieved with the Karhunen-Loeve transform as given in (8.6.1) specialized to the case of a Gaussian input vector. □

8.7 Performance Gain of Transform Coding

The key to the performance gain of transform coding comes from the effective use of bit allocation. We consider the special case where the input vector $\mathbf{X}$ is a block of k samples from a weakly stationary Gaussian process. Then each X_i has the common variance σ_X^2. and the alternative to transform coding with B bits per vector is simply to use PCM quantization of each sample with a bit allocation of $\bar{b}$ bits.

Using the Karhunen-Loeve transform and optimal bit allocation with a quota of B bits, we obtain the transform coding performance:

$$
D_{\text{tc}} = kh_g 2^{-2\bar{b}}(\prod_{i=1}^{k}\lambda_i)^{1/k},
$$

where $\bar{b} = B/k$.

Now for comparison, consider the case where no transform is used and each component of the input vector $\mathbf{X}$ is separately quantized with a scalar quantizer having $\bar{b}$ bits. This keeps the total bit quota, B, the same in both cases. Then, we have the overall PCM distortion (corresponding to the special case of transform coding where the transform matrix $\mathbf{T}$ is the identity matrix):

$$D_{\text{pcm}} = k h_g 2^{-2\bar{b}} \sigma_X^2.$$

The *coding gain* of transform coding can be defined as the ratio

$$\frac{D_{\text{pcm}}}{D_{\text{tc}}} = \frac{\sigma_X^2}{(\prod_{i=1}^{k} \lambda_i)^{1/k}}.$$

To interpret this result, we note that

$$\text{Tr}(\mathbf{R}_X) = k\sigma_X^2 = \sum_{i=1}^{k} \lambda_i.$$

Thus, we have

$$\frac{D_{\text{pcm}}}{D_{\text{tc}}} = \frac{\frac{1}{k} \sum_{i=1}^{k} \lambda_i}{(\prod_{i=1}^{k} \lambda_i)^{1/k}}. \tag{8.7.1}$$

Hence, the coding gain is the ratio of the arithmetic mean to the geometric mean of the eigenvalues of $\mathbf{R}_X$. This ratio is always greater or at least equal to unity. The larger the ratio, the more correlated are the vector components. In fact, it can be shown (See Problem 8.4) that as the dimension of the block goes to infinity, the ratio approaches the limiting value G_p which is the infinite-past least squared error prediction gain as defined in Chapter 4.

Suppose now that the input random variables X_i are *not* samples of a weakly stationary random process and have different variances, σ_i^2, although they are still jointly Gaussian. What advantage is there in transform coding in this case over direct quantization of these variables with optimal bit allocation? From the expression for the average distortion for optimal bit allocation (8.3.5) without transform coding and the corresponding expression (8.6.1) for the case of optimal transform coding with the Karhunen-Loeve transform, it can be seen that the performance gain is given by

$$\frac{D_{\text{direct}}}{D_{\text{tc}}} = \frac{(\prod_{i=1}^{k} \sigma_i^2)^{1/k}}{(\prod_{i=1}^{k} \det \mathbf{R}_X)^{1/k}},$$

which is always greater than unity when the X_i are not uncorrelated, according to the covariance determinant bound. Thus, we see that transform coding can indeed be beneficial for coding a set of random variables that are not necessarily samples of a weakly stationary process.

In practice some of the benefit of transform coding comes from the fact that several of the transform coefficients have a sufficiently small variance that they can be ignored completely, that is, they can be quantized with zero bits of resolution. What this means is that the transform has "concentrated" the information needed to reconstruct the information into a reduced number (less than k significant parameters). When a subset of transform coefficients is chosen for quantization with zero bits, the quantized value is taken as zero (assuming zero mean random variables) and the remaining coefficients are quantized as usual. This situation is sometimes called *zonal sampling*.

8.8 Other Transforms

There are a large variety of orthogonal transforms that can be used for transform coding. For a comprehensive treatment of transform coding with application to image coding, see Clarke [71]. The most popular transform in terms of systems actually implemented and in established and proposed standards is the *discrete-cosine transform* (DCT). The DCT is an orthogonal transform having some of the features of a transformation to the frequency domain. It is equivalent to a discrete Fourier transform (DFT) of a symmetricized extension of the input set of samples and fast algorithms are available for efficiently computing the DCT. The DCT has an advantage over the DFT in that it is a purely real transform if the input vector is real.

The DCT is specified by t_{pm}, the pmth element of the $k \times k$ discrete-cosine transform matrix $\mathbf{T}$, and is given by:

$$t_{pm} = \begin{cases} \sqrt{\frac{2}{\pi}} \cos(\frac{\pi}{k} p(m + \frac{1}{2})) & p = 1, 2, \cdots, k-1 \\ \sqrt{\frac{1}{\pi}} \cos(\frac{\pi}{k} p(m + \frac{1}{2})) & p = 0 \end{cases}$$

where for convenience we have indexed p and m over the range $0, 1, \cdots, k-1$,

For signals whose samples are naturally organized into a two-dimensional array, such as a sampled image, two-dimensional orthogonal transforms are convenient and workable, and correspond closely to two-dimensional Fourier transforms. The DCT is amenable to fast computation just as the FFT performs the discrete-Fourier transform in a computationally efficient manner.

It has frequently been reported that the DCT achieves performance very close to that of the optimal Karhunen-Loeve transform and yet has the advantage that it is a fixed, signal-independent transform matrix. Problems 8.7–8.8 provide a proof of the equivalence of the DFT and the Karhunen-Loeve transform in the limiting case of large bit-rate and large dimension.

Most of the applications of transform coding have focused on the DCT. Today, several commercial integrated circuit chips intended for video coding applications are available which perform the two-dimensional DCT for arrays as high as 16×16 pixels. One such transform code is an integral part of the CCITT standards for still frame image compression (JPEG) and video compression (MPEG).

Other orthogonal transforms can also be used for transform coding. They often lack the decorrelating properties of the Karhunen-Loeve transform, but they can be simpler or have other useful properties. Binary transforms such as the Walsh or Hadamard transform have been used with success in image compression [255]. The Hartley transform has been proposed as an alternative to the DCT for doing real valued transforms related to the Fourier transform [35]. Recently various wavelet transforms have been proposed and tried, although it is not yet clear that they provide definite advantages over the simpler DCT [70][230][231]. Significant compression ratios have been claimed, but the tradeoffs among performance, bit rate, and complexity and the relative advantages and disadvantages with other approaches are not yet well understood.

8.9   Sub-band Coding

Transform coding belongs to a larger family of coding techniques where the signal is decomposed or *analyzed* into components that in some sense offer a more fundamental or more primitive representation of the signal. The quantized components are then used to synthesize a reproduction of the original signal. Such coding schemes are called *analysis-synthesis coding systems.* The components most commonly represent some form of spectral decomposition of the signal as with the DCT. An important and widely used type of analysis-synthesis coding system is *sub-band coding.* In sub-band coding, a bank of filters operates on the input signal to generate a set of narrowband signals each representing a different sub-band of the input spectrum. The narrow bandwidth of each sub-band signal allows subsampling to be performed, reducing the bit-rate needed to code each sub-band. Interpolation is then used to synthesize the reproduction of the original signal. We first present the basic theory of sub-band coding in an idealized setting in order to convey the fundamental ideas without being encumbered by the very important but more complicated issues of realistic filter designs. Then some of the key developments in sub-band coding filter banks will be outlined.

Suppose we have a stationary random Gaussian process $\{X_n\}$ with zero mean, variance (or power) σ_X^2, sampling rate f_s, and spectral density $S_X(f)$. We define a bank of *ideal* bandpass filters

$$H_m(f) = \begin{cases} 1 & \text{for } \frac{(m-1)f_s}{2M} < f < \frac{mf_s}{2M} \\ 0 & \text{otherwise .} \end{cases} \qquad (8.9.1)$$

Thus, each of the M filters has bandwidth f_s/M. When X_n is applied to each filter, the outputs obtained are the sub-band signals X_{mn} for $m = 1, 2, \cdots, M$. The ideal nature of the filters implies that

$$X_n = \sum_{m=1}^{M} X_{mn} \qquad (8.9.2)$$

Also, it is easy to show that the sub-band signals are uncorrelated with one another, i.e., that

$$E[X_{mn}X_{ls}] = 0 \text{ for } m \neq l \qquad (8.9.3)$$

so that for all integers n and s

$$X_n = \sum_{m=1}^{M} X_{mn} \qquad (8.9.4)$$

$$\sigma_X^2 = \sum_{m=1}^{M} \sigma_m^2 \qquad (8.9.5)$$

where σ_m^2 is the variance (or power) of the mth sub-band signal. We have therefore decomposed the input signal into M uncorrelated sub-band signals which can be used to exactly reconstruct X_n according to (8.9.2).

A simplistic and idealized sub-band coding system is illustrated in Figure 8.3 where the ideal filter bank defined above is used to obtain the M sub-band signals. In order to code the sub-band signals, a quantizer Q_m with b_i bits operates on each successive sample X_{mn} of the mth sub-band signal producing the reproduction values $\hat{X}_{mn} = Q_m(X_{mn})$. The final reproduction value $\hat{X}_n$ for the composite signal is given by the "synthesis" or "reconstruction" operation:

$$\hat{X}_n = \sum_{m=1}^{M} \hat{X}_{mn}. \qquad (8.9.6)$$

Combining (8.9.2) and (8.9.6), and using (8.9.3) gives the mean squared error for the coding scheme

$$D = E[(\hat{X}_n - \hat{X}_n)^2] = \sum_{m=1}^{M} E[(\hat{X}_{mn} - \hat{X}_{mn})^2], \qquad (8.9.7)$$

which shows that the overall average distortion is the sum of the average distortions in each sub-band. With optimal bit allocation among the sub-bands using the high resolution approximation, the minimum achievable average distortion from (8.3.5) is given by

$$D_{\text{sbc}} = M h_g \rho_S^2 2^{-2\bar{b}} \qquad (8.9.8)$$

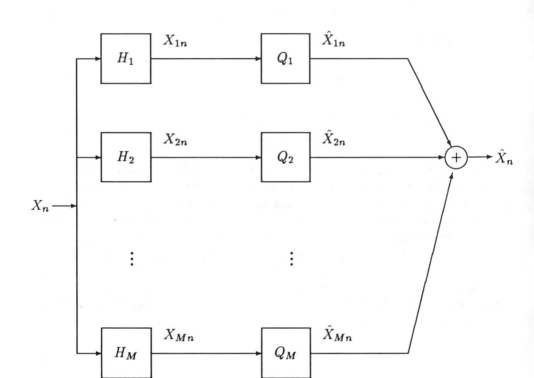

Figure 8.3: Sub-Band Coding System without Subsampling

where $\bar{b}$ is the average of the bit allocations for the sub-band samples. Thus, $\bar{b} = B/M$ where $B = \sum_{m=1}^{M} b_i$, ρ_S^2 is the geometric mean of the sub-band variances, and h_g is the Gaussian quantization constant given by (8.2.4).

The coding scheme descibed so far is conceptually a simple way to introduce the idea of sub-band coding; however, it requires a large increase in the total number of samples to be quantized so that it is not a meaningful signal compression scheme in its present form. Instead of the original rate of f_s samples per second the sub-band signals collectively consist of $M f_s$ samples per second. To make a transition to a more meaningful scheme, we note that each sub-band signal has a bandwidth of $f_s/2M$ so that it can be subsampled by a factor of M while allowing perfect recovery of of the original sub-band signal from the reduced rate sequence of samples. A simple band-pass version of the sampling theorem applies here. (See Problem 8.9.) Specifically, we extract every Mth sample of each sub-band to form the subsampled sub-band signals given by

$$Y_{mk} = \hat{X}_{mn} \text{ where } n = kM, \qquad (8.9.9)$$

where k is now the time index for the subsampled signals. Figure 8.4 illustrates a complete but idealized sub-band coding system that uses subsampling and interpolation to achieve *critical sampling*, the condition where the total number of samples per unit time generated by the analysis scheme is equal to the number of samples per unit time of the original signal.

The blocks with downward pointing arrows represent the operation of subsampling by a factor of M; similarly the blocks with upward arrows represent the first stage of interpolation where $M - 1$ zeros are inserted between each pair of incoming samples thereby increasing the sampling rate by the factor M. The ideal bandpass filters are then used to complete the interpolation process. In the absence of quantization errors, the output of the mth interpolation filter would be the full-rate sub-band signal X_{mn}. In other words, the analysis-synthesis scheme has no reconstruction error in the absence of quantization. This condition is called *perfect reconstruction*. The effect of quantization is to produce a modified sub-band signal $\hat{X}_{mn}$ after the interpolation filter $H_m(f)$.

The same quantizers Q_m are used as before but now they operate on the subsamples Y_{mk}, producing quantized values $\hat{Y}_{mk}$. Note that X_{mn} and Y_{mk} have the same variance and that $\hat{Y}_{mk}$ and $\hat{X}_{mn}$ have the same variance. Hence, the same bit allocation and the same average overall distortion D_{sbc} results as in the simplistic scheme of Figure 8.3 where subsampling was not performed. The only difference is that now the number of samples per second to be quantized is exactly equal to the rate at which input samples arrive. Since B bits are needed for quantization of all sub-band samples

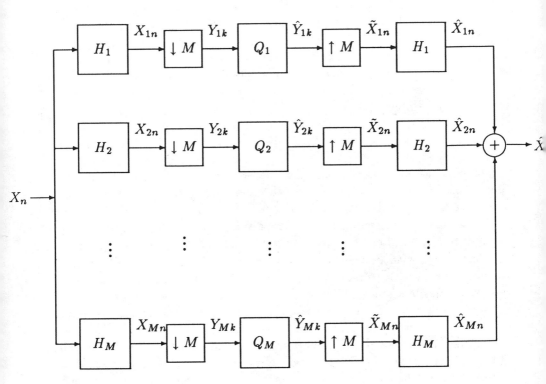

Figure 8.4: Ideal Sub-Band Coding System with Critical Sampling and Perfect Reconstruction

at any particular time instant, the reduced rate of f_s/M implies that the overall bit rate is $R = \bar{b}f_s$ bits per second (b/s).

We now examine the performance of this idealized sub-band coding system which has both critical sampling and perfect reconstruction. Suppose PCM is performed on the input signal X_n with optimal quantization at the rate of $\bar{b}f_s$ b/s. Then, the minimum mean squared error achievable is given by

$$D_{\mathrm{pcm}} = h_g \sigma_X^2 2^{-2\bar{b}}. \tag{8.9.10}$$

Now, the performance advantage of sub-band coding compared with PCM (or direct sample by sample waveform coding) can be measured by the ratio

$$\frac{D_{\mathrm{pcm}}}{D_{\mathrm{sbc}}} = \frac{h_g \sigma_X^2 2^{-2\bar{b}}}{M h_g \rho_S^2 2^{-2\bar{b}}} = \frac{\sigma_X^2}{M \rho_S^2} = \frac{\frac{1}{M}\sum_{m=1}^{M}\sigma_m^2}{(\prod_{m=1}^{M}\sigma_m^2)^{1/M}} \tag{8.9.11}$$

where we have used (8.9.8), (8.9.10) and (8.9.5). Thus the performance gain is simply the ratio of arithmetic to geometric means of the sub-band variances. Consequently, the gain is largest when there is a wide variation in the power levels of the different sub-bands. Clearly for a widely varying input spectral density $S_X(f)$ and a sufficiently large number of bands M, a substantial performance gain is achieved. It is not difficult to show (see Problem 8.10) that in the limit as M approaches infinity, the performance gain approaches the infinite memory prediction gain G_p. Exactly the same asymptotic performance for transform coding was noted in Section 8.7 and is considered in Problem 8.4.

Thus, highly correlated signals with high prediction gains are amenable to efficient removal of redundancy by sub-band coding with the same ultimate performance as achieved in transform coding. For a finite number of sub-bands and a finite order transform coder, the performance comparison between the two coding systems is determined by examining (8.7.1) and (8.9.11). Since the numerators in these two expressions are identical, the comparison is determined by the ratio of the geometric mean of the eigenvalues of the autocorrelation matrix to the geometric mean of the sub-band variances. A fair comparison should consider both the relative computational complexity of the two methods and the effect of relaxing the idealizations made here to obtain explicit performance results.

Practical sub-band coding systems generally maintain the constraint of critical sampling while replacing the ideal bandpass filters with physically realizable filters whose transfer functions are rational functions of $z = e^{j2\pi f/f_s}$ and usually FIR filters. Inevitably, frequency components in one sub-band will leak into other sub-bands since the filters cannot have sharp cutoffs at the edges of the passband. This generally introduces aliasing effects that degrade the reconstructed signal even in the absence of quantization. Various filter design techniques have been developed to reduce or eliminate this aliasing by achieving nearly perfect or perfect reconstruction. These include in particular, quadrature mirror filter (QMF) banks [84] and various extensions of QMF techniques. More recently perfect reconstruction QMF techniques were introduced [298], [316]. The QMF techniques are intimately connected with wavelet transformation techniques. Although neither approach is a strict special case of the other, a large class of QMF sub-band coders are equivalent to a class of wavelet transform coders . (See, e.g., [230].)

Sub-band coding has been used for some time in speech coding and recently it was selected for use in wideband audio compression for the MPEG standard. It has also become popular for image coding [330][340] where two-dimensional sub-band filtering is performed. Sub-band coding is readily generalized to multidimensional signals for use in image and video coding. Often separable sub-band filters are used where the two-dimensional transfer function has the form $G_{mn}(f_1, f_2) = H_m(f_1)H_n(f_2)$. The simplest case of practical interest is a four band decomposition where the four bands are often described as "low-low", "low-high", "high-low", and "high-high" where the component one-dimensional filters consist of a lowpass and a highpass filter each covering half the frequency range. For a general treatment of multidimensional sub-band coding see [321] and for its application to image coding see [339].

8.10   Problems

8.1. Evaluate the high resolution quantization constant h of (8.2.3) for the case of a doubly exponential pdf.

8.2. Prove the arithmetic/geometric mean inequality. *Hint:* Take logarithms and apply Jensen's inequality. It may also be proved by formulating it as a constrained optimization problem with a Lagrange multiplier.

8.3. Prove the optimal bit allocation formula (8.3.5) for the high resolution, weighted average distortion case.

8.4. Show that as the block size goes to infinity the coding gain ratio given by transform coding using the optimal bit allocation high resolution distortion approximation approaches a limiting value of G_p, the infinite-past least squared error prediction gain.

8.5. Suppose that you wish to quantize a Gaussian two-dimensional vector $\mathbf{X} = (X_1, X_2)$ formed by taking two samples from a stationary Gauss Markov source $\{X_n\}$ with unit variance and $R_X(1) = .9$. Either by tables or simulations, design Lloyd-Max quantizers for each component separately, for all possible individual integer bit allocations that result in one bit per sample overall. Find the average distortion for each such quantizer pair. What is the optimal bit allocation? Next look at the vector $\mathbf{Y}$ formed by taking the Karhunen-Loeve transform of $\mathbf{X}$, that is, $\mathbf{Y} = \mathbf{UX}$, where $\mathbf{U}$ is an orthogonal matrix and $\mathbf{Y}$ has uncorrelated components. Find $\mathbf{U}$. Describe the Lloyd-Max quantizers for the components of $\mathbf{Y}$ for all possible bit allocations that result in one bit per sample overall. What is the optimal bit allocation?

Compare the results for the scalar quantized and the transform coded. Compare the results also with those predicted by the optimal bit allocation high resolution approximations.

8.6. Suppose that X is a complex-valued Gaussian random variable. Such a random variable can be considered as a two-dimensional real-valued random vector $X = (X_r, X_i)^t$ where X_r is the real part and X_i is the imaginary part of X, respectively, and X is considered to be a column vector. Using ordinary complex notation we can also write $X = X_r + jX_i$. Both representations will be used as convenient. Assume that the real random variables X_r and X_i have zero mean and variances $\sigma_r^2 = E(X_r^2)$ and $\sigma_i^2 = E(X_i^2)$ and correlation $E(X_r X_i) = \rho \sigma_r \sigma_i$.

We wish to quantize X using B bits, where B is assumed to be large.

If the quantized version of X is $Q(X)$ (also complex valued), then the average distortion is measured by the mean squared error which is defined using the complex number notation as

$$D = E(|X - Q(X)|^2) = E[(X - Q(X))^*(X - Q(X))],$$

where a^* denotes the complex conjugate of a (scalar) complex number a. This can be expanded as

$$D = E[(X_r - Q(X)_r)^2] + E[(X_i - Q(X)_i)^2]$$

where $Q(X)_r$ denotes the real and $Q(X)_i$ the imaginary parts of $Q(X)$, respectively. Thus the overall distortion is the sum of the distortion in the two coordinates. This can be written in the vector notation as

$$D = E(\|X - Q(X)\|^2)$$

where

$$||x|| = \sqrt{\sum_l |x_l|^2}$$

is the Euclidean norm of a vector $x = \{x_l\}$.

(a) First suppose that the two coordinates of X are simply quantized separately, e.g., there are two scalar quantizers q_r and q_i so that

$$Q(X)_r = q_r(X_r)$$

and

$$Q(X)_i = q_i(X_i)$$

or, equivalently, $Q(X) = (q_r(X_r), q_i(X_i))$. Using the Bennett approximations find the average distortion resulting if both quantizers have the same number of bits, $B/2$ each. Call this D_1. Your answer should depend on the system parameters σ_r, σ_i, ρ. (Any integrals should be evaluated.)

(b) Now repeat the problem allowing optimal bit allocations between the real and imaginary parts (again using high rate theory). Call the resulting average distortion D_2 and show that

$$D_2 \leq D_1.$$

(c) Next suppose that instead we use a two-dimensional Karhunen-Loeve transform code. Using vector notation we form a new complex random variable $V = TX$ where we assume that $T^* = T^{-1}$. We now apply a quantizer Q to V to form a reproduction $Q(V)$ and then define $\hat{X} = T^{-1}Q(V)$. Suppose that the quantization is accomplished by two scalar quantizers operating on the coordinates of V. What is the optimal bit allocation and what is the resulting average distortion $D_3 = E[||X - \hat{X}||^2]$? Prove that $D_3 \leq D_2$.

(d) Suppose now that the $\sigma_r^2 = \sigma_i^2$ and hence that the real and imaginary parts are identically distributed. Specialize the previous results to this special case and interpret the results.

8.7. Suppose that we wish to quantize a complex k-dimensional vector $\mathbf{X} = (X_0, X_1, \cdots, X_{k-1})$ drawn from a stationary complex Gaussian random process $\{X_n\}$ using an average of $\bar{b}$ bits per (complex) coordinate. Assume a mean

$$E(X_n) = 0$$

and autocorrelation

$$R_X(k,n) = E[X_k X_n^*] = R_X(n-k).$$

The power spectral density is given by the discrete time Fourier transform of the autocorrelation

$$S_X(f) = \sum_{k=-\infty}^{\infty} R_X(k)e^{j 2\pi f k}.$$

One way to code the k-dimensional vector would be to simply repeat the codes of problem 8.6 k times. In this case the resulting average distortion per (complex) coordinate would be the same as found there.

Suppose that instead the complex vector $\mathbf{X}$ is transformed into $\mathbf{Y} = \mathbf{WX}$ with

$$\mathbf{W} = \frac{1}{\sqrt{k}}\{e^{j 2\pi \frac{il}{k}} \; i, l = 0, 1, \cdots, k-1\},$$

that is, the Fourier transform (here the DFT) is used.

Assume as in the previous part that the simple two-dimensional scalar quantizer of (a) of the previous problem is used for each coordinate of $\mathbf{Y}$, thus yielding for the ith coordinate an average distortion proportional to $E[|Y_i|^2]$ using the results of part (a).

(a) Find an expression for the overall average distortion (per coordinate), D_{FFT}, resulting from the optimal bit allocation (among the complex coordinates) for this FFT transform code in terms of the $E[|Y_i|^2]$. Note that the bits are allocated optimally among the Y_i, but that as in part (a) of the previous problem for each i, the real and imaginary parts of Y_i get an equal number of bits.

(b) Show that

$$E[|Y_i|^2] = \sum_{l=-(k-1)}^{k-1} R_X(l)(1 - \frac{|l|}{k})e^{j 2\pi i l / k}.$$

(c) It can be shown (and it should seem reasonable) that if both i and k are large and i/k is fixed, then

$$\sum_{l=-(k-1)}^{k-1} R_X(l)(1 - \frac{|l|}{k})e^{j 2\pi i l / k} \approx S_X(\frac{i}{k}),$$

that is, $E[|Y_i|^2] \approx S_X(\frac{i}{k})$. Use this fact and some calculus approximations to evaluate the average distortion in the limit of asymptotically large k.

(*Hint:* Your answer should contain the one-step prediction error σ_p^2 appearing in equation (4.9.22).)

8.8. In this part the Karhunen-Loeve transform coding ideas are extended to complex vectors and compared with the previous problem which treated DFT transform coding. As in the previous problem, suppose that we wish to quantize a complex k-dimensional vector $\mathbf{X} = (X_0, X_1, \cdots, X_{k-1})$ drawn from a stationary complex Gaussian random dom process $\{X_n\}$ using $\overline{b}$ bits per (complex) coordinate. Assume a mean, autocorrelation, and power spectral density as in the previous problem. Since an autocorrelation matrix is Hermitian and nonnegative definite $(\mathbf{a}^*\mathbf{R}_X\mathbf{a} \geq 0)$, the eigenvalues of $\mathbf{R}_X$ are real and nonnegative, say

$$\lambda_1 \geq \lambda_2 \geq \cdots \geq \lambda_k \geq 0.$$

Suppose that the complex vector $\mathbf{X}$ is transformed into $\mathbf{Y} = \mathbf{U}^*\mathbf{X}$ where $\mathbf{U}^*$ is the complex Karhunen-Loeve transform, i.e., $\mathbf{UU}^* = \mathbf{I}$, the $k \times k$ identity matrix, and

$$\mathbf{R}_Y = \begin{bmatrix} \lambda_1 & 0 & \cdots & 0 \\ 0 & \lambda_2 & \cdots & 0 \\ \vdots & \vdots & \ddots & \vdots \\ 0 & \cdots & 0 & \lambda_k \end{bmatrix}.$$

Suppose that each of the individual complex Y_i is quantized as in part (a) of problem 8.7 and as in the previous problem, that is, using separate quantizers without further transforming and using equal bit allocation among the coordintates. Find an expression for the average distortion per coordinate having the form

$$D = C(\det \mathbf{R}_X)^{1/k},$$

where C is a constant that does not depend on k.

It can be shown using results from the theory of Toeplitz matrices that

$$\lim_{k \to \infty} (\det \mathbf{R}_X)^{1/k} = e^{\int_{-1/2}^{1/2} \ln S_X(f)\, df} = \sigma_p^2,$$

where $S_X(f)$ is the power spectral density of $\{X_n\}$ (the discrete time Fourier transform of $R_X(k)$) and σ_p^2 is the one-step prediction error or (4.9.22). (See problem 8.4.) [172] This provides a description of the performance achievable with Karhunen-Loeve transform coding when the vector dimension is large.

Compare the result to that of the previous problem. Your conclusion should be that the FFT transform works as well as the Karhunen-Loeve transform in the limit of large dimension and large bit rate.

8.9. Consider the sub-band signal X_{mn} produced by the ideal bandpass filter $H_m(f)$ as defined in Section 8.9 for the sub-band coding system of Figure 8.3 and show that X_{mn} can be exactly recovered from its samples taken at instants $n = kM$.

8.10. Show that the asymptotic MSE distortion of ideal sub-band coding as the number of bands M goes to infinity approaches the infinite memory least mean squared prediction error σ_p^2 as given by Equation (4.9.26). *Hint:* Express the sub-band variances in terms of the spectral density $S_X(f)$ and use suitable limiting arguments as $M \to \infty$.

Chapter 9

Entropy Coding

9.1   Introduction

All of the coding systems considered thus far have been *fixed rate* codes in the sense that a fixed number of channel bits per time unit is produced by the encoder and processed by the decoder. In some communication and storage systems, fixed rate operation is not desirable because the data source may display wide variations in activity. For example, sampled speech may change very little during long periods of silence and then exhibit very complex behavior during plosives. Ideally, one would like to waste few bits coding the silence and preserve them for coding the highly informative transients. Such a strategy requires a *variable rate* code, a code which can adjust its own bit rate to better match local behavior. In order to use fixed rate communication and storage links, however, the long term average bit rate must be constant. Thus buffers are usually required as an interface when variable rate codes are used on fixed rate communication or storage media. The buffers will hold bits arriving at a variable rate from the encoder until they are accepted by the fixed rate channel for transmission. Such buffers add complexity to a system and can also add errors when they overflow, which occurs when the data source produces bits faster than the buffer can accept them. Similarly, errors can be introduced when the buffers underflow, which occurs when the data source produces bits slower than the rate at which the buffer is releasing bits. To combat this problem, a technique known as *buffer feedback* is commonly used, where the occupancy level of the buffer is fed back to the source encoder to suitably adjust the quantizer data rate. This added complexity is often justified, however, by the potentially significant performance gains possible with variable rate strategies. Entropy codes are often used in conjunction with scalar quantizers (to con-

259

serve the average bit rate) and are often fairly simple to implement when the input alphabets are of reasonable size. The overall variable rate code is then a simple cascade of a scalar quantizer, which performs the analog-to-digital conversion in a fixed rate manner, and a variable length noiseless code, which maps the quantizer output into a variable length binary index in a way that can be perfectly decoded by the receiver. It can be shown that in the high rate case, coupling simple scalar uniform quantizers with noiseless variable rate codes of vectors can achieve performance within approximately 1/4 bit of the Shannon optimum as given by the Shannon rate distortion function. This result is sketched in Section 9.9 and considered in detail in more general form in [353], [244], and Chapter 5 of [159].

Communication and storage systems that are inherently variable rate are increasing in importance and variable length codes can be well matched to such systems. For example, variable rate codes cause no problems in off-line storage (the bits are accepted as they come until the file is complete) and variable rate codes are no more complicated than fixed rate codes for use in packet communication environments.

The emphasis of this book is on fixed rate systems, but two chapters on variable rate systems are included for completeness and because these systems can offer significant performance improvements in some applications. This chapter concentrates on simple variable rate noiseless codes, a technique for encoding discrete data into variable length codewords in an invertible fashion. This coding technique is commonly called *entropy coding* for reasons that will be seen. It is also often referred to as *noiseless coding*, *lossless coding*, and *data compaction coding*. It is also referred to simply as *data compression* in the computer science literature, but we avoid this nomenclature as entropy coding is a very special case of what we have defined to be data compression. The narrow use of the term by computer scientists is perhaps understandable because of the disastrous consequences that can result from even rare bit errors if the compressed file is a binary executable file. When bit errors cause catastrophe, lossy codes are not useful for compression (except possibly as a component of an overall lossless code).

Entropy coding has been extensively studied in the literature and detailed accounts may be found in many books and papers. For example, various aspects are treated in Blahut [34], Gallager [134], Lynch [226], Rice [266], and Storer [303]. As a result, our treatment will be brief and leave deeper developments to the references.

The goal of noiseless coding is to reduce the average number of symbols sent while suffering no loss of fidelity. A classical example is the Morse code where short binary codewords are used for more probable letters and long codewords used for less probable letters. The Morse code in fact is a very good code for its age and, when applied to English text, results in many fewer bits on the average than would the use of one byte ASCII codes for each letter. A more recent but still venerable example is the run-length code used to code sources which tend to repeat symbols for long periods of time. For example, a binary source such as facsimile may produce long runs of zeros and, occasionally, ones. Hence one means of compression is to sequentially send a symbol followed by the number of its repetitions, the *run length*. This will result in compression on the average if the source tends to produce such runs. It will not compress a memoryless source. (See Chapter 10 of [195] for a history and general development of run-length codes.) In this chapter we will emphasize three of the most popular techniques for entropy coding: Huffman codes, arithmetic codes, and Ziv-Lempel codes.

9.2 Variable-Length Scalar Noiseless Coding

Suppose that $\{X_n\}$ is a stationary sequence of random variables with a finite alphabet $A = \{a_0, \cdots, a_{M-1}\}$ with a marginal probability mass function $p(a) = p_X(a) = \Pr(X_n = a)$. The case of primary interest for our present purposes is that where the X_n are quantized versions of a continuous alphabet sequence W_n; that is, $X_n = Q(W_n)$, with Q an ordinary scalar quantizer. Another important example to be considered later is the case of a binary source such as a computer file.

A *variable length scalar noiseless code* consists of an encoder α, which maps a single input symbol x in A into a binary vector $\alpha(x)$ of dimension or length $l(x)$, and a decoder β, which maps binary vectors u of differing length into an output $\beta(u)$ so that $\beta(\alpha(x)) = x$; that is, the encoding/decoding operation is *lossless* or *noiseless* or *transparent*. The goal of the code is to keep the average number of bits transmitted for each source symbol as small as possible, that is, to minimize the *average length*

$$\bar{l}(\alpha) = El(X_n) = \sum_{a \in A} p(a)l(a). \tag{9.2.1}$$

If (9.2.1) is accepted as a definition of quality of a noiseless source code, then it is of interest to quantify how small $\bar{l}(\alpha)$ can be made and hence what the optimal achievable performance is. It is also of interest to construct actual codes that perform very near to the optimal quantity.

Unfortunately, the given definition of a code is not enough to ensure that it is useful. Suppose, for example, that the input alphabet has four letters, $A = \{a_0, a_1, a_2, a_3\}$, possibly the output of a 2 bit per sample quantizer. Consider the code given in Table 9.1. Although this is a noiseless code by

Input Letter	Codeword
a_0	0
a_1	10
a_2	101
a_3	0101

Table 9.1: A Non-Uniquely Decodable Code

the above definition, it cannot always be decoded in a noiseless fashion when the code is applied to a sequence of inputs. For example, if the receiver gets the sequence $0101101010\cdots$, it could have been produced by the input sequence $a_0a_2a_2a_0a_1\cdots$ or by $a_3a_2a_0a_1\cdots$. To make matters worse, the ambiguity can never be resolved regardless of future received bits. Hence for a code to be useful, it must be *uniquely decodable* in the sense that if the decoder receives a valid encoded sequence of finite length, there is only one possible input sequence that could have produced the encoded sequence. This effectively extends the idea of a noiseless or transparent code from a single letter to a sequence. Note that we could accomplish this by inserting punctuation in the binary sequence between codewords, e.g., add a third letter "," and send the sequence $0, 101, 101, 0, 10\cdots$. While this disambiguates the sequence, it also increases the average length of the encoding as well as the required channel alphabet. This may be a simple fix, but it is not an efficient use of symbols. An alternative and less restrictive approach is to require that the code satisfy a *prefix condition* in the sense that no codeword be a prefix of any other codeword. In the previous example, a_0 is a prefix of a_3 and a_1 a prefix of a_2. An example of a code satisfying the prefix condition is given in Table 9.2.

Input Letter	Codeword
a_0	0
a_1	10
a_2	110
a_3	111

Table 9.2: Code Satisfying the Prefix Condition

Binary prefix codes can be depicted as a binary *tree* as in Figure 9.1. As we shall often encounter code trees, we pause to point out several relevant

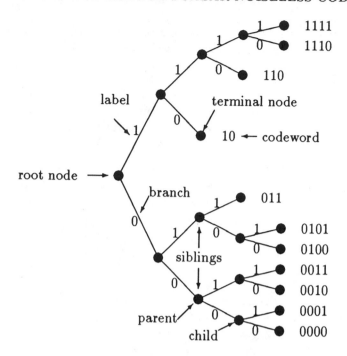

Figure 9.1: A Binary Code Tree

definitions. The binary tree starts with a *root* or *root node* which has two *branches* extending from it. Each such branch ends in a *node*, which can be thought of as first level nodes or depth one nodes. The branches are *labeled* by a 1 or 0 (for a binary tree). By convention, we often put the label 1 on the upper branch in a horizontally drawn tree and a 0 on the lower branch. Nodes either have further branches leading to more nodes, or they are *terminal nodes* or *leaves* with no extending branches. This tree is depicted as growing from left to right, but they are often drawn in vertical fashion with the root on the bottom (like most biological trees) or with the root on top and the branches extending downward. A level $n+1$ node connected by a branch from a level n node is said to be a *child* of the latter node, which is called the *parent* of the level $n+1$ node. Children of a common parent are called *siblings*. There is a one-to-one correspondence between paths from the root node to the leaves and the codewords. The codewords are for this reason sometimes called "path maps." Reading the branch labels from the root on the left to the leaf on the right yields a binary codeword. By construction of the tree, no codeword can be a prefix of another codeword since codewords terminate in leaves, i.e., no other codewords begin with the same binary sequence. Conversely, given any prefix code we can represent it as a tree. An encoder is a means of assigning one of the codewords to a source symbol. It might (or might not) take advantage of the tree structure.

The Kraft Inequality

The most important fundamental property of a uniquely decodable noiseless source code is given by the *Kraft inequality* described in the following two theorems.

Theorem 9.2.1 *(The Kraft Inequality: Necessity)*

   A necessary condition for unique decodability of a noiseless source code with input alphabet $A = \{a_0, \cdots, a_{M-1}\}$, encoder α, and codeword lengths $l_k = l(a_k)$, $k = 0, 1, \cdots, M - 1$, is

$$\sum_{k=0}^{M-1} 2^{-l_k} \leq 1.$$

   Proof: Given K, an input sequence of K symbols, say $\mathbf{b} = b_0, \cdots, b_{K-1}$ has a total length $l(\mathbf{b}) = l(b_0) + l(b_1) + \cdots + l(b_{K-1})$. Let $N(L)$ denote the total number of such input sequences $\mathbf{b}$ of length K that result in a total code sequence length of $l(\mathbf{b}) = L$. Let $l_{\max} = \max\{l_k; k = 0, 1, \cdots, M - 1\}$ be the maximum possible codeword length and observe that $l(\mathbf{b}) \leq K l_{\max} = L_{\max}$. If the code is uniquely decodable, then all $N(L)$ input sequences yielding codeword sequences of overall length L must yield *distinct* codeword sequences of that length. Since there are 2^L distinct binary sequences of length L, we must have that

$$N(L) \leq 2^L. \tag{9.2.2}$$

To apply this equality to the sum considered in the theorem, consider the power

$$\left(\sum_{k=0}^{M-1} 2^{-l_k}\right)^K = \left(\sum_{a \in A} 2^{-l(a)}\right)^K$$

$$= \sum_{b_0 \in A} \sum_{b_1 \in A} \cdots \sum_{b_{K-1} \in A} 2^{-(l(b_0) + l(b_1) + \cdots + l(b_{K-1}))}$$

$$= \sum_{L=1}^{L_{\max}} N(L) 2^{-L}. \tag{9.2.3}$$

Using the bound of (9.2.2) in (9.2.3) yields

$$\left(\sum_{k=0}^{M-1} 2^{-l_k}\right)^K \leq L_{\max} = K l_{\max}$$

and therefore

$$\sum_{k=0}^{M-1} 2^{-l_k} \leq (Kl_{\max})^{\frac{1}{K}} = 2^{\frac{1}{K}\log K + \frac{1}{K}\log l_{\max}}.$$

Since the bound is true for all integer K, we can take the limit as $K \to \infty$, in which case the right hand side converges to 1, completing the proof. $\square$

Theorem 9.2.2 (*The Kraft Inequality: Sufficiency*) *If the collection of lengths l_k; $k = 0, 1, \cdots, M - 1$ satisfies the Kraft inequality*

$$\sum_{k=0}^{M-1} 2^{-l_k} \leq 1, \tag{9.2.4}$$

then there exists a uniquely decodable code for an alphabet $\{a_0, \cdots, a_{M-1}\}$ having these lengths.

Proof: We can assume without loss of generality that the codeword lengths are ordered as $l_0 \leq l_1 \leq \cdots \leq l_{M-1}$. To develop a collection of codewords satisfying the prefix condition, consider the binary tree of depth l_{M-1} with $M = 4$ depicted in Figure 9.2.

Binary codewords of length l_{M-1} and shorter can be considered as paths through the tree or, equivalently, as the terminal nodes of such a path. In the figure, a complete tree is depicted with each branch being labeled by a 0 or 1. The code is represented by the *subtree* consisting of the branches from the root of the tree to the terminal nodes (leaves) of the subtree denoted by the circles. The codewords correspond to the sequences of branch labels from the root of the tree to the leaf. The lengths of the codewords in the figure are $\{1,2,3,4,4\}$. The codewords corresponding to the leaves of the subtree are given in the boxes near the leaves.

In a general binary tree of arbitrary depth, a codeword of length l corresponds to a path of l branches in the tree beginning at the root node (depth 0) and finishing at a terminal node of depth l in the tree. The codeword is the sequence of binary labels of the branches read from the first branch to the branch at depth l.

Given a collection of lengths satisfying the Kraft inequality, pick an arbitrary node of depth l_0 and hence an arbitrary length l_0 binary sequence as the first codeword. In the figure this first choice is the single symbol sequence 0 corresponding to the downward branch emanating from the root node. Since no other code word can have this first codeword as a prefix, we *prune* the tree at the terminal node of this first codeword at depth l_0 in the tree. This removes all of the deeper nodes emanating from the terminal

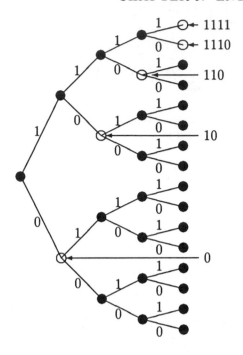

Figure 9.2: A Variable-Length Binary Tree Code

node of the first codeword from consideration as terminal nodes for other codewords.

Next pick one of the remaining available depth l_1 nodes and hence the corresponding binary l_1-tuple as the second codeword. In the figure, this is the length 2 sequence 10. Observe that there are $2^{l_1} - 2^{l_1-l_0}$ available nodes at this depth. In the example, this is $2^2 - 2^{2-1} = 2$. Then prune the tree at this terminal node and continue in this fashion. For the third word we have available $2^{l_2} - 2^{l_2-l_0} - 2^{l_2-l_1}$ available nodes. This is $2^3 - 2^2 - 2^1 = 2$.

For the mth codeword we choose one of the

$$n(m) = 2^{l_m} - \sum_{k=0}^{m-1} 2^{l_m-l_k} = 2^{l_m}(1 - \sum_{k=0}^{m-1} 2^{-l_k}). \qquad (9.2.5)$$

From (9.2.4), if $m < M$

$$n(m) = 2^{l_m}(1 - \sum_{k=0}^{M-1} 2^{-l_k} + \sum_{k=m}^{M-1} 2^{-l_k})$$

$$\geq 2^{l_m} \sum_{k=m}^{M-1} 2^{-l_k}$$

$$= 1 + \sum_{k=m+1}^{M-1} 2^{l_m - l_k},$$

which is always at least one. (The last summation is zero when $m = M-1$.) Thus for each $m \leq M - 1$ there is always a terminal node and hence a codeword of the desired depth available. This proves that a prefix code exists with the given lengths. Since a prefix code is uniquely decodable, the theorem is proved. $\qquad\square$

The Kraft inequality provides the basis for simple lower and upper bounds to the average length of uniquely decodable variable length noiseless codes. The remainder of this section is devoted to the development of the bound and some of its properties.

Entropy

We have from the Kraft inequality that

$$\bar{l}(\alpha) = \sum_{a \in A} p(a) l(a)$$

$$= -\sum_{a \in A} p(a) \log 2^{-l(a)}$$

$$\geq -\sum_{a \in A} p(a) \log \frac{2^{-l(a)}}{\sum_{b \in A} 2^{-l(b)}}, \qquad (9.2.6)$$

where the logarithm is base 2. The bound on the right-hand side has the form

$$\sum_{a \in A} p(a) \log \frac{1}{q(a)}$$

for two pmf's p and q. The following lemma provides a basic lower bound for such sums that depends only on p.

Theorem 9.2.3 *The Divergence Inequality:*
Given any two pmf's p and q with a common alphabet A, then

$$D(p\|q) \equiv \sum_{a \in A} p(a) \log \frac{1}{q(a)} \geq H(p) \equiv \sum_{a \in A} p(a) \log \frac{1}{p(a)}. \qquad (9.2.7)$$

Equality holds in (9.2.7) if and only if $p(a) = q(a)$ for all $a \in A$.

$D(p\|q)$ is called the *divergence* or *relative entropy* or *cross entropy* of the pmf's p and q. $H(p)$ is called the *entropy* of the pmf p or, equivalently, the entropy of the random variable X described by the pmf p. Both notations $H(p)$ and $H(X)$ are common, depending on whether the emphasis is on the distribution or on the random variable.

Proof: Change the base of logarithms and use the elementary inequality $\ln r \leq r - 1$ (with equality if and only if $r = 1$) to write

$$
\begin{aligned}
\sum_{a \in A} p(a) \log \frac{1}{p(a)} - \sum_{a \in A} p(a) \log \frac{1}{q(a)} &= \sum_{a \in A} p(a) \log \frac{q(a)}{p(a)} \\
&= \frac{1}{\ln 2} \sum_{a \in A} p(a) \ln \frac{q(a)}{p(a)} \\
&\leq \frac{1}{\ln 2} \sum_{a \in A} p(a)(\frac{q(a)}{p(a)} - 1) \\
&= \frac{1}{\ln 2}(\sum_{a \in A} q(a) - \sum_{a \in A} p(a)) = 0.
\end{aligned}
$$

The inequality is an equality if and only if $q(a) = p(a)$ for all a, which completes the proof. □

Combining Theorem 9.2.3 with (9.2.6) immediately yields the following lower bound:

Theorem 9.2.4 *Given a uniquely decodable scalar noiseless variable length code with encoder α operating on a source X_n with marginal pmf p, then the resulting average codeword length satisfies*

$$\bar{l}(\alpha) \geq H(p); \tag{9.2.8}$$

that is, the average length of the code can be no smaller than the entropy of the marginal pmf. The inequality is an equality if and only if

$$p(a) = 2^{-l(a)} \text{ for all } a \in A. \tag{9.2.9}$$

Note that the equality (9.2.9) follows when both (9.2.7) and (9.2.6) hold with equality. The latter equality implies that $q(b) = 2^{-l(b)}$.

Because the entropy provides a lower bound to the average length of noiseless codes and because, as we shall see, good codes can perform near this bound, uniquely decodable variable length noiseless codes are often called *entropy codes*.

To achieve the lower bound, we need to have (9.2.9) satisfied. Obviously, however, this can only hold in the special case that the input symbols all have probabilities that are powers of $1/2$. In general $p(a)$ will not have this form and hence the bound will not be exactly achievable. The practical design goal in this case is to come as close as possible. The following result uses Theorem 9.2.2 to balance Theorem 9.2.4 by showing that codes exist that are not too far above the lower bound.

Theorem 9.2.5 *There exists a uniquely decodable scalar noiseless code for a source with marginal pmf p for which the average codeword length satisfies*

$$\bar{l}(\alpha) < H(p) + 1. \tag{9.2.10}$$

Proof: Suppose that the marginal probabilities of the source are

$$p_0, p_1, \cdots, p_{M-1}.$$

For $k = 0, 1, \cdots, M - 1$ choose l_k as the integer satisfying

$$2^{-l_k} \le p_k < 2^{-l_k+1} \tag{9.2.11}$$

or, equivalently,

$$-\log p_k \le l_k < -\log p_k + 1. \tag{9.2.12}$$

Thus l_k is the smallest integer greater than or equal to $-\log p_k$. From (9.2.11) the lengths must satisfy the Kraft inequality since

$$\sum_{k=0}^{M-1} 2^{-l_k} \le \sum_{k=0}^{M-1} p_k = 1$$

and hence there is a prefix (and hence a uniquely decodable) code with these lengths. From (9.2.12), the average length must satisfy the bound of the theorem. □

9.3 Prefix Codes

Prefix codes were introduced in Section 9.2 as a special case of uniquely decodable codes wherein no codeword is a prefix of any other codeword. Assuming a known starting point, decoding a prefix code simply involves scanning symbols until one sees a valid codeword. Since the codeword cannot be a prefix for another codeword, it can be immediately decoded. Thus each codeword can be decoded as soon as it is complete, without having to wait for further codewords to resolve ambiguities. Because of this property, prefix codes are also referred to as *instantaneous codes*.

Although prefix codes appear to be a very special case, the following theorem demonstrates that prefix codes can perform just as well as the best uniquely decodable code and hence no optimality is lost by assuming the prefix code structure. The theorem follows immediately from Theorems 9.2.1 and 9.2.2 since the lengths of a uniquely decodable code must satisfy the Kraft inequality and hence there must exist a prefix code with these lengths.

Theorem 9.3.1 *Suppose that* (α, β) *is a uniquely decodable variable length noiseless source code and that* $\{l_m; \ m = 0, 1, \cdots, M - 1\} = \{l(a); \ a \in A\}$ *is the collection of codeword lengths. Then there is a prefix code with the same lengths and the same average length.*

The theorem implies that the optimal prefix code, where here *optimal* means providing the minimum average length, is as good as the optimal uniquely decodable code. Thus we lose no generality by focusing henceforth on the properties of optimal prefix codes. The following theorem collects the two most important properties of optimal prefix codes.

Theorem 9.3.2 *An optimum binary prefix code has the following properties:*

(i) *If the codeword for input symbol* a *has length* $l(a)$, *then* $p(a) > p(b)$ *implies that* $l(a) \leq l(b)$; *that is, more probable input symbols have shorter (at least, not longer) codewords.*

(ii) *The two least probable input symbols have codewords which are equal in length and differ only in the final symbol.*

Proof:

(i) If $p(a) > p(b)$ and $l(a) > l(b)$, then exchanging codewords will cause a strict decrease in the average length. Hence the original code could not have been optimum.

(ii) Suppose that the two codewords have different lengths. Since a prefix of the longer codeword cannot itself be a codeword, we can delete the final symbol of the longer codeword without the truncated word being confused for any other codeword. This strictly decreases the average length of the code and hence the original code could not have been optimum. Thus the two least probable codewords must have equal length. Suppose that these two codewords differ in some position other than the final one. If this were true, we could remove the final binary symbol and shorten the code without confusion. This is true since we could still distinguish the shorter codewords and since the prefix condition precludes the possibility of confusion with another codeword. This, however, would yield a strict decrease in average length and hence the original code could not have been optimum. □

The theorem provides an iterative design technique for optimal codes, as will be seen in the next section.

9.4 Huffman Coding

In 1952 D. A. Huffman developed a scheme which yields performance quite close to the lower bound of Theorem 9.2.2 [187]. In fact, if the input probabilities are powers of $1/2$, the bound is achieved. The design is based on the ideas of Theorem 9.3.2. Suppose that we order the input symbols in terms of probability, that is, $p(a_0) \geq p(a_1) \geq \cdots \geq p(a_{M-1})$. Depict the symbols and probabilities as a list $\mathcal{L}$ as in Table 9.3.

Symbol	Probability
a_0	$p(a_0)$
a_1	$p(a_1)$
$\vdots$	$\vdots$
a_{M-2}	$p(a_{M-2})$
a_{M-1}	$p(a_{M-1})$

Table 9.3: List $\mathcal{L}$

The input alphabet symbols can be considered to correspond to the terminal nodes of a code tree which we are to design. We will design this tree from the leaves back to the root in stages. Once completed, the codewords can be read off from the tree by reading the sequence of branch labels encountered passing from the root to the leaf corresponding to the input symbol.

The theorem implies that the two least probable symbols have codewords of the same length which differ only in the final binary symbol. Thus we begin a code tree with two terminal nodes with branches extending back to a common node. Label one branch 0 and the other 1. We now consider these two input symbols to be *tied* together and form a single new symbol in a reduced alphabet A' with $M - 1$ symbols in it. Alternatively, we can consider the two symbols a_{M-2} and a_{M-1} to be merged into a new symbol (a_{M-2}, a_{M-1}) having as probability the sum of the probabilities of the original two nodes. We remove these two symbols from the list $\mathcal{L}$ and add the new merged symbol to the list. This yields the modified list of Table 9.4.

We next try to find an optimal code for the reduced alphabet A' (or modified list $\mathcal{L}$) with probabilities $p(a_m)$; $m = 0, 1, \cdots, M - 3$ and $p(a_{M-1}) + p(a_{M-2})$. A prefix code for A' implies a prefix code for A by adjoining the final branch labels already selected. Furthermore, if the prefix code for A' is optimal, then so is the induced code for A. To prove this, observe that the

Symbol	Probability
a_0	$p(a_0)$
a_1	$p(a_1)$
$\vdots$	$\vdots$
(a_{M-1}, a_{M-2})	$p(a_{M-1}) + p(a_{M-2})$

Table 9.4: List $\mathcal{L}$ After One Huffman Step

lengths of the codewords for a_m; $m = 0, 1, \cdots, M - 3$ in the two codebooks are the same. The code book for A' has a single word of length l_{M-2} for the combined symbol (a_{M-2}, a_{M-1}) while the codebook for A has two words of length $l_{M-2} + 1$ for the two input symbols a_{M-2} and a_{M-1}. This means that the average length of the codebook for A is that for the codebook for A' plus $p(a_{M-1})$, a term which does not depend on either codebook. Thus minimizing the average length of the code for A' also minimizes the average length of the induced code for A.

We continue in this fashion. The probability of each node is found by adding up the probabilities of all input symbols connected to the node. At each step the two least probable nodes in the tree are found. Equivalently, the two least probable symbols in the ordered list $\mathcal{L}$ are found. These nodes are tied together and a new node is added with branches to each of the two low probability nodes and with one branch labeled 0 and the other 1. The procedure is continued until only a single node remains (the list contains a single entry). The algorithm can be summarized in a concise form due to Gallager [135] as in Table 9.5.

An example of the construction is depicted in Figure 9.3.

Observe that a prefix code tree combined with a probability assignment to each leaf implies a probability assignment for every node in the tree. The probabilities of two children sum to form the probability of their parent. The probability of the root node is 1.

We have demonstrated that the above technique of constructing a binary variable length prefix noiseless code is optimal in the sense that no binary variable length uniquely decodable scalar code can give a strictly smaller average length. Smaller average length could, however, be achieved by relaxing these conditions. First, one could use nonbinary alphabets for the codebooks, e.g., ternary or quaternary. Similar constructions exist in this case. (See, e.g., Gallager [134], Chapter 3.) Second, one could remove the scalar constraint and code successive pairs or larger blocks or vectors of input symbols, that is, consider the input alphabet to be vectors of input

---

Table 9.5: **Huffman Code Design**

---

1. Let $\mathcal{L}$ be a list of the probabilities of the source letters which are considered to be associated with the leaves of a binary tree.

2. Take the two smallest probabilities in $\mathcal{L}$ and make the corresponding nodes siblings. Generate an intermediate node as their parent and label the branch from parent to one of the child nodes 1 and label the branch from parent to the other child 0.

3. Replace the two probabilities and associated nodes in the list by the single new intermediate node with the sum of the two probabilities. If the list now contains only one element, quit. Otherwise go to step 2.

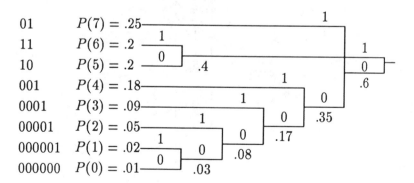

Figure 9.3: A Huffman Code

symbols instead of only single symbols. This approach is considered in Section 9.5.

The Sibling Property

In this section we describe a structural property of Huffman codes due to Gallager [135]. This provides an alternative characterization of Huffman codes and is useful in developing the adaptive Huffman code to be seen later.

A binary code tree is said to have the *sibling property* if

1. Every node in the tree (except for the root node) has a sibling.

2. The nodes can be listed in order of decreasing probability with each node being adjacent to its sibling.

The list need not be unique since distinct nodes may possess equal probabilities.

The code tree of Figure 9.3 is easily seen to have the sibling property. Every node except the root has a sibling and if we list the nodes in order of decreasing probability we have Table 9.6. Each successive pair in the ordered stack of Table 9.6 is a sibling pair.

.6
.4
.35
.25
.2
.2
.18
.17
.09
.08
.05
.03
.02
.01

Table 9.6: List of Nodes in Order of Probability

Theorem 9.4.1 *A binary prefix code is a Huffman code if and only if it has the sibling property.*

Proof: First, assume that we have a binary prefix code that is a Huffman code. We modify Gallager's version of the Huffman code design algorithm as follows:

Step 1. Let $\mathcal{L}$ be a list of the probabilities of the source letters which are considered to be associated with the leaves of a binary tree. Let $\mathcal{W}$ be a list of nodes of the code tree; initially $\mathcal{W}$ is empty.

Step 2. Take the two smallest probabilities in $\mathcal{L}$ and make the corresponding nodes siblings. Generate an intermediate node as their parent and label the branch from parent to one of the child nodes 1 and label the branch from parent to the other child 0.

Step 3. Replace the two probabilities and associated nodes in the list $\mathcal{L}$ by the new intermediate node with the sum of the two probabilities. Add the two sibling nodes to the top of the list $\mathcal{W}$, with the higher probability node on top. If the list $\mathcal{L}$ now contains only one element, quit. Otherwise go to Step 2.

The list $\mathcal{W}$ is constructed by adding siblings together, hence siblings in the final list are always adjacent. The new additions to $\mathcal{W}$ are chosen from the old $\mathcal{L}$ of Step 2 by choosing the two smallest probability nodes. Thus the two new additions have smaller probabilities (at least no greater) than all of the remaining nodes in the old $\mathcal{L}$. This in turn implies that these new additions have smaller probabilities than all of the nodes in the new $\mathcal{L}$ formed by merging these two nodes in the old $\mathcal{L}$. Thus in the next iteration the next two siblings to be added to the list $\mathcal{W}$ must have probability no smaller than the current two siblings since the next ones will be chosen from the new $\mathcal{L}$. Thus $\mathcal{W}$ has adjacent siblings listed in the order of descending probability and therefore the code has the sibling property.

Next suppose that a code tree has the sibling property and that $\mathcal{W}$ is the corresponding list of nodes. The bottom (smallest probability) nodes in this list are therefore siblings. Suppose that one of these nodes is an intermediate node. It must have at least one child which in turn must have a sibling from the sibling property. As the probabilities of the siblings must sum to that of the parent, this means that the children must have smaller probability than the parent, which contradicts the assumption that the parent was one of the lowest probability nodes. (This contradiction assumes that all nodes have nonzero probability, which we assume without any genuine loss of generality.) Thus these two bottom nodes must in fact be leaves of the code tree and they correspond to the two lowest probability source letters. Thus the Huffman algorithm in its first pass will in Step 2 assign siblings in the original code tree to these two lowest probability symbols and it can label the siblings in the same way that the siblings are labeled in the original tree. Next remove these two siblings from the code tree and remove the corresponding bottom two elements in the ordered list. The reduced code tree still has the sibling property and corresponds to the reduced code tree $\mathcal{L}$ after a complete pass through the Huffman algorithm. This argument can be applied again to the reduced lists: At each pass the Huffman algorithm chooses as siblings two siblings from the original prefix code tree and labels the corresponding branches exactly as the original siblings were labeled in the original tree. Continuing in this manner shows that the Huffman algorithm "guided" by the original tree will reproduce the same tree.   □

9.5   Vector Entropy Coding

All of the entropy coding results developed thus far apply immediately to the "extended source" consisting of successive nonoverlapping N-tuples of the original source. In this case the entropy lower bound becomes the entropy of the input vectors instead of the marginal entropy. For example, if p_{X^N} is the pmf for a source vector $X^N = (X_0, \cdots, X_{N-1})$, then a uniquely decodable noiseless source code for successive source blocks of length N has an average codeword length no smaller than

$$H(p_{X^N}) = \sum_{x^N} -p_{X^N}(x^N) \log p_{X^N}(x^N),$$

the Nth order entropy of the input. This is often written in terms of the average codeword length per input symbol and entropy per input symbol:

$$\bar{l} \geq \frac{1}{N} H(X^N). \qquad (9.5.1)$$

This bound is true for all N. Similarly, Theorem 9.2.5 can be applied to N-tuples to obtain an upper bound. The bounds are summarized in Theorem 9.5.1.

Theorem 9.5.1 *For any integer N a prefix code has average length satisfying the lower bound of (9.5.1). Furthermore, there exists a prefix code for which*

$$\bar{l} < \frac{1}{N} H(X^N) + \frac{1}{N}. \qquad (9.5.2)$$

It can be shown that if the input process is stationary, then one can take the minimum over N on the right hand side to achieve a lower bound for all N and that this minimum is $\overline{H}$, the *entropy rate* of the source defined by

$$\overline{H} = \lim_{N \to \infty} \frac{H(X^N)}{N}.$$

(See, e.g., [134].) If the source is iid, then in fact $\overline{H} = H(p_{X_0})$. Eq. (9.5.1) then implies that the average length of a uniquely decodable code can never be smaller than the entropy rate of the code, even though input blocks of arbitrarily large length are encoded.

The construction of Huffman codes extends in principle to such block inputs, but obviously the technique becomes far more complicated as the number of input symbols grows. Furthermore, if we are going to code groups of input symbols into groups of code symbols, then the previous approach of coding fixed length input blocks into variable length output blocks is not the only possible code structure. While a Huffman code may be optimal for this structure, other code structures may provide superior performance, that is, smaller average length with comparable or less complexity. One can consider codes that map variable length blocks into fixed length blocks and codes that map variable length blocks into variable length blocks. We next turn to alternative noiseless coding techniques.

9.6 Arithmetic Coding

Arithmetic coding is a direct descendent of an unpublished coding technique of P. Elias (reported in Abramson [1]) that was developed by Pasco [252] and Rissanen [271] and subsequently improved by Rissanen and Langdon [273], Jones [197], and others. A good tutorial overview and reference list may be found in the paper by Witten et al. [337].

We demonstrate the basic idea by focusing on an example of the Elias code itself. Arithmetic codes can be viewed as Elias codes with finite precision arithmetic, that is, codes which do not assume arbitrary arithmetic precision. To simplify the description we also restrict interest to a memoryless binary source. Extensions involve similar ideas with added complexity. Since we wish to compress a source with only two symbols, clearly we will have to code groups of input symbols. Unlike the vector entropy coders, however, now the number of input symbols grouped together will vary.

Once again the code will be described by a tree, a binary tree for the case of designing a code for a binary source. A good way to think of the structure of the tree is as a *classification tree* for points in the unit interval $[0, 1)$. The tree will have as an input a real number $r \in [0, 1)$ and each node will make a binary decision based on r. Based on this decision the classifier will either output a 1 and advance along the (say) right branch

emanating from the node or output a 0 and advance along the left. The tree
will not much resemble its eventual application of a noiseless code while we
construct it, instead it will look like a means of assigning a binary sequence
to a real number, reminiscent of the ordinary binary expansion of numbers
in the unit interval:

$$r = \sum_{i=1}^{\infty} u_i 2^{-i}, \qquad (9.6.1)$$

where the u_i are 0 or 1. This expansion will play a key role in using the
tree as a code, but the tree itself will not try to produce a binary sequence
$\{u_i\}$ for a real number r for which (9.6.1) is true, instead it will try to
trace a path through the tree for a given "input" r with the following
property: If r is selected at random according to a uniform distribution,
then the probability of having the classifier produce a given binary k-tuple
after k decisions is the same as the probability of the original binary source
producing that binary k-tuple. The tree can be thought of as a *model*
for the source, a means for producing binary sequences having the same
probabilities as the original source by making a sequence of deterministic
decisions on a uniform random variable. Stated in another way, there is
a one-to-one correspondence between source binary k-tuples and tree path
maps from the root to depth k. We show how such a classifier/model can
be constructed and then demonstrate how the resulting tree can be used to
noiselessly encode the original source.

Suppose that the input is a memoryless binary source with alphabet
$\{0,1\}$ and pmf $p(0) = q$, $p(1) = 1 - q$ and entropy $H(p) = -q \log q - (1 - q) \log(1 - q)$. Consider the unit interval $[0, 1)$ to consist of two subintervals with lengths proportional to the two input letter probabilities: $[0, q)$
and $[q, 1)$. We can then subdivide each of these intervals into two subintervals with lengths proportional to the two letter probabilities. Now the
four subintervals have lengths corresponding to the probabilities of all 2-
dimensional source blocks: $q^2, q(1-q), (1-q)q, (1-q)^2$. From the modeling
standpoint, a uniform input to this two-level tree will produce four binary
pairs with the same probability as will the original source. There is a one-
to-one correspondence between pairs (binary two-tuples) from the source
and these four subintervals having length equal to the probabilities of all
possible binary two-tuples. Hence we could "code" the input pairs into
subintervals in an invertible manner; that is, we could assign a subinterval to each binary pair and infer without error the original pair from the
subinterval.

We can continue this idea and recursively subdivide the unit interval
into smaller subintervals so that after the nth subdivision there would be
2^n subintervals with lengths equal to the probabilities of all possible 2^n

binary n-tuples. The tree at two levels is depicted in Figure 9.4.

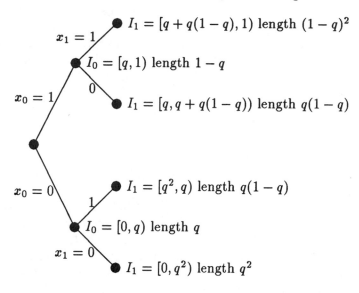

$I_1 = [q + q(1 - q), 1)$ length $(1 - q)^2$

$x_1 = 1$

$I_0 = [q, 1)$ length $1 - q$

0

$x_0 = 1$

$I_1 = [q, q + q(1 - q))$ length $q(1 - q)$

$x_0 = 0$

$I_1 = [q^2, q)$ length $q(1 - q)$

1

$I_0 = [0, q)$ length q

$x_1 = 0$

$I_1 = [0, q^2)$ length q^2

Figure 9.4: Beginning Elias Code Tree

As with pairs, in principle we could assign to each input k-tuple the one of the possible 2^k intervals having as length the probability of the k-tuple. This coding is invertible, but it is not yet a practical noiseless coding scheme because the endpoints of the intervals are in general real numbers; it is not clear how to convert these "codewords" into binary codewords for communication purposes.

At this point recall the binary expansion of (9.6.1) for $r \in [0, 1]$. The endpoints of any of the intervals can be expressed in such a fashion. This clearly assigns at least an infinite binary sequence to each interval, but in fact we can find a finite binary sequence which will tell us if we are in a particular interval or not. This finite binary sequence will serve as the codeword that specifies the interval and therefore the original source binary sequence. The idea is equivalent to viewing the source sequence as a specification of the random number r with increasing resolution as successive source symbols arrive. For a given number of source digits the encoder generates a codeword that represents the conventional binary expansion of that number r to the best precision possible from the available number of source digits. We next show how to generate this code.

Suppose we wish to encode a binary sequence $x_0, x_1, \cdots$, first look at x_0 and the corresponding interval I_0 in $[0, 1)$. If both ends of I_0 have the same first term in the binary expansion (which means the interval is entirely in $[0, 1/2)$ or $[1/2, 1)$, then the encoder will release that common bit. This becomes the first bit of the output codeword. If the first two binary symbols of the binary expansion of the endpoints of the interval corresponding to x_0 agree, then the encoder will check the next binary symbol. As many such symbols that agree are released to the channel. When no more binary symbols match, the encoder proceeds to the next input symbol x_1 and inspects its corresponding interval. If the decoder gets a bit at time 0, it will know which of the intervals was seen and hence what x_0 was.

If, on the other hand, the interval endpoints of the first interval do not have a common first binary symbol, the encoder sends nothing and instead immediately inspects the second input symbol x_1. The observed pair x_0, x_1 now corresponds to a specific subinterval of $[0, 1)$ with length equal to the probability of x_0, x_1. The encoder again checks the endpoints of this interval to see if the first binary term agrees (which, as before, would specify that the subinterval lay either entirely on the left or entirely on the right of $1/2$). If the first binary symbol agrees, the encoder can release that binary symbol to the decoder. On seeing that symbol, the decoder will be able to determine x_0 as above. The encoder can then check the second binary symbol in the expansion for the endpoints of the second level subinterval. If they agree, the common symbol is released. If not, a new symbol is checked.

If even the first binary symbols of the endpoints do not agree, then the encoder must look at more input symbols before releasing any binary symbols. The initial steps are depicted in Figure 9.5 for the case where $q = \Pr(x_i = 0) = 1/4$.

The encoder continues in this way: at each time it views a new input symbol and then looks at the endpoints of the corresponding subinterval. The subintervals are shrinking with each new input symbol. If there are any new binary symbols in accord (symbols not already sent), then they are released to the decoder. As the decoder receives these binary symbols, it can determine with ever increasing accuracy a subinterval (having length a power of $1/2$) which contains the "code" subinterval and hence can reconstruct the input sequence.

To illustrate how compression occurs, consider the special case of Figure 9.5. Consider only the effects of coding the first three source symbols, x_0, x_1, x_2. Table 9.6 shows the source symbols, the resulting binary words produced by the Elias code after the three source symbols have been encoded, the resulting length of the codeword, and the probability of seeing that source sequence and hence having that length. The average channel

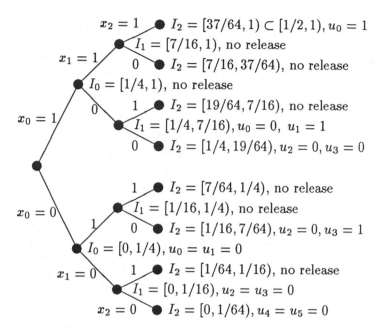

Figure 9.5: Initial Elias Code Encoding

x_0, x_1, x_2	$u_0, u_1, u_2, u_3, u_4, u_5$	l	$p(x_0, x_1, x_2)$
111	1	1	27/64
110		0	9/64
101	01	2	9/64
100	0100	4	3/64
011	00	2	9/64
010	0001	4	3/64
001	0000	4	3/64
000	000000	6	1/64

Table 9.7: Elias Code Example

codeword length considering only the encoder up to three symbols is

$$\bar{l} = \frac{27 + 18 + 12 + 18 + 12 + 12 + 6}{64} = \frac{105}{64} \approx 1.83,$$

slightly better than 3:2 compression. In fact an arithmetic code would be applied to a very long input sequence and the above analysis is applicable to only a single application of the code and not to a sequence of applications as considered with the Huffman code. In particular, the above code is not uniquely decodable and it does not meet the prefix condition. The code could be improved slightly by realizing that for a single use, we could shorten several of the words and still be able to successfully decode, e.g., 0100 could be replaced by 010. The example is intended simply to show how an arithmetic code achieves compression, not to provide a practical code.

To achieve compression, one codes long strings of input symbols. This, however, places impossible demands on the precision of the arithmetic as the length of the input sequence grows. Modifying the algorithm to incorporate occasional rescaling and finite precision arithmetic in a consistent way yields an arithmetic code.

We now describe in somewhat more detail the workings of the basic Elias code. The encoder at each time n will look at an input symbol x_n and then, given its past actions, determine a subinterval $I_n = [a_n, b_n)$. It may or may not then output code symbols, depending on what I_n is. Beginning at time $n = 0$, if the first input symbol $x_0 = 0$, set $I_0 = [0, q)$. If the first symbol is a 1, set $I_0 = [q, 1)$. Thus the time 0 subinterval has length equal to the probability of the symbol seen. Note that if we know the first subinterval, we also know the first input symbol x_0. We then divide the selected subinterval I_0 into two further subintervals proportional to the input probabilities: $[a_0, a_0 + q(b_0 - a_0))$ (having length $q(b_0 - a_0)$) and $[a_0 + q(b_0 - a_0)), b_0)$ (having length $(1 - q)(b_0 - a_0)$). If $x_1 = 0$, then I_1 is the first subinterval. Otherwise it is the second. Knowing the second subinterval tells us the second symbol x_1. In addition, it specifies the first subinterval since it is a subset of the first subinterval. Thus it specifies also the first input symbol. This procedure is then continued, each time dividing the previous subinterval into two subintervals proportional to the input probabilities. In general the new endpoints are given in terms of the previous endpoints by

$$a_n = \begin{cases} a_{N-1} & \text{if } x_n = 0 \\ a_{N-1} + q(b_{N-1} - a_{N-1}) & \text{if } x_n = 1 \end{cases}$$

$$b_n = \begin{cases} a_{N-1} + q(b_{N-1} - a_{N-1}) & \text{if } x_n = 0 \\ b_{N-1} & \text{if } x_n = 1 \end{cases}.$$

The algorithm produces at time n a subinterval of length equal to the probability of the input sequence produced up to that time. Knowing the subinterval I_n is sufficient to completely determine the original input sequence up to the nth symbol, i.e., $x_0, \cdots, x_{n-1}$.

The sequence of subintervals I_n is itself used to produce the code symbol sequence as follows. For each n, the subinterval endpoints a_n and b_n of I_n both have binary expansions. At time 0 check to see if the first term in the binary expansions of a_0 and b_0 agree. This will be the case if either $I_0 \subset [0, 1/2)$ or $I_0 \subset [1/2, 1)$. If this is the case, the encoder produces the common symbol in the binary expansion, $u_0 = 0$ if $I_0 \subset [0, 1/2)$ and $u_0 = 1$ if $I_0 \subset [1/2, 1)$. If further binary symbols agree, then these too are released.

If the first symbols in the binary expansions of the interval endpoints do not agree, then no encoder symbol is output and the same test is conducted on I_1. The encoder repeats the test for I_n for increasing n until an n is found for which the first symbol of the binary expansions of a_n and b_n agree, at which point the common binary symbol is released to the channel. The encoder then tests to see if the second symbols in the binary expansion agree. If yes, the second symbol is output to the channel. If no, the next subinterval is tested and so on until an n is achieved for which the first two binary symbols in the binary expansion of the endpoints of a_n and b_n of I_n agree.

In general, at time n the encoder will have found the largest k for which the first k binary symbols in the binary expansions for a_n and b_n agree and it will have produced the common symbols as the output code symbols $u_0 u_1 \cdots u_k - 1$. This condition is equivalent to

$$I_n \subset [\sum_{i=0}^{k-1} u_i 2^{-i}, \sum_{i=0}^{k-1} u_i 2^{-i} + 2^{-k}) \qquad (9.6.2)$$

with k being the largest integer for which the formula holds.

The decoder upon encountering k symbols from the encoder will know that the above inclusion is true and will be able to reconstruct the corresponding n input symbols. For example, denote the interval on the right-hand side by J_k. The decoder tests to see if $J_k \subset [0, q)$ or $J_k \subset [q, 1)$. If the former is true, then $x_0 = 0$. If the latter is true, $x_0 = 1$. One of them must be true since the encoder released symbols. The decoder continues in this way, checking to see if J_k belongs to one of the possible I_j for increasing j (the possible I_j are found by the same recursion used at the input) until it is found that J_k is not a subset of one of the possible I_j, at which point the decoder must wait for more encoded symbols.

The code is noiseless and must look at a variable number of input symbols for each code symbol produced. Although we will not prove it, it can be shown that for an iid input the average number of code symbols produced for each input symbol will converge to the entropy of the source as the encoded sequence becomes long. Unfortunately, however, the code as described is impracticable because the precision required to specify the interval endpoints grows without bound. This defect can be surmounted by modifying the encoding algorithm to use fixed precision arithmetic. Roughly speaking, one simply computes the needed intervals approximately to within the accuracy of the fixed precision arithmetic. In order to avoid overlapping intervals due to the approximation of the endpoints, a rule is needed to adjust the endpoints so as to produce disjoint intervals. This can be accomplished by suitable scaling and rounding. Although the resulting code no longer yields an average word length exactly equal to the entropy, it can be made arbitrarily close by using sufficiently high precision arithmetic. Furthermore, the approach can be extended to sources with memory by carving up the unit interval according to conditional probabilities instead of marginal probabilities. The conditional probabilities can, in turn, be estimated from the source itself while coding is going on [273].

Arithmetic coding is more complicated to implement than Huffman coding, but its compression is typically greater and hence it is a popular approach for entropy coding where the extra compression justifies the extra complexity.

9.7 Universal and Adaptive Entropy Coding

Both Huffman codes and arithmetic codes assume *a priori* knowledge of the input probabilities. This information is often not known in practice. Furthermore, the probabilities may change with time because of the non-stationarities of real data, e.g., different computer files may have differing probabilities of 0 and 1, varying from equally distributed (for programs) to highly skewed (for facsimile data). Hence better performance will usually be achieved if a code is flexible or robust in the sense of being able to change according to the local statistical behavior of the input being compressed. In other words, a smart code should *adapt* to the source at hand.

Perhaps the earliest approach to adaptive entropy coding was that developed by Robert Rice of JPL and subsequently called the "Rice machine." (See, e.g., [265].) In his example the input process tended to have two distinct modes with corresponding distributions. The modes would remain in effect for long periods of time relative to the codeword sizes. Hence his simple but elegant solution was to design two entropy codes, one for each mode. A long block of input symbols could be encoded by simultaneously encoding the input with both codes and seeing which code yielded the most compression. The encoder then sent one bit describing which code was used followed by the long encoded sequence to the decoder. The lead bit told the decoder which decoder to use to produce the original sequence. The single bit overhead describing which code to use could be made small (in its contribution to the overall bit per sample) by making the "superblock" length large. This approach to noiseless coding was an early example of what later came to be known as *universal codes*: have a collection of codes matched to different input modes and choose the code which yields the best compression. Alternatively one can observe the input sequence and guess (or estimate or identify) which mode is in effect, possibly by looking at the histograms or relative frequencies of symbol occurrences, and then choose the code designed for that mode. This latter encoder tends to be simpler than the universal encoder if there are many modes, but it might not choose the best code for the input sequence (that is, the code designed for the mode guessed to be in effect might not yield the best compression on the current input sequence). This latter approach, estimating the input statistics and using a code matched to those statistics, is referred to as *adaptive entropy coding*. Clearly the universal and adaptive techniques are intimately related. We shall describe a few adaptive techniques here. A good survey of universal entropy coding may be found in [94] [96]. Universal extensions of arithmetic codes are considered in [273].

Lynch-Davisson and Enumerative Coding

One of the earliest adaptive codes was a simple and natural means of encoding binary vectors observed by Lynch and Davisson [226] [93] and generalized by Cover [81]. The idea is this: Given an input vector of dimension N, first count the number of 1's and call this number w (the *weight* of the vector). The entropy encoded vector then consists of a prefix giving this count (possibly in binary notation for a binary code) followed by an index specifying which of all of the

$$\binom{N}{w} = \frac{N!}{w!(N-w)!}$$

possible weight w vectors is input. The references provide simple algorithms for computing this index or *enumerating* the collection of all weight w vectors. The decoder then reverses the procedure. Note that the determination of w can be viewed as an estimate of the underlying binary symbol probability since w/N is the relative frequency of 1's in the input sequence. It can be shown that as the vector size becomes large, this approach performs near the entropy bound if the source is memoryless. In other words, the code is nearly optimal even though its underlying statistics are not known in advance.

Adaptive Huffman Coding

In principle it is straightforward to make the Huffman code adaptive by combining the ideas seen thus far. One approach is to observe a large block of N symbols and estimate the underlying probability distribution from the relative frequencies of the symbols. The estimate of symbol a occurring would then be simply $n_N(a)/N$, where $n_N(a)$ is the count of a's appearance in the block of length N. This probability vector would serve as a prefix to a long codeword. Given this vector, both encoder and decoder could design an identical Huffman code which would then be used to actually encode the entire data block. If the block is large enough, the overhead information in the prefix would contribute only a small amount to the average bit rate. Although straightforward, this scheme has the obvious drawback of a large delay. In addition, it implicitly assumes that the underlying distributions remain constant over the large block size.

Another approach is more adaptive in spirit. Suppose that we have an initial Huffman code based on *a priori* statistics, but we wish to modify the estimates of these probabilities as more data arrive and to adapt the code correspondingly. A strategy similar to the previous construction would be as follows. Suppose that at time $N - 1$ we have probability estimates $p_{N-1}(a_i)$ for all of the source symbols $a_0, a_1, \cdots, a_{M-1}$,

$$p_{N-1}(a_i) = \frac{n_{N-1}(a_i)}{N - 1},$$

along with the corresponding Huffman code. The Nth input symbol $x_N = a$ is then encoded and decoded using this Huffman code and all of the probabilities are updated using the new relative frequencies: Since only the count for the symbol a is changed,

$$p_N(a) = \frac{n_N(a)}{N} = \frac{(N - 1)p_{N-1}(a) + 1}{N}$$

$$p_N(a_i) = \frac{n_N(a_i)}{N} = \frac{(N - 1)p_{N-1}(a_i)}{N} \text{ if } a_i \neq a.$$

These new and improved probabilities are known to both the encoder and decoder, which can then design a new Huffman code for use on the next input symbol.

Although theoretically a reasonable way to adapt, this code has complexity bordering on the ridiculous. For each new symbol a brand new code must be designed. Furthermore, the increasing precision demanded of the arithmetic as N grows large is not practicable. A more practical approach is found by using the structure of Huffman codes and emphasizing the counts rather than the relative frequencies. The main idea is simple. Because of the sibling property of Huffman codes, we need not redesign the entire tree at each step. If the probability estimates are modified by a new symbol so that a change in the ordering results and the sibling property is violated, we only have to do some minor surgery on the tree to regain the sibling property and hence have a Huffman code for the modified probability estimates.

Suppose now that instead of saving the probabilities, we save something proportional to probabilities. For the moment suppose that at each step we associate a set of *weights* w_i with the source symbols a_i. The weights are nonnegative and we could form explicit estimates of the probabilities by normalizing; that is,

$$p(a_i) = \frac{w_i}{\sum_{k=0}^{M-1} w_k},$$

but in fact we will not do this. The weights can be used in place of probabilities to design a Huffman code and to find the corresponding ordered tree $\mathcal{W}$ of siblings guaranteed by the sibling property.

Suppose now that at time N we have (as before) a Huffman code together with its ordered list $\mathcal{W}$ (now containing the weights instead of the probabilities, but the order of listing nodes is the same). As before we see a new symbol and we begin to encode it with the currently available tree. We first select the leaf corresponding to the symbol in the current tree. Before continuing through the tree, however, we increment the weight of this node by one; that is, the new weight of the node is 1 plus the old weight. We then check the ordered list to see if the weight of this node is still no greater than the node above it in the ordered list. If this is the case, then this part of the tree has not changed and we advance to the parent node of the leaf, producing the branch label on the way. If, on the other hand, the weight of the current node now exceeds that of the node above it on the list, we no longer have an ordered list and the old tree is no longer a Huffman code for the new set of weights. Thus we must rearrange things before advancing. Look above the current node and find the highest (largest weight) node in the list that has a weight that is less than the current node. Exchange the current node with that higher node, forcing its weight to be listed correctly in the ordered list. This exchange carries with it all of the corresponding subtree information; that is, the current node loses its old parent and inherits the parent of the former occupant of the higher place in the list. Likewise the dethroned higher node exchanges parents. (Tracking the details of the tree exchanges requires an appropriate data structure for storing the trees and is treated in some detail in Gallager [135].) Now that the weightier current node has moved up to its appropriate (higher weight) position in the ordered list $\mathcal{W}$, we once again have a Huffman code for the new weights (at least up to the current node in its new position). Thus we can advance to the (new) parent node and produce an output symbol, the label of the branch.

We have now either arrived at a new node in the original tree, or exchanged nodes and arrived at a new node in a modified tree. Either way we have produced a symbol and arrived at a new node. The process now repeats. Augment the weight of the new node and check its position in $\mathcal{W}$. Either the tree is not changed and we can advance, or we must exchange nodes and advance in a modified tree. This process continues until we arrive at the root, at which time a complete word is produced for the input symbol. Note that the weights of all the nodes visited have been augmented by one.

An obvious difficulty with this scheme is that as the number of inputs observed increases, the weights grow without bound. This can be resolved by periodically dividing the weights by some constant to reset them. The period of this rescaling and the size of the constant effectively determine how slowly or rapidly the adaptation algorithm forgets the past.

The program called "compact" on Unix systems is an adaptive Huffman code based on Gallager's algorithm. As described in the on-line manual, the algorithm compresses text by 38%, Pascal source code by 43%, C source code by 36%, and binary by 19%. Thus the compression is typically less than 2:1.

A final adaptive code which makes no assumptions on the input statistics and which does not explicitly estimate those statistics is of sufficient importance to merit its own section. The next section is devoted to a simple and elegant technique for noiseless coding due to Ziv and Lempel.

9.8   Ziv-Lempel Coding

The final noiseless code that we consider is inherently universal in its operation. It is called the Ziv-Lempel code after its inventors [354], [355]. The code shares the property of the arithmetic codes that variable numbers of input symbols are required to produce each code symbol. Unlike both Huffman and arithmetic codes, however, the code does not use any knowledge of the probability distribution of the input. As with the Elias code, the Ziv-Lempel code achieves the entropy lower bound when applied to a suitably well-behaved source. Also as with the Elias code, the Ziv-Lempel code is not practicable in its simplest form and variations are required for real applications that require some loss of optimality.

Once again we sketch the basic idea of the code by means of a binary example. We use a variation of the Ziv-Lempel algorithm suggested by Welch [327], who corrected an error in the original algorithm and whose paper was largely responsible for popularizing the algorithm in the computer science community.

The basic idea is that an input sequence is recursively parsed into nonoverlapping blocks of variable size while constructing a dictionary of blocks seen thus far. The dictionary is initialized with the available single symbols, here 0 and 1. Each successive block in the parsing is chosen to be the largest (longest) word ω that has appeared in the dictionary and hence the word ωa formed by concatenating ω and the following symbol is not in the dictionary. Before continuing the parsing ωa is added to the dictionary and a becomes the first symbol in the next block. Before detailing the parsing and dictionary construction, we show how the parsing alone works. Suppose that we see the sequence

$$01100110010110000100110.$$

Parsing this sequence using the above rule yields the following segmentation, where the parsed blocks are enclosed in parentheses and the new dictionary word (the parsed block followed by the first symbol of the next block) written as a subscript:

$$(0)_{01}(1)_{11}(1)_{10}(0)_{00}(01)_{011}(10)_{100}(01)_{010}(011)_{0110}$$

$$(00)_{000}(00)_{001}(100)_{1001}(11)_{110}(0).$$

The parsed data implies a code by indexing the dictionary words in the order in which they were added and sending the indices to the decoder. We now consider these operations in more detail.

For an input sequence $x_0 x_1 \cdots$ we will produce an output binary sequence $u_0 u_1 \cdots$. The encoder will map a variable number of input symbols into a fixed number, say N, of coded symbols. The N coded symbols should be thought of as an integer between 0 and $2^N - 1$. Both encoder and decoder will construct a code table of 2^N entries as the data is processed. The table will consist of variable length input strings, each assigned an integer (or binary N-tuple) codeword. At time 0 the encoder and decoder both have identical initial tables assigning integers to single input symbols. At step n in the encoding process the encoder will look at the $k(n)$th input symbol, where the pointer $k(n)$ will be made explicit shortly. The encoder then finds the largest integer l for which the sequence $x_{k(n)}, x_{k(n)+1}, \cdots, x_{k(n)+l-1}$ is contained in its table and hence the sequence $x_{k(n)}, x_{k(n)+1}, \cdots, x_{k(n)+l}$ is not contained in the table. The encoder then produces as a codeword the integer index of the existing word $x_{k(n)}, x_{k(n)+1}, \cdots, x_{k(n)+l-1}$ and also adds the new word $x_{k(n)}, x_{k(n)+1}, \cdots, x_{k(n)+l}$ to the table, assigning it the next available index. The encoder then sets the pointer for the next step at $k(n+1) = k(n) + l$, that is, it will begin the next step on the last input symbol considered in the current step.

As an example of encoding, suppose we begin with Table 9.8 and we

Input String	Index
0	0
1	1

Table 9.8: Initial Code Table

wish to encode the binary string 0110011001011000010011 0. The steps are depicted in Tables 9.9 through 9.15.

Input String	Index
0	0
1	1
01	2

Table 9.9: Step 1: Longest string in table: 0, Output: 0, Add to table: 01, pointer: 2

Input String	Index
0	0
1	1
01	2
11	3

Table 9.10: Step 2: Longest string in table: 1, Output: 1, Add to table: 11, pointer: 3

Input String	Index
0	0
1	1
01	2
11	3
10	4

Table 9.11: Step 3: Longest string in table: 1, Output: 1, Add to table: 10, pointer: 4

Input String	Index
0	0
1	1
01	2
11	3
10	4
00	5

Table 9.12: Step 4: Longest string in table: 0, Output: 0, Add to table: 00, pointer: 5

Input String	Index
0	0
1	1
01	2
11	3
10	4
00	5
011	6

Table 9.13: Step 5: Longest string in table: 01, Output: 2, Add to table: 011, pointer: 7

Input String	Index
0	0
1	1
01	2
11	3
10	4
00	5
011	6
100	7

Table 9.14: Step 6: Longest string in table: 10, Output: 4, Add to table: 100, pointer: 9

Input String	Index
0	0
1	1
01	2
11	3
10	4
00	5
011	6
100	7
010	8

Table 9.15: Step 7: Longest string in table: 01, Output: 2, Add to table: 010, pointer: 11

Looking through the sequence of tables, it can be seen that things start slowly as the table builds. Observe that the input words in the table are always prefixes of longer words as the longer words are added. At the completion of all the steps in the tables, the encoder has produced the sequence of integers (or N-dimensional binary vectors)

$$0110242.$$

Before describing the operation of the decoder we observe an important property of the encoder:

The Last-First Property: The *last* symbol of the most recent word added to the table is the *first* symbol of the next parsed vector.

The decoder starts with the same table and hence immediately decodes the first symbol as 0 and the second as 1. The decoder recognizes the new string 01 and adds it to its table with index 2. The next symbol is a 1, which means that an input of 1 was seen by the encoder. The decoder sees the string 11 terminating with the current symbol and adds it to its table with index 3. On seeing the fourth symbol 0, the decoder again knows the encoder saw a 0 and it recognizes the sequence 10 terminating with the current symbol as a new one for its table with index 4. On encountering the code symbol 2, the decoder knows that the encoder saw the pattern 01 and decodes this pattern. It also recognizes the previously decoded 0 and the first 0 of this new decoded pattern as a 00, which must be added to the table with index 5. The next symbol is a 4, which means that the decoder produces a 10. The previous pattern 01 combined with the first symbol of the new pattern produces a 011, which is added to the table with index 6. In this way the decoder builds its copy of the encoder table. This procedure can be summarized as in Table 9.16. The question marks denote symbols that cannot yet be determined.

In	Out	Reconstructed Sequence	Add to Table
0	0	$(0)_{0,?}$	
1	1	$(0)_{0,1}(1)_{1,?}$	$(0,1)$ (as 2)
1	1	$(0)_{0,1}(1)_{1,1}(1)_{1,?}$	$(1,1)$ (as 3)
0	0	$(0)_{0,1}(1)_{1,1}(1)_{1,0}(0)_{0,?}$	$(1,0)$ (as 4)
2	$(0,1)$	$(0)_{0,1}(1)_{1,1}(1)_{1,0}(0)_{0,0}(01)_{0,1,?}$	$(0,0)$ (as 5)
4	$(1,0)$	$(0)_{0,1}(1)_{1,1}(1)_{1,0}(0)_{0,0}(01)_{0,1,1}(10)_{1,0,?}$	$(0,1,1)$ (as 6)

Table 9.16: Decoder

Unfortunately, there is a problem with the algorithm as described. There exists a type of sequence which at first glance appears confusing to the decoder. Consider, for example, a ternary source with symbols $\{0, a, b\}$ and an input sequence

$$a000ba0a \cdots .$$

The initial table is given in Table 9.17.

Input String	Index
0	0
a	1
b	2

Table 9.17: Initial Code Table

The encoder will parse this sequence as

$$(a)_{a,0}(0)_{0,0}(0,0)_{0,0,b}(b)_{b,a}(a0)_{a,0,a} \cdots$$

and take the actions depicted in Table 9.18.

Time	Send	New Entry	Index
1	1 (for 0)	$(a,0)$	3
2	0 (for 0)	$(0,0)$	4
3	4 (for $(0,0)$	$(0,0,b)$	5
4	2 (for b)	$(b,0)$	6
5	3 (for $(a,0)$)	$(a,0,a)$	7

Table 9.18: Ternary Encoder

The decoder then begins as previously described and performs the actions of Table 9.19. The problem is immediate. The decoder receives an index for table entry 4, but the entry does not yet exist in the table! Without additional guidance, the decoder is stuck. It turns out that this behavior can arise whenever one sees a pattern of the form $x\omega x\omega x$, where x is a single symbol and ω is either empty or a sequence of symbols such that $x\omega$ already appears in the encoder and decoder table, but $x\omega x$ does not. In our example $x = 0$ and ω is empty. The encoder will send the codeword for $x\omega$ and add $x\omega x$ ((0,0) in the example) to the table with a new index i (4 in the example). Next it will parse $x\omega x$ and send the new index i

In	Out	Reconstructed Sequence	Add to Table
1	a	$(a)_{a,?}$	
0	0	$(a)_{a,0}(0)_{0,?}$	(a,0) (as 3)
4	?	$(a)_{a,0}(0)_{0,?}$	?

Table 9.19: Ternary Decoder

corresponding to the just added word. The decoder will receive the index i for $x\omega x$, but will not yet have this word in its table because it has not yet determined the character that terminates the previously received string $x\omega$ to complete the previously added table entry. In other words, the decoder knows that i corresponds to a new table entry of the form $x\omega?$, where ? is the input signal that followed the $x\omega$ previously decoded. But the final symbol ? of the previously added table entry must be the first symbol in the current block being decoded using the Last-First Property, which the decoder knows is an x. This tells the decoder that the new encoder entry is $x\omega x$ and the decoder can proceed.

In the example, the decoder knows that entry 4 must be an extension of the previously decoded string, that is, have the form $x?$ where x is the last encoded string, 0 in the current example, and ? is a single letter. The Last-First Property, however, implies that the last letter (? here) of the last entry must be the first letter of the current decoded string, $x = 0$ in the current example. Thus 4 must be $(0,0)$ and decoding can proceed. This is Welch's fix to the basic Ziv-Lempel algorithm.

The code is noiseless and can be shown to be optimum in the limit of unbounded table size. The disadvantage of the algorithm is that unmanageably large tables may be required in some applications in order to achieve the desired performance. Any real application must necessarily have a bound on table size. Once reached, the encoder can no longer add codewords and must simply use the existing dictionary. This dictionary can be used in a static fashion to encode the remaining inputs, or it can dynamically adapt to track varying input behavior. Bunton and Borriello [39] evaluate practical Ziv-Lempel dictionary adaptation schemes with respect to their resource consumption and effects on compressor performance.

9.9 Quantization and Entropy Coding

Until now we have been concerned in this chapter with noiseless coding of arbitrary discrete sources and we have focused on the noiseless coding algorithms themselves. One of the most common discrete alphabet sources is the output of a quantizer and hence in this section we focus on the natural combination of quantization and entropy coding. In part there is nothing new here, we could simply combine the separate design techniques treated in Chapter 6 for quantizers and in this chapter for entropy codes. Three issues merit further investigation, however.

- Can Bennett-style high resolution approximations be used to predict the tradeoffs between average distortion and the final rate in bits per second, that is, can we include the additional compression provided by entropy coding into the performance figures of Chapters 5 and 6?

- If a quantizer is to be followed by an entropy code, then should the quantizer be designed as previously done in Chapter 6? Perhaps the use of a quantizer in cascade with an entropy coder has an effect on the design philosophy.

- How far is a combined quantizer/entropy code from the optimal achievable rate-distortion tradeoff over all possible coding structures?

In this section we develop and interpret the following answers to these questions.

- If one follows a quantizer by an entropy coder, then the resolution of the quantizer is more meaningfully measured by the entropy of the output rather than the log of the codebook size (since subsequent noiseless coding can achieve performance near the entropy). The approximations of the Bennett theory can be used to approximate the first order entropy of the quantizer output as well as the average distortion. Subsequent noiseless coding can achieve an average number of bits per input symbol close to the entropy and hence the high resolution theory indeed provides a rate-distortion analysis for the combined quantization and entropy coding.

- Using the high resolution approximations for both entropy and average distortion, the optimization of Chapter 6 can be modified to minimize the average distortion for a fixed entropy instead of for a fixed codebook size. The resulting minimum is achieved by a uniform quantizer and not by the same quantizer that was optimal subject to a codebook size constraint. Thus the theory of Chapter 6 provides the approximately optimal quantizer when entropy coding is not used, but a uniform quantizer is approximately optimal if entropy coding is used. In both cases "optimal" means the minimum average distortion for a fixed resolution or rate, the difference being in how resolution is measured.

- The high resolution theory in combination with Shannon's rate distortion theory shows that simple uniform quantization followed by an entropy coder can yield a system which for a given average distortion yields a rate within 0.255 bits of the optimal achievable performance for memoryless sources. Thus at least for large bit rates, the variable rate scheme comprising a uniform quantizer followed by an entropy coder is nearly optimal for memoryless sources.

We begin by assuming that Q is a quantizer with quantization cells and output levels $\{R_i, y_i;\ i = 1, \cdots, N = 2^R\}$, where R is the resolution of the quantizer. We do not assume for the moment that the quantizer is optimum in the sense of minimizing the mean squared error for the given R, i.e., it need not satisfy the Lloyd conditions. The original way of encoding the quantizer levels for communication over a channel was simply to send a binary R-tuple giving the binary representation for the index i of the reproduction level nearest to the input. This results in sending R bits for each input sample or symbol. If the input is a stationary and ergodic process $\{X_n\}$, then the resulting quantized process $\{Q(X_n)\}$ is also stationary and ergodic and, in addition, has a finite alphabet. Thus we can apply all of the noiseless coding techniques of this chapter to this quantized source. In particular, we should be able to construct a noiseless code for each quantized symbol so that the average bits per sample required is not too much larger than the marginal entropy of the quantized process defined by

$$H_Q = H(Q(X_0)) = -\sum_{i=1}^{2^R} P(R_i) \log P(R_i), \qquad (9.9.1)$$

where

$$P(R_i) = \Pr(Q(X_0) = y_i) = \int_{R_i} f_X(x)\, dx, \qquad (9.9.2)$$

where f_X is the pdf for X_n. From the divergence inequality (see Theorem 9.2.3) it follows that

$$H_Q \leq R, \qquad (9.9.3)$$

(see Problem 9.1) and hence noiseless coding should provide further compression. The questions remain: how much further compression can be achieved and should the quantizer design be modified? Before trying to respond to these questions, it is worth making a few observations.

We know that more compression can be achieved by coding vectors instead of scalars, hence it is natural to wonder why we are focusing on entropy coding only single quantized samples at a time. The answer is simplicity; it is easier to entropy code scalars and entropy codes can be quite complicated if the input alphabets are too large. We already have an alphabet of size 2^R and this would grow exponentially fast by coding several quantized samples at a time. It is of practical interest to know how well a sensibly designed simple system can work. In later chapters we will explore a variety of techniques for coding vectors; here we consider only scalars. It is true that only coding scalars may result in a code that is as far as 1 bit away from H_Q, but we will focus on the high-resolution Bennett approximations where one more bit is not a serious concern.

To continue we again invoke the high-resolution approximation of Chapter 5 and develop optimality results similar to the high-resolution results of Chapter 6. As before we assume that 2^R is large, that only granular noise is important, and that the maximum cell length is small. Thus the approximation of (5.6.9) for average distortion D is valid. In particular, if $N = 2^R$ and $\lambda(x)$ is the quantizer point density function, then

$$D \approx \frac{1}{12} \frac{1}{N^2} \int f_X(x)\lambda(y)^{-2}\, dy = \frac{1}{12} E\left(\frac{1}{(N\lambda(X))^2}\right), \qquad (9.9.4)$$

where the final form will prove useful shortly. Recall also the approximation of (5.6.3) for the cell probabilities

$$P_i = P(R_i) \approx f_X(y_i)\Delta_i$$

and, from (5.6.7), the relation $\Delta_i = \frac{1}{(N\lambda(X))}$. Hence,

$$P_i \approx \frac{f_X(y_i)}{N\lambda(y_i)}. \qquad (9.9.5)$$

This provides a Bennett-style approximation to the quantizer output entropy:

$$
\begin{aligned}
H_Q &= -\sum_{i=1}^{N} P_i \log P_i \\
&= -\sum_{i=1}^{N} \frac{f_X(y_i)}{N\lambda(y_i)} \log \frac{f_X(y_i)}{N\lambda(y_i)} \\
&= -\sum_{i=1}^{N} \Delta_i f_X(y_i) \log f_X(y_i) - \sum_{i=1}^{N} \Delta_i f_X(y_i) \log \frac{1}{N\lambda(y_i)} \\
&\approx -\int f_X(y) \log f_X(y)\, dy - \int f_X(y) \log \frac{1}{N\lambda(y)}\, dy \\
&\equiv h(X_0) - E[\log \frac{1}{N\lambda(X)}], \qquad (9.9.6)
\end{aligned}
$$

where the final approximations follow the same reasoning as that used to obtain the distortion integral (5.6.9) and where $h(X_0)$ is Shannon's *differential entropy* of the random variable X_0, the continuous alphabet analog of the entropy.

In Section 6.3, Hölder's inequality was used to find the optimal quantizer point density function in the sense of minimizing D subject to a constraint on the total number of quantization levels N. In particular, if we required that $N \leq 2^R$ for a fixed number R, the minimum D and the λ achieving that D were found. Now suppose that we do similar analysis with the alternative constraint that H_Q be required to be less than or equal to a fixed value, that is, now we will require that $H_Q \leq R$ and ask how low can the average distortion D be made and what λ achieves this. Toward this end write

$$H_Q \approx h(X_0) - \frac{1}{2}E[\log \frac{1}{(N\lambda(X))^2}]. \qquad (9.9.7)$$

We can then apply Jensen's inequality which states that if ϕ is a convex function, then $E[\phi(X)] \geq \phi(E[X])$ with equality if and only if X is a constant with probability one. In (9.9.7) the function $-\log$ is convex and hence we have that

$$H_Q \geq h(X_0) - \frac{1}{2}\log E\left(\frac{1}{(N\lambda(X))^2}\right) \qquad (9.9.8)$$

$$\approx h(X_0) - \frac{1}{2}\log 12D, \qquad (9.9.9)$$

where the second approximation is (9.9.4). Equality is achieved above if and only if $\lambda(X)$ is a constant, that is, if the quantization point density is uniform. Thus we have shown that a uniform quantizer achieves the minimum entropy for a fixed average distortion and that the entropy is given approximately by

$$H_Q \approx h(X_0) - \frac{1}{2}\log 12D. \qquad (9.9.10)$$

Equivalently, for a constrained output entropy the minimum average distortion is achieved similarly by the uniform quantizer and

$$D \approx \frac{1}{12}2^{-2(H_Q - h(X))}. \qquad (9.9.11)$$

The surprise here is that if one wishes to measure bit rate by entropy rather than resolution, that is, by the achievable bit rate using entropy coding and not just by log of the codebook size, then the optimum quantizer is the uniform quantizer and not the Lloyd quantizer. We have proved this

approximation only for the high resolution case and the resulting quantizer
may require an enormous number of levels, but the entropy will be as low as
possible. But how low? How small an average distortion can be achieved?
It is not obvious from (9.9.11) that the average distortion is in fact smaller
than (6.3.2). While it can be shown that indeed the former is smaller
than the latter if the two rates are constrained by the same number, it
is fairly easy to see that the general behavior of the two formulas is the
same. Suppose that we set $H_Q = R$ for the entropy coded quantizer and
let $N = 2^R$ for the ordinary quantizer. Then both bounds yield an average
distortion proportional to 2^{-2R}, where the proportionality constants differ
(that for the entropy coder can be shown to be better).

On a more absolute scale, the approximate average distortion achievable
by uniform quantization and entropy coding can be compared with the
Shannon optimum performance as given by the rate-distortion or distortion-
rate bound[134][32][159]. In particular, if the source is memoryless, then
the Shannon lower bound to the rate-distortion function is given by

$$R_{\mathrm{SLB}}(D) = h(X_0) - \frac{1}{2}\log(2\pi e D). \qquad (9.9.12)$$

The converse coding theorem states that no scheme with average distortion
D can be coded at fewer bits than the rate-distortion function and hence
at fewer bits that the Shannon lower bound (which usually either equals
or closely approximates the actual rate-distortion function for small D).
Thus the uniform quantizer/entropy coding scheme results in a bit rate
approximately H_Q which exceeds this unbeatable optimum by

$$H_Q - R_{\mathrm{SLB}}(D) \approx \frac{1}{2}\log(\frac{\pi e}{6}) \approx 0.255. \qquad (9.9.13)$$

Thus the average rate is only approximately 0.255 bits from the Shannon
optimum, a result that is good in the sense of promising nearly optimum
performance with a simple system, but discouraging in suggesting that more
sophisticated techniques are not worth the effort. This point merits further
comment. We have proved the 0.255 bit result only for very large rates, but
it has been shown by Farvardin and Modestino [109] to remain true for a
variety of memoryless sources even at low rates. In addition, one can use an
iterative algorithm similar to the Lloyd algorithm to numerically find the
quantizer that is optimum subject to an entropy constraint. Even at low
rates (such as 0.5 bits per symbol) this quantizer is very near to a uniform
quantizer [109]. If this simple techique is so near the theoretical optimum,
why bother reading Part III of this book?

It is true that in some cases the combination of uniform quantization
and entropy coding works quite well, is well understood, and is easy to

implement. In such cases it is likely not worth the effort of considering alternative coding techniques. It would seem clear that one such case is that of memoryless sources with high bit rates. This we have proved, but even here one must be careful because high bit rates mean large alphabets which can mean entropy codes that are not necessarily simple. While it is difficult to make precise exactly what constitutes a "high" bit rate, we have argued that the near optimality of the scheme remains valid even at low bit rates for many memoryless sources. Here the large alphabet issue vanishes, but other points arise. First, if one is operating at a fractional bit rate, then being within 0.255 of the rate distortion function may not be all that good. For example, if the rate distortion function is itself 0.25 and the code technique has a bit rate of 0.5, then it meets the 0.255 bound, yet it is double the optimum.

Second, all that has been argued is that the first order entropy of the uniformly quantized source will be within 0.255 bits of the rate distortion function, but an actual noiseless code such as a Huffman code which is constrained to operate on scalars is only guaranteed to come within 1 bit per symbol of the entropy. In order to guarantee better performance, one would have to noiselessly encode vectors, which we are not now permitting. This extra fractional bit required to actually accomplish the compression can make a significant difference.

Thirdly, the nonasymptotic versions such as [109] hold only for memoryless sources, which are the least interesting to compress. The high rate versions, however, do extend to sources with memory with the same 0.255 bit result *provided* one allows vector entropy coding of possibly large dimension. Again our restriction to scalar quantization and scalar entropy coding precludes this. Ziv [353] has shown that at all bit rates one can obtain performance within .75 bits of the rate distortion function for sources with memory, but again this requires vector entropy coding and again .75 bits is a lot if the rate distortion function is itself a fraction of a bit.

Fourthly, this simple scheme is still a variable-rate scheme and requires buffers. In some applications variable-rate codes are undesirable and entropy coding is not suitable due to the complexity of using buffer feedback. In this case fixed rate schemes are necessary, but the extra constraint will mean inferior performance if one is limited to coding scalars.

In summary, one should always consider the uniform quantization/entropy coding combination if

- one is dealing with memoryless sources,

- the bit rate is moderate to large, and

- variable bit rates are permitted.

At fractional bit rates (rates below 1 bit per sample), the performance may still be within a quarter bit of the optimum, but that quarter bit is now a significant quantity. Of course this simple code may be used in a predictive loop or following a transform in order to make the quantizer inputs appear more like memoryless symbols. These techniques form the workhorses of traditional analog-to-digital conversion and data compression for good reason—they often work quite well. Beginning with the next chapter we will investigate a different approach wherein vectors instead of scalars are quantized. As in the entropy coding of this chapter, better performance can always be achieved in this manner. The question will be whether or not it is worth the effort, that is, what is the additional cost in complexity for the improvement in quality. In addition to the obvious tradeoffs of rate and distortion, however, some of the vector codes to be seen will have useful structural properties that can make them a good match to certain specific applications. Most of Part III will be devoted to fixed rate codes, codes primarily suited for fixed rate environments where entropy coding may be undesirable. In Chapter 17 we will return to the issue of variable-rate codes and consider several alternatives to the simple quantizer/entropy coder combination.

The reader interested in pursuing a deeper study of the 0.255 bit result and the relative merits of uniform scalar quantization plus vector entropy codes vs. vector quantization is referred to Neuhoff [244] who considers the relative complexity vs. performance issues in more detail.

9.10   Problems

9.1. Use the divergence inequality to show that if the alphabet for a random variable X has K symbols, then

$$H(X) \leq \log K$$

with equality if and only if the pmf for X is uniform.

9.2. Consider the ternary source X_n formed by applying a Lloyd-Max quantizer with three levels to a zero mean, unit variance Gaussian sequence. Call the output alphabet $\{0, a, b\}$. Describe for this source a Huffman code, an Elias code, and a Ziv-Lempel code. Encode and decode the sequence

$$aa0bbab000ba$$

Keep the values for $\Pr(0)$, $\Pr(a)$ and $\Pr(b)$ to two decimal places.

9.3. Design a Huffman code for a source with 8 letters a_k, $k = 0, 1, \cdots, 7$ with $p(a_k)$ proportional to 2^{-k}. Is the entropy bound achieved with equality?

9.4. Can a source with K letters having distinct probabilities (no two the same) satisfy the entropy bound for noiseless coding with equality?

9.5. Construct a pmf for a source having 8 letters which satisfies the entropy bound for noiseless coding with equality.

9.6. How would you choose the scaling period and constant in order to have an adaptive Huffman code adapt quickly to rapid changes? Slowly for long term changes?

9.7. A dual to the Huffman code was introduced by Tunstall [313] (see also [272]). Tunstall argued that if we wish to noiselessly encode a binary memoryless source into a fixed nonbinary alphabet A by encoding variable length input blocks into single symbols in A, then a procedure dual to the Huffman procedure can be used. The dual procedure will optimally assign variable length input binary vectors to single symbols in A, where here "optimally" means that the average length of the input vectors is the maximum possible while preserving a noiseless coding into A. This is a "variable-to-fixed length code" as opposed to the "fixed-to-variable length code" obtained by the Huffman procedure. Show how to construct such a tree if A consists of all binary 3-tuples and the source is memoryless with probability $P(1) = p < 1/2$. *Hint:* Here instead of combining nodes with the smallest weights, split nodes with the largest weights.

9.8. Suppose a memoryless binary source has pmf: $p(0) = 0.6$, $p(1) = 0.4$ and is coded with the Elias code. Compute a table corresponding to that of Table 9.9 for the eight possible input sequences of length 3.

9.9. Suppose that $\{X_n\}$ is an iid source with $\Pr(X_n = 1) = 1 - \Pr(X_n = 0) = 2/3$. Construct a vector Huffman code for triples produced by this source, e.g., for the vector (X_0, X_1, X_2). Find the average length and compare it to the entropy rate of the source. Describe the Elias code for this source.

9.10. Construct a Ziv-Lempel code for the sequence

$$1011101100101110101101111101011100111101.$$

What is the average length of the code? Can you compare this example with that of the previous problem?

9.11. (a) Design a Huffman code for the source with 8 symbols with probabilities:

$$0.04, 0.07, 0.09, 0.10, 0.10, 0.15, 0.2, 0.25$$

(b) Compute the average word length and compare with the entropy.

9.12. Generate a sequence of indexes in the range of $\{0, 2, \cdots, 15\}$ to encode the binary sequence

1110110101000101 0101011010101010101001

01110101011010011001 1001010101010111

1010101010111010101011101 01010100100010001101

0110101110101110 101011101011101011101000

111000111011010101010

(listed with gaps for visual clarity only) using Ziv-Lempel coding and specify the code table that you generated. Assume the first two entries of the table are the strings $\{0\}$ and $\{1\}$ assigned indices 0 and 1 respectively. (Check your work by trying to decode the sequence of indices you obtain without looking at the binary data above.)

9.13. The unix utility called "compact" is an adaptive Huffman entropy coder. The utility "compress" is an adaptive Lempel-Ziv encoder. Try using these utilities to compress a few computer files. Try to use different file types, e.g., binary (e.g., compiled programs), ascii, postscript, etc. Report your sample statistics for how long the compression takes and what compression is achieved (use the appropriate Unix utilities to measure execution time and file size).

9.14. In this problem we consider noiseless source coding for a source with an infinite alphabet. Suppose that $\{X_n\}$ is an iid source with pmf $p_X(k) = C2^{-k}$; $k = 1, 2, \cdots$.

(a) Find C and $H(X_0)$, the entropy of the random variable X_0 (which is also the entropy rate of the process).

(b) Assume that there is no constraint on the length of binary noiseless codewords. Construct a noiseless source code for this source using something like a Huffman code. It is not an obvious Huffman code because you cannot start by linking the least probable source symbols. It should, however, be a prefix code and have the smallest possible average codeword length. What is the average codeword length? Is the code optimum?

(c) Suppose now that you have a maximum codeword length of 100 binary symbols. Hence it is clearly no longer possible to noiselessly encode the sequence. Suppose you simply truncate the code of the previous part by encoding input symbols 1–100 using that code and all other input symbols get mapped into the same binary codeword as does symbol 100. What is the resulting probability of error and what is the resulting average codeword length in bits?

9.15. Suppose that a random variable X has an exponential pdf

$$f_X(x) = \lambda e^{-\lambda x}; \ x \geq 0$$

and that q is a uniform quantizer with stepsize Δ and an infinite number of levels. In other words, if $x \in [k\Delta, (k+1)\Delta)$, then $q(x) = (k + 1/2)\Delta$.

(a) Find the mean square error $E[(q(X) - X)^2]$.

(b) Find the entropy $H(q(X))$.

(c) Find the differential entropy $h(X)$.

(d) Compare $H(q(X))$ to the Shannon lower bound of (9.9.12). What happens as $\Delta \to 0$?

Part III

Vector Coding

Chapter 10

Vector Quantization I: Structure and Performance

10.1 Introduction

Vector quantization (VQ) is a generalization of scalar quantization to the quantization of a vector, an ordered set of real numbers. The jump from one dimension to multiple dimensions is a major step and allows a wealth of new ideas, concepts, techniques, and applications to arise that often have no counterpart in the simple case of scalar quantization. While scalar quantization is used primarily for analog-to-digital conversion, VQ is used with sophisticated digital signal processing, where in most cases the input signal already has some form of digital representation and the desired output is a compressed version of the original signal. VQ is usually, but not exclusively, used for the purpose of data compression. Nevertheless, there are interesting parallels with scalar quantization and many of the structural models and analytical and design techniques used in VQ are natural generalizations of the scalar case.

A vector can be used to describe almost any type of *pattern*, such as a segment of a speech waveform or of an image, simply by forming a vector of samples from the waveform or image. Another example, of importance in speech processing, arises when a set of parameters (forming a vector) is used to represent the spectral envelope of a speech sound. Vector quantization can be viewed as a form of pattern recognition where an input pattern is "approximated" by one of a predetermined set of standard patterns, or in other language, the input pattern is matched with one of a stored set of templates or codewords. Vector quantization can also be viewed as a front end to a variety of complicated signal processing tasks, including classification and linear transforming. In such applications VQ can be viewed as a complexity reducing technique because the reduction in bits can simplify the subsequent computations, sometimes permitting complicated digital signal processing to be replaced by simple table lookups. Thus VQ is far more than a formal generalization of scalar quantization. In the last few years it has become an important technique in speech recognition as well as in speech and image compression, and its importance and application are growing.

Our treatment of VQ in this book is motivated primarily by its value as a powerful technique for data compression. We hope, however, that the treatment presented here will provide a foundation for applications in pattern recognition as well.

The topics presented in this chapter and the next closely parallel those of Chapters 5 and 6 on scalar quantization. Many of the basic definitions and properties immediately generalize from scalars to vectors, while some do not generalize at all. These similarities and differences will be emphasized in the development.

We first present the basic definition of VQ and the structural properties that are independent of any statistical considerations or distortion measures. The structure and basic ideas for software or hardware implementation of VQ are considered for both the general case and the special case of uniform quantizers based on lattices. Basic complexity considerations are also presented. The presentation here assumes the reader has read or reviewed the basic material on scalar quantization presented in Chapters 5 and 6. We shall occasionally refer to "quantizers" or "quantization" implying the generality of VQ but without specifically attaching the modifier "vector." Of course, scalar quantization is always a special case of VQ and all results can so be specialized. In Chapter 11 we continue the treatment of VQ by focusing on the optimality properties of vector quantizers and their implications for quantizer design. In Chapter 12 we consider special structures for VQ that often help to improve distortion-rate performance when a complexity constraint is placed on the encoding operation.

Basic Definitions

A vector quantizer Q of dimension k and size N is a mapping from a vector (or a "point") in k-dimensional Euclidean space, $\mathcal{R}^k$, into a finite set $\mathcal{C}$ containing N output or reproduction points, called *code vectors* or *codewords*. Thus,

$$Q : \mathcal{R}^k \to \mathcal{C},$$

where $C = (\mathbf{y}_1, \mathbf{y}_2, \cdots, \mathbf{y}_N)$ and $\mathbf{y}_i \in \mathcal{R}^k$ for each $i \in \mathcal{J} \equiv \{1, 2, \cdots, N\}$. The set C is called the *codebook* or the *code* and has size N, meaning it has N distinct elements, each a vector in $\mathcal{R}^k$. The *resolution, code rate*, or, simply, *rate* of a vector quantizer is $r = (\log_2 N)/k$, which measures the number of bits per vector component used to represent the input vector and gives an indication of the accuracy or precision that is achievable with a vector quantizer if the codebook is well-designed. It is important to recognize that for a fixed dimension k the resolution is determined by the size N of the codebook and not by the number of bits used to numerically specify the code vectors stored in the codebook. The codebook is typically implemented as a table in a digital memory and the number of bits of precision used to represent each component of each code vector does not affect the resolution or the bit-rate of the vector quantizer; it is of concern only in connection with storage space limitations and with the question of adequate precision in describing a well-designed codebook.

Associated with every N point vector quantizer is a *partition* of $\mathcal{R}^k$ into N regions or *cells*, R_i for $i \in \mathcal{J}$. The ith cell is defined by

$$R_i = \{\mathbf{x} \in \mathcal{R}^k : Q(\mathbf{x}) = \mathbf{y}_i\}, \tag{10.1.1}$$

sometimes called the *inverse image* or *pre-image* of $\mathbf{y}_i$ under the mapping Q and denoted more concisely by $R_i = Q^{-1}(\mathbf{y}_i)$.

From the definition of the cells, it follows that

$$\bigcup_i R_i = \mathcal{R}^k \text{ and } R_i \bigcap R_j = \emptyset \text{ for } i \neq j, \tag{10.1.2}$$

so that the cells form a partition of $\mathcal{R}^k$. A cell that is *unbounded* is called an *overload cell* and the collection of all overload cells is called the *overload region*. A bounded cell, i.e., one having finite (k-dimensional) volume, is called a *granular cell*. The collection of all granular cells is called the granular region.

An important property of a set in $\mathcal{R}^k$ is *convexity*. Recall that in two or three dimensions, a set is said to be convex if given any two points in the set, the straight line joining these two points is also a member of the set. This familiar idea remains applicable in $\mathcal{R}^k$. A set $\mathbf{S} \in \mathcal{R}^k$ is convex if a and $b \in \mathbf{S}$ implies that $\alpha a + (1 - \alpha)b \in \mathbf{S}$ for all $0 < \alpha < 1$.

Definition: A vector quantizer is called *regular* if

a) Each cell, R_i, is a convex set, and

b) For each i, $\mathbf{y}_i \in R_i$.

It is also convenient to define a *polytopal vector quantizer* as a regular quantizer whose partition cells are bounded by segments of hyperplane surfaces in k dimensions. Equivalently, each partition region is a regular polytope and consists of an intersection of a finite number of half spaces of the form $\{\mathbf{x} \in \mathcal{R}^k : \mathbf{u}_\nu \cdot \mathbf{x} + \beta_\nu \geq 0\}$. For a thorough treatment of polytopes, the generalization of polyhedra, see Coxeter [82]. The *faces* of a polytopal cell are hyperplane segments of dimension less than k that bound the cell, so that every point on one side of the face is inside the cell and every point on the other side of the face is outside the cell. Usually, a face refers to a $k-1$ dimensional hyperplane segment for a cell in k dimensions. Note that the definition of a regular vector quantizer is consistent with the scalar case and that a regular quantizer in the one-dimensional case is always polytopal.

A vector quantizer is said to be *bounded* if it is defined over a bounded domain, $B \subset \mathcal{R}^k$, so that every input vector, $\mathbf{x}$, lies in this set. The *volume* of the set B, denoted by $V(B)$ and given by

$$V(B) = \int_B d\mathbf{x}, \qquad (10.1.3)$$

is therefore finite. A bounded quantizer does not have any overload regions in its partition.

A vector quantizer can be decomposed into two component operations, the vector *encoder* and the vector *decoder*. The encoder $\mathcal{E}$ is the mapping from $\mathcal{R}^k$ to the index set $\mathcal{J}$, and the decoder $\mathcal{D}$ maps the index set $\mathcal{J}$ into the reproduction set $\mathcal{C}$. Thus,

$$\mathcal{E} : \mathcal{R}^k \to \mathcal{J} \text{ and } \mathcal{D} : \mathcal{J} \to \mathcal{R}^k. \qquad (10.1.4)$$

It is important to note that a given partition of the space into cells fully determines how the encoder will assign an index to a given input vector. On the other hand, a given codebook fully determines how the decoder will generate a decoded output vector from a given index. The task of the encoder is either implicitly or explicitly to identify in which of N geometrically specified regions of k space the input vector lies. Contrary to popular belief, the encoder does not fundamentally need to know the codebook to perform its function. On the other hand, the decoder is simply a table lookup which is simply and fully determined by specifying the codebook. The decoder does not need to know the geometry of the partition to perform its job. Later we shall see that for most vector quantizers of practical interest, the codebook provides sufficient information to characterize the partition and in this case, the encoder operation can be performed by using the codebook as the data set which implicitly specifies the partition.

The overall operation of VQ can be regarded as the cascade or composition of two operations:

$$Q(\mathbf{x}) = \mathcal{D} \cdot \mathcal{E}(\mathbf{x}) = \mathcal{D}(\mathcal{E}(\mathbf{x})). \qquad (10.1.5)$$

Occasionally it is convenient to regard a quantizer as generating both an index, i, and a quantized output value, $Q(x)$. The decoder is sometimes referred to as an "inverse quantizer." Figure 10.1 illustrates how the cascade of an encoder and decoder defines a quantizer.

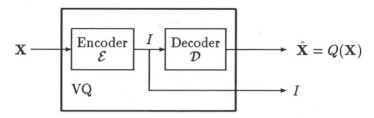

Figure 10.1: A Vector Quantizer as the Cascade of an Encoder and Decoder

In the context of a digital communication system, the encoder of a vector quantizer performs the task of selecting (implicitly or explicitly) an appropriately matching code vector $\mathbf{y}_i$ to approximate, or in some sense to describe or represent, an input vector $\mathbf{x}$. The index i of the selected code vector is transmitted (as a binary word) to the receiver where the decoder performs a table-lookup procedure and generates the reproduction $\mathbf{y}_i$, the quantized approximation of the original input vector. If a sequence of input vectors is to be quantized and transmitted, then the *bit-rate* or *transmission rate* R, in bits per vector, is given by $R = kr$, where r is the resolution and k the vector dimension. If we let f_v denote the *vector rate*, or the number of input vectors to be encoded per second, then the *bit-rate*, R_s, in bits per second is given by $R_s = krf_v$.

Of particular interest is the case of a scalar waveform communication system where each vector represents a block of contiguous samples of the waveform. The vector sequence, called a *blocked scalar process*, then corresponds to consecutive blocks of the waveform and the vector dimension, k, and the sampling rate, f_s, measured in samples per second, together determine the vector rate, in this case $f_v = f_s/k$ vectors per second. The bit-rate in bits per second is therefore, $R_s = rf_s$ which is independent of the dimension k. We distinguish between resolution and rate since there are important applications of VQ where the vectors are extracted parameters of a signal rather than blocked samples of a waveform.

Generality of Vector Quantization

VQ is not merely a generalization of scalar quantization. In fact, it is the "ultimate" solution to the quantization of a signal vector. No other coding technique exists that can do better than VQ. This appears to be a sweeping statement that may leave the reader somewhat skeptical at first. Yet it is readily justified. The following theorem shows that VQ can at least match the performance of any arbitrarily given coding system that operates on a vector of signal samples or parameters.

Theorem 10.1.1 *For any given coding system that maps a signal vector into one of N binary words and reconstructs the approximate vector from the binary word, there exists a vector quantizer with codebook size N that gives exactly the same performance, i.e., for any input vector it produces the same reproduction as the given coding system.*

Proof: Enumerate the set of binary words produced by the coding system as indexes $1, 2, \ldots, N$. For the ith binary word, let the decoded output of the given coding system be the vector $\mathbf{y}_i$. Define the codebook $\mathcal{C}$ as the ordered set of code vectors $\mathbf{y}_i$. Then a VQ decoder achieves equivalent performance to the decoder of the given coding system and a VQ encoder can be defined to be identical to the encoder of the given coding system. $\square$

In general, an *ad hoc* or heuristically designed coding technique that codes a set of k signal samples or parameters with b bits, is a suboptimal way to map a k dimensional input vector into one of $N = 2^b$ index values and into one of N reproduction vectors. Since VQ encompasses all such possible schemes, it is natural to seek the optimal encoding scheme by starting with the most general way of modeling the encoding problem, namely VQ. In the chapter that follows we consider this problem, that is, the joint optimization of the encoder and decoder for VQ with a given performance measure and given input statistics. At this point, it should be clear that if we can find the optimal vector quantizer for a given performance objective, no other coding system will be able to achieve a greater performance.

Examples of Vector Quantization

It is convenient to view the operation of a vector quantizer geometrically, using our intuition for the case of two- or three-dimensional space. Thus, a 2-dimensional quantizer assigns any input point in the plane to one of a particular set of N points or locations in the plane. As a simple illustration, consider a map of a city that is divided into school districts and the codebook is simply the location of each school on the map. The "input" is the location of a particular child's residence and the quantizer is simply the rule that assigns each child to a school according to the child's location.

Figure 10.2 demonstrates an example of a two-dimensional quantizer that is neither polytopal nor regular since the cells have faces (one-dimensional boundaries) that are not segments of hyperplanes (straight line segments) and the cells are not convex. The dots represent code vectors in a

2 dimensional space (the plane) and the region containing each codevector is a partition cell. Figure 10.3 shows a two-dimensional regular quantizer

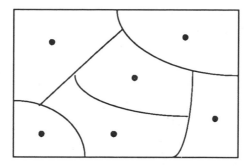

Figure 10.2: A Nonregular Quantizer

whose bounded cells are polygons (closed polytopes in two dimensions).

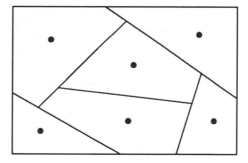

Figure 10.3: A Regular Quantizer

As an illustration of the generality of VQ, consider the coding scheme where two random variables x_1 and x_2 are each quantized by a scalar quantizer. As indicated by Theorem 10.1.1 on the generality of VQ, this case can be considered as a degenerate special case of vector quantization of the vector $\mathbf{x} = (x_1, x_2)$ where the vector quantizer is given by

$$Q(\mathbf{x}) = (Q_1(x_1), Q_2(x_2)) \qquad (10.1.6)$$

where Q_1 and Q_2 are the scalar quantizers for x_1 and x_2, respectively. Figure 10.4 shows the resulting vector quantizer corresponding to a particular choice of scalar quantization for each variable. We note in passing that this is an example of a *product VQ* because the overall VQ is formed as a Cartesian product of smaller dimensional VQs. Product codes are studied in more detail in Chapter 12. It is evident that the VQ defined by separately

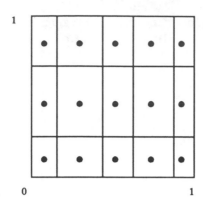

Figure 10.4: VQ Based Upon Scalar Quantization

quantizing the components of a vector must *always* result in quantization cells that are rectangular. In contrast, a more general vector quantizer is freed from these geometrical restrictions and can have arbitrary cell shapes as indicated in the examples of Figures 10.2 and 10.3. In higher dimensions the same idea is clearly applicable. Thus, in three dimensions, scalar quantization of the three components of a vector always results in cells that have rectangular box-like shapes where each face is a plane parallel to one of the coordinate axes. On the other hand, regular quantizers in three dimensions will have polyhedral cells. (A polyhedron is a polytope in three dimensions.) Extending this idea to k dimensions, it is clear that scalar quantization of the components of a vector always generates a very restricted class of vector quantizers where the faces are $(k-1)$-dimensional hyperplanes each parallel to a coordinate axis in the k-dimensional space. The inherent superiority of VQ is thereby evident simply because of the greater structural freedom it allows in quantization of a vector.

As another example of a special but restricted case of VQ, consider transform coding of a k dimensional vector as treated in Chapter 8. Since transform coding with an orthogonal matrix is equivalent to a rotation of the coordinate system, the use of scalar quantization on the transform coefficients is equivalent to a partition of the input vector space into rectangular cells which are parallel to the axes of a rotated coordinate system. From Theorem 10.1.1, transform coding can be considered as a special case of VQ whose performance is limited by the imposed coding structure. While general VQ is based on a partition of the input space into N arbitrarily shaped regions, the partition cells for transform coding are restricted to shapes that are rotations of those induced by scalar coding.

10.2 Structural Properties and Characterization

A structural decomposition of a vector quantizer is particularly valuable for obtaining insight needed to find effective algorithms for implementing vector quantizers [138]. It is also valuable for analytical studies including vector quantizer optimization. As in the scalar case, it is desirable to find a decomposition into simple building blocks and to identify the primitive building blocks that can be used to construct a model of a vector quantizer. We begin with the primary structural decomposition which is a direct generalization of the scalar case. Then we present the secondary structure, a nontrivial and important decomposition of a vector quantizer that reveals the intriguing richness of the multidimensional case. Following naturally after this background is an examination of the encoding complexity and encoding techniques for VQ. An important special class of vector quantizers, lattice-based quantizers, is treated next.

Primary Structure of a Vector Quantizer

The encoding task of a quantizer is to examine each input vector $\mathbf{x}$ and identify in which partition cell of the k-dimensional space $\mathcal{R}^k$ it lies. The vector encoder simply identifies the index i of this region and the vector decoder generates the code vector $\mathbf{y}_i$ that represents this region.

To model the operation of the encoder, we define as in the scalar case the *selector function*, $S_i(\mathbf{x})$, as the indicator or membership function $1_{R_i}(\mathbf{x})$ for the cell R_i of the partition, that is,

$$S_i(\mathbf{x}) = \begin{cases} 1 & \text{if } \mathbf{x} \in R_i \\ 0 & \text{otherwise.} \end{cases} \tag{10.2.1}$$

The operation of a vector quantizer can then be represented as:

$$Q(\mathbf{x}) = \sum_{i=1}^{N} \mathbf{y}_i S_i(\mathbf{x}). \tag{10.2.2}$$

Note that for any given value of the input vector $\mathbf{x}$, only one term of the sum is nonzero.

Figure 10.5 shows this model of a vector quantizer, which we call the *primary structural decomposition*. Each multiplier, indicated by circles, simply multiplies a particular code vector by either one or zero to produce as its output either the code vector or the zero vector. Alternatively, the multiplier can be viewed as a memory cell that stores a code vector whose value is retrieved when the binary input has value 1. The summation of

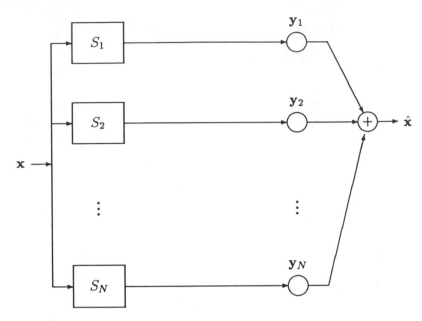

Figure 10.5: Primary Structural Decomposition of VQ

course is only symbolic since only one of its inputs is nonzero. Each S box is a memoryless (in terms of vectors) nonlinear operation which examines all components of the input vector simultaneously in order to evaluate whether or not the input belongs to the particular cell. The task of computing the binary output of an S box can indeed be a rather complex operation depending on the particular geometrical character of the cell in k-dimensional Euclidean space. The internal operation of the S box, however, is not of explicit interest in the primary decomposition. It is often convenient to abbreviate $S_i(\mathbf{x})$ to simply S_i and consider it the binary valued output of the ith selector box.

As in the scalar case, the encoder and decoder can be separately modeled with a corresponding primary structure. Let A denote the *index generator*, as the mapping $A : \mathcal{B}_N \rightarrow \mathcal{J}$ and its inverse, the *index decoder*, $A^{-1} : \mathcal{J} \rightarrow \mathcal{B}_N$, where $\mathcal{B}_N$ denotes the set of binary N-tuples of the form $\mathbf{b} = (b_1, b_2, \cdots, b_N)$ with $b_i \in \{0, 1\}$ and $\mathbf{b}$ is restricted to have exactly one nonzero component. Then $A(\mathbf{b}) = i$ if $b_i = 1$ and $b_j = 0$ for $j \neq i$. Then the encoder operation can be written as

$$\mathcal{E}(\mathbf{x}) = A(S_1(\mathbf{x}), S_2(\mathbf{x}), \cdots, S_N(\mathbf{x})) \qquad (10.2.3)$$

and the corresponding decoder operation is given by

$$\mathcal{D}(i) = \sum_{l=1}^{N} \mathbf{y}_l A^{-1}(i)_l. \tag{10.2.4}$$

The structural diagrams corresponding to the encoder and decoder are shown in Figure 10.6. In the context of a communications system, the

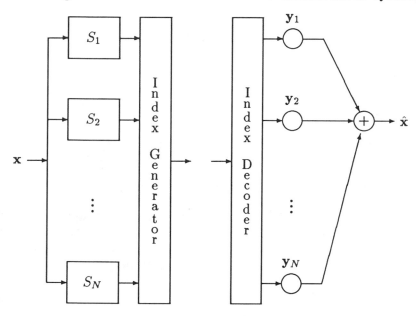

Figure 10.6: Primary Structure of a Vector Encoder and Vector Decoder

index i may be regarded as the identifier of a channel symbol whose physical nature can have many forms. The index generator and index decoder correspond respectively to the address encoder and address decoder as used in digital memory circuits.

Secondary Structure of Polytopal Vector Quantizers

The selector function for a particular cell of the vector quantizer can be a rather complex operation and it does not fit with the notion of a primitive building block for modeling a quantizer. Further decomposition of the selector function is indeed feasible, but it is dependent on the particular classification of the vector quantizer. Since the cells that partition $\mathcal{R}^k$ for a vector quantizer can have arbitrarily intricate geometric shapes, there are no natural "elementary" geometrical components that can be used as universal building blocks to model the selector operation. We now focus on a restricted but extremely important and widely used class of quantizers, the polytopal quantizers, in order to obtain a secondary decomposition of the structure of VQ.

We have defined a regular vector quantizer as *polytopal* if each cell in the partition is a convex polytope, that is, the $(k-1)$-dimensional faces of the cells consist of hyperplane segments. To be more explicit, consider the *half-space* denoted by

$$H_\nu = \{\mathbf{x} \in \mathcal{R}^k : \mathbf{u}_\nu \cdot \mathbf{x} + \beta_\nu \geq 0\}, \qquad (10.2.5)$$

where the dot product denotes the usual scalar product or inner product

$$\mathbf{u} \cdot \mathbf{x} = \mathbf{u}^t \mathbf{x} = \sum_{i=0}^{k-1} u_i x_i$$

between two vectors. The half-space H_ν is characterized by the vector parameter $\mathbf{u}_\nu \in \mathcal{R}^k$ and the scalar value β_ν. Let

$$\Lambda_\nu = \{\mathbf{x} \in \mathcal{R}^k : \mathbf{u}_\nu \cdot \mathbf{x} + \beta_\nu = 0\}, \qquad (10.2.6)$$

be the hyperplane that bounds H_ν. Then, any polytopal region can be represented by a finite intersection of half planes and the associated hyperplanes contain the "faces" or boundaries of the polytope. In particular, for any polytopal vector quantizer the partition cells can be written as

$$R_i = \bigcap_{\nu=1}^{L_i} H_\nu, \qquad (10.2.7)$$

where L_i is equal to the number of $(k-1)$-dimensional faces of the cell R_i. Let $u(v)$ denote the unit step function, where $u(v) = 1$ for $v \geq 0$ and $u(v) = 0$ for $v < 0$. Then the indicator function $T_\nu(\mathbf{x}) = 1_{H_\nu}(\mathbf{x})$ for the half-space H_ν is defined by

$$T_\nu(\mathbf{x}) = u(\mathbf{u}_\nu \cdot \mathbf{x} + \beta_\nu). \qquad (10.2.8)$$

Finally, we obtain

$$S_i(\mathbf{x}) = \prod_{\nu=1}^{L_i} T_\nu(\mathbf{x}), \qquad (10.2.9)$$

an explicit decomposition of the cell selector function into a product of binary-valued hyperplane decision functions. Observe that we could replace the product by a logical AND; the key idea is that S_i is 1 if and only if all of the T_ν are 1. A hyperplane decision function, simply answers the question:

"Is **x** on the positive side of hyperplane Λ_ν?"

It is quite reasonable to consider this as a primitive operation in k-dimensional space. Figure 10.7 illustrates this decomposition of the selector function into a product of such primitive hyperplane decision functions.

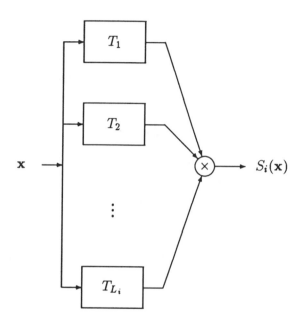

Figure 10.7: Structural Decomposition of a Selector Function

It should be noted that the complexity of the VQ encoder can be very large since the number of hyperplanes that typically bound a polytope grows rapidly with the dimension of the space.

The above representation assumed that the cells are convex as well as polytopal. This condition is indeed apparent from the decomposition since any half-space as defined above is readily shown to be convex and each cell is the intersection of convex sets and is thereby convex.

The primitive elements of this structure, which we have called hyperplane decision functions, can also be recognized as fundamental elements in the field of pattern recognition where they are typically known as *linear decision functions*. These elements are closely related to perceptrons and arise in neural network modeling. In the one-dimensional case, the hyperplane decision functions reduce to the binary threshold elements discussed in Chapter 5.

Tertiary Structure of a Polytopal Vector Quantizer

If indeed the hyperplane decision functions are appropriately considered as primitive elements, the secondary structure of a polytopal quantizer offers the last word in decomposition into elementary building blocks. However, there is an alternative approach that leads to a different structure which has in some sense greater simplicity, at least on a conceptual level. By further decomposing the hyperplane decision function, we obtain a *tertiary decomposition* of a polytopal vector quantizer.

Consider one particular selector function, say $S_i(\mathbf{x})$. Form the $L_i \times k$ matrix $\mathbf{M}_i$ whose νth row is the vector $\mathbf{u}_\nu$ as defined for the secondary structure. Then the linear transformation

$$\mathbf{w} = \mathbf{M}_i \mathbf{x} \qquad (10.2.10)$$

maps the k dimensional input vector $\mathbf{x}$ into the L_i-dimensional transformed vector $\mathbf{w}$. Define the scalar encoders,

$$\mathcal{E}_\nu(w_\nu) = u(w_\nu + \beta_\nu), \qquad (10.2.11)$$

where w_ν is the νth component of the transformed vector $\mathbf{w}$. Observe that if the original input is $\mathbf{x}$, then $\mathcal{E}_\nu(w_\nu)$ is just $T_\nu(\mathbf{x})$, the indicator function for the half plane H_ν. The one-bit outputs of these scalar encoders can be applied to an AND gate whose output is a logical one if and only if $\mathbf{x}$ lies in the regions S_j. An input vector can therefore be encoded by performing N linear transformations on the input vector $\mathbf{x}$ followed by one-bit scalar encoding of the transformed vector components. Finally, some combinational logic operations will produce the binary representation of the correct index.

The set of N linear transformations can be combined into one higher dimensional linear transformation matrix whose order is $L \times k$ where $L = \sum_{j=1}^{N} L_j$. Then by performing one-dimensional scalar encodings of all the transformed variables, followed by combinational logic operations the encoder can be implemented with the general structure is illustrated in Figure 10.8, where the comparator inputs are the transformed input vector coordinates and $b_k = -\beta_k$. Therefore, *any polytopal vector encoder can be decomposed into a linear transformation followed by scalar encodings and logical operations.* It is interesting to note the similarity with the operation of transform coding as described in Chapter 8. Although transform coding leads to a very restricted structure of the partition, nevertheless any vector quantizer can be modeled in a manner similar to transform coding. The key distinction here, is that the size of the transform matrix needed to model a given vector quantizer can be much larger than the vector dimension.

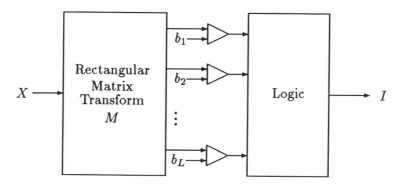

Figure 10.8: Tertiary Structural Decomposition of a VQ based on a Linear Transformation, Scalar Encoders, and Logical Operations

10.3 Measuring Vector Quantizer Performance

A distortion measure d is an assignment of a nonnegative cost $d(\mathbf{x}, \hat{\mathbf{x}})$ associated with quantizing any input vector $\mathbf{x}$ with a reproduction vector $\hat{\mathbf{x}}$. Given such a measure we can quantify the performance of a system by an average distortion $D = Ed(\mathbf{X}, \hat{\mathbf{X}})$ between the input and the final reproduction. Generally, the performance of a compression system will be good if the average distortion is small.

In practice, the overall measure of performance is the long term sample average or time average

$$\bar{d} = \lim_{n \to \infty} \frac{1}{n} \sum_{i=1}^{n} d(\mathbf{X}_i, \hat{\mathbf{X}}_i), \qquad (10.3.1)$$

where $\{\mathbf{X}_\nu\}$ is a sequence of vectors to be encoded. If the vector process is stationary and ergodic, then with probability one the above limit exists and equals the statistical expectation, i.e., $\bar{d} = D$. Stationarity and ergodicity, however, are not necessary and similar properties can hold under more general conditions. (See, for example, Chapter 1 of [159].) For the moment we will assume that such conditions are met and that such long term sample averages are given by expectations. Later remarks will consider the general assumptions required and their implications for practice.

Let $\mathbf{X}$ denote a continuously distributed random vector in $\mathcal{R}^k$ with a specified pdf (in this case the pdf refers to a joint probability density function of the vector components, X_i, for $i = 1, 2, \cdots, k$). Then the average distortion can be expressed as

$$D = Ed(\mathbf{X}, \mathbf{Y}) = \int_{\mathcal{R}^k} d(\mathbf{x}, Q(\mathbf{x})) f_{\mathbf{X}}(\mathbf{x}) \, d\mathbf{x} \qquad (10.3.2)$$

and using the partition and codebook for the given quantizer Q, we get

$$D = \sum_{j=1}^{N} \int_{R_j} d(\mathbf{x}, \mathbf{y}_j) f_{\mathbf{X}}(\mathbf{x}) \, d\mathbf{x}, \qquad (10.3.3)$$

or,

$$\begin{aligned} D &= \int \sum_{j=1}^{N} P_j d(\mathbf{X}, \mathbf{y}_j) f_{\mathbf{X}|j}(\mathbf{x}) \, d\mathbf{x} \\ &= \sum_{j=1}^{N} P_j E[d(\mathbf{x}, \mathbf{y}_j) | \mathbf{X} \in R_j], \end{aligned} \qquad (10.3.4)$$

where $P_j = P(\mathbf{X} \in R_j)$, and $f_{\mathbf{X}|j}(\mathbf{x}|j)$ is the conditional probability density of $\mathbf{X}$ given that $\mathbf{X} \in R_j$. (Note that $f_{\mathbf{X}|j}(\mathbf{x}) = 0$ for $\mathbf{x} \notin R_j$, so that the integral in (10.3.4) is taken over the entire space $\mathcal{R}^k$). Usually we omit any indication of the range of integration when the integral is taken over the entire k dimensional space.

Occasionally, it is preferable to use a *worst-case* distortion as a measure of performance rather than an average value. With respect to an arbitrary measure $d(\mathbf{x}, \mathbf{y})$ of distortion, this is simply defined as the maximum attainable distortion for a given quantizer:

$$D_{\max} = \max_{\mathbf{x} \in B} d(\mathbf{x}, Q(\mathbf{x})) \qquad (10.3.5)$$

where B is the closed subset of $\mathcal{R}^k$ to which the input $\mathbf{X}$ is confined, that is, B is the domain over which the pdf of $\mathbf{X}$ is nonzero. In some cases, this maximum may not exist since the distortion can become arbitrarily large and this measure is then not meaningful. Use of the maximum distortion as a performance measure is limited to the case of bounded random vectors or quantizers with countably infinite codebook sizes with no overload regions. For the most part, we shall focus on the statistical average of the distortion as a performance measure.

Particular Distortion Measures of Interest

Ideally a distortion measure should be tractable to permit analysis and design and it should be computable so that it can be evaluated in real time for guiding the actual encoding process for encoders which select a nearest neighbor or minimum distortion output (we shall see that as with scalar quantizers, this form of encoder is optimal for a given codebook). It should also be subjectively meaningful so that large or small average distortion values correlate with bad and good subjective quality as perceived by the ultimate user of the reproduced vector sequence.

The most convenient and widely used measure of distortion between an input vector $\mathbf{X}$ and a quantized vector $\hat{\mathbf{X}} = Q(\mathbf{X})$, is the *squared error* or squared Euclidean distance between two vectors defined as

$$
\begin{aligned}
d(\mathbf{X}, \hat{\mathbf{X}}) &= ||\mathbf{X} - \hat{\mathbf{X}}||^2 \\
&\equiv (\mathbf{X} - \hat{\mathbf{X}})^t (\mathbf{X} - \hat{\mathbf{X}}) \\
&= \sum_{i=1}^{k} (X_i - \hat{X}_i)^2.
\end{aligned}
\tag{10.3.6}
$$

If the input and reproduction vectors are complex, then this becomes

$$
\begin{aligned}
d(\mathbf{X}, \hat{\mathbf{X}}) &= ||\mathbf{X} - \hat{\mathbf{X}}||^2 \\
&\equiv (\mathbf{X} - \hat{\mathbf{X}})^* (\mathbf{X} - \hat{\mathbf{X}}) \\
&= \sum_{i=1}^{k} |X_i - \hat{X}_i|^2.
\end{aligned}
\tag{10.3.7}
$$

The *average squared error distortion* or, more briefly, the *average distortion* (when no confusion with other distortion measures arises) is defined as

$$
D = Ed(\mathbf{X}, \hat{\mathbf{X}}) = E(||\mathbf{X} - \hat{\mathbf{X}}||^2).
\tag{10.3.8}
$$

This measure is frequently associated with the energy or power of an error signal and therefore has some intuitive appeal in addition to being an analytically tractable measure for many purposes.

Numerous alternative distortion measures may also be defined for assessing dissimilarity between the input and reproduction vectors. (See, for example, Chapter 2 of [159].) Many of the measures of interest for VQ have the form

$$
d(\mathbf{X}, \hat{\mathbf{X}}) = \sum_{i=1}^{k} d_m(X_i, \hat{X}_i)
\tag{10.3.9}
$$

where $d_m(x, \hat{x})$ is a scalar distortion measure (often called the *per-letter distortion*) as in one-dimensional quantization. Any distortion measure having this additivity property with the same scalar distortion measure used for each component is called an *additive* or *single letter* distortion measure and is particularly appropriate for waveform coding where each vector component has the same physical meaning, being a sample of a waveform.

Of particular interest is the case where the scalar distortion measure is given by $d_m(x, \hat{x}) = |x - \hat{x}|^m$ for positive integer values of m. When $m = 1$, this specializes to the l_1 *norm* of the error vector, $\mathbf{X} - \hat{\mathbf{X}}$. When $m = 2$, we obtain the squared error measure already discussed. The mth root of d_m is the l_m norm of the error vector $\mathbf{X} - \hat{\mathbf{X}}$.

Another distortion measure of particular interest is the *weighted squared error* measure

$$d(\mathbf{x}, \mathbf{y}) = (\mathbf{x} - \mathbf{y})^t \mathbf{W} (\mathbf{x} - \mathbf{y}), \qquad (10.3.10)$$

where $\mathbf{W}$ is a symmetric and positive definite weighting matrix and the vectors $\mathbf{x}$ and $\mathbf{y}$ are treated as column vectors. Note that this measure includes the usual squared error distortion in the special case where $\mathbf{W} = \mathbf{I}$, the identity matrix. In the case where $\mathbf{W}$ is a diagonal matrix with diagonal values $w_{ii} > 0$, we have

$$d(\mathbf{x}, \mathbf{y}) = \sum_{i=1}^{k} w_{ii}(x_i - y_i)^2 \qquad (10.3.11)$$

which is a simple but useful modification of the squared error distortion that allows a different emphasis to be given to different vector components.

It should be noted that the weighted squared error measure (10.3.10) can be viewed as an unweighted squared error measure between linearly transformed vectors, $\mathbf{x}' = \mathbf{A}\mathbf{x}$, and $\mathbf{y}' = \mathbf{A}\mathbf{y}$ where $\mathbf{A}$ is obtained by the factorization $\mathbf{W} = \mathbf{A}^t \mathbf{A}$. In some applications, the matrix $\mathbf{W}$ is selected in accordance with statistical characteristics of the input vector, $\mathbf{X}$, being quantized. For example, in the *Mahalanobis distortion measure*, $\mathbf{W}$ is chosen to be the inverse of the covariance matrix of the input vector $\mathbf{X}$. In particular, if the components of $\mathbf{X}$ are uncorrelated, the Mahalanobis weighting matrix reduces to the diagonal matrix discussed above with $w_{ii} = \sigma_i^{-2}$, where σ_i^2 is the variance of X_i.

All of the distortion measures discussed so far are symmetric in their arguments $\mathbf{x}$ and $\mathbf{y}$. It is sometimes convenient and effective to choose a weighting matrix $\mathbf{W}(\mathbf{x})$ (assumed to be symmetric and positive definite for all $\mathbf{x}$) that depends explicitly on the input vector $\mathbf{x}$ to be quantized in order to obtain perceptually motivated distortion measures for both speech and image compression. In this case, the distortion

$$d(\mathbf{x}, \mathbf{y}) = (\mathbf{x} - \mathbf{y})^t \mathbf{W}(\mathbf{x})(\mathbf{x} - \mathbf{y}) \qquad (10.3.12)$$

is in general asymmetric in $\mathbf{x}$ and $\mathbf{y}$.. As an example of such a distortion measure, let $\mathbf{W}(\mathbf{x})$ be $||\mathbf{x}||^{-2}\mathbf{I}$, where $\mathbf{I}$ is the identity matrix. Here the distortion between two vectors is the noise energy to signal energy ratio:

$$d(\mathbf{x}, \hat{\mathbf{x}}) = \frac{||\mathbf{x} - \hat{\mathbf{x}}||^2}{||\mathbf{x}||^2}.$$

This allows one to weight the distortion as being more important when the signal is small than when it is large. Note the distortion measure is not well defined unless $||\mathbf{x}|| > 0$.

Finally, we define the *maximum* or l_∞ *norm* distortion measure, by:

$$d_{\max}(\mathbf{x}, \hat{\mathbf{x}}) = \max_i |x_i - \hat{x}_i| \qquad (10.3.13)$$

where the distortion is determined by the component of the error vector $\mathbf{x} - \hat{\mathbf{x}}$ that contributes the largest absolute error. It is a well-known mathematical result that the l_m norm approaches the l_∞ norm as $m \to \infty$. Thus,

$$\lim_{m \to \infty} [d_m(\mathbf{x}, \hat{\mathbf{x}})]^{1/m} = d_{\max}(\mathbf{x}, \hat{\mathbf{x}}). \qquad (10.3.14)$$

10.4 Nearest Neighbor Quantizers

An important special class of vector quantizers that is of particular interest, called *Voronoi* or *nearest neighbor* vector quantizers, has the feature that the partition is completely determined by the codebook and a distortion measure. We shall see in the next chapter that for a given codebook such an encoder is in fact optimal in the sense of minimizing average distortion, as was the case for scalar quantization. In fact, the term "vector quantizer" is commonly assumed to be synonymous with "nearest neighbor vector quantizer.".An advantage of such an encoder is that the encoding process does not require any explicit storage of the geometrical description of the cells. Instead, a conceptually simple algorithm can encode by referring to the stored codebook.

Suppose that $d(\mathbf{x}, \mathbf{y})$ is a distortion measure on the input/output vector space, for example the ubiquitous *squared error distortion measure* defined by the squared Euclidean distance between the two vectors:

$$d(\mathbf{x}, \mathbf{y}) = ||\mathbf{x} - \mathbf{y}||^2 = \sum_{i=1}^{k} (x_i - y_i)^2. \qquad (10.4.1)$$

This is the simplest measure and the most common for waveform coding. While not subjectively meaningful in many cases, generalizations permitting input-dependent weightings have proved useful and often only slightly more complicated.

We define a *Voronoi* or *nearest neighbor* (NN) vector quantizer as one whose partition cells are given by

$$R_i = \{\mathbf{x} :\ d(\mathbf{x}, \mathbf{y}_i) \leq d(\mathbf{x}, \mathbf{y}_j) \text{ all } j \in \mathcal{J}\}. \qquad (10.4.2)$$

In other words, with a nearest neighbor (NN) encoder, each cell R_i consists of all points $\mathbf{x}$ which have less distortion when reproduced with code vector $\mathbf{y}_i$ than with any other code vector. To avoid excessive mathematical formality we have been a bit careless in defining the cells. In order for the cells to constitute a partition, each boundary point must be uniquely assigned to one cell. This modification of the above definition is readily handled by assigning $\mathbf{x}$ to be a member of R_m where m is the smallest index i for which $d(\mathbf{x}, \mathbf{y}_i)$ attains its minimum value.

The most direct encoding algorithm for a NN encoder is given by the following simple algorithm.

Nearest Neighbor Encoding Rule

Step 1. Set $d = d_0$, $j = 1$, and $i = 1$.

Step 2. Compute $D_j = d(\mathbf{x}, \mathbf{y}_j)$.

Step 3. If $D_j < d$, set $D_j \rightarrow d$. Set $j \rightarrow i$.

Step 4. If $j < N$, set $j + 1 \rightarrow j$ and go to step 2.

Step 5. Stop. Result is index i.

The resulting value of i gives the encoder output $\mathcal{C}(\mathbf{x})$ and the final value of d is the distortion between $\mathbf{x}$ and $\mathbf{y}_i$. The initial value, d_0, must be larger than any expected distortion value and is usually set to the largest positive number that can be represented with the processor's arithmetic.

The key feature of the above encoding algorithm is that no geometrical description of the partition is needed to perform the encoding rule. Thus the encoder operation may be described by x

$$\mathcal{E}(\mathbf{x}) = c(\mathbf{x}, \mathcal{C}) \qquad (10.4.3)$$

where the functional operation described by $c(\cdot, \cdot)$ is independent of the specific quantizer and depends only on the distortion measure. This important concept is central to the implementation of virtually all vector quantizers in use today. Figure 10.9 illustrates the form of the NN encoder where the

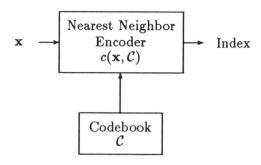

Figure 10.9: Nearest Neighbor Encoder with a Codebook ROM

codebook $\mathcal{C}$ is a read-only memory that is accessed by the NN encoder box that implements the function $c(\cdot, \cdot)$. Virtually any computable distortion measure can be used for a NN encoder. Later, however, we shall see that other properties of a distortion measure are required to yield a statistically based performance measure that is amenable to optimal codebook design.

In general, the NN cells defined above need not be polytopal or even convex. In the important special case where the distortion measure is the squared error measure, however, we shall see that the cells are indeed polytopal so that the secondary and tertiary decompositions described in Section 10.2 are applicable. Furthermore, the NN encoder has the additional feature that the parameters defining the hyperplane decision functions that characterize the cells can be conveniently specified from the values of the code vectors.

Applying (10.4.1) to (10.4.2), we find that the NN cells can be expressed as:

$$R_i = \bigcap_{\substack{j=1 \\ j \neq i}}^{N} H_{ij} \tag{10.4.4}$$

where

$$H_{ij} = \{\mathbf{x} : ||\mathbf{x} - \mathbf{y}_i||^2 \leq ||\mathbf{x} - \mathbf{y}_j||^2\}. \tag{10.4.5}$$

Expanding the squared norms and simplifying, we get

$$H_{ij} = \{\mathbf{x} : \mathbf{u}_{ij} \cdot \mathbf{x} + \beta_{ij} \geq 0\} \tag{10.4.6}$$

where

$$\mathbf{u}_{ij} = 2(\mathbf{y}_i - \mathbf{y}_j), \tag{10.4.7}$$

and

$$\beta_{ij} = ||\mathbf{y}_j||^2 - ||\mathbf{y}_i||^2. \tag{10.4.8}$$

Thus, we see that the NN cells of a Voronoi quantizer are formed by the intersection of half-spaces so that every Voronoi quantizer is polytopal and its half-spaces are explicitly determined from the code vectors $\mathbf{y}_i$ of its codebook. This convenient property is a direct consequence of the fact that the squared error distortion measure is a quadratic function of the code vector components. Other more complex distortion measures which depend on higher order powers of the input vector components will not give rise to polytopal NN quantizers.

For the squared error distortion measure or squared Euclidean distance, the partition cells of the NN encoder not only are polytopal but have an explicitly determined specification derived from the code vectors. This allows a simplification of the general tertiary structure.

Every hyperplane that determines a face of a cell is characterized by two code vectors to which this hyperplane is equidistant. (The converse, on the other hand, is not true: it is not necessary for every pair of code vectors to define a hyperplane that is a face of a cell.) We see that the hyperplane parameter vectors $\mathbf{u}_\nu$ that determine the rows of the transform matrix are given by (10.4.7) for NN cells. Consequently, the only vector-based operation that needs to be performed on the input vector is a linear transform with an $N \times k$ matrix $\mathbf{Y}$ whose rows are the code vectors themselves! In this way, a new N-dimensional vector $\mathbf{a}$ is generated, given by $\mathbf{a} = \mathbf{Y}\mathbf{x}$. Subsequent processing requires selected pairwise subtractions to generate scalar variables of the form $w_{ij} = a_i - a_j$. Finally these values are applied to a one-bit encoder as previously described for the tertiary decomposition. What have we achieved by doing this? The main distinction here is that the order of the linear transformation has been reduced and the matrix itself has an immediate direct interpretation: it is simply a representation of the codebook $\mathcal{C}$ itself.

Finally, we point out a simple and important variation of the nearest neighbor encoding algorithm described earlier. It follows either from the preceding discussion or directly from the squared Euclidean distance measure that the nearest neighbor search can be performed by evaluating scalar products rather than taking squared norms. Since

$$\min_i{}^{-1}\|\mathbf{x} - \mathbf{y}_i\|^2 = \max_i{}^{-1}[\mathbf{x}^t\mathbf{y}_i + \alpha_i]$$

where $\alpha_i = \|\mathbf{y}_i\|^2/2$ and the values of α_i can be precomputed and stored along with the codebook, we have the following algorithm.

Alternate Nearest Neighbor Encoding Rule

Step 1. Set $f = f_0$, $j = 1$, and $i = 1$.

Step 2. Compute $F_j = \mathbf{x}^t \mathbf{y}_j + \alpha_j$

Step 3. If $F_j > f$, set $F_j \to f$. Set $j \to i$.

Step 4. If $j < N$, set $j + 1 \to j$ and go to step 2.

Step 5. Stop. Result is index i.

The resulting value of i gives the encoder output $C(\mathbf{x})$ and the final value of f determines the distortion between $\mathbf{x}$ and $\mathbf{y}_i$ according to $d(\mathbf{x}, \mathbf{y}_i) = ||\mathbf{x}||^2 - 2f$. The initial value, f_0, must be smaller than any expected distortion value and is usually set to zero.

In real-time implementations of VQ encoders, the preferred choice of search algorithm depends on the particular processor architecture. The earliest reported hardware implementation of VQ [306] implemented the nearest neighbor search algorithm and used off-the-shelf components and the Z80A, an 8-bit microprocessor. Most commercial applications of VQ for speech compression algorithms today are implemented on single-chip programmable signal processors. For example, one engineer using a Texas Instruments TMS32010, a first generation signal processor chip, found that the basic nearest neighbor algorithm is most efficient. Another engineer used the AT&T DSP32 processor (which has floating point arithmetic) and found that the alternate nearest neighbor algorithm was more efficient. Generally, the code vectors are read from ROM into the processor's RAM and the input vector is also stored in RAM. Each component of each vector is stored in a separate memory location in RAM. With current generation processors, one component of the code vector and the corresponding component of the input vector can be simultaneously copied to the two input registers of a hardware multiplier, then the multiplication is performed and the product is added to the accumulator. This sequence of events provides the basic operation needed for the alternate nearest neighbor algorithm and can be performed in a single instruction cycle on most DSP chips. If an instruction cycle requires τ seconds, neglecting overhead operations, the encoding of an input vector for a codebook of N k-dimensional vectors takes place in $kN\tau$ seconds Advanced processors of 1991 vintage operate with an instruction cycle as fast as 35 ns so that for a 10-dimensional codebook of size 1024 vectors. one codebook search can be performed in 360 μs, for a coding rate of 2800 input vectors per second or 28,000 input samples per second.

As an alternative to programmable processors, VQ can be performed with application-specific integrated circuits (ASICs). Several experimental VLSI chips have been proposed or implemented that were specifically designed for codebook searching with the squared Euclidean distance algorithm [89], [92], [51], [99], [90]. In particular, a pipeline architecture chip based on the nearest neighbor algorithm was implemented in [89] and applied to a single-board speech coder [92]. A systolic chip set based on the alternate nearest neighbor algorithm was designed and implemented in [51], [90]. In [99], the alternate nearest neighbor algorithm was modified to implement the weighted squared error measure of equation 10.3.11. To our knowledge, there is no commercial use so far of custom or semi-custom VLSI implementations of VQ. This appears to be due to the rapid advance of programmable processor technology and the large variety of limited volume applications where custom chips are not cost-effective. Nevertheless, future high volume applications based on standardized speech coding algorithms are likely to lead to ASIC implementations of sophisticated algorithms that include VQ.

Encoding Complexity and Efficient Search Algorithms

The NN encoding algorithm for a Voronoi quantizer given earlier can be viewed as an *exhaustive search* algorithm where the computation of distortion is performed sequentially on every code vector in the codebook, keeping track of the "best so far" and continuing until every code vector has been tested. For a codebook of size N, this requires N distortion evaluations and each such evaluation involves k multiplications and $k - 1$ additions for the squared error distortion. Other distortion measures may have much higher computational demands. Frequently the codebook size needed in applications is very large. Furthermore, the vector rate f_v is rather high in typical communications systems and the number of distortion calculations that must be performed per unit time, given by $N f_v$, implies a very demanding computational complexity, typically involving many millions of arithmetic operations per second.

These considerations motivate serious study of more efficient algorithms that yield the nearest neighbor code vector without requiring an exhaustive search through the codebook. Several approaches have been taken to circumvent the complexity problem in VQ. Some solutions are based on designing a vector quantizer whose codebook has a special structure that allows faster encoding while paying the price that the quantizer is suboptimal for the given input statistics, vector dimension, and resolution. In

Chapter 12 some of these techniques will be presented.

At this point, we consider only the issue of finding fast search algorithms for an arbitrarily specified VQ without compromising performance. The idea here is that the codebook is assumed to be prespecified and designed to be optimal for a given performance objective. Does there exist an efficient nearest neighbor encoding algorithm that can achieve reduced complexity compared to the exhaustive search algorithm while guaranteeing that it will always find the nearest neighbor code vector? The answer is indeed yes for polytopal VQs as we shall see. The viewpoint that leads to efficient algorithms stems from the secondary and tertiary structure of polytopal quantizers as discussed in Section 10.2. Recall, in particular, that a decision that an input lies in a particular cell can be based on a number of binary hyperplane tests.

Consider the two-dimensional quantizer shown in Figure 10.10. In the figure the six code vectors are indicated with dots and the labeled "hyperplane" decision boundaries (line segments) are indicated with thick lines which separate neighbor regions. Thin lines indicate extensions of the hyperplane segments to help visualize the efficient encoding operation. Each

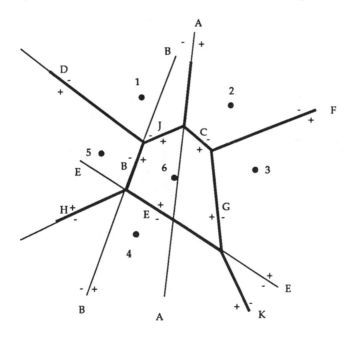

Figure 10.10: Two-Dimensional Quantizer

decision boundary is a segment of a hyperplane and is labeled with the letters $A, B, C, \cdots$ and the two half-spaces separated by each hyperplane are labeled ' + ' and '−'. We now describe a tree structure for an efficient successive approximation procedure to locate an input vector by a series of hyperplane decision tests. Suppose the initial step of a search algorithm is to compute the binary decision function for the hyperplane A. If the input **x** is on the right (labeled +) side of A, then code vector 1 and 5 are immediately eliminated as candidates for the nearest neighbor since they are contained entirely on the left side of A. Similarly, code vectors 2 and 3 are eliminated as candidates if the input is on the left side (−) of A. Depending on the result of the first test, we choose one of two new tests to perform. Thus if the input lies on the + side of A, we then test to see on which side of the hyperplane C it lies. If the input is on the − side of A, we then determine on which side of hyperplane B it lies. Each test eliminates one or more candidate code vectors from consideration. This kind of procedure corresponds to a tree structure for the search algorithm, as shown in Figure 10.11. In the figure each node is labeled with a letter corresponding to a particular binary decision test for the hyperplane (line) labeled in Figure 10.10. The tree is said to be *unbalanced* since different paths through the

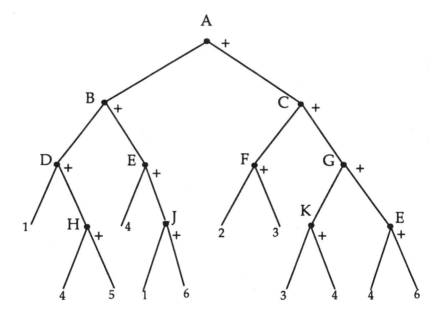

Figure 10.11: Tree Structure for efficient nearest neighbor search.

tree do not all have the same number of nodes. Each node of the tree where a hyperplane test occurs is labeled to identify the corresponding hyperplane in Fig. 10.10. The terminal nodes of the tree identify the final code vectors.

Note that this particular tree always leads to a final determination of the nearest neighbor code vector after at most 4 binary decisions. Hence, the complexity of the search has been reduced from 6 distance computations to 4 scalar product operations (neglecting the cost of performing comparisons). In less trivial cases where the codebook size is much larger than 6, it often turns out that a single hyperplane decision can eliminate a large subset of candidate code vectors. The maximum depth of the tree can be much less than the codebook size N. Furthermore, often the average complexity is much less than the worst case complexity of the tree search. In two dimensions, it is quite simple to see how to design the search tree from examination of the Voronoi cells for the given codebook. In higher dimensions, however, it is not so simple. In [65], a technique is described for finding an efficient tree structure for any given VQ codebook with any vector dimension. It was found that significant complexity savings can be achieved for searching an arbitrary codebook if the codebook size satisfies $N >> 2^k$, i.e., as long as the resolution r is somewhat larger than one bit per vector component. The reader should be warned, however, that complexity is not adequately measured by counting multiplies and ignoring comparison operations. Often simple logic operations can be more costly than multiply-adds.

Alternative methods exist for fast VQ encoding, however, they frequently require that the codebook itself be specially designed to allow a tree structure and hence, the performance achievable is not quite as good as with an optimal codebook. Alternative fast search methods will be described in Chapter 12. A systematic design technique for codebooks with balanced tree structures is given in Chapter 12 and for unbalanced tree structures is given in Chapter 17.

10.5 Lattice Vector Quantizers

Until this point we have retained considerable generality in discussing the nature and structure of VQ. A special class of vector quantizers that are of particular interest because of their highly regular structure are the so-called lattice quantizers. In one-dimensional quantization we have found that the codebook for a uniform quantizer is formed by truncating a lattice, where a lattice was defined as a set of equidistant points extending over the entire real line. Here we wish to consider multidimensional lattices. The concept of a lattice in R^k is identical to that presented in Chapter 3 on multidimensional sampling. The basic definitions are repeated here for the reader's convenience.

A lattice is a regular arrangement of points in k-space that includes the origin or zero vector, $\mathbf{0}$. The word "regular" means that each point "sees" the same geometrical environment as any other point. In other words, any translation of the set of points by subtracting one lattice point from all points in the lattice will yield the same lattice. A lattice Λ in $\mathcal{R}^k$ is a collection of all vectors of the form

$$\sum_{i=1}^{n} m_i \mathbf{u}_i, \tag{10.5.1}$$

where m_i are arbitrary integers and $\mathbf{u}_1, \mathbf{u}_2, \cdots, \mathbf{u}_n$ are a linearly independent set of $n \leq k$ vectors. The most common case of interest is $n = k$, which we call *nondegenerate*. The set of vectors $\{\mathbf{u}_i\}$ is called a *basis* or *generating set* for the lattice Λ. Observe that if Λ is a lattice, then it has the property that if any two points $\mathbf{x}$ and $\mathbf{y}$ are in Λ, then $m\mathbf{x} + n\mathbf{y}$ is in Λ for any pair of integers (positive, negative, or zero) m and n. It can be shown (see Problem 10.12) that this is an equivalent definition of a lattice. It is convenient to define the *generating matrix* of a lattice, $\mathbf{U} = [\mathbf{u}_1, \mathbf{u}_2, \cdots, \mathbf{u}_n]$. Since the vectors are linearly independent, $\mathbf{U}$ is nonsingular in the nondegenerate case where $n = k$. A well-known result of multidimensional geometry is that the volume of the parallelopiped formed by the basis vectors of a nondegenerate lattice is given by $|\det\mathbf{U}|$, the absolute value of the determinant of $\mathbf{U}$. If we view a lattice Λ as the codebook for a vector quantizer (with $N = \infty$), the Voronoi or NN cell of the lattice point 0 is given by $V = \{\mathbf{x} : \|\mathbf{x} - \mathbf{0}\| \leq \|\mathbf{x} - \mathbf{y}\|$ for each $\mathbf{y} \in \Lambda\}$. Since the NN cells form a partition of the space with the same density (cells per unit volume) as the partition formed by the parallelopiped cells, it follows that the volume of the NN cell is also equal to $|\det\mathbf{U}|$.

The reciprocal or dual lattice to a given nondegenerate lattice Λ with generating matrix $\mathbf{U}$ is that lattice whose generating function is given by $(\mathbf{U}^{-1})^t$, the transpose of the inverse matrix of $\mathbf{U}$.

Consider in particular the two-dimensional lattice Λ^2 whose generating matrix is given by

$$\mathbf{U} = \begin{bmatrix} 0 & \sqrt{3} \\ 2 & 1 \end{bmatrix} \tag{10.5.2}$$

so that each point $\mathbf{x} = (x_1, x_2)$ in Λ has coordinates of the form

$$x_1 = \sqrt{3}m_2$$

$$x_2 = 2m_1 + m_2$$

where m_1 and m_2 take on all integer values. Figure 10.12 shows the resulting arrangement of lattice points in the plane. Observe that the Voronoi cells are regular hexagons and the parallelopiped cell is a parallelogram.

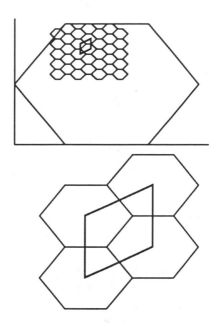

Figure 10.12: Two-Dimensional Hexagonal Lattice

In 3 dimensions there are several interesting and geometrically distinct lattice patterns which can be characterized by specifying the Voronoi cell shape. Many such patterns have been studied in crystallography. In particular, an important lattice in $\mathcal{R}^3$ has for its Voronoi cell the truncated octahedron which, as we shall see in Chapter 11, has certain optimal properties for quantization of uniform random vectors in 3 dimensions. The dodecahedron defines another three-dimensional lattice which occurs in the honeycombs of bees. A *lattice vector quantizer* is simply a vector quantizer whose codebook is either a lattice or a coset of a lattice or a truncated version of a lattice or its coset so that the codebook size is finite. (A coset of a lattice is the set of points formed by adding a fixed vector to each point in the lattice.) Clearly lattice quantizers are natural choices when uniform quantization is desired.

An interesting feature arises when the tertiary structure is applied to NN encoding of a lattice VQ. Let $\mathbf{n}_i$ denote the k-dimensional vector of integer components used to form the code vector $\mathbf{y}_i$,

$$\mathbf{y}_i = \mathbf{U}\mathbf{n}_i \tag{10.5.3}$$

for a lattice Λ. Then the $N \times k$ transform matrix $\mathbf{Y}$ in the tertiary structure can be written as

$$\mathbf{Y} = \mathbf{N}\mathbf{U}^t \tag{10.5.4}$$

where $\mathbf{N}$ is the $N \times k$ matrix whose ith row is $\mathbf{n}_i$. Thus the transformed variables to be scalar encoded in the tertiary structure are given by

$$a = \mathbf{Yx} = \mathbf{NU}^t\mathbf{x}. \qquad (10.5.5)$$

However, $\mathbf{N}$ is merely an integer matrix, so that the only scalar product operation that requires nontrivial arithmetic for computation is the formation of the k vector $\mathbf{U}^t\mathbf{x}$. This means that the main computational task for a NN lattice vector quantizer is to perform k scalar products of the input vector $\mathbf{x}$ with the basis vectors of the lattice. Once this is done, only integer linear combinations of the scalar product values must be formed and then scalar encoded with one-bit encoders and finally combinational logic produces the correct index. The great simplification in the encoding complexity that results from lattices is of course due to the regularity of the code vector locations and specifically due to the fact that each NN cell has the same set of faces apart from a translation of coordinates.

The preceding discussion is not meant to offer a specific encoding algorithm but rather merely to demonstrate the reduction in complexity due to the regularity of the code vector locations and also to show that the tertiary structure can be conceptually helpful. Efficient explicit algorithms for lattice encoding must depend on the specific lattice. See Problem 10.4 for a two-dimensional case of some interest. For a more extensive discussion of lattice encoding algorithms see [76] [75] [78] [79] [283] [148].

As a final note, we point out that the encoding algorithms for lattices are not directly applicable to vector quantizers based on truncated lattices. The difficulty arises in the boundary regions where the regularity of the lattice structure is no longer followed.

10.6   High Resolution Distortion Approximations

In this section we sketch the extension of the high resolution distortion approximations from scalar quantization to vector quantization. Here the results are more complicated and we only briefly describe the types of results that are available, confining interest to the squared error distortion. Additional details and developments for more general distortion measures may be found in [347] [348] [349] [137] [345] [159].

Given a vector quantizer with partition cells R_i and reproduction vectors $\mathbf{y}_i$, $i = 1, \cdots, N$, the average distortion can be expressed as in (10.3.3) by

$$D = \sum_{i=1}^{N} \int_{R_i} f_\mathbf{X}(\mathbf{x}) \|\mathbf{x} - \mathbf{y}_i\|^2 \, d\mathbf{x},$$

where $||\cdot||$ denotes the Euclidean distance. As in the scalar high resolution development, we confine interest to the granular region and assume that the total contribution to the average distortion from the overload region is small. If N is large, the cells R_i small, and the probability density smooth, this permits us to make the approximation

$$D \approx \sum_{i=1}^{N} f_i \int_{R_i} ||\mathbf{x} - \mathbf{y}_i||^2 \, d\mathbf{x}$$

where we are assuming that $f_X(x) \approx f_i$ for $\mathbf{x} \in R_i$. We define

$$P_i = \Pr(X \in R_i) \approx f_i V(R_i),$$

where $V(R)$ is the volume of the set R (which is finite since we are in the granular region). Thus

$$D \approx \sum_{i=1}^{N} \frac{P_i}{V(R_i)} \int_{R_i} ||\mathbf{x} - \mathbf{y}_i||^2 \, d\mathbf{x}. \tag{10.6.1}$$

Unlike the scalar high resolution results, (10.6.1) is not immediately useful because there is no easy way to approximate the integral for the arbitrary shapes possible for the cells R_i. The integral can be immediately recognized as the moment of inertia of the set R_i about the point $\mathbf{y}_i$. This at least provides a simple lower bound to the average distortion since the moment of inertia of an arbitrary set about an arbitrary point can be bounded below by the moment of inertia of a k-dimensional sphere of the same volume about its center of gravity. This approach leads to a family of bounds which can themselves be optimized over possible densities of quantizer levels. By generalizing the idea of a quantizer point density of Chapter 5 to vectors, applying multidimensional approximation arguments analogous to those used to obtain (5.6.9), and optimizing over the point density function it can be shown that in the high resolution regime,

$$D \geq \frac{k}{k+2} \left(\frac{2\pi^{\frac{k}{2}}}{k\Gamma(\frac{k}{2})} \right)^{-2/k} \left[\int f_X(\mathbf{x})^{\frac{k}{k+2}} \, d\mathbf{x} \right]^{\frac{k+2}{k}}, \tag{10.6.2}$$

where $\Gamma(r)$ is the usual gamma function and k is the vector dimension.

This results will occasionally be quoted for comparison with the actual performance of some of the design examples to be presented in later chapters.

If the quantizer is a lattice quantizer and hence can be considered as a multidimensional uniform quantizer, then all of the cells are congruent.

If we further assume that the reproduction vectors are the centers of gravity (centroids) of the cells (the optimum choice), then all of the moments of inertia are the same. Assuming that the zero vector $\mathbf{0}$ is one of the reproduction vectors, (10.6.1) becomes

$$D \approx \frac{1}{V(R_0)} \int_{R_0} ||\mathbf{x}||^2 \, d\mathbf{x} = \frac{1}{|\det \mathbf{U}|} \int_{R_0} ||\mathbf{x}||^2 \, d\mathbf{x}, \qquad (10.6.3)$$

where R_0 is the nearest neighbor or Voronoi cell about the point $\mathbf{0}$ and $\mathbf{U}$ is the generator matrix for the lattice. This quantity has been tabulated for many popular lattices [78] [76] [75] [79] and hence the high resolution approximation does give simple expressions for the average distortion for lattices.

10.7   Problems

10.1. A three point (N=3) quantizer in two dimensions has the codebook given below. If the input vectors are two-dimensional random variables uniformly distributed on the unit square: $0 < X < 1, 0 < Y < 1$, determine the mean squared quantizing error by Monte-Carlo generating 150 independent random vectors with this distribution and averaging the squared error obtained in each case. You should turn in a printout of your source code and data output.

Point	x-comp	y-comp
1	0.25	0.25
2	0.75	0.50
3	0.40	0.75

Table 10.1: Codebook

10.2. A pattern classifier operates on 3-dimensional input vectors $\mathbf{x} = (x_1, x_2, x_3)$ and performs the following classification rule: If $x_1 > 0$ and $x_2^2 + x_3^2 < 1$, then the pattern is in class 1; if $x_1 < 0$, then the pattern is in class 2; if neither of these conditions is satisfied, the pattern is in class 3. Explain why this classifier is a valid example of a VQ encoder. Is it regular? Explain.

10.3. A VQ encoder is defined by a table consisting of N pairs of vectors $(\mathbf{y}_i, \mathbf{z}_i)$ for $i = 1, 2, \ldots, N$. The encoder computes the distortion values $d_i = ||\mathbf{x} - \mathbf{y}_i||^2$ for $i = 1, 2, \ldots, N$ and $d_{N+j} = ||\mathbf{x} - \mathbf{z}_j||^2$

for $j = 1, 2, \ldots, N$. Then the encoding rule assigns index m, for $m = 1, 2, \ldots, N$, to the given input vector $\mathbf{x}$ when either $d_m \leq d_i$ for $i = 1, 2, \ldots, 2N$ or $d_{N+m} \leq d_i$ for $i = 1, 2, \ldots, 2N$ and a suitable tie-breaking rule is used.

(a) Sketch a typical shape of the partition for a simple 2-dimensional example.

(b) Is this VQ regular? Explain.

10.4. (a) Derive an expression for the computational complexity needed to encode a k-dimensional random variable with a b bit codebook (containing 2^b k-dimensional code vectors), each component of each vector is a real number stored to a suitably high accuracy. The result of an encoding is simply the index identifying which vector in the codebook is the best representation of the input vector. Assume the cost of additions or comparisons is negligible but a scalar multiplication of two real numbers is the unit for measuring complexity.

(b) Repeat part (a) but now suppose that an addition costs 3 units, a scalar multiplication costs 23 units, and a comparison between two real numbers costs 1 unit.

In both parts you must explain how you derive your result.

(c) If one unit in part (b) requires 0.1 microseconds of a processor's time, and only one processor is used to implement the encoder, how many vectors can be encoded per second when $k = 8$ and $b = 16$?

10.5. A digital signal processor (DSP) with a multiply-add cycle time of 100 ns is used to quantize speech samples arriving at 8 kHz rate. The DSP sends out an encoder bitstream at 8 kb/s. Ignoring whatever overhead might be necessary to implement an encoder and considering just VQ complexity, give an estimate of the maximum block size that you can use for encoding consecutive blocks of k contiguous samples by VQ with an arbitrary unstructured codebook. Explain.

10.6. A two-dimensional random vector $\mathbf{X}$ is uniformly distributed in the X_1, X_2 plane over the square centered at the origin, with sides parallel to the coordinate axes, and of length 10. A nearest neighbor vector quantizer $Q_{\mathbf{X}}$ for $\mathbf{X}$ has codebook given by

$$\mathcal{C}_{VQ} = \{(2,5), (-2,3), (1,4), (3,4), (-3,-1), (2,0)\}$$

Alternatively a pair of nearest neighbor scalar quantizers Q_1 and Q_2 for separately quantizing X_1 and X_2 are given by the codebooks $C_1 = \{-2, +2\}$ and $C_2 = \{-3, 0, +3\}$.

(a) If the input to the vector quantizer is the vector $(-1, 4)$ find the quantized output vector and the mean squared error.

(b) If x_1 is quantized with the scalar quantizer whose output levels are (-2, +2) and x_2 is quantized with a scalar quantizer whose output levels are (-3, 0, +3), where $\mathbf{x} = (x_1, x_2)$ is as given in part (a), find the reconstructed output of the quantizer when the input is as given in part (a). Explain.

(c) Find the equivalent VQ codebook for the scalar quantization case of part (b). In other words, find the set of all possible reconstructed vectors that can be reproduced by scalar quantizing the components according to (b).

(d) In general (without reference to the specific numbers given in this problem), compare qualitatively the performance achievable for the two quantization methods for the input random vector. Assume the same total number of bits for both methods.

10.7. Each code vector of a two-dimensional quantizer is defined by $x = m\sqrt{3}$ and $y = m+2n$ where m and n take on all possible integer values so that the number of output points is infinite. (This approximation is made to neglect boundary effects.)

(a) Sketch the Voronoi cells in the plane for a few points near the origin.

(b) Find a simple algorithm that will identify the nearest code vector to any given input point in the plane.

(c) Find the nearest lattice point to $x = (374.23, -5384.71)$.

10.8. The following technique leads to a table that has been proposed for fast search through a VQ codebook. Let x_1 denote the first component of the k dimensional vector $\mathbf{x}$ in the k dimensional space $\mathcal{R}^k$. For a given vector quantizer with partition regions R_k, find the smallest value s_i and biggest value b_i of the x_1 component of all points lying in region R_i. Then sort all the values of s_i and b_i into a sequence of increasing numbers: $u_1 \leq u_2 \leq u_3 \leq \ldots \leq u_m$ and for each pair of adjacent values u_j, u_{j+1} record the indices of all regions of the partition which contain point with x_1 component lying in this interval. Once this data is computed and stored, a fast search algorithm can be achieved. Describe a fast search algorithm that will make use of these tables to find the nearest neighbor region for a given input vector without performing an exhaustive search through the codebook. Illustrate the main idea of this method by sketching a VQ partition in 2 dimensions and indicating the role of the values u_i in the figure. Explain.

10.9. Consider a k-dimensional nearest neighbor vector quantizer with a given codebook of size N. Prove the following: The encoder can be implemented by augmenting the input vector $\mathbf{x}$ (dimension k) by adding a $(k+1)$th component with value 1 and then simply finding the maximum scalar product between the augmented input vector and each vector in a modified codebook where each vector has dimension $k+1$ and the codebook size N remains the same. Specify how the new codebook is generated from the old codebook.

10.10. Define the taxicab distance measure as

$$d(\mathbf{x}, \mathbf{y}) = \sum_{i=1}^{k} |x_i - y_i|$$

(a) Sketch the region in the plane for the 2-dimensional VQ case consisting of all points having a taxicab distance measure less than 2 from the origin.

(b) Sketch the boundary separating the set A of all points in the plane that are closer (in the sense of the taxicab distance measure) to (-1,0) than to (3,0) from the set B of all points closer to (3,0) than to (-1,0).

(c) In general, for a given VQ codebook, how would you compare the partition regions for a nearest neighbor partition for the case of squared error distance and taxicab distance?

10.11. For VQ based on the Euclidean distance measure, let two codevectors $\mathbf{y}_1$ and $\mathbf{y}_2$ be separated by a distance H and suppose an input vector $\mathbf{x}$ is distance A from $\mathbf{y}_1$ and distance B from $\mathbf{y}_2$. Show that if $H > 2A$, then $A < B$, so that it is not necessary to compute the value of B in order to do a nearest neighbor search for the input $\mathbf{x}$.

10.12. Prove the statement made in Section 10.5 that a collection of vectors $\Lambda = \{\mathbf{x}_i\}$ is a lattice if and only if for any $i \neq j$, the points $m\mathbf{x}_i + k\mathbf{x}_j \in \Lambda$ for all integers m and n.

10.13. Evaluate the high resolution approximation (10.6.3) for the lattice of (10.5.2) (see also Problem 10.7). Repeat for the dual lattice.

Chapter 11

Vector Quantization II: Optimality and Design

11.1 Introduction

We now turn to the optimality properties of vector quantizers, the vector extensions of the properties developed in Chapter 6. Such properties aid in the design of good quantizers by giving simple conditions that a quantizer must satisfy if it is to be optimal and by giving a simple iterative technique for improving a given quantizer. This technique yields the generalized Lloyd algorithm for codebook design, the most widely used method for designing good vector quantizers. We also explore the additional structure of a class of quantizers that is particularly amenable to low complexity implementation: polytopal quantizers. Unlike Chapter 6 the approximations to the optimal performance for high resolution quantizers are not explored in depth. A few of the ideas are sketched, but the detailed developments are left to the literature. See in particular Chapter 5 of [159] and the original references such as [347], [348], [349], [137], [345], [38].

Statistical Advantages of Vector Quantization

Before proceeding with the topic of optimality, it is helpful to gain some initial insight into the superiority of VQ over scalar quantization for random input vectors. In Chapter 10, we noted the generality of VQ as shown in Theorem 10.1.1 as is evident from the increased freedom in choosing the partition geometry for VQ compared to the very restrictive geometry in the case where each vector component is scalar quantized and the resulting quantization cells are rectangles. These results indicate the greater

345

flexibility of VQ and make very clear that scalar quantization is simply a restricted special case of VQ. Now we wish to go further and demonstrate that indeed VQ can offer a performance gain that is tied to the degree of correlation between the vector components. At this point, the arguments will be intuitive and informal. Later, after we formulate statistical measures of performance for VQ, we can be more explicit in demonstrating the advantage of VQ over scalar quantization.

Consider the two-dimensional random variable $\mathbf{X} = (X_1, X_2)$, which has pdf uniformly distributed over the shaded regions within the unit square as shown in Figure 11.1(a).

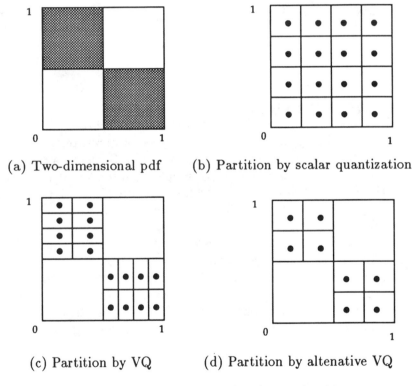

(a) Two-dimensional pdf   (b) Partition by scalar quantization

(c) Partition by VQ   (d) Partition by altenative VQ

Figure 11.1: Two-dimensional quantization

X_1 and X_2 each has a marginal distribution that is uniform over the interval $(0, 1)$. Notice that the two-component random variables are statistically dependent since the probability that $X_1 < 0.5$ conditioned on $X_2 = \alpha$ depends very strongly on whether α is less than or greater than 0.5. Since each component is uniformly distributed, scalar quantization of each component is optimal for a uniform quantizer. Figure 11.1(b) shows a partition of the domain of **X** based on individual quantization of the components in an optimal manner (uniform step size in this case) into 16 cells, whereas Figure 11.1(c) shows another partition into 16 cells available if VQ is used rather than scalar quantization of individual components. Clearly, the latter partition will be superior to the former for any reasonable choice of performance measure. In fact, it is easy to see that the worst-case quantization error measured by Euclidean distance between **X** and $Q(\mathbf{X})$ is $\sqrt{2}/8 \approx 0.18$ in case (b) and is $\sqrt{5}/16 \approx 0.13$ in case (c). The total number of bits (4) is the same in both cases. Thus, in this example, VQ is clearly superior.

It is easy to extend this intuitive idea to recognize that whenever some correlation exists between the components of a vector, some performance gain can be expected with VQ and stronger correlation produces greater gain. Another feature of this example is that the statistical dependence includes a *linear* dependence (correlation), but it is not limited to a linear dependence. A linear transformation cannot be found to map (X_1, X_2) into a pair of statistically independent random variables. VQ has the appealing feature that it efficiently quantizes vectors which have *nonlinear* statistical dependence between the components.

On the other hand, if the dependence between two random variables were purely linear (as in the case of a joint Gaussian pdf), we would be able to "decorrelate" them by a linear transformation leading to two new independent random variables. This might suggest that VQ has nothing to offer in cases of purely linear dependence. A linear transformation followed by scalar quantization of the components would appear to be optimal in this case. Not so! This is a common misconception; in fact, this coding technique, which is simply transform coding [186] as described in Chapter 8, is theoretically *suboptimal* even in the case of a Gaussian vector. The fallacy lies in the assumption that nothing is to be gained by applying VQ to statistically independent random variables.

In fact, VQ can indeed give superior performance over scalar quantization even when the components of a random vector are statistically independent of each other! Although this follows from the basic Shannon source coding theorems [134],[32],[159], it is also readily seen by considering the following example. Suppose that the two-dimensional vector **X** is uniformly distributed on the square domain shown in Figure 11.2(a). In Figure 11.2(b), a rectangular partition is shown corresponding to scalar quantization of each component. For the example to work we assume that the number of quantization points is large, even though only a modest number are shown in the figure to avoid clutter. In Figure 11.2(c), a hexagonal partition is shown. Except possibly at the edge regions, it is easily verified that the hexagonal partition has a lower worst-case error based on Euclidean

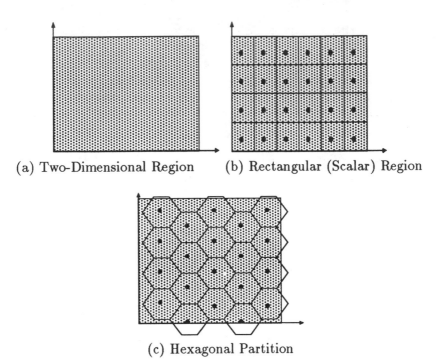

(a) Two-Dimensional Region   (b) Rectangular (Scalar) Region

(c) Hexagonal Partition

Figure 11.2: Two-Dimensional Quantization

distance than the square partition, for the same number of partition cells. It can be shown that for a statistical error criterion (average squared Euclidean distance) the hexagonal partition performs better than does the rectangular partition if the edge effects are assumed negligible. The point here is that by using a vector quantizer, one can choose cells whose shape more efficiently fills space than do the multidimensional rectangles forced by scalar quantization. These potential advantages of VQ:

- the ability to exploit linear and nonlinear dependence among the vector coordinates and

- the extra freedom in choosing the multidimensional quantizer cell shapes

are discussed in depth from an intuitive point of view in Makhoul et al. [229] and from a more quantitative point of view using multidimensional high rate quantizer theory in Lookabaugh et al. [221], where one can weigh the relative gains due to each advantage.

Another advantage offered by VQ is the flexibility of codebook sizes N that can be used. If scalar quantization of the components of a k-

dimensional vector uses N_1 levels, then the codebook size of the equivalent vector quantizer is N_1^k. On the other hand, direct use of VQ on the vector allows arbitrary partitions with any integer number N of cells. Thus in the example of Figure 11.1, the random vector could be coded with $N = 8$ cells as shown in Figure 11.1(d). Note that the maximum error achieved with this 8-point vector quantizer is exactly the same as in the case of the 16-point rectangular quantizer of Figure 11.1(b). The resolution in case (b) is 2 bits/component whereas it is $(\log_2 8)/2 = 1.5$ bits/component in case (c). An important advantage of VQ over scalar quantization is that VQ can achieve *fractional* values of resolution (measured in bits per sample or bits per vector component). This feature is particularly important where low resolution is required due to limited bit allocations for low bit-rate applications.

11.2 Optimality Conditions for VQ

The principal goal in design of vector quantizers is to find a codebook, specifying the decoder, and a partition or encoding rule, specifying the encoder, that will maximize an overall measure of performance considering the entire sequence of vectors to be encoded over the lifetime of the quantizer. The overall performance can be assessed by either a *statistical average* of a suitable distortion measure or by a *worst-case* value of distortion. We focus here only on statistical criteria. The statistical average of the distortion for a vector quantizer $Q(\cdot)$, can be expressed as

$$D = Ed(\mathbf{X}, Q(\mathbf{X})) = \int d(\mathbf{x}, Q(\mathbf{x}))f_{\mathbf{X}}(\mathbf{x})\, d\mathbf{x}, \qquad (11.2.1)$$

where $f_{\mathbf{X}}(\mathbf{x})$ is the (joint) pdf of the vector $\mathbf{X}$ and the integration above is understood to be a multiple integral over the k-dimensional space. When the input vector has a *discrete* distribution, a case that will also be of interest to us later, it is more convenient to avoid the use of a pdf made up of delta functions and instead describe the distribution by the probability mass function, (pmf), denoted by $p_{\mathbf{X}}(\mathbf{x})$. Then we have

$$D = Ed(\mathbf{X}, Q(\mathbf{X})) = \sum_i d(\mathbf{x}_i, Q(\mathbf{x}_i))p_{\mathbf{X}}(\mathbf{x}_i), \qquad (11.2.2)$$

where $\{\mathbf{x}_i\}$ are the values of $\mathbf{X}$ that have nonzero probability.

We assume that the codebook size N is given, the k-dimensional input random vector, $\mathbf{X}$, is statistically specified, and a particular distortion measure $d(\cdot, \cdot)$ has been selected. We wish to determine the necessary conditions for a quantizer to be optimal in the sense that it minimizes the average distortion for the given conditions. As in the scalar case, we proceed by finding the necessary condition for the encoder to be optimal for a given decoder. Then for a given encoder we find the necessary condition for the decoder to be optimal. Recall that the encoder is completely specified by the partition of $\mathcal{R}^k$ into the cells $R_1, R_2, \cdots, R_N$, and the decoder is completely specified by the codebook, $\mathcal{C} = \{\mathbf{y}_1, \mathbf{y}_2, \cdots, \mathbf{y}_N\}$. The optimality conditions presented next will explicitly determine the optimal partition for a given codebook and the optimal codebook for a given partition.

We first consider the optimization of the encoder for a fixed decoder. For a given codebook, an optimal partition is one satisfying the *nearest neighbor condition* (NN condition): for each i, all input points closer to code vector $\mathbf{y}_i$ than to any other code vector should be assigned to region R_i. Thus:

Nearest Neighbor Condition

For a given set of output levels, $\mathcal{C}$, the optimal partition cells satisfy

$$R_i \subset \{\mathbf{x} : d(\mathbf{x}, \mathbf{y}_i) \leq d(\mathbf{x}, \mathbf{y}_j); \text{ for all } j\}, \qquad (11.2.3)$$

that is,

$$Q(x) = \mathbf{y}_i \text{ only if } d(\mathbf{x}, \mathbf{y}_i) \leq d(\mathbf{x}, \mathbf{y}_j) \text{ all } j.$$

Thus, given the decoder, the encoder is a minimum distortion or nearest neighbor mapping, and hence

$$d(\mathbf{x}, Q(\mathbf{x})) = \min_{\mathbf{y}_i \in \mathcal{C}} d(\mathbf{x}, \mathbf{y}_i). \qquad (11.2.4)$$

The astute reader will note that the condition is the same as that for scalar quantizers. The condition states that if $\mathbf{x}$ has $\mathbf{y}_j$ as its *unique* nearest neighbor among the set of code vectors, then it must be assigned to R_j. If an input vector $\mathbf{x}$ is equally distant from two or more code vectors, the assignment of an index is not unique and we adopt the convention that it should always be assigned to the code vector with the smallest index among all the nearest neighbor code vectors. Any input point which does not have a unique nearest neighbor can be arbitrarily assigned to any of its nearest neighboring code vectors without altering the average distortion of the resulting quantizer.

Proof (General Case): For a given codebook, $\mathcal{C}$, the average distortion given by (11.2.1) can be lower bounded according to

$$D = \int d(\mathbf{x}, Q(\mathbf{x})) f_{\mathbf{X}}(\mathbf{x}) \, d\mathbf{x} \geq \int [\min_{i \in \mathcal{I}} d(\mathbf{x}, \mathbf{y}_i)] f_{\mathbf{X}}(\mathbf{x}) \, d\mathbf{x}, \qquad (11.2.5)$$

where $\mathcal{I}$ is the index set $\{1, 2, \cdots, N\}$. This lower bound is attained if $Q(\cdot)$ assigns each $\mathbf{x}$ to a code vector that has the lowest distortion, in other words, if the nearest neighbor condition is satisfied.

We next consider the optimality of the codebook for a given partition. This leads to the *centroid condition* for specifying the code vector associated with each partition region. We define the *centroid*, cent(R), of any set $R \in \mathcal{R}^k$ with nonzero probability as that vector $\mathbf{y}$ (if it exists) which minimizes the distortion between a point $\mathbf{X}$ in R and $\mathbf{y}$ averaged over the probability distribution of $\mathbf{X}$ given that $\mathbf{X}$ lies in R. Thus,

$$\mathbf{y}^* = \text{cent}(R) \text{ if } E[d(\mathbf{X}, \mathbf{y}^*)|\mathbf{X} \in R] \leq E[d(\mathbf{X}, \mathbf{y})|\mathbf{X} \in R] \qquad (11.2.6)$$

for all $\mathbf{y} \in \mathcal{R}^k$. It is convenient to write this as an *inverse minimum*, that is, (11.2.6) is equivalent to

$$\text{cent}(R) = \min_{\mathbf{y}}^{-1} E[d(\mathbf{X}, \mathbf{y})|\mathbf{X} \in R]$$

The notation *argmin* is also used sometimes instead of $\min^{-1}$. Thus the centroid is in some sense a natural representative or "central" vector for the set R and the associated probability distribution on R. It should be noted at this point that for some distortion measures, the centroid is not always uniquely defined. If the set R has zero probability, then the centroid can be defined in an arbitrary fashion.

For the squared error distortion measure, the centroid of a set R is simply the minimum mean squared estimate of $\mathbf{X}$ given that $\mathbf{X} \in R$. Similar to the proof of Theorem 4.2.1 this is easily seen to be

$$\text{cent}(R) = E(\mathbf{X}|\mathbf{X} \in R), \qquad (11.2.7)$$

so that the general definition of the centroid reduces to the usual use of the term "centroid" as in mechanics and, in this case, the centroid is uniquely defined.

For the case of a finite set $R \in \mathcal{R}^k$, the centroid definition (11.2.7) remains applicable and can be evaluated if we know the pmf (probability mass function) for the input distribution. In particular, we shall be interested later in the case where each point in R has equal probability. For example, suppose the probability distribution is a sample distribution based on a *training set* of data. A training set is a finite collection of sample vectors generated from the source distribution in order to represent the statistics of the source with a finite set of data. When there is a natural ordering to the data, the training set is also called the *training sequence*. In the statistical literature training sets are also called *learning sets*. Usually, the training vectors are generated independently from the source. This gives rise to a discrete model for the source where each of the L vectors in the training set has a probability of $1/L$. For the squared error measure the centroid then reduces to the arithmetic average

$$\text{cent}(R) = \frac{1}{\|R\|} \sum_{i=1}^{\|R\|} \mathbf{x}_i \qquad (11.2.8)$$

for $R = \{\mathbf{x}_i;\ i = 1, 2, 3, \cdots, \|R\|\}$, where $\|R\|$ is the *cardinality* of the set R, that is, the number of elements in R.

With this background, we state the condition for codebook optimality when the partition is given.

Centroid Condition

For a given partition $\{R_i; i = 1, \cdots, N\}$, the optimal code vectors satisfy

$$\mathbf{y}_i = \text{cent}(R_i). \qquad (11.2.9)$$

The result is an exact generalization of the centroid condition for scalar quantizers of Chapter 6. It is proved in three different ways below. Each proof gives a somewhat different insight.

First Proof (General Case): The average distortion is given by

$$D = \sum_{i=1}^{N} \int_{R_i} d(\mathbf{x}, \mathbf{y}_i) f_{\mathbf{X}}(\mathbf{x})\, d\mathbf{x} = \sum_{i=1}^{N} P_i \int d(\mathbf{X}, \mathbf{y}_i) f_{\mathbf{X}|i}(\mathbf{x})\, d\mathbf{x},$$

where we use the notation $f_{\mathbf{X}|i}(\mathbf{x})$ to mean the conditional pdf for $\mathbf{X}$ given $\mathbf{X} \in R_i$ and where $P_i = P[\mathbf{X} \in R_i]$. Since the partition is fixed, each term can be separately minimized by finding $\mathbf{y}_i$ that will minimize the expected distortion, or

$$E[d(\mathbf{x}, \mathbf{y}_i) | \mathbf{X} \in R_i] = \int d(\mathbf{X}, \mathbf{y}_i) f_{\mathbf{X}|i}(\mathbf{x})\, d\mathbf{x}. \qquad (11.2.10)$$

By definition, the centroid $\mathbf{y}_i = \text{cent}(R_i)$ minimizes this conditional distortion.   □

The following two proofs are restricted to the case where the squared error distortion measure is used as the performance measure. They are based on the view that the decoder receives partial information about the input vector and must perform a statistical estimate of the input given the transmitted index. The second proof finds the optimal decoder by finding the optimal (not necessarily linear) estimator of X and the third proof is based on the recognition that the decoder always performs a linear operation and the optimal estimate is linear.

As background for these proofs, we introduce the binary vector **S** of selector functions following the treatment in [138] for modeling vector quantizers and defined in a manner similar to the presentation of scalar quantization in Chapter 5. Since the partition is fixed, the selector functions, $S_i(\mathbf{x}) = 1_{R_i}(\mathbf{x})$, defined in Chapter 10, are well-defined and the encoder output can be expressed in terms of the binary vector

$$\mathbf{S} = (S_1(\mathbf{X}), \cdots, S_N(\mathbf{X})) \tag{11.2.11}$$

where for each $i \in \mathcal{I}$, $S_i = S_i(\mathbf{X})$ has value unity when $\mathbf{X} \in R_i$ and value zero otherwise. Then the N-dimensional vector $\mathbf{S} = (S_1, S_2, \cdots, S_N)$ fully describes the output of the encoder. This vector can only take on values in which all components but one are zero and the remaining component has value unity. With this background, the remaining proofs can be presented.

Second Proof (Squared Error Distortion): The decoder can be regarded as a vector-valued function of $\mathbf{S}$, with output $\mathbf{Y} = \mathbf{F}(\mathbf{S})$ where $\mathbf{Y}$ is a k-dimensional reproduction vector. To minimize $D = E[\|\mathbf{X} - \mathbf{F}(\mathbf{S})\|^2]$, we have from Theorem 4.2.1 that $F(\mathbf{S}) = E[\mathbf{X}|\mathbf{S}]$ or, in terms of sample values,

$$\mathbf{F}(\mathbf{z}) = E[\mathbf{X}|\mathbf{S} = \mathbf{z}]. \tag{11.2.12}$$

Now, recognizing that only one component of the binary vector $\mathbf{z}$ can be nonzero, there are only N distinct values that can be taken on by $\mathbf{z}$. In particular, suppose $\mathbf{z} = \mathbf{u}_i$, where $\mathbf{u}_i$ has a 1 in the ith coordinate and 0 elsewhere. Then

$$\mathbf{F}(\mathbf{u}_i) = \mathbf{y}_i = E[\mathbf{X}|S_i = 1],$$

which again proves the necessity of the centroid condition for optimal quantization. □

Third Proof (Squared Error Distortion): From the primary decomposition of a vector quantizer, we can express the reproduction vector, $\mathbf{Y}$, as

$$\mathbf{Y} = Q(\mathbf{X}) = \sum_{j=1}^{N} \mathbf{y}_j S_j. \tag{11.2.13}$$

It may be seen that this is a *linear* combination of the observable random variables S_j. Since we wish to minimize the average distortion $E(\|\mathbf{X} - \mathbf{Y}\|^2)$, we seek the best estimate of $\mathbf{X}$ using the estimator $\mathbf{Y}$, which is constrained to be a linear combination of well-defined random variables, S_j. By the orthogonality principle (Theorem 4.2.4), the condition for optimality is that the estimation error $\mathbf{X} - \mathbf{Y}$ must be orthogonal to each of the observable variables, S_i. Thus we have

$$E(\mathbf{X}S_i) = \sum_{j=1}^{N} \mathbf{y}_j E(S_j S_i). \tag{11.2.14}$$

Note that $E(S_j S_i)$ is zero for $j \neq i$ and is equal to $ES_i^2 = P_i$ when $j = i$. Also, the expectation $E(\mathbf{X}S_i)$ can be expressed in the form

$$E(\mathbf{X}S_i) = E[\mathbf{X}|S_i = 1]P_i. \qquad (11.2.15)$$

From these observations we see that (11.2.14) simplifies and yields:

$$\mathbf{y}_i = \frac{E(\mathbf{X}S_i)}{ES_i^2} = E[\mathbf{X}|S_i = 1] = \text{cent}(R_i). \qquad (11.2.16)$$

$\square$

This result assumes that the partition is *nondegenerate* in the sense that each region has nonzero probability of containing the input vector, i.e., $P_i \neq 0$. In the degenerate case where $P_i = 0$, we have what is called the *empty cell problem*. For such a partition region, the centroid is undefined and clearly it makes no sense to have a code vector dedicated to representing this cell. This situation is not of theoretical concern since a quantizer with a code vector representing an empty cell will not be optimal: the cell can be omitted from the partition and another cell with positive probability can be split into two cells thereby keeping the codebook size unchanged while reducing the average distortion. In practice however, the empty cell problem must be considered in codebook design as will be discussed later in this chapter.

It is noteworthy that the decoder of any vector quantizer always performs a *linear estimate* of the input vector based on partial information about that input as given by the binary observables S_i. The optimal decoder for a given encoder is therefore always an *optimal linear estimate* of the input vector. The restriction to the squared error distortion was needed only to obtain a simple and explicit solution for the optimal linear estimator (or equivalently, for the optimal code vectors). For other distortion measures, the optimal linear estimator is computable, but may be less tractable analytically. Later we shall consider the solution for other important distortion measures.

The two necessary conditions are direct generalizations of the scalar case as derived by Lloyd and others. These conditions are so fundamental to quantization and more generally to source coding, that they recur in similar forms in many other contexts. It is essential that the reader clearly grasp the two necessary conditions for optimality.

We shall see in the next section that these two necessary conditions are not sufficient for optimality. It is hence of interest to know if there are additional necessary conditions that might be used as further tests for optimality. There is a third condition, due also to Lloyd in the scalar case, which is useful in the case of discrete alphabets.

Given a vector quantizer satisfying the nearest neighbor condition and the centroid condition, let $\{y_i\}$ denote the reproduction points, $\{R_j\}$ the partition cells, and $\{R'_j\}$ the sets

$$R'_j = \{x : d(x, y_j) \le d(x, y_i), \text{ for all } i \in \mathcal{I}\}.$$

Thus R_j consists of the nearest neighbor cell R_j augmented by its boundary points. From the NN Condition $R_j \subset R'_j$. Call the set $B_j = R'_j - R_j$ of points in R'_j but not in R_j the *boundary* of R_j. The boundary consists of points which are not in R_j, but which are equally close to both y_j and to some other y_i and hence do not have a unique nearest neighbor. This leads to the following necessary condition for optimality.

Zero Probability Boundary Condition

A necessary condition for a codebook to be optimal for a given source distribution is

$$P(\bigcup_{j=1}^{N} B_j) = 0,$$

that is, the boundary points occur with zero probability. An alternative way to give this condition is to require that the collection of points equidistant from at least two code words has probability 0, that is,

$$P(x : d(x, y_i) = d(x, y_j) \text{ for some } i \ne j) = 0.$$

Proof: Suppose that $P(B_j) \ne 0$ for some j and hence there is at least one input point, say x_0, which is encoded into y_j but which is equally distant from another code vector y_i. Thus a new partition formed by breaking the tie in a different way, that is, mapping x_0 into the other reproduction value y_i, will yield a code with the same average distortion. This results, however, in moving a nonzero probability input point into the cell of a different reproduction vector, y_i, which must move the centroids of both R_j and R_i, which means that the codebook is no longer optimal for the new partition. $\qquad\qquad\square$

The zero-probability boundary condition is useless (i.e., automatically satisfied) when the input is a continuous random variable since for all the distortion measures that we have considered, the boundary will have zero volume and hence zero probability. The condition can, however, be useful in the discrete case since then probability can easily be placed on boundary points, e.g., one of the training samples may be equally distant from two reproduction vectors. The above condition states that the resulting quantizer cannot be optimal even if the nearest neighbor and centroid conditions are satisfied. This suggests that a better quantizer can be found by breaking the tie in a different manner and proceeding with more iterations.

Sufficiency of the Optimality Conditions

Suppose one has a vector quantizer that satisfies the centroid condition, the nearest neighbor condition, and the zero-probability boundary condition. Does this guarantee that it is globally or even locally optimal? First, we must be more explicit about the meaning of "global" and "local" optimality. We are trying to minimize the average distortion, D, which is a function of both the partition and the codebook. Since the NN condition is necessary for optimality, let us assume that it is satisfied for any given codebook C. Then we may regard the performance measure D as an explicit function of the codebook; the distortion is a multivariate function of each component of each vector. Thus the average distortion is simply a function of kN real variables. Now, the idea of optimality is quite simple. The quantizer is *locally optimal* if every small perturbation of the code vectors does not lead to a decrease in D. (Any such perturbation implies the nearest neighbor regions are correspondingly perturbed.) It is *globally optimal* if there exists no other codebook that gives a lower value of D. If we have a codebook that satisfies both necessary conditions of optimality, it is widely believed that it is indeed locally optimal. No general theoretical derivation of this result has ever been obtained. For the particular case of a discrete input distribution such as a sample distribution produced by a training sequence, however, it can be shown that under mild restrictions, a vector quantizer satisfying the necessary conditions is indeed locally optimal. (See Appendix C of [167].) This is intuitively evident since in the discrete input case, a slight perturbation of a code vector will not alter the partitioning of the (countable) set of input vectors as long as none of the input values lies on a partition boundary; once the partition stays fixed, the perturbation causes the centroid condition to be violated and the average distortion can only increase. At least under these conditions, a vector quantizer that satisfies the necessary conditions will be locally optimal.

The designer should be aware that locally optimal quantizers can, in fact, be very suboptimal. Examples of such quantizers may be found in [166], but one example is fairly easy to visualize. Suppose one has an optimal 3 bit scalar quantizer for the unit interval. One can design a two-dimensional quantizer for the unit square, also having 3 bits, by using the scalar quantizer on one coordinate and ignoring the other. This produces a partition of the plane into eight parallel stripes whose widths correspond to the quantizing intervals of the optimal scalar quantizer. The resulting quantizer is clearly poor since it completely ignores one coordinate, yet it satisfies the Lloyd conditions.

Implications of Optimal Vector Quantizers

The following two properties follow from the optimality conditions for the squared error distortion measure.

Lemma 11.2.1 *A vector quantizer which satisfies the necessary conditions for optimality with the squared error distortion measure is regular.*

Proof: We have seen in Chapter 10 that the partition cells are convex for vector quantizers satisfying the NN condition with the squared error distortion measure. If the pdf of a random vector is distributed over a convex set in $\mathcal{R}^k$, its mean must lie within the set. (See Problem 11.10.) Hence the centroid condition implies that each code vector lies within its corresponding partition cell. Therefore, the quantizer is regular. □
    Lemma 11.2.1 can readily be generalized to the case of any distortion measure that is a monotonic function of Euclidean distance. (See Problem 11.8.) The next lemma generalizes the results of Lemma 6.2.2 to vectors.

Lemma 11.2.2 *A vector quantizer which satisfies the centroid condition for the squared error distortion measure has*

(a) $EQ(\mathbf{X}) = E\mathbf{X}$,

(b) $E(\mathbf{X}^t Q(\mathbf{X})) = E(\|Q(\mathbf{X})\|^2)$, *and*

(c) $E(\|Q(\mathbf{X})\|^2) = E(\|\mathbf{X}\|^2) - E(\|\mathbf{X} - Q(\mathbf{X})\|^2)$.

Proof: Since the decoder is optimal for the given encoder, the orthogonality condition is satisfied, i.e., $E[(\mathbf{X} - Q(\mathbf{X}))S_i] = 0$ for each i. Summing these equations over i gives $E[(\mathbf{X} - Q(\mathbf{X}))(\sum_i S_i)] = 0$. But the sum $\sum_i S_i$ is always unity due to the special character of the variables S_i defined earlier. Hence, $E(\mathbf{X} - Q(\mathbf{X})) = 0$ and part (a) is proved. Now, recalling that Y is a linear combination of the S_i's, the orthogonality condition implies that

$$E[(\mathbf{X} - Q(\mathbf{X}))^t Q(\mathbf{X})] = 0,$$

showing that the quantization error is uncorrelated with the output vector. This result immediately implies part (b), that $E(\mathbf{X}^t Q(\mathbf{X})) = E(\|Q(\mathbf{X})\|^2)$. Also, it follows that

$$D \equiv E(\|\mathbf{X} - Q(\mathbf{X})\|^2) = E[(\mathbf{X} - Q(\mathbf{X}))^t \mathbf{X}].$$

Hence

$$
\begin{aligned}
E(\|\mathbf{X} - Q(\mathbf{X})\|^2) &= E(\|\mathbf{X}\|^2) - E[\mathbf{X}^t Q(\mathbf{X})] \\
&= E(\|\mathbf{X}\|^2) - E(\|Q(\mathbf{X})\|^2).
\end{aligned}
$$

□

Lemma 11.2.2(a) demonstrates that the study of optimal quantization can focus on zero mean random input vectors without loss of generality. Given the optimal quantizer for a zero mean random vector, assume we wish to modify it for input vectors having a constant mean μ added to them. We simply add μ to each code vector and the resulting codebook is again optimal.

Lemma 11.2.2(b) shows that the quantization error $Q(\mathbf{X}) - \mathbf{X}$ is *always* correlated with the input $\mathbf{X}$ if the vector quantizer has an optimal decoder for its encoder. This implies that a model of vector quantization as the addition of an independent "noise" vector to the input vector cannot be valid or at least cannot be strictly correct. In high resolution VQ, however, it may offer a useful approximation (see, e.g., [159], Chapter 5).

Since the quantization error energy can never be zero for a continuously distributed input vector, Lemma 11.2.2(c) shows that the quantized vector $Q(\mathbf{X})$ always has less energy than the input vector $\mathbf{X}$. This statement also implies that the additive independent noise vector model is not valid, since such a model would imply that $Q(\mathbf{X})$ has larger energy than $\mathbf{X}$. At high resolution, the average distortion is very small so that the input and output energies are almost equal.

11.3   Vector Quantizer Design

The necessary conditions for optimality provide the basis for iteratively improving a given vector quantizer. If the iteration continues to convergence, a good (hopefully close to optimal) quantizer can be found. The iteration begins with a vector quantizer consisting of its codebook and the corresponding optimal (NN) partition and then finds the new codebook which is optimal for that partition. This new codebook and its NN partition are then a new vector quantizer with average distortion no greater (and usually less) than the original quantizer. Although each of these steps of optimizing a partition for a codebook and a codebook for a partition is simple and straightforward, the simultaneous satisfaction of both conditions is not easy to achieve. There are no known closed-form solutions to the problem of optimal quantization. The repeated application of the improvement step, however, yields an *iterative* algorithm which at least reduces (or leaves unchanged) the average distortion at each step. In effect, we design a sequence of vector quantizers, which continually improve in performance if the algorithm is effective.

We begin with the problem of obtaining the initial codebook for improvement since this, too, is a problem of vector quantizer design. In fact, if the initial codebook is good enough, it may not be worth the effort to run further improvement algorithms. There are a variety of techniques for generating a codebook that have been developed in cluster analysis (for pattern recognition) and in vector quantization (for signal compression). We survey several of the most useful.

Random Coding

Perhaps the simplest conceptual approach towards filling a codebook of N code words is to randomly select the code words according to the source distribution, which can be viewed as a Monte Carlo codebook design. The obvious variation when designing based on a training sequence is to simply select the first N training vectors as code words. If the data is highly correlated, it will likely produce a better codebook if one takes, say, every Kth training vector. This technique has often been used in the pattern recognition literature and was used in the original development of the k-means technique [227]. One can be somewhat more sophisticated and randomly generate a codebook using not the input distribution, but the distribution which solves the optimization problem defining Shannon's distortion-rate function. In fact, the Shannon source coding theorems imply that such a random selection will on the average yield a good code [134][32][159]. Unfortunately, the codebook will have no useful structure and may turn out quite awful.

Observe that here "random coding" means only that the codebook is selected at random: once selected it is used in the usual deterministic (nearest neighbor) fashion. This is the same sense that "random coding" is used in information theory.

Pruning

Pruning refers to the idea of starting with the training set and selectively eliminating (pruning) training vectors as candidate code vectors until a final set of training vectors remains as the codebook. In once such method, a sequence of training vectors is used to populate a codebook recursively as follows: put the first training vector in the codebook. Then compute the distortion between the next training vector and the first code word. If it is less than some threshold, continue. If it is greater than the threshold, add the new vector to the codebook as a codeword. With each new training vector, find the nearest neighbor in the codebook. If the resulting distortion is not within some threshold, add the training vector to the codebook. Continue in this fashion until the codebook has enough words. For a given finite set of training vectors, it is possible that the resulting codebook will have fewer than the desired number of code vectors. If this happens, the threshold value must be reduced and the process repeated. A typical choice of threshold value for the MSE distortion measure is proportional to $N^{-\frac{2}{k}} = 2^{2r}$ where r is the rate of the code. (See Problem 11.11.) This technique is well known in the statistical clustering literature. (See, e.g., Tou and Gonzales [308].)

Pairwise Nearest Neighbor Design

A more complicated, but better, means of finding a codebook from a training sequence is the pairwise nearest neighbor (PNN) clustering algorithm proposed by Equitz [106][107]. Similar algorithms have also been used in the clustering literature [326][9]. This is also a form of pruning as it begins with the entire training sequence of L vectors, and ends with a collection of N vectors. Unlike the previous design technique, however, the final vectors need not be in the training sequence. The technique involves more computation than the preceding methods, but it is faster than the generalized Lloyd algorithm which attempts to optimize the codebook.

Suppose that the training sequence has L vectors, each of which is considered to be a separate cluster containing a single vector. The goal will be to merge vectors together into groups or clusters until we have the desired number, say N. The codebook will then contain the centroids of these clusters. In this manner we will have a partition of the training sequence into the correct number of cells and have the optimal codebook for this partition. The partition might not be a nearest neighbor partition, however. The induced vector quantizer would then replace this partition by the NN partition. The partition is obtained as follows. First compute the distortion between all pairs of vectors. The two training vectors having the smallest distortion are combined into a single cluster and represented by their centroid. We now have $L - 1$ clusters, one containing two vectors and the rest containing a single vector. Henceforth at each step clusters may have more than one vector. Suppose now that we have K clusters with $N < K \leq L-1$ and we wish to merge two of the clusters to get a good set of $K - 1$ clusters. This single step merging can be done in an optimal fashion as follows: For every pair of clusters drawn from the full collection of K clusters, compute the increase in average distortion resulting if the two clusters and their centroids are replaced by the merged two clusters and the corresponding centroid. This computation is much easier than it sounds in the case of squared error. When the best pair of clusters for merging is found, they are merged to form a codebook with $K - 1$ vectors. Continue in this way until only N vectors remain. Thus, for example, each of the K^2 pairs of clusters consists of two clusters of vectors $R_i = \{x_i(l);\ l = 1, 2, \cdots, L_i\}$ and

$R_j = \{x_j(l); \; l = 1, 2, \cdots, L_j\}$. The contribution of these two clusters to the average distortion if they are not merged is

$$\Delta_{i,j} = \sum_{l=1}^{L_i} d(x_i(l), \text{cent}(R_i)) + \sum_{l=1}^{L_j} d(x_j(l), \text{cent}(R_j))$$

while the contribution if they are merged is

$$\Delta'_{i,j} = \sum_{l=1}^{L_i} d(x_i(l), \text{cent}(R_i \bigcup R_j)) + \sum_{l=1}^{L_j} d(x_j(l), \text{cent}(R_i \bigcup R_j)) \geq \Delta_{i,j}.$$

The pair of clusters R_i, R_j for which $\Delta'_{i,j} - \Delta_{i,j}$ is the smallest is merged. That is, the two clusters are merged which thereby cause the least increase in the overall average distortion.

Note that each merge is optimal, but the overall procedure need not be optimal, that is, need not produce the optimal codebook of the given size.

Product Codes

In some cases a product codebook may provide a good initial guess. For example, if one wishes to design a codebook for a k-dimensional VQ with codebook size 2^{kR} for some integral resolution R, then one can use the product of k scalar quantizers with 2^R words each. Thus if $q(x)$ is a scalar quantizer, then $Q(x_0, \cdots, x_{k-1}) = (q(x_0), \cdots, q(x_{k-1}))$, the Cartesian product of the scalar quantizers, is a vector quantizer. This technique will not work if R is not an integer. In general other product structures can be used, e.g., one could first design a one-dimensional quantizer q_1 from scratch (perhaps using a uniform quantizer as an initial guess). One could then use $(q_1(x_0), q_1(x_1))$ as an initial guess to design a good two-dimensional quantizer $q_2(x_0, x_1)$. One could then initiate a three-dimensional VQ design with the product $(q_1(x_0), q_2(x_1, x_2))$ as an initial guess. One could continue in this way to construct higher dimensional quantizers until the final size is reached.

Splitting

Linde et al. introduced a technique that resembles the product code initialization in that it grows large codebooks from small ones, but differs in that it does not require an integral number of bits per symbol [215]. The method is called the *splitting algorithm* and it produces increasingly larger codebooks of a fixed dimension. The globally optimal resolution 0 codebook of a training sequence is the centroid of the entire sequence. The one code word, say y_0, in this codebook can be "split" into two code words, y_0 and $y_0 + \epsilon$, where ϵ is a vector of small Euclidean norm. One choice of ϵ is to make it proportional to the vector whose ith component is the standard deviation of the ith component of the set of training vectors. Another choice is to make it proportional to the eigenvector corresponding to the largest eigenvalue of the covariance matrix of the training set. This new codebook has two words and can be no worse than the previous codebook since it contains the previous codebook. The iterative improvement algorithm can be run on this codebook to produce a good resolution 1 code. When complete, all of the code words in the new codebook can be split, forming an initial guess for a resolution 2 codebook. One continues in this manner, using a good resolution r codebook to form an initial resolution $r + 1$ codebook by splitting. This algorithm provides a complete design technique from scratch on a training sequence and will later be seen to suggest a vector analog to successive approximation quantizers.

The Generalized Lloyd Algorithm

We now return to the iterative codebook improvement algorithm based on the necessary conditions for optimality, the generalization of Lloyd's Method I for designing scalar quantizers. We have just surveyed a collection of techniques which produce VQ codebooks, but none of them will in general produce a codebook meeting Lloyd's necessary conditions for optimality. Any of the codebooks can, however, be used as an initial codebook for application of the Lloyd iterative improvement algorithm. We now delve more deeply into this algorithm for designing vector quantizers.

The generalized Lloyd algorithm for VQ design is sometimes known as the *k-means algorithm* after MacQueen [227] who studied it as a statistical clustering procedure. The idea had been described earlier by Forgey in 1965 [122] in a clustering context. It has since led to a variety of extensions and applications in in the statistical literature. (See, e.g., Diday and Simon [101].) It is also sometimes referred to as the LBG algorithm in the data compression literature after [215] where a detailed treatment of the algorithm for data compression applications is presented, although this name is perhaps more appropriate for the splitting algorithm variation of the Lloyd algorithm. We shall refer to it as the generalized Lloyd or GL algorithm (or the GLA) since it is indeed a direct generalization of Lloyd's treatment in 1957 [218]. For design based on empirical data, it was first used for VQ in 1977 by Chen for the design of two-dimensional quantizers [60] and by Hilbert [178] for the design of multispectral image compression systems.

The algorithm is based on the iterative use of the codebook modification operation, which generalizes the Lloyd Iteration for scalar quantization. We first state the generalization for the vector case when the joint pdf of the input vector is assumed to be known and continuously distributed.

**The Lloyd Iteration for Codebook Improvement
Known Statistics:**

(a) Given a codebook, $\mathcal{C}_m = \{\mathbf{y}_i; \ i = 1, \cdots, N\}$, find the optimal partition into quantization cells, that is, use the Nearest Neighbor Condition to form the nearest neighbor cells:

$$R_i = \{\mathbf{x} : d(\mathbf{x}, \mathbf{y}_i) < d(\mathbf{x}, \mathbf{y}_j); \text{ all } j \neq i\}.$$

If $\mathbf{x}$ yields a tie for distortion, e.g., if $d(\mathbf{x}, \mathbf{y}_i) = d(\mathbf{x}, \mathbf{y}_j)$ for one or more $j \neq i$, then assign $\mathbf{x}$ to the set R_j for which j is smallest.

(b) Using the Centroid Condition, find $\mathcal{C}_{m+1} = \{\text{cent}(R_i); \ i = 1, \cdots, N\}$, the optimal reproduction alphabet (codebook) for the cells just found.

Lemma 11.3.1 *Each application of the Lloyd iteration must reduce or leave unchanged the average distortion.*

Proof: From the necessary conditions for optimality it follows that Step (a) in the algorithm can only improve or leave unchanged the encoder for the given decoder. Similarly Step (b) can only improve or leave unchanged the decoder for the given encoder. Hence, the average distortion of the vector quantizer cannot increase. □

Note that the lemma does not actually prove that the sequence of codebooks actually converges to a local optimum, nevertheless practical implementations of the algorithm have been found to be very effective. The above form of the Lloyd Iteration requires that the input pdf be known and that the geometry of the partition be specified in order to compute the centroids. If the input pdf happens to have regions of zero probability it is possible that for some choices of codebook $\mathcal{C}_m$ the empty cell problem can arise and one or more centroids cannot be computed in Step (b). The computation of the centroids in Step (b), by evaluating a multiple integral over a complicated region in $\mathcal{R}^k$, is generally impossible by analytical methods even if the pdf has a simple and tractable expression. In practice, a numerical integration would be essential to find the centroids. However, an adequate analytical description of the input pdf is generally not available in most applications. Instead, a sample distribution based on empirical observations of the input vector is used to generate the improved codebook. This approach has become a standard technique for quantizer design in recent years. In fact, this approach to be described next is equivalent to a Monte Carlo method for evaluating the needed integrals that determine the centroids.

Suppose we have a set of observations of the signal to be quantized, called the *training set*,

$$T = \{\mathbf{v}_1, \mathbf{v}_2, \cdots, \mathbf{v}_M\}, \qquad (11.3.1)$$

where M is the size of the training set. We define the *training ratio* as $\beta \equiv M/N$, where N is the codebook size. This parameter indicates how effectively the training set can describe the true pdf of the source and how well the Lloyd algorithm can yield a codebook that is nearly optimal for the true pdf. The training set can be used to define statistically a random vector $\mathbf{U}$ by assigning the probability mass function (pmf) to have mass $1/M$ for each value of $\mathbf{v}_i$ in T. Although a pdf does not exist for a discrete distribution (unless we use delta functions), the cumulative distribution function is well defined for $\mathbf{U}$. Specifically, the *empirical cdf* of $\mathbf{U}$ is given by

$$F^{(M)}(\mathbf{x}) = \frac{1}{M} \sum_{i=1}^{M} \hat{u}(\mathbf{x} - \mathbf{v}_i) \qquad (11.3.2)$$

where $\hat{u}(\mathbf{w})$ denotes the unit vector-step function which is one if *all* components of $\mathbf{w}$ are nonnegative and zero otherwise.

The strong law of large numbers or the pointwise ergodic theorem implies that the empirical cdf converges with probability one to the actual cdf of the input random variable (at all points where the cdf is continuous) as the training size M approaches infinity. (See, e.g., [251] for a discussion of the convergence of empirical distributions for general asymptotically stationary processes.) If we find an N point vector quantizer that is optimal for input vectors whose distribution is given by the empirical cdf of the training set and if the training set is large enough, we can expect the quantizer to be very nearly optimal for the actual input vector (which generally has a continuous cdf). Conditions under which such convergence can be mathematically guaranteed are developed in [167] [279].

For a discrete input, the selector functions $S_i(\mathbf{x})$ (defined in Chapters 5 and 10) are now defined for a given partition of T as $S_i(\mathbf{x}) = 1$ if $\mathbf{x} \in R_i$.

Hence the Centroid Condition for a cell R_i is given by

$$\mathbf{Y}_j = E[\mathbf{X}|\mathbf{X} \in R_j] = \frac{E[\mathbf{X}S_j(\mathbf{X})]}{E[S_j(\mathbf{X})]}. \qquad (11.3.3)$$

Thus we have

$$\mathbf{Y}_j = \frac{\frac{1}{M}\sum_{i=1}^{M} \mathbf{v}_i S_j(\mathbf{v}_i)}{\frac{1}{M}\sum_{i=1}^{M} S_j(\mathbf{v}_i)} \qquad (11.3.4)$$

for $j = 1, 2, \cdots, N$. The index i counts training points and the index j counts partition cells. Again, we note that the empty cell problem will arise if a region R_j has zero probability, i.e., if there are no training points that lie in this set. In this case, the centroid (11.3.4) is undefined.

The nearest neighbor condition for the discrete input case give the partition regions:

$$R_j = \{\mathbf{v} \in \mathcal{T} : ||\mathbf{v} - \mathbf{y}_j|| < ||\mathbf{v} - \mathbf{y}_i|| \text{ all } i\}. \qquad (11.3.5)$$

For the discrete case, the average distortion can be expressed as

$$D = \frac{1}{M}\sum_{i=1}^{M}\sum_{j=1}^{N} d(\mathbf{v}_i, \mathbf{y}_j)S_j(\mathbf{v}_i) \qquad (11.3.6)$$

where the quantizer input is confined to values in the finite set $\mathcal{T}$, and $\mathcal{C} = \{\mathbf{y}_j\}$ is the codebook.

The Lloyd iteration can now be directly applied to the discrete input distribution defined from the training set $\mathcal{T}$ to obtain a locally optimal quantizer for this distribution. The general conclusion here is that if we design an optimal N point quantizer for a finite training ratio, it is reasonable to expect that for M sufficiently large, it will be very nearly optimal for the true distribution of the input random vector.

The Lloyd Iteration for Empirical Data

(a) Given a codebook, $\mathcal{C}_m = \{\mathbf{y}_i\}$, partition the training set into cluster sets R_i using the Nearest Neighbor Condition:

$$R_i = \{\mathbf{x} \in \mathcal{T} : d(\mathbf{x},\mathbf{y}_i) \le d(\mathbf{x},\mathbf{y}_j); \text{ all } j \ne i\}$$

(and a suitable tie-breaking rule).

(b) Using the Centroid Condition, compute the centroids for the cluster sets just found to obtain the new codebook, $\mathcal{C}_{m+1} = \{\mathrm{cent}(R_i)\}$. If an empty cell was generated in Step (a), an alternate code vector assignment is made (in place of the centroid computation) for that cell.

A variety of heuristic solutions have been proposed for handling the empty cell problem. In some programs a cell is declared "empty" if it has 3 or fewer training vectors. In one method, the cell with the highest number of training vectors is assigned a second code vector by splitting its centroid into two vectors and the empty cell is deleted. Some researchers use the cell with the highest partial distortion as the one whose centroid is to be split. If c cells are empty in a given iteration, the c cells with highest partial distortion can have their centroids split.

Now that we have defined the codebook improvement iteration, the actual design algorithm is stated concisely in Table 11.1 Note that Step 2 should include a suitable heuristic for handling the empty cell problem. Although not usually included in most implementations of the GL algorithm, one could check before terminating if the zero-probability boundary condition is satisfied, i.e., that no training vector is equidistant from two (or more) code vectors. If this is not satisfied and the algorithm terminates then the resulting codebook can in principle be improved by re-assigning such a training vector to a different cell (i.e., one not consistent with the tie-breaking rule) and then performing an additional Lloyd iteration. The check for ties is easily done as part of Step (a) of the Lloyd Iteration. In practice, this additional step will ordinarily offer only a negligible improvement when a reasonable large training ratio is used. For very low training ratios this step can be an important one.

The flow chart for the generalized Lloyd algorithm is shown in Figure 11.3. Although various stopping criteria can be used, it is common and effective to test if the fractional drop in distortion, $(D_m - D_{m+1})/D_m$,

Table 11.1: **The Generalized Lloyd Algorithm**

Step 1. Begin with an initial codebook C_1. Set $m = 1$.

Step 2. Given the codebook, C_m, perform the Lloyd Iteration to generate the improved codebook C_{m+1}.

Step 3. Compute the average distortion for C_{m+1}. If it has changed by a small enough amount since the last iteration, stop. Otherwise set $m + 1 \rightarrow m$ and go to Step 2.

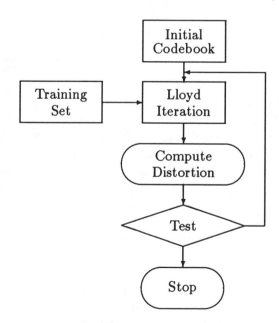

Figure 11.3: Lloyd Algorithm Flow Chart

is below a suitable threshold. When this happens and the zero-probability boundary condition is satisfied, then the algorithm is halted. With a threshold value of zero, the algorithm will necessarily produce a sequence of codebooks with monotone nonincreasing values of average distortion. If the algorithm converges to a codebook in the sense that further iterations no longer produce any changes in the reproduction values, then the resulting codebook must simultaneously satisfy both necessary conditions for optimality.

Lemma 11.3.2 *For a finite training set, the GL algorithm always produces a sequence of vector quantizers whose average distortions converge in a finite number of iterations.*

Proof: There is a finite number of ways to partition the training set into N subsets. For each partition, the centroid condition yields a codebook with a particular value of average distortion. The monotone nonincreasing average distortion implies that the algorithm cannot return in subsequent iterations to a partition yielding a higher average distortion. Hence, the average distortion of the sequence of vector quantizers produced must converge to a fixed value in a finite number of steps.   □

The GL algorithm can be initialized by selecting a codebook that has code vectors reasonably well scattered over the space and somehow covers the principal region of interest, where the input pdf is fairly large. All of the previously described VQ design techniques can be used to generate initial guesses, but among the most popular are the random code, splitting, and pairwise nearest neighbor techniques.

Although we have focused on the GL algorithm as the centerpiece of VQ design, a number of alternative approaches to codebook design exist. In this section we briefly sketch the main idea of some of these alternatives. The special attention to the GL algorithm is, however, justified, not only because of its prevalence in the engineering literature but also because it can always be used as an add-on to any other design technique. Since the GL algorithm can only improve (or at worst leave unchanged) the performance of any given initial codebook, any alternative VQ design method can always be regarded as a way to generate an initial codebook. Of course, any truly effective design algorithm ought to leave little or no room for improvement by the GL algorithm.

Before discussing specific methods, it is important to emphasize that the codebook design problem for a nearest neighbor quantizer is equivalent to the minimization of a multimodal performance measure over a high dimensional space. The GL algorithm is a *descent* algorithm, meaning that each iteration always reduces (or at least never increases) average distortion. Furthermore, each Lloyd iteration generally corresponds to a *local* change in the codebook, that is, the new codebook is generally not drastically different from the old codebook. These properties suggest that once an initial codebook is chosen, the algorithm will lead to the nearest local minimum to the initial codebook in the space of all possible codebooks. Since a complex cost function will generally have many local minima and some can be much better than others, it is clear that the GL algorithm does not have the ability to locate an *optimal* codebook.

Stochastic Relaxation

By introducing randomness into each iteration of the GL algorithm it becomes possible to evade local minima, reduce or eliminate the dependence of the solution on the initial codebook, and locate a solution that may actually be a global minimum of the average distortion as a function of the codebook. A family of optimization techniques called *stochastic relaxation* (SR) is characterized by the common feature that each iteration of a search for the minimum of a cost function (e.g., the average distortion) consists of perturbing the *state*, the set of independent variables of the cost function (e.g., the codebook) in a random fashion. The magnitude of the perturbations generally decreases with time, so that convergence is achieved. A key feature of SR algorithms is that increases in the value of the cost function are possible in each iteration. An important family of SR algorithms known as *simulated annealing* (SA) are considered below and treated in [210].

The earliest use of randomness for VQ design was proposed in [215] where noise is added to the training vectors prior to a Lloyd iteration and the variance of the noise is gradually decreased to zero. This is indeed an SR algorithm although it was not so labelled. A convenient and effective SR algorithm for codebook design was introduced in [350] and is a simpler version of earlier methods proposed in [319] and [52]. This technique modifies the Lloyd iteration in the GL algorithm in the following very simple manner: After each centroid calculation, zero-mean noise is added to each component of each code vector. The variance of the noise is reduced with successive iterations according to a predetermined schedule. Corresponding to the simulated annealing literature, the terminology *temperature* is used to refer to the noise variance and a *cooling schedule* to refer to the sequence $\{T_m\}$ of temperature values as a function of the iteration number m. An effective cooling schedule was found to be

$$T_m = \sigma_x^2 (1 - \frac{m}{I})^3,$$

where σ_x^2 is the average variance of the components of the source vector and I is the total number of iterations to be performed. The noise samples were generated as independent uniform random variables (whose range is determined by the variance). In one example, an improvement of 0.8 dB in SNR was reported in [350] over the standard GL algorithm, although typically the improvements achieved are more modest. In some applications, even a gain of 0.5 dB in SNR can be of importance and since the codebook design is off-line, it may well be worth the increase in design computation to obtain even a slightly enhanced codebook. A more extensive discussion of SR algorithms for VQ design is given in [351].

Simulated Annealing

Simulated annealing is a stochastic relaxation technique in which a randomly generated perturbation to the state (the codebook in our context) at each iteration is accepted or rejected probabilistically where the probability depends on the change in value of the cost function resulting from such a perturbation. The terminology comes from observations in physical chemistry that certain chemical systems, particularly glass and metals, can be driven to low-energy states with useful physical properties by the process of annealing, which is a very gradual reduction in temperature. Simulated annealing as a combinatorial optimation technique comes from the Metropolis algorithm [237] for atomic simulations. The set of possible states is finite but the number of states is generally so large or astronomic that exhaustive search is not feasible.

Specifically, let $H(\mathbf{s})$ denote the cost function or "energy"' as a function of the state $\mathbf{s}$. In our case H is the average distortion for a nearest neighbor or Voronoi quantizer with a given codebook and the state $\mathbf{s}$ is the codebook or equivalently the set of labels associating each training vector with a particular cluster membership. In each iteration a candidate perturbation of the partition is randomly generated and then either accepted or rejected with a probability that depends on the cost of the new codebook. If the cost decreases, then the change is always accepted. If the cost increases it is accepted with probability

$$P = e^{-\beta \Delta H}$$

where $\beta = T^{-1}$, T is a parameter called the *temperature*, and ΔH is the increase in cost. If a sufficient number of iterations are made for a given temperature, the sequence of states approaches the condition of *thermal equilibrium* where the probability distribution of the states is stationary. If thermal equilibrium is reached at a very low temperature, then states that are globally optimal or very nearly so have very high probability. The time to approach thermal equilibrium increases with the temperature and so annealing is achieved by very gradually decreasing the temperature. Geman and Geman [136] have shown that convergence in probability to a global minimum can be achieved with SA under the condition that the cooling schedule has temperatures decreasing in proportion to $(\log m)^{-1}$ where m is the iteration count. Generally, this cooling schedule is impractically slow and other faster schedules are chosen. Nevertheless, effective solutions have been found with this approach to many nonconvex optimization problems.

In VQ design from a training set, we indeed have a finite number of possible codebook designs since, as noted in the proof of Lemma 11.3.2, there is a finite number of ways to partition the training set into N clusters. The cost function is given by the distortion (11.3.6) incurred in coding the training set with the current codebook. Several reports of successful use of SA for VQ codebook design have been reported. In [319] SA is combined with the Lloyd iteration; the perturbation is implemented by randomly moving training vectors between neighboring partition regions until thermal equilibrium is reached, then a Lloyd iteration is performed before proceeding to the next temperature. In [119] the method of [319] is modified by omitting the Lloyd iteration and using a distortion measure that is more easily computed. In [52] the state perturbation is taken as a random perturbation to the codebook and and usual SA method of accepting or rejection the change is performed with a cooling schedule that is exponentially decreasing.

Fuzzy Clustering and Deterministic Annealing

It would be of great value if we could find a codebook design algorithm that leads to a global optimum without the typically very time consuming process that is introduced by the use of SA with effective cooling schedules. Recent work of Rose et al. [275] [276] [274] offers a novel approach to nonconvex optimization called *deterministic annealing* (DA), that appears to capture the benefits of SA for VQ codebook design without any randomness in the design process. In method, it is conceptually similar to the technique of *fuzzy clustering* described in [104] [33].

In designing a VQ codebook from empirical data, we encountered the formulas (11.3.4) and (11.3.5) for the centroid of a cluster (partition region) and and for the average distortion, respectively, where both were based on the selector functions. These functions can be regarded as *membership functions*, i.e., $S_j(\mathbf{v}_i)$ has value one if the training vector $\mathbf{v}_i$ is a member of cluster j and value zero otherwise. A cluster is said to be a *fuzzy* set if we may assign to each element of the training set a degree of membership or partial membership value between zero and one which indicates to what extent the particular vector is to be regarded as belonging to that set. In this way, we can generalize the selector functions to become partial membership functions. Then, we define a *fuzzy distortion* to be

$$D_f = \frac{1}{M} \sum_{i=1}^{M} \sum_{j=1}^{N} d(\mathbf{v}_i, \mathbf{y}_j)[S_j(\mathbf{v}_i)]^q$$

where q is a parameter that controls the "fuzziness" of the distortion. If $q = 1$, then D_f is simply the average distortion where the contribution of each training set is weighted by its degree of membership in the cluster. In the latter case, the task of finding the best codebook and best partial memberships for the training vectors leads to the usual (nonfuzzy) solution. By choosing values of q greater than one a fuzzy solution is found. This suggests that a search procedure that begins with a large value of q and gradually reduces q to unity might lead to a global minimum. Actually no such procedure is known but DA is somewhat similar in concept.

In DA, instead of making random perturbations on the surface of a given cost function, the statistical description of the randomness (that would arise in a corresponding SA scheme) is incorporated into the cost function. A sequence of effective cost functions is parametrized by the temperature T. At infinite temperature the cost function will be convex; at high temperatures, the cost functions will be very smooth so that it is easy to move from one local minimum to another. As the temperature drops, the cost functions become more ragged and as $T \rightarrow \infty$, the effective cost function converges to the original cost function. In effect, DA finds the global minimum at a very high temperature and then tracks the changing location of this minimum as the temperature drops while using simpler convex optimization methods at each temperature. For brevity, we omit the mathematical description of DA.

11.4   Design Examples

We now turn to a variety of examples illustrating the generalized Lloyd algorithm. The examples range from artificially simple codes, which illustrate the basic operations, to codes for real sampled speech and image data. The quality of the basic "vanilla" VQ on such real sources is not high for very low bit rates. As we shall discuss in the remainder of the book, simple memoryless VQ is limited because of the rapid growth of complexity and memory with bit rate or dimension and large dimensions or rate are usually needed for high quality. Nonetheless the examples are important because they illustrate how the algorithm works and what quality is possible with basic VQ. All of the variations to be seen will still have the Lloyd algorithm at the core of their design. Hence they cannot be understood without first becoming familiar with the unadorned version.

IID Gaussian Source

The first example is a highly artificial one taken from [215] [153]. The source is a two-dimensional Gaussian vector with independent components each with zero mean and unit variance. As the training set for the design of a codebook containing $N = 4$ code vectors, twelve pairs of real numbers are selected at random from an iid Gaussian source having zero mean and unit variance. Thus, the training ratio is 3. As an initial guess for the codebook, we use a product code, taking a scalar quantizer with output levels $\{-1, +1\}$ for each of the two dimensions, that is, we use four points of a rectangular lattice in two dimensions. The training sequence is depicted by x's and the initial codewords by circles in Figure 11.4. The first step

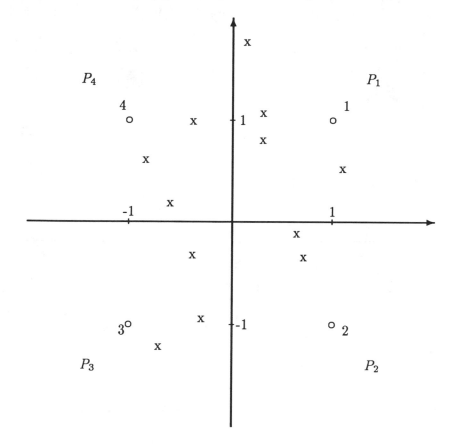

Figure 11.4: Training Sequence and Initial Code Book

of the Lloyd iteration is to partition the training sequence by performing a nearest neighbor encoding into the initial codebook. Assuming a squared error distortion measure, this means assigning each training vector to the nearest code word, resulting in the partition corresponding to the four quadrants of the figure. Now that the training vectors are grouped, it is clear that the code words are not the best representatives for their groups. For example, in the upper right quadrant a good representative (in the sense of minimizing the squared error between the code word and the training vectors in the group) should lie within the cluster, near its visual center. The initial codeword is well outside of the cluster, producing a larger than necessary squared error. The second step of the Lloyd iteration fixes this: for each group of input vectors, the old code word is replaced by the group centroid, the vector which *by definition* minimizes the group's squared error. For the squared error distortion, this is the sample mean of the vectors in the group. The new code words are displayed in Figure 11.5. The code

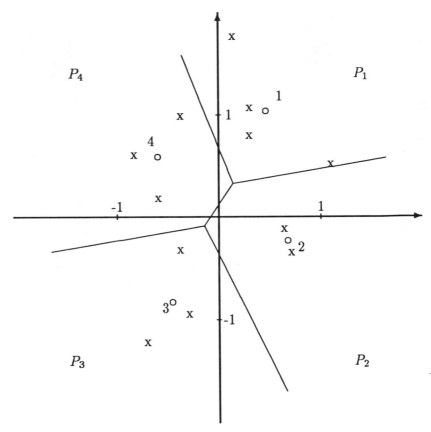

Figure 11.5: Centroids of Nearest Neighbor Regions

words have "moved" to the center of gravity of the vectors they are to represent. This provides the first new codebook.

The quadrants now no longer correspond to the nearest neighbor regions of the new codebook. Hence the next Lloyd iteration begins by partitioning the space into nearest neighbor regions for the current codebook. These regions or cells are indicated by the thin lines.

In this simple example we are now done since the new cells are the nearest neighbor regions for the codewords, the codewords are the centroids of the cells, and there are no ties in the training set. Thus the Lloyd conditions are all satisfied.

Here we have only reported the average distortion of the code when it is used on the training sequence for which it is designed. If our intent were to design a good codebook and sell it, this would be cheating. Codes will often work unrealistically well on the training sequence. The real test is if they also work well on other data produced by the same source. We previously argued that if the training sequence is large enough, then the ergodic theorem implies that all of our sample averages are good estimates of the true expectations and that the training sequence numbers should be good indications of the performance on other long sequences. Here, however, our training ratio is ridiculously small and there is no such guarantee that the code will be robust. For the Gaussian examples we will focus only on the distortion incurred in quantizing the training set with the resulting codebook. This distortion is called for short the *design distortion* and is distinguished from distortions measured by testing performance on vectors that are not in the training set, the latter distortion is called *test distortion*. These two ways of assessing performance correspond to the commonly used phrases of coding "inside the training set" or "outside the training set," respectively. In the statistics literature, the performance inside the training set is called the "resubstitution" performance. Later we shall concern ourselves with test distortion when we turn to the more important real data sources such as speech and images.

A more complete design can be run on this simple case which illustrates the properties of a codebook produced by the Lloyd algorithm. Figure 11.6 shows the centroids and Voronoi (nearest neighbor) regions of a two-dimensional quantizer designed using the Lloyd algorithm on an iid zero mean unit variance Gaussian sequence. The codebook has 64 words so that the rate is $\frac{1}{2} \log_2 64 = 4$ bits per sample. The figure shows the polytopal shapes of the Voronoi regions with the codewords (centroids) (the dots) visually at the center of the regions.

As a more realistic example, the GL algorithm was used to design VQs of one bit per sample and dimensions $k = 1, 2, 3, 4, 5, 6$ (so that the codebook size is 2^k). The source is again assumed to have independent Gaussian components with zero mean and unit variance. Using a training sequence of

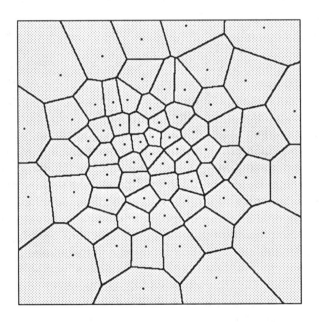

Figure 11.6: Voronoi Regions and Centroids: Two-Dimensional Gauss iid

100,000 samples and a distortion threshold of 0.01, the algorithm converged in fewer than 50 iterations and the resulting performance in terms of SNR vs. vector dimension is plotted as the solid line in Figure 11.7.

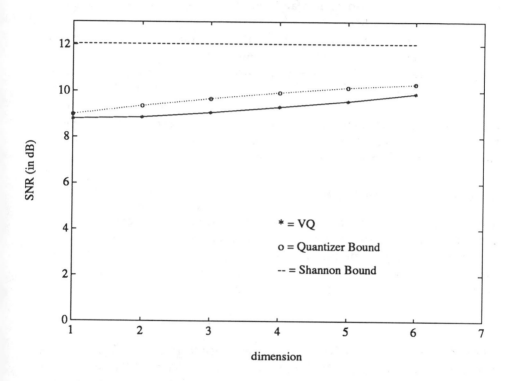

Figure 11.7: SNR vs. Blocklength: IID Gaussian

As a benchmark, two theoretical performance bounds are also plotted in the figure. The straight line corresponds to Shannon's distortion-rate function

$$D(R) = 2^{-2R},$$

where R is the rate in bits per sample [290] [134] [32] [159]. Here $R = 1$ and hence on a dB scale the distortion-rate function is $20\log_{10}0.25 = 12.04$ dB. The distortion-rate function provides an unbeatable lower bound to the average distortion of a rate R VQ of *any* dimension and arbitrary complexity. Furthermore, the Shannon bound is achievable in the limit as $k \to \infty$; that is, there exist codes that perform $D(R)$ (although Shannon theory does not say how to construct them). Clearly the performance of the actual VQs is not close to the optimum over all dimensions. A more realistic bound is provided by the high resolution quantizer bound of (10.6.2), which is plotted in circles. This bound is a vector generalization of the high resolution approximation for scalar quantizers of Chapter 5. Although strictly speaking these bounds are only accurate for large resolution, they are reasonably close to the actual performance and can be trusted at larger dimensions, e.g., when $k = 6$ and hence $N = 64$.

In this example a natural way of generating initial guesses for dimension k is to use the product code (to be defined more carefully in the next chapter) formed by using the $k - 1$ dimensional quantizer just designed for the first $k - 1$ samples and a one bit scalar quantizer on the final coordinate. This provides a one bit per sample dimension k codebook which provides a good start for the dimension k codebook. If, however, one only wishes to design a dimension k codebook and has no need for smaller dimensions, then the splitting algorithm works well.

Gauss Markov Sources

We have previously argued that one might expect VQ to show more improvement in comparison with scalar quantization when the source has significant memory. A simple example of a source with memory and a common model for a variety of sources in signal processing applications is the Gauss Markov or Gauss autoregressive source $\{X_n\}$ defined by

$$X_{n+1} = aX_n + W_n,$$

where the regression coefficient a has magnitude less than 1 and where $\{W_n\}$ is a zero mean, unit variance, iid Gaussian source. (The variance is chosen as unity for convenience; the extension to arbitrary finite variance is straightforward.) The autocorrelation of this source has the familiar form

$$R_X(k) = a^{|k|}$$

for all integers k. We consider the case of fairly high correlation: $a = 0.9$. In this case Shannon's distortion-rate function provides an optimal performance (over all block lengths) of 13.2 dB.

Vector quantizers were designed for this source at a rate of 1 bit per sample based on a training sequence of 60,000 samples and dimensions of 1 to 7. The performance is plotted in Figure 11.8. The improvement over scalar quantization is now more striking as expected. The codes were also tested on other sequences outside of the training sequence (but produced by the same random number generator) and the differences were under 0.25 dB, suggesting that the training sequence is large enough in this example to be trustworthy. For comparison, Arnstein's optimized one bit per sample

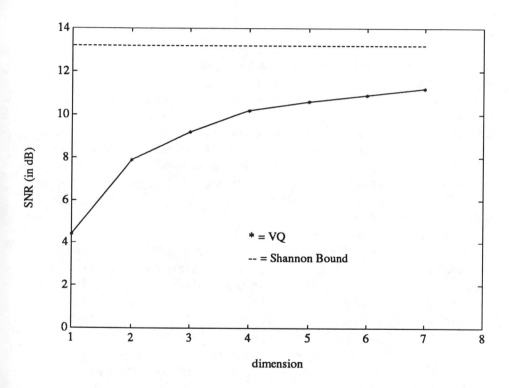

Figure 11.8: SNR vs. Blocklength: Gauss Markov Source

predictive quantization system [18] yields an SNR of 10 dB, close to that of a dimension 4 VQ. Three additional dimensions provide approximately 1 dB further improvement. We shall later see that a vector predictive quantizer can outperform the dimension 7 VQ by using only dimension 2 vectors.

Clearly the same basic approach works for any Monte Carlo generated random process and several papers have appeared reporting the performance of VQs on a variety of sources, especially on Laplacian and gamma densities. A variety of results along these lines may be found in the IEEE reprint collection [2].

Speech Waveform Coding

As a first source of practical importance we consider sampled speech. We should begin with the admission that plain VQ does not provide outstanding quality when applied directly to speech; more sophisticated VQs are required. It does, however, provide some interesting comparisons with traditional techniques and a straw dog for comparisons with smarter systems later on. Another caveat is that squared error is a notoriously poor measure of subjective quality for speech, but is often used anyway because of its simplicity. Good quality systems commonly use weighted squared error distortion measures, where the weightings are in the frequency domain and reflect knowledge of the human auditory system.

A training sequence of 640,000 samples of ordinary speech from four different speakers sampled at 6.5 kHz was generated. One and two bit per sample codes were designed in exactly the same way as those for the Gaussian sources. Figure 11.9 presents the SNRs for a test sequence consisting of 76,800 samples from a speaker not in the training sequence. The improvement in performance with dimension is largest at the smaller dimensions, but maintains a slope of 0.6 dB (per dimension) for dimensions 5 and above. The two bit per sample codes were designed for dimensions 1 through 4. These codes were designed using the splitting technique.

It is worth observing at this point that the SNRs reported in such experiments are of primary interest for relative comparisons among different code design techniques run on the same training set and compared on common test sets. For example, the SNRs reported here are uniformly lower by from 0.5 dB to 2 dB than those reported in [139] for basic vector quantization of speech waveforms. In addition to using different training and test sets, the sampling rate of the speech waveform results reported here was 6.5 kHz in comparison with the 8 kHz of [139]. The higher the sampling rate, the less variable are the vector shapes for the same dimension and the simpler the needed codebooks. (This is essentially the idea behind oversampling analog-to-digital conversion.)

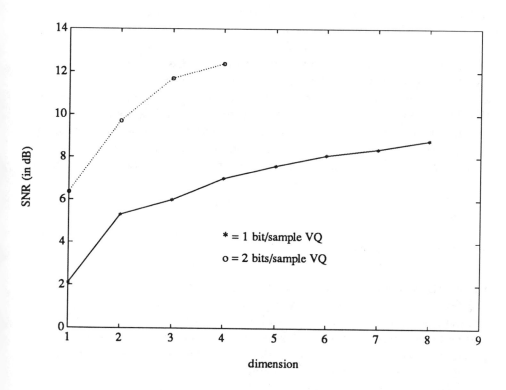

Figure 11.9: SNR vs. Blocklength: Sampled Speech

To gain a feeling for the SNR comparison with traditional techniques, the two bit per sample codes can be compared with the Lloyd-Max scalar quantizer design based on a gamma model density yielding an SNR of 6.04 dB on real speech [249]. The Lloyd algorithm gave slightly better performance of 6.4–6.5 dB for a scalar quantizer, but significantly better performance (more than 5 dB better) at higher dimensions. A scalar Lloyd-Max quantizer optimized for a Laplace model and imbedded in a predictive quantization system led to an SNR of 11.67 dB, a performance slightly lower than with a 3 dimensional VQ. The low rate Karhunen-Loeve transform coder of Campanella and Robinson [44] results in an SNR of 12.6 dB at 14 kbits/sec (their Fourier transform coder is about 4 dB worse). They used a transform vector dimension of $k = 16$. Their performance is close to that of the dimension 4 VQ. The adaptive predictive PCM (ADPCM) system of Crochiere et al. [83] yielded a performance of 10.9 dB at 2 bits per sample and of 11.1 dB for their subband coder of the same rate, comparable to the dimension 3 VQ. (Of course the subband coder has some perceptual advantage not reflected by SNR values.) The point of these comparisons is that simple memoryless VQ is at least competitive with several traditional scalar-quantizer based systems employing prediction or transforms in order to increase the effective vector size.

Simulated Speech

Unlike the Gaussian examples previously considered, the performance of VQ on speech cannot be compared to theoretical optima because no such optima are known. A natural possibility is to compute the optimal performance for random process models of speech, that is, well-defined random processes that exhibit similarity with speech. This is of interest because it might both suggest performance bounds and indirectly shed light on how good such models of real data sources are. It is also of interest in the Gaussian case because it shows even better than the Gauss Markov source how the performance improvement with dimension increase can be greater for more highly correlated sources. For this experiment an autoregressive or all-pole model for speech was used (see Section 2.8 for the terminology and Section 4.10 on random process modeling). This model is a gross simplification of reality since speech is known to exhibit a far more complex statistical behavior. Nevertheless, autoregressive modeling is very commonly used for modeling the short term behavior of speech for an individual phonetic unit whose duration is a small fraction of a second. More sophisticated models use a different autoregressive model for successive segments.

Here we model speech as a stationary random process $\{X_n\}$ of the form

$$X_n = \sum_{k=1}^{M} a_k X_{n-k} + \sigma W_n,$$

where the W_n are produced by an iid zero mean random process and the parameters a_k are chosen so that the filter with z-transform

$$\frac{1}{A(z)} = \frac{1}{1 + \sum_{k=1}^{M} a_k z^{-k}}$$

is stable. The parameter σ^2 is the squared gain or one-step prediction error of the process $\{X_n\}$ as developed in Chapter 4. An autoregressive source having the same autocorrelation as a speech sequence to lag M is called a *matching autoregressive source* or *MARS* for speech [12]. The same sampled speech training sequence with $L = 640,000$ can be used to estimate the autocorrelation function of speech to a lag of M by computing the sample autocorrelations

$$r(k) = \sum_{l=1}^{L-k} x_l x_{l+k}$$

for $k = 1, 2, \cdots, M$. The results are described in Table 11.2 where the values $r(k)$ are normalized by the energy $r(0) = 9662.57$. It can be shown that

Lag k	$r(k)/r(0)$	a_k
0	1.00000	1.00000
1	0.40094	-0.35573
2	-.04745	0.29671
3	-0.16734	0.09716
4	-0.18298	0.13798
5	-0.19850	0.10661
6	-0.26476	0.18392
7	-0.25939	0.17867
8	-0.12955	0.07244
9	-0.01075	0.11004
10	0.10457	0.03551

Table 11.2: Speech Autocorrelation and Filter Parameters

because of the autocorrelation matching property of linear prediction [234], the filter parameters a_k which yield an Mth order autoregressive process having exactly the given autocorrelation to lag M are the a_k produced by the Levinson-Durbin algorithm used in Chapter 4 to find the optimal linear predictor based on the finite past for a process given M autocorrelation values. Intuitively, if one can build a good predictor for a process, then one also can construct a good model for the process. Also intuitively, if $P(z) = -\sum_{k=1}^{M} a_k z^{-k}$ is the z-transform of a good finite impulse response linear predictor for a process, then ideally the error that equals the difference between the original process and its linear prediction, that is, the signal produced by passing the original process through a linear filter with z transform $A(z) = 1 + \sum_{k=1}^{M} a_k z^{-k}$, should be white. (Recall that the prediction coefficients were the negative a_k.) Thus if one wants to produce a process with the same autocorrelation function as the original process, one can reverse the system and drive the inverse filter (an autoregressive filter) $1/A(z)$ with a white process. See the discussion of Section 4.10 on simulation of random processes for further details. This produces an output with power spectrum $1/|A(e^{-2\pi jf})|^2$, which turns out to have an autocorrelation function which agrees with the given autocorrelation function. Thus we can run the Levinson-Durbin algorithm on the autocorrelation of Table 11.2 to find the filter parameters of Table 11.2. This filter will arise later when considering the LPC-VQ vocoder. The remaining issue is how to choose the probability density function for the iid process which drives the model. We choose a popular density function for this purpose, the Laplacian or doubly exponential density given by

$$f_Z(z) = \frac{1}{\sqrt{2}} e^{-\sqrt{2}|z|}.$$

Figures 11.10 and 11.11 depict the performance of VQs designed for real speech and for a training set of 60,000 samples from the matching autoregressive source with the Laplacian driver (produced by a random number generator). The training set for the fake speech is relatively short and the reported distortions are design distortions rather than independent test distortions. This use of design distortion was simply for convenience because these early experiments were performed several years ago on relatively small and slow computers. Nevertheless, they still provide an interesting comparison provided one realizes that the reported design distortions are generally optimistic. Also plotted are the high resolution quantization bound of (10.6.2) and Shannon's distortion rate bound. (Details of the bounds as well as similar experiments for K_0 marginals may be found in [3]. The distortion rate bounds are treated in [159].)

Several observations are worth making. First, the relative performance improvements achieved by real speech are much larger than those achieved

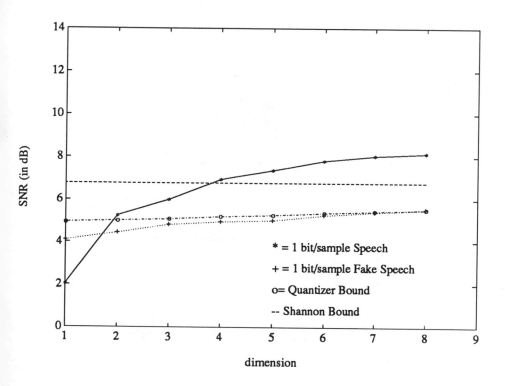

Figure 11.10: SNR vs. Blocklength: Speech and Fake Speech at 1 b/s

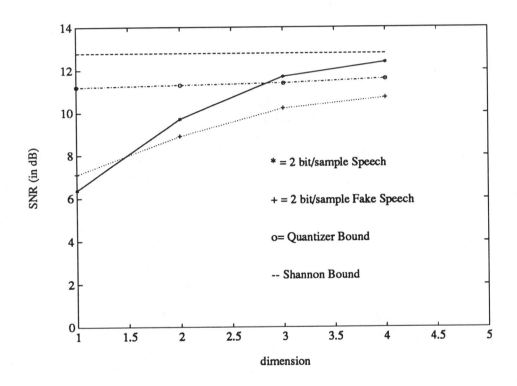

Figure 11.11: SNR vs. Blocklength: Speech and Fake Speech at 1b/s

for the model. This suggests that real speech has a much richer structure than the model and that a code trained directly on the real speech is more likely to capture and use this structure than is a code designed on a model of speech. Second, the actual performance of the designed VQs is better than the theoretical Shannon "optimum" in the 1 bit per sample case. This is not a violation of theory, the bounds are based on the model, not on real speech. It again suggests that the model is not very good, especially not at predicting the performance for real speech. Third, the performance on the model compares quite well to the high resolution quantization bound at the higher dimensions (and hence the higher rates in bits per vector). This suggests that in some applications at realistic dimensions VQ can perform nearly optimally for a fixed dimension and rate. Note in particular that the highly correlated speech model has performance much closer to the theoretical bound than the less correlated Gauss Markov and iid Gauss sources.

We should not leave speech waveform coding without commenting on the subjective quality of simple VQ. For a rate of 1 bit per second (b/s) the quality goes from very noisy and only barely intelligible scalar codes (effectively hard limiters) to easily intelligible but still noisy at dimension 8. The quality improves steadily with increased dimension. For 2 b/s, the quality is significantly better at all dimensions. A dimension 4, 2 b/s code exhibits little quantization noise on an ordinary high fidelity audio tape system. When these tests are run on a test sequence outside of the training sequence, there is a clear increase in quantization noise and buzziness, especially at the smaller dimensions and 1 b/s.

LPC-VQ

An alternative to coding a speech waveform directly is to construct a model of a short segment of speech, say 25 ms or so, quantize the parameters of the model, and then synthesize the reproduction using the quantized model parameters. Although this analysis/synthesis approach to speech compression is quite old, its major impact for good quality practical systems came with the development of the PARCOR or linear predictive coding (LPC) approaches to analysis/synthesis due to Itakura and Saito [189] [190] and Atal and Hanauer [19]. The basic idea is to model speech as an all-pole or autoregressive filter driven by either a white noise source (for unvoiced sounds such as "sss" and "sh") or a periodic pulse train (for voiced sounds such as vowels). The parameters describing the filter, voicing pitch and period are quantized and communicated. The user then synthesizes the speech by simulating the filter and driver. The technique, commonly called LPC, is treated in detail in numerous texts such as [234] [195] [247]. Here

we extract a few key points necessary to describe the combination of LPC
and VQ. The name "linear predictive coding" is appropriate because, as
noted previously, the construction of an all-pole model for speech turns out
to be equivalent to the construction of an optimal one-step finite memory
linear predictor for speech. The model can be designed by first estimating
the sample autocorrelation of the speech and then running the Levinson-
Durbin algorithm to obtain the filter parameters describing the predictor
or, equivalently, the model. The voiced/unvoiced decision and the determination of pitch are completely separate (and difficult) issues and will be
ignored here but are amply discussed in the literature. See for example
[247].

Traditional LPC systems used scalar quantization of the various parameters, although a variety of distortion measures, parameter representations,
and bit allocation strategies were used. (See, e.g., [171].) If a VQ is to be
used, clearly a distortion measure is required and it is not clear that the
squared error is appropriate. An alternative and natural distortion measure was implicit in Itakura and Saito's original development. Rather than
using linear prediction arguments, they used techniques from estimation
theory to show that the selection of the best autoregressive model given a
vector (or frame) of speech could be viewed as a minimum distortion selection procedure using what they called the "error matching measure." The
error matching measure has since come to be known as the Itakura-Saito
distortion. Suppose that we consider the input vector $\mathbf{x}$ to be a vector of N
speech samples, typically from 50 to 200 samples. The reproduction to be
produced by the vector quantizer however, will not be a quantized speech
sequence. Instead it will be an Mth order ($M \ll N$) autoregressive model
for speech of the form $\hat{x} = \sigma/A(z)$ with

$$A(z) = 1 + \sum_{k=1}^{M} a_k z^{-k}.$$

Typically N is from 8 to 15. The final reproduced speech segment will
then be synthesized by using this filter on either zero mean white noise or
a periodic pulse train. In fact, the filter is determined by linear prediction
arguments (Atal and Hanauer) or maximum likelihood arguments (Itakura
and Saito), both of which lead to running the Levinson-Durbin algorithm
on the speech frame sample autocorrelation (for the autocorrelation method
of LPC). In principle, however, this is equivalent to choosing $\sigma/A(z)$ (or
any equivalent set of parameters) over all stable, minimum phase filters so
as to minimize the distortion measure

$$d(\mathbf{x}, \hat{\mathbf{x}}) = \frac{\mathbf{a}^t \mathbf{R}(\mathbf{x}) \mathbf{a}}{\sigma} - \ln \frac{\alpha_M(\mathbf{x})}{\sigma} - 1, \qquad (11.4.1)$$

where $\mathbf{a}^t = (1, a_1, \cdots, a_M)$, where

$$\mathbf{R}(\mathbf{x}) = \{r_x(k - j); \ k, j = 0, 1, \cdots, M\}$$

is the $(M + 1) \times (M + 1)$ matrix of sample autocorrelations from 0 to M lag for the input sequence of N samples, that is,

$$r_x(k) = \frac{1}{N - k} \sum_{j=0}^{N-k-1} x(j)x(j + k).$$

Finally,

$$\alpha_M(\mathbf{x}) = \min_{\mathbf{y}:b_0=1} \mathbf{b}^t \mathbf{R}(\mathbf{x})\mathbf{b} \tag{11.4.2}$$

is the quantity D_M of the Levinson-Durbin algorithm of Section 4.5, the minimum mean squared prediction error achievable by a linear predictor with M stages. That d is nonnegative and hence is indeed a distortion measure follows from the $\ln x \leq x - 1$ inequality and (11.4.2) since

$$\frac{\mathbf{a}^t \mathbf{R}(\mathbf{x})\mathbf{a}}{\sigma} - 1 - \ln \frac{\alpha_M(\mathbf{x})}{\sigma} \geq \ln \frac{\mathbf{a}^t \mathbf{R}(\mathbf{x})\mathbf{a}}{\sigma} - \ln \frac{\alpha_M(\mathbf{x})}{\sigma}$$

$$= \ln \frac{\mathbf{a}^t \mathbf{R}(\mathbf{x})\mathbf{a}}{\alpha_M(\mathbf{x})}$$

$$\geq 0$$

The distortion of (11.4.1) appears strange at first: it is not symmetric and is not simply related to a metric. The asymmetry should not cause a problem since the input and output are very different; one is a sequence of real sampled speech and the other describes a model for synthesizing speech. The distortion measure is called the *modified Itakura-Saito distortion* or often simply the *Itakura-Saito distortion* without the extra adjective. Strictly speaking, the original Itakura-Saito distortion, which they called the "error-matching measure" [189][190], has α_∞, the infinite one-step prediction error σ_p^2 of Section 4.9 in place of $\alpha_M(x)$ in (11.4.1). The modification yields an equivalent nearest neighbor rule and is computationally simpler (See, e.g., [165].)

The Itakura-Saito distortion has many interesting properties for computation, theory, and applications [161] [296] [165]. For example, it can be expressed in terms of the sample power spectrum of the input sequence and the power spectrum of the autoregressive model. It can be shown to be a special case of a relative entropy or Kullback-Leibler information rate. It can be put into a more computationally transparent form by writing

$$\mathbf{R}(\mathbf{x}) = \{r_x(k - j); \ k, j = 0, 1, \cdots, M\}$$

and by defining the inverse filter parameter autocorrelations

$$r_a(k) = \sum_{l=0}^{M} a_l a_{l+k}$$

and then doing some matrix algebra to write

$$d(\mathbf{x}, \hat{\mathbf{x}}) = \frac{\alpha}{\sigma} - \ln \frac{\alpha_M(\mathbf{x})}{\sigma} - 1, \qquad (11.4.3)$$

where

$$\alpha = r_x(0)r_a(0) + 2 \sum_{k=1}^{M} r_x(k)r_a(k), \qquad (11.4.4)$$

an inner product between the autocorrelation of the input sequence (up to M lags) and the autocorrelation of the inverse filter coefficients.

Since the Itakura-Saito distortion measure is implicit in the selection of the perfect (unquantized) all-pole model for speech, it is reasonable to use the same distortion measure when selecting one of a finite number of possible reproductions, that is, when using a VQ which has a speech sequence as an input and an all-pole model as an output. In order to use the Lloyd algorithm, however, one must be able to find a centroid with respect to a distortion measure. Recall that for any collection $R = \{\mathbf{x}_1, \cdots, \mathbf{x}_K\}$, where $K = \|R\|$, we must be able to find a vector $\mathbf{y} = \sigma/A(z)$ which minimizes the sample distortion

$$\sum_{i=1}^{K} d(\mathbf{x}_i, \mathbf{y}) = \sum_{i=1}^{K} [\frac{\mathbf{a}^t R(\mathbf{x}_i)\mathbf{a}}{\sigma} - \ln \frac{\alpha_M(\mathbf{x}_i)}{\sigma} - 1].$$

Equivalently, we must find the $\mathbf{y}$ that minimizes the above sum normalized by $1/K$, and hence we must find the σ and the $\mathbf{a}$ which minimize

$$\frac{\mathbf{a}^t \overline{R} \mathbf{a}}{\sigma} - \frac{1}{K} \sum_{i=1}^{K} \ln \frac{\alpha_M(\mathbf{x}_i)}{\sigma} - 1 \qquad (11.4.5)$$

where $\overline{\mathbf{R}}$ is the average sample autocorrelation

$$\overline{\mathbf{R}} = \frac{1}{K} \sum_{i=1}^{K} R(\mathbf{x}_i).$$

Only the leftmost term of (11.4.5) depends on $\mathbf{a}$ and its minimization over all $\mathbf{a}$ with $a_0 = 1$ is exactly the minimization encountered in Chapter 4 (in

(4.3.12)). It is solved by the Levinson-Durbin algorithm which gives the a vector portion of the centroid. If $\overline{\sigma}$ is the resulting minimum of $\mathbf{a}^t \overline{\mathbf{R}} \mathbf{a}$, then the σ minimizing (11.4.5) is simply $\sigma = \overline{\sigma}$ since, ignoring terms not depending on σ, we have that

$$\frac{\overline{\sigma}}{\sigma} + \ln \sigma \geq 1$$

with equality if $\sigma = \overline{\sigma}$. Note that in order to find the nearest neighbor and the centroids with respect to the Itakura-Saito distortion, we do not actually have to compute the individual $\alpha(\mathbf{x}_i)$ for the input vectors, a computation which can be time consuming. It is necessary, however, to compute these mean squared errors if one wishes to quantitatively measure the Itakura-Saito distortion.

We now have all the pieces: to design an LPC-VQ we run the Lloyd algorithm (here the splitting method is a natural initialization) on the speech training sequence. The same basic training sequence is first parsed into long vectors (frames) of 128 samples each. Each of these vectors has its sampled autocorrelation matrix $\mathbf{R}(\mathbf{x}_i)$ computed and this can be considered henceforth as the training sequence. Here the splitting algorithm is natural. One first finds the centroid of the entire training sequence. In fact we have already done this in the previous example when we constructed the simulated speech model by finding the autoregressive source that matched the speech autocorrelation to lag 10. This model is split into two close models (e.g., just change the gain term σ slightly) and the training sequence is encoded into one of the two reproduction vectors using the Itakura-Saito distortion. Note that we do not need to compute the entire distortion to find the best fit. For each input vector $\mathbf{x}$ we need only find the model for which the σ and $\mathbf{a}$ minimize

$$\frac{\mathbf{a}^t \mathbf{R}(\mathbf{x}) \mathbf{a}^t}{\sigma} + \ln \sigma;$$

that is, we need not evaluate all the $\alpha_M(\mathbf{x})$ terms. After partitioning the input space, we replace the codewords by the centroids, found using the Levinson-Durbin algorithm. We then continue to convergence at a rate of 1 bit per vector, split, and so on. The results are depicted in Figure 11.12. Shown are the rate in bits per vector (128 input symbols) and the design and test SNRs (here ten times the logarithm of the ratio of the Itakura-Saito distortion of the code to that of the optimal zero rate code.) Since the number of samples is fixed and the vector size has increased, the training sequence now has only 5000 vectors in it. The test sequence has 600 vectors. Note that the design performance is noticeably optimistic, often excelling the test performance by 1 dB or more.

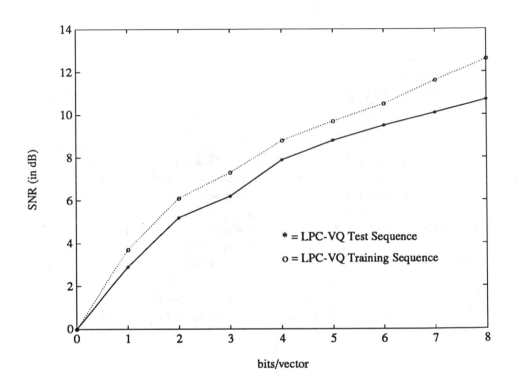

Figure 11.12: SNR vs. Bits per vector: LPC-VQ

The LPC VQ was one of the earliest successful applications of VQ because it brought the bit rate of traditional LPCs using scalar quantization for various parameterizations of the LPC coefficients from 4800 bits per second down to 1200 bits per second with little loss of quality. The problem remained that LPC itself has questionable quality, often tinny and sensitive to high and low pitched voices as well as multiple speakers. Nonetheless, LPC VQ forms a core part of most of the best current low to very low rate (2.4 kb/s or below) voice coding systems. Several of these variations will be sketched later. The preceding example, based on the Itakura-Saito measure, is just one of many alternative methods available today for coding the linear prediction parameters. Efficient coding of linear prediction coefficients has become an important issue in almost all advanced speech coders studied today since the autoregressive model is widely used in waveform coding of speech as well as in low rate vocoders.

Image VQ

A natural way to apply basic VQ to images is to decompose a sampled image into rectangular blocks of fixed size and then use these blocks as the vectors. For example, each vector might consist of a 4 × 4 square block of picture elements or *pixels*. Thus each vector has 16 coordinates. A typical digital image has a resolution of 8 bits per pixel (bpp). The goal of VQ is to reduce this to less than 1 bit per pixel without perceptible loss of picture quality. We shall see that this is not possible with basic VQ, but will be possible with some variations. The application of VQ to coding images was independently proposed in [344], [140], and [23] and subsequently research activity in this area expanded greatly. The earliest study, by Yamada, Fujita, and Tazaki [344], appeared in Japanese in 1980. Subsequently commercial video teleconferencing coders emerged in the U.S. and Japan based on VQ.

We shall consider two different types of images for variety. The first consists of ordinary still frame images, for which we will use a training and test set taken from an image data base prepared at the University of Southern California. Actually the USC image data base comes in full 24 bpp color, but we here consider only gray scale compression by taking only the luminance component of the image, that is, the so-called Y component of the YIQ transformation from the red, green, blue components of the original color image. Several of the images from this data base are shown in Figure 11.13. The lower right image, called variously "Lenna" and "the girl with the hat," was not in the training sequence and is used as our test image for this data base.

The Lloyd algorithm was run on the training sequence using the simple

Figure 11.13: USC Images

squared error distortion measure and 4×4 blocks. The quality measure for ordinary still images is commonly taken as the peak signal-to-noise ratio or PSNR which is defined as 10 times the log to the base 10 of the ratio of peak input amplitude squared to the mean square error (MSE). In the case of an 8 bpp original this is

$$\mathrm{PSNR} = 10\log_{10}\Big(\frac{256^2}{\mathrm{MSE}}\Big).$$

The PSNR is plotted as a function of the rate in bits per pixel (bpp) in Figure 11.14.

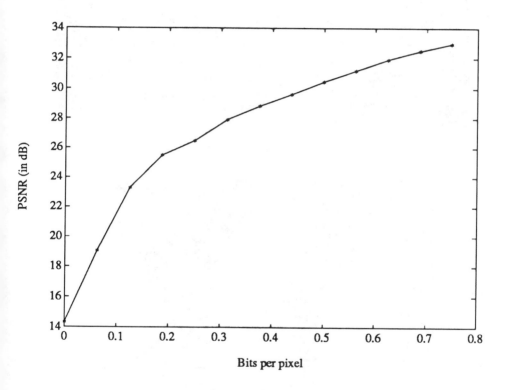

Figure 11.14: PSNR vs. Bits per pixel: Lenna

The resulting codebook for .5 bits per pixel (bpp) is shown in Figure 11.15. There are 256 different patterns (8 bits) of 4×4 pixel squares so that the rate is indeed $8/15 = 0.5$ bpp.

In order to get a feeling for the corresponding subjective quality and the operation of the code, Figure 11.16 shows the coded Lenna at 0.5 bpp and Figure 11.17 shows the resulting error image, the difference between the

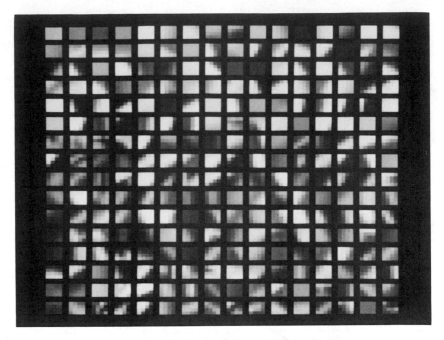

Figure 11.15: Image VQ Code Book

Figure 11.16: Lenna at 0.5 bpp

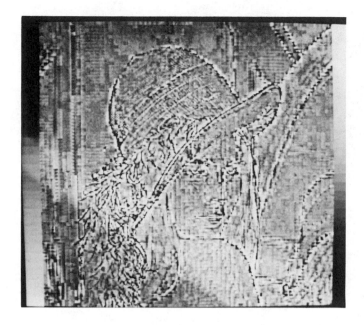

Figure 11.17: Quantization Error

original and the coded image. Figure 11.18 shows a closeup of the eye of the original Lenna image along with the coded eye. One can see that even the original is not perfect when magnified enough; high resolution digital images are blocky when viewed at this scale. The VQ image is, however, much worse: the blockiness and the sawtooth effect along diagonal edges are apparent. Figure 11.19 shows a similar closeup of the original image along with the coded version of the shoulder. This shows the inability of a simple VQ to well code diagonal edges.

As a distinct image example, a training sequence of magnetic resonance (MR) images of sagittal brain slices were used to design a 1 bit per pixel (bpp) VQ with 4×4 blocks. The original 256×256 image was 8 bpp and the Lloyd algorithm was used to design a VQ. (In fact the original MRI images produced had 9 bits, but only the highest order bit was truncated for this study because our monitor could only display to 8 bit accuracy.) The original and VQ image of a person not in the training sequence (but the same view) are shown in Figure 11.20. We refer to this image as "Eve" as it shows the brain of Dr. Eve Riskin who did much of the original work applying vector quantization to medical images. Once again the block and staircase effects can be seen in the edges, but the overall appearance is fairly good for a simple code. When dealing with medical images, ordinary

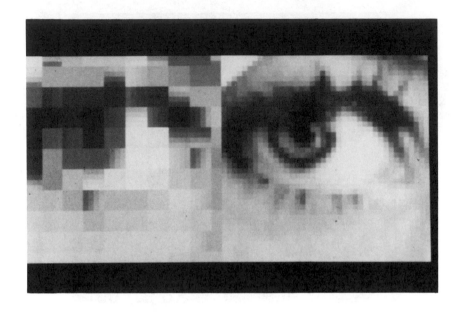

Figure 11.18: VQ Coded Eye and Original

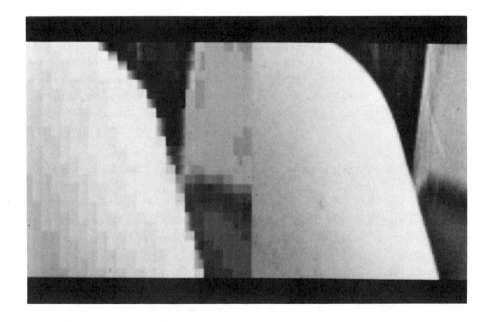

Figure 11.19: VQ Coded Shoulder and Original

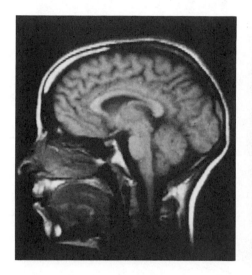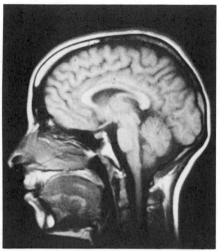

Figure 11.20: VQ MR Image at 1.5 bpp

SNR is more common than PSNR as a quality measure. The SNR for the test MRI image is plotted as a function of bit rate in Figure 11.21.

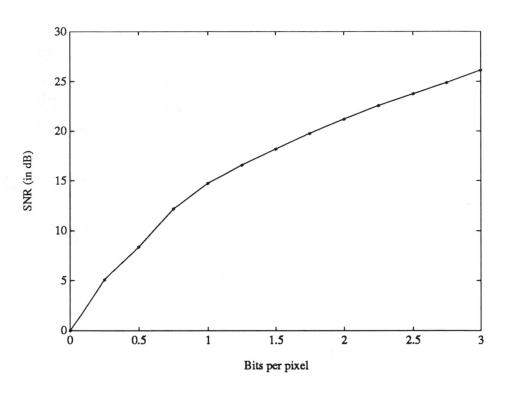

Figure 11.21: PSNR vs. Bits per pixel: Eve

11.5 Problems

11.1. Design a good codebook of size 4 for a Gaussian random vector $\mathbf{X}$ in the plane ($k = 2$). Assume each component of $\mathbf{X}$ has zero mean, unit variance, and the correlation coefficient is 0.9. After computing the codebook, plot the code vectors (indicate with an 'x' or other distinctive sign) on a graph containing the region $-2 \leq X_1 \leq +2$, $-2 \leq X_2 \leq +2$. Plot the first 100 training vectors with a dot ($\cdot$) on the same graph as the code vectors. (Use the generalized Lloyd I algorithm, generating a training set of 1000 random vectors.) Pick any reasonable initial codebook and use a stopping criterion of 1% threshold in relative drop in distortion AND a maximum of 20 Lloyd iterations. Specify the final distortion (averaged over your training set) achieved for your codebook and specify the location of the four code vectors.

11.2. A two-dimensional random vector is uniformly distributed on the unit circle $\{x, y : x^2 + y^2 \leq 1\}$. Find a VQ satisfying the Lloyd conditions for a codebook of size 2. Repeat for codebook sizes 3 and 4. (You are to do this by trial and error rather than by computer, although you can program a solution as a guide. You must prove your quantizers meet the Lloyd conditions.)

Are the codebooks of size 2 and 3 essentially unique, that is, is there only one codebook satisfying these conditions except for rotations and permutations? (Say what you can about this, do not be discouraged if you do not come up with a solution.)

Show that there are clearly distinct codebooks of size 4 by finding one with a vector at the origin and the other without. Which is better?

11.3. A vector quantizer is to be designed to minimize the weighted distortion measure given by

$$d(\mathbf{x}, \mathbf{y}) = E[(\mathbf{x} - \mathbf{y})^t \mathbf{W}_{\mathbf{x}}(\mathbf{x} - \mathbf{y})]$$

where for each $\mathbf{x} \neq 0$ $\mathbf{W}_{\mathbf{x}}$ is a symmetric positive definite matrix (that depends on $\mathbf{x}$) and $E(\mathbf{W}_{\mathbf{x}})$ and $E(\mathbf{W}_{\mathbf{x}}\mathbf{X})$ are finite. For a given partition, find the necessary condition for the code vectors to be optimal.

11.4. Consider the weighted squared error distortion measure of the previous problem. Find extensions of Lemma 11.2.2 to codes satisfying the centroid condition with respect to this distortion measure. (The results will be different in form, but similar in appearance. If you are stuck, track down [166].)

11.5. A coding system performs some processing on an incoming signal and then generates a random vector **U** and a scalar random variable G that is statistically dependent on **U**. (But G is *not* the gain of **U**; it depends on other random variables derived from the input signal that are not contained in the vector **U**.) Specifically, G is typically the rms level of a long segment of the input signal, whereas **U** is a vector formed from a small number of consecutive samples of this input segment.) The scalar G provides a rough estimate of the rms level of U (without actually requiring the use of bits to identify the true rms level of **U**) and is used to normalize **U** before applying it to a vector quantizer, **Q**. The quantizer output is rescaled (multiplied by G) to produce the final estimate $\hat{U}$ of **U**. Assume G is transmitted to the receiver with perfect fidelity. Assume that the joint probability distribution of **U** and G are known so that all expectations can be determined.

   (a) Given the codebook for **Q**, derive the optimal partition rule for **Q**.

   (b) Given a fixed partition for **Q**, derive a formula for the optimal codevectors in the codebook for **Q** which minimize the average squared Euclidean distance between **U** and $\hat{U}$. *Hint:* one way to do this it to use a natural generalization of the orthogonality condition; alternatively, you can express the desired codevector in terms of its (scalar) components and optimize separately for each component.

11.6. (a) Derive an expression for the computational complexity needed to encode a k-dimensional random variable with a b bit codebook (containing 2^b k-dimensional code vectors, where each component of each vector is a real number stored to a suitably high accuracy). The result of an encoding is simply the index identifying which vector in the codebook is the best representation of the input vector. Assume the cost of additions or comparisons is negligible but a scalar multiplication of two real numbers is the unit for measuring complexity.

   (b) Repeat part (a) but now suppose that an addition costs 3 units, a scalar multiplication costs 23 units, and a comparison between two real numbers costs 1 unit.

     In both parts you must explain how you derive your result.

   (c) If one unit in part (b) requires 0.1 microseconds of a processor's time, and only one processor is used to implement the encoder, how many vectors can be encoded per second when $k = 8$ and $b = 16$?

11.7. Design and sketch a good 2-dimensional quantizer with 3 levels for an iid sequence of uniform random variables (uniform on [-1,1]) and the mean squared error fidelity criterion using the Lloyd algorithm and a training sequence of at least 5000. What is the final mean squared error? What is the mean squared error if your code is run on a separate test sequence (produced by the same random number generator with a different seed or initial value)?

11.8. Show that the conclusions of Lemma 11.2.1 remain true for any distortion measure of the form $d(\mathbf{x}, \mathbf{y}) = \rho(||\mathbf{x} - \mathbf{y}||)$, where $\rho(\cdot)$ is a monotonically nondecreasing function and $|| \cdot ||$ is the Euclidean distance.

11.9. Use the Lloyd algorithm to design a scalar ($k = 1$) quantizer for an iid Gaussian source for codebooks of size $N = 1, 2, 3, 4, 6, 8$ using a training sequence of 10,000 samples. Compare your results with those reported by Max [235].

11.10. Given that a random vector $\mathbf{X}$ lies in a convex set R, show that its mean value lies in R.

11.11. Consider a convex bounded region R in k-dimensional space with volume V. Assume that the region contains N training vectors that are very roughly uniformly distributed in R and that the nearest neighbor region for each vector is roughly approximated as a hypersphere. Find a formula for the radius of each hypersphere and indicate how this can be used for selecting a threshold value for the pruning method of designing an initial VQ codebook.

11.12. You are given the following 4-dimensional binary training set:

$$T = \{(1,1,1,1),(1,1,1,0),(1,1,1,0),(0,0,0,1),(1,0,0,1),$$

$$(0,0,0,1),(1,0,0,0),(0,0,1,0),(0,0,0,1),(1,1,0,1)\}.$$

Use the generalized Lloyd algorithm with the Hamming distance distortion measure to design a 2-codeword VQ. Use the initial codebook

$$Y_1 = \{(1,1,0,0),(0,0,1,1)\}.$$

Assume in the case of a tie in the distortion between an input vector and the two codewords, that the training vector is assigned to the codeword with the lower index, that is, to the cell R_1. Also assume that in the case of a tie in the centroid computation (a majority rule) that a 0 is chosen.

No computer is necessary.

(a) What is the final codebook you get?

(b) What is the average Hamming distortion between the training sequence and the final codebook?

(Problem courtesy of E. Riskin.)

11.13. Use the splitting algorithm to design a 2-word (rate 1) codebook for the training sequence

$$T = \{(0,\frac{1}{4}),(\frac{1}{4},\frac{1}{4}),(\frac{3}{4},\frac{1}{4}),(1,\frac{1}{4}),(\frac{1}{4},\frac{1}{2}),$$

$$(\frac{3}{4},\frac{1}{2}),(0,1),(\frac{1}{2},1),(1,1),(\frac{1}{4},\frac{3}{4})\}.$$

You may assume that the perturbation is done with $\epsilon = 0.01$, that is, split the rate 0 codebook y_0 into y_0 and $y_0 + 0.01$.

(a) What is the initial rate 0 codebook?

(b) What is the final rate 1 codebook?

(c) What is the distortion between the training set and the rate 1 codebook?

(Problem courtesy of E. Riskin.)

11.14. Design a nearly optimal codebook of size $N = 5$ for a Gaussian random vector $\mathbf{X}$ in the plane ($k = 2$). Assume each component of $\mathbf{X}$ has zero mean, unit variance, and the correlation coefficient is 0.95. After computing the codebook, plot the codevectors (indicate with an 'x' or other distinctive sign) on a graph containing the region $-2 \le X_1 \le +2$, $-2 \le X_2 \le +2$. Plot the first 100 training vectors with a dot ($\cdot$) or other distinctive marker on the same graph as the codevectors. Use the generalized Lloyd I algorithm, generating a training set of 1000 random vectors. [First figure out how to generate independent uniform random variables over $(0, 1)$ using a modulo congruence relation or an existing library routine in C or Fortran, then find a way of generating independent Gaussian variables from the uniform variables (lots of literature available on this) and then use take a linear transformation to produce the desired random variables. See also Problem 2.10.] Pick any reasonable initial codebook and use a stopping criterion of 1% threshold in relative drop in distortion AND a maximum of 20 Lloyd iterations. Specify the final mean squared distortion per vector (averaged over your training set) achieved for your codebook and specify the location of the five codevectors.

11.15. Obtain a training sequence for speech and design VQs for 1 bit per sample and dimension 8 using both squared error distortion and the input weighted squared error distortion defined by

$$d(\mathbf{x}, \hat{\mathbf{x}}) = \frac{||\mathbf{x} - \hat{\mathbf{x}}||^2}{||\mathbf{x}||^2},$$

that is, the short term noise to signal ratio. Compare the resulting subjective quality.

11.16. Suppose that a a distortion measure is defined to depend on the sample variance of the input in the following manner. Given a k-dimensional input vector $\mathbf{x} = (x_1, \cdots, x_k)^t$, define the sample mean

$$m(\mathbf{x}) = \frac{1}{k} \sum_{l=1}^{k} x_l$$

and the sample variance

$$\sigma^2(\mathbf{x}) = \frac{1}{k} \sum_{l=1}^{k} (x_l - m(\mathbf{x}))^2.$$

An input vector $\mathbf{x}$ is classed as being in C_1 if $\sigma^2(\mathbf{x}) > \delta$ and as being in class C_2 if $\sigma^2(\mathbf{x}) \leq \delta$. The threshold $\delta > 0$ is fixed. Intuitively, the C_1 vectors are more "active" since they possess a larger sample variance. The distortion measure is now defined as

$$d(\mathbf{x}, \hat{\mathbf{x}}) = \begin{cases} \lambda ||\mathbf{x} - \hat{\mathbf{x}}||^2 & \text{if } \mathbf{x} \in C_1 \\ (1 - \lambda) ||\mathbf{x} - \hat{\mathbf{x}}||^2 & \text{if } \mathbf{x} \in C_2 \end{cases},$$

where $\lambda \in (0, 1)$ is a fixed weight. This distortion measures allows us to weight active input vectors as more (or less) important. Describe the details of the generalized Lloyd algorithm for designing a VQ with respect to this distiortion measure.

11.17. The partial distortion theorem (Theorem 6.3.1) for scalar quantizers suggests that in addition to the two known Lloyd conditions for optimality, the optimal vector quantizer should perhaps force all of the partial distortions

$$D_i = E[||\mathbf{X} - Q(\mathbf{X})||^2 | \mathbf{X} \in R_i] \Pr(\mathbf{X} \in R_i)$$

to be equal, regardless of i. Is this true? Explain.

Chapter 12

Constrained Vector Quantization

12.1 Introduction

There is no better way to quantize a single vector than to use VQ with a codebook that is optimal for the probability distribution describing the random vector. However, direct use of VQ suffers from a serious complexity barrier that greatly limits its practical use as a complete and self-contained coding technique.

Unconstrained VQ is severely limited to rather modest vector dimensions and codebook sizes for practical problems. In speech coding, for example, the largest codebook sizes used typically lie in the range of 1024 to 4096 and the largest vector dimensions used are typically between 40 and 60. More commonly, codebook sizes range from 64 to 256 and vector dimensions from 4 to 10. Reducing the dimension of a vector often sacrifices the possibility of effectively exploiting the statistical dependency known to exist in a set of samples or signal parameters. Thus, as designers of signal compression systems, we tend to push toward the complexity limits of VQ to obtain the best coding gain achievable.

Several techniques have been developed which apply various constraints to the structure of the VQ codebook and yield a correspondingly altered encoding algorithm and design technique. As a result, higher vector dimensions and larger codebook sizes become feasible without hitting a "brick wall" complexity barrier. These methods generally compromise the performance achievable with unconstrained VQ, but often provide very useful and favorable trade-offs between performance and complexity. In other words, it is often possible to reduce the complexity by orders of magnitude while paying only a slight penalty in average distortion. Moreover, the constrained VQs can often be designed for larger dimensions and rates. Hence quality that is simply not possible for unconstrained VQ becomes practicable and hence no performance is "lost" at all.

The clever design of such constrained VQ schemes is part of what makes engineering an art that goes beyond the realm of straightforward use of mathematics to analyze engineering problems. This chapter focuses on some of the most important constrained VQ techniques. Rather than provide an encyclopedic listing of techniques, we concentrate on several basic forms. It is our hope that this presentation will provide the basis for readers to find innovations suited to their own applications. The IEEE Reprint collection of H. Abut provides a thorough catalogue of such variations [2].

12.2   Complexity and Storage Limitations

Consider the task of quantizing a picture. A typical image can be sampled with a 512 by 512 square lattice of sample points and specified as a single vector consisting of 512^2 or $262,144$ sample amplitudes, called *pixels* or *pels* (for *picture elements*). Suppose the probability distribution of this rather high dimensional vector represents the set of all "natural" images that could ever be seen by a camera. It might be reasonable to estimate that very high quality reproduction can be achieved with a bit rate of 0.1 bits per pixel (or *bpp*). However, this rate implies a codebook size of $N = 2^{26214}$ or 10^{7891} code vectors which no doubt exceeds the size of the largest read-only memory (ROM) chip likely to be available in the next ten years. Furthermore, the complexity required of a processor that must perform a nearest neighbor search is on the order of N operations. Needless to say, no realizable processor is capable of performing this search within a few years' time, even if it had a cycle time of less than a picosecond. Even a massively parallel processor capable of handling such a task in reasonable time is not yet on the horizon. Incidentally, it is interesting to note that, loosely speaking, the number N represents the set of all possible pictures that are distinguishable from one another.

While the above example is obviously extreme, the complexity barrier does present a very real and practical limitation on the applicability of VQ. Suppose we consider the coding of a signal that can be modeled as a stationary random sequence and we focus on the question of how to partition this sequence into vectors for direct application of VQ to each vector. In general, if we constrain the resolution r measured in bits per vector component to a fixed value, we know that the performance of VQ can only increase as the dimension k of the vector increases. This performance improvement with dimension can be explained by recognizing that we are exploiting longer term statistical dependency among the signal samples (and the extra freedom to choose higher dimensional quantizer cell shapes). On the other hand, the required codebook storage space in words and the search complexity (number of operations per input vector in an exhaustive codebook search) are both proportional to kN. Thus both time and space complexity are given by

$$kN = k2^{rk},$$

which grows exponentially with dimension. A word of storage represents one component of one code vector and one operation refers to either a multiplication of two scalars or a squaring operation. For simplicity, additions and comparisions are assumed to have negligible complexity. (In many applications, this is not a good approximation and the actual complexity can be much higher than indicated here.)

To see clearly how the complexity barrier is reached, consider a speech waveform that is sampled at $f_s = 8$ kHz. Suppose we wish to use VQ with a resolution of 1 bit per sample for a bit rate of 8 kb/s. Blocking k consecutive samples into a vector then requires a codebook size of $N = 2^k$ so that the search complexity and storage space are each $k2^k$. The number of operations per unit time for an exhaustive codebook search with the squared error performance measure indicates the required processor speed and is given by $Nf_s = 2^k f_s$ and the reciprocal of this speed is the maximum time available per operation, i.e., the instruction cyle time of the processor. Table 12.1 shows the complexity needed to perform real-time speech coding, the minimum number of operations per unit time and the maximum cycle time required by a processor that is devoted to perform the codebook search. Note that processor speed and ROM storage limitations impose an upper limit on vector dimension achievable. In this example, the limit is approximately $k = 12$ corresponding to a codebook size $N = 4096$ for the best digital signal processors of 1991 vintage (with a 30 ns cyle time).

12.3 Structurally Constrained VQ

One approach to mitigate the complexity barrier is to impose certain structural constraints on the codebook. This means that the code vectors cannot have arbitrary locations as points in k-dimensional space but are distributed in a restricted manner that allows a much easier (lower complexity) search for the nearest neighbor. Lattice VQs and polytopal VQs are examples already seen.

Complexity vs. Dimension			
Dimension	Complexity	Operation Speed	Cycle Time
1	2	16 kHz	62.5 μs.
4	64	128 kHz	8 μs.
8	2048	2 MHz	500 ns.
10	10,240	8 MHz	125 ns.
12	49,152	33 MHz	31 ns.
20	10^{13}	8 GHz	100 ps.
40	10^{25}	8.7 10^9 GHz	0.1 ps.

Table 12.1: Complexity as a function of dimension for waveform VQ at 1 bit per sample and 8000 samples per second. Complexity indicates both codebook storage in words and operations per input vector. Operation speed is the minimum number of multiplies per unit time. Cycle time indicates the maximum time available per operation.

In some of these techniques, the encoder, using a search procedure associated with the new structure, will still find the nearest neighbor, but the performance will be degraded compared to that of conventional VQ with an unconstrained codebook. Any constraints imposed on the codebook as part of the design process can generally be expected to lead to an inferior codebook for a given rate and dimension.

Some techniques that impose structural constraints on the codebook use a search procedure that may not find the nearest neighbor, but will generally find one of the collection of approximately nearest neighbors. As a result, the performance compared to conventional VQ is degraded due to both the suboptimal search procedure and the suboptimal codebook. Nevertheless, the degradation can often be kept quite small and tolerable in exchange for a substantial complexity reduction.

12.4   Tree-Structured VQ

One of the most effective and widely-used techniques for reducing the search complexity in VQ is to use a tree-structured codebook search. In tree-structured VQ (TSVQ), the search is performed in stages. In each stage a substantial subset of candidate code vectors is eliminated from consideration by a relatively small number of operations. In an m-ary tree search with a balanced tree, the input code vector is compared with m predesigned test vectors at each stage or node of the tree. The nearest (minimum dis-

tortion) test vector determines which of m paths through the tree to select
in order to reach the next stage of testing. At each stage the number of
candidate code vectors is reduced to $1/m$ of the previous set of candidates.
Such a tree-structured search is a special case of a *classification tree* or
decision tree [36]. In many applications $m = 2$ and we have a binary tree.

We have already encountered codes with a tree structure: noiseless prefix codes such as the Huffman code of Chapter 9 have a tree structure
(although they are not quantizers in the sense of using a minimum distortion selection of tree branches), and the successive approximation scalar
quantizers of Chapter 6 (which are one-dimensional quantizers). Furthermore, in Section 10.4 we presented an example of an unbalanced tree that
efficiently performs a nearest neighbor search for a given codebook. In this
section we focus on jointly designing the tree and the VQ codebook in a
manner that leads to a balanced tree, where each path from the root to a
terminal node traverses the same number of nodes. Although the resulting
codebook may not be optimal, it often gives a performance that is quite
close to optimal while allowing an encoding complexity that is much less
than that of exhaustive search through an arbitrary codebook of the same
size.

If the codebook size is $N = m^d$, then d m-ary search stages are needed
to locate the chosen code vector. An m-ary tree with d stages is said to
have *breadth* m and *depth* d. The basic structure of the tree search is shown
in Figure 12.1, where each node in the tree (except for terminal nodes) is
associated with a set of m test vectors, forming a *node codebook*. The figure
shows the tree to depth three. Each set of test vectors or node codebook
are depicted as being associated with a node. The tree can also be drawn
to associate the individual test vectors as labels of the branches emanating
from the node, that is, each test vector labels the branch that the encoder
follows if that test vector is chosen by the nearest neighbor selection. A
particular search leads to a particular path down the tree till a terminal
node is reached. Each terminal node corresponds to a particular vector in
the codebook. The encoder first searches the root codebook $\mathcal{C}^*$ and finds
the minimum distortion codeword (test vector). If the vector has m-ary
index i, then i becomes the first symbol in the m-ary channel vector and
the encoder advances along the ith branch emanating from the root node to
node (i), where each node is represented by the m-ary *path map* describing
how the encoder went from the root node to the current node. Given
that the search has reached the current node (i), the encoder searches the
corresponding node codebook $\mathcal{C}_i$ to find the minimum distortion test vector,
which determines the next branch, and so on. For notational convenience,
each node codebook is labeled with a subscript that is the binary vector or
path map that leads to this node from the root. The encoding algorithm

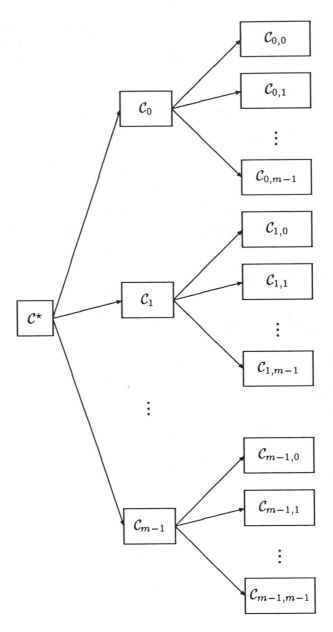

Figure 12.1: Basic Structure of Tree-Structured VQ Encoding

of an m-ary TSVQ is summarized in Table 12.2.

Table 12.2: TSVQ Encoder

Step 0. *Given:* Depth d, breadth m, rate $\log_2 m^d$ bits per vector, input vector $\mathbf{x}$.

Step 1. Root node: Find the codeword $\mathbf{y} \in C^\star$ minimizing $d(\mathbf{x}, \mathbf{y})$. Let $u_0 \in \{0, 1, \cdots, m-1\}$ be the index of this minimum distortion word. Set the one-dimensional channel m-ary codeword to $u^1 = u_0$ and advance to node (u^1). Set the current tree depth $k = 1$.

Step 2. Given the k-dimensional channel codeword $u^k = (u_0, u_1, \cdots, u_{k-1})$ and the current node (u^k), find the codeword $\mathbf{y} \in C_{u^k}$ minimizing the distortion $d(\mathbf{x}, \mathbf{y})$. Let u_k denote the index of the minimum distortion codeword. Set the $(k+1)$-dimensional channel m-ary codeword u^{k+1} equal to the concatenation of u^k and u_k:

$$u^{k+1} = (u^k, u_k) = (u_0, u_1, \cdots, u_k).$$

Step 3. If $k + 1 = d$, halt with the final channel codeword u^d (corresponding to a reproduction vector in $C_{u^{d-1}}$). Otherwise set $k + 1 \to k$ and go to Step 2.

Since a tree-structured codebook only affects the search strategy, the decoder does not need the test vectors and is in fact identical to that of the conventional VQ. There is, however, one situation where the test vectors are also used by the decoder. In *progressive* reconstruction (or *successive approximation*) of the vector, it is desired to rapidly provide successively better approximations to the input vector as the digital information arrives rather than wait till the vector is completely specified. In this application the transmitter can send an m-ary symbol or index to the receiver as soon as the search passes through each stage of the tree. The decoder can use the indicated test vector as a temporary reproduction vector and as each new symbol arrives, the reproduction is replaced with a better approximation using a test vector from a deeper level in the tree. Finally when the last symbol is transmitted, a particular code vector can be identified and used as the reproduction vector.

The number of search operations at each stage is proportional to m since each test is an exhaustive nearest neighbor search through a set of m test vectors. Thus, the total search complexity is proportional to md rather than m^d where the proportionality constant depends on the vector dimension k. On the other hand, the storage requirement of TSVQ is increased compared to unstructured VQ. In addition to storing the m^d code vectors (the leaves of the tree), the test vectors for each node of the tree must also be stored. There is one node at the first stage, m nodes at the second stage, m^2 nodes at the third stage, etc. Hence, the total number of nonterminal nodes is $1 + m + m^2 + \cdots + m^{d-1} = (m^d - 1)/(m - 1)$. Since each nonterminal node stores m code or test vectors, the total number of k-dimensional vectors to be stored including code vectors is $m(m^d - 1)/(m - 1)$. Alternatively, one must store the m^d code vectors plus

$$\sum_{l=1}^{d-1} m^l = \frac{m(m^{d-1} - 1)}{m - 1}$$

additional test vectors.

The simplest tree structure to implement is the binary tree where $m = 2$. In this case the complexity is reduced by a factor of $2d/2^d$ compared to ordinary VQ and the total storage is $2(2^d - 1)$ vectors, slightly less than double that of ordinary VQ with the same codebook size $N = 2^d$. It is simple to show that for a given codebook size the complexity increases in proportion to $m/\log m$ and the storage decreases in proportion to $m/(m-1)$ as the tree breadth m increases. For a binary tree, the complexity is lowest and the storage is highest.

An important question for the designer of a coding system naturally arises at this point. How much degradation in performance is there compared to ordinary VQ for a fixed codebook size and for various choices of the breadth m? Unfortunately, this is difficult to answer in general since the performance depends very much on the particular signal statistics. Usually the only way to answer this question is to design a tree-structured codebook that satisfies the complexity and storage limitations for the particular application and measure the performance by computer simulation. It is often difficult or impossible to compare the performance with ordinary full search VQ, since the complexity for the same codebook size and vector dimension may be too high even for computer simulation studies. Typically, the degradation is quite modest but often it is not negligible. In speech or image coding applications, it can result in a perceptually noticeable quality difference.

The tree-structured search does not in general find the nearest neighbor code vector. This is easily seen in the example of a 2 bit quantizer in two dimensions illustrated in Figure 12.2. Line A is the hyperplane boundary associated with the root node and lines B and C correspond to the two nodes at level 2 in the tree. The corresponding test vectors, labeled 0 and 1 are used to determine on which side of A the input lies. The input **X** shown lies closer to code vector 1,0 than to code vector 0,1, yet the first stage of the search eliminates code vector 1,0 from further consideration and code vector 0,1 is chosen to quantize **X**. Figure 12.3 illustrates the corresponding tree structure for this quantizer.

Design of Tree-Structured VQ

A standard method for designing the tree structure (consisting of the set of test vectors for each node and the associated codebook) is based on application of the generalized Lloyd algorithm (GLA) to successive stages using a training set. The procedure, which was proposed in [215], is a variation on the standard splitting method of designing full search VQ. Similar methods have been used in the field of pattern classification. To motivate the design, consider the two-dimensional example of Figure 12.4. The first step is identical to that of a standard Lloyd design with the splitting technique: the optimum resolution 0 bit code is found, that is, the code having only a single codeword, the centroid of the entire training set (Figure 12.4(a)). This codeword is then split into two, shown by the open circles in Figure 12.4(b), and the Lloyd algorithm is run to produce a good resolution 1 bit code, as depicted by the arrows pointing from the initial 1 bit seed to the final discs depicting the two codewords.

The ordinary Lloyd (full search) design would now proceed by splitting the two codewords and running the Lloyd algorithm on the full training set to obtain a new codebook of four words. Instead, we replace the nearest neighbor (optimum) encoder by an encoder that does the following. The encoder first looks at the 1 bit codebook and selects the nearest neighbor. Given that selection, it then confines its search to the descendants of the selected codeword, that is, to the two words formed by splitting the selected word. It encodes the training set and then replaces all four codewords (the labels of the terminal nodes or leaves in a depth two tree) by the centroid of all training vectors mapped into those nodes (Figure 12.4(c)).

The replacement encoder has substituted two binary searches for a single quaternary search, but the remainder of the algorithm is the same. This depth 2 binary tree of course offers no savings in complexity as the two encoders have comparable complexity and the double binary search is clearly not an optimum search (it need not find the nearest neighbor). It does,

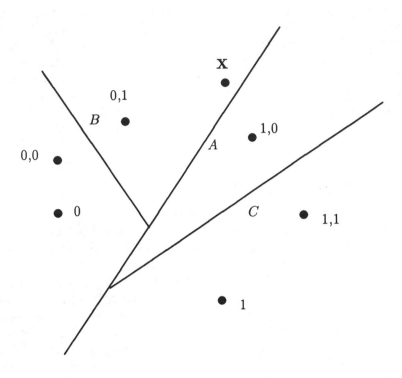

Figure 12.2: Quantizer resulting from a binary tree with 2 stages. The input vector **X** is quantized with code vector 0,1 although its nearest neighbor is code vector 1,0.

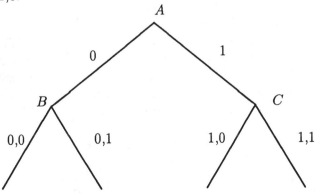

Figure 12.3: Corresponding Tree

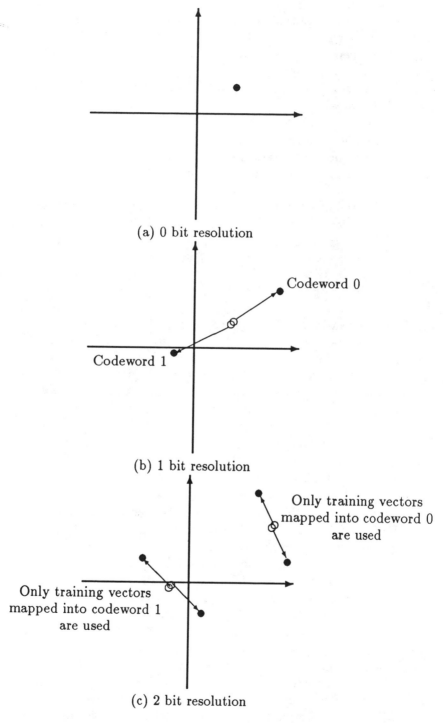

(a) 0 bit resolution

(b) 1 bit resolution

(c) 2 bit resolution

Figure 12.4: Splitting TSVQ Design

however, have the potentially useful progressive reconstruction property: the first channel codeword bit itself specifies a 1 bit reconstruction. The second bit then improves upon the first. The ordinary full search VQ has no such inherent property.

One can continue to build layers of the tree in this fashion. At each step the terminal nodes are split and the training vectors which reach each node are used to design the node codebook.

For each node in the tree other than the terminal nodes, the set of test vectors which determine which branch of the tree is to be followed is a VQ codebook of size m. However, these "code vectors" are to be used as test vectors and should not be confused with the code vectors at the terminal nodes that actually form the codebook that the decoder will use. Each of these test codebooks for a particular level of the tree are designed based on a suitably generated training set, that makes use of the test codebooks designed for the higher level nodes of the tree. An ordinary training set

$$T = \{\mathbf{v}_1, \mathbf{v}_2, \cdots, \mathbf{v}_M\},$$

with training ratio $\beta = M/N$ is used to represent the statistics of the input vectors. We can now describe the basic m-ary TSVQ design algorithm, which is best viewed as a simple variant of the splitting version of the generalized Lloyd algorithm. The algorithm is given in Table 12.3.

In Step 1, an input vector will be routed down to the next node according to which of the test vectors it is closest. Hence it is natural to apply this operation to each training vector thereby partitioning the training set into m new subsets $T_0, T_1, \cdots, T_{m-1}$. The training set T_i then represents the statistics of all input vectors that have passed the test of being closer to the ith test vector than to any other test vector in the set of test vectors at the root node.

As noted earlier, this codebook design procedure results in a suboptimal codebook and in a tree search method that does not necessarily find the nearest neighbor in that codebook. The above method can also be generalized to unbalanced trees which do not have a fixed number of branches emanating from each node. This somewhat complicates the tree design process, but allows greater flexibility in finding "natural" structures based on the input statistics. If by chance the statistics of the source vectors have the property of being multimodal with m modes so that any input vector has a natural characterization as being distinctly of one type, i.e., belonging to one mode, then a partitioning of the training set into m subsets (generally not of equal size) occurs quite naturally and a separate codebook can be designed by the Lloyd algorithm for each of the subsets of the training set. The selection of the mode and hence codebook is a form of classification

---

Table 12.3: **TSVQ Design Procedure**

Step 1. Use the training set T to generate a codebook $C^\star$ of size m test vectors for the root node (level 1) of the tree. Partition the training set into m new subsets $T_0, T_1, \cdots, T_{m-1}$.

Step 2. For each i, design a test codebook C_i of size m using the GLA applied to T_i. (We thus obtain the test codebooks for the m nodes at level 2 of the tree.)

Step 3. Partition each training set T_i into m training subsets T_{ij} and use these new training sets to design the m^2 test codebooks C_{ij} for level 3.

Step 4. Continue this process until level $d - 1$ is reached. (The test vectors in the test codebooks obtained for the nodes at this level are actually code vectors for the terminal nodes of level d. The collection of all these test vectors at this level constitute the codebook.)

and this structure will be further studied later in the section on classified VQ.

Single-Node Splitting

Another approach to designing a tree-structured VQ follows from viewing it as a classification tree. Breiman, Friedman, Olshen, and Stone [36] provide a variety of algorithms for "growing" such trees. Their algorithms are "greedy" in the sense that one tries to grow a tree one step further by finding the node whose split will yield the best possible improvement of the overall quality measure for the tree. That node is then split and two new branches and two new nodes are formed. In the VQ context using average distortion as the quality (or lack thereof) measure for the tree, this is similar to the technique proposed by Makhoul, Roucos, and Gish [229]. Their idea was to split only the one terminal node in the current tree which has the largest contribution to the overall average distortion. The Lloyd algorithm then runs on the resulting tree as before, except now the trees are in general unbalanced: terminal nodes occur at different levels. This clearly yields at the interim steps a very different type of code that resembles the noiseless codes of Chapter 9: the codewords have variable lengths. Permitting a variable length code removes a constraint from the VQ design so that it is possible for such codes to outperform even full search VQ of the same rate (because full search VQ must devote the same number of bits to every vector while the incomplete TSVQ can devote more bits to important vectors and fewer to unimportant vectors, mimicking the behavior of good noiseless codes). Of course, "rate" here refers to the *average* of a code. We will later return to variations of this technique when we consider variable rate codes in Chapter 17. The primary modification that will be made in deciding which node to split is to select that node which results in the biggest decrease in average distortion after the split, rather than select the node that currently has the highest average distortion. Yet another strategy for choosing a node to split is to perform a principal components analysis and to split in the direction of maximium variance [343].

One can obtain a collection of fixed-length codewords from an unbalanced tree such as that produced by single-node splitting design techniques. Makhoul et al. simply relabel the leaves with fixed length binary vectors and thereby obtain a fixed rate code with a tree-search encoding algorithm. Unfortunately, however, the successive approximation property of the binary codewords is thereby lost. The resulting code does have a simple encoder because of the tree structure and the resulting codes performed better than the TSVQ designed with the splitting technique.

The Makhoul-Roucos-Gish design algorithm is not a special case of a greedy growing algorithm with respect to average distortion since splitting the node with the largest partial distortion need not produce the best split in the sense of causing the largest reduction of total average distortion. Such greedy growing design algorithms will be considered in Chapter 17.

TSVQ Design Examples

Figures 12.5 and 12.6 provide a comparison of full search VQ with binary TSVQ for two of the design examples previously considered in Section 11.4: a Gauss Markov source and sampled speech. As usual, in these examples a vector is formed from a block of k consecutive samples of the waveform. As in Section 11.4, the figures include for comparison Shannon's distortion-rate function for the source considered.

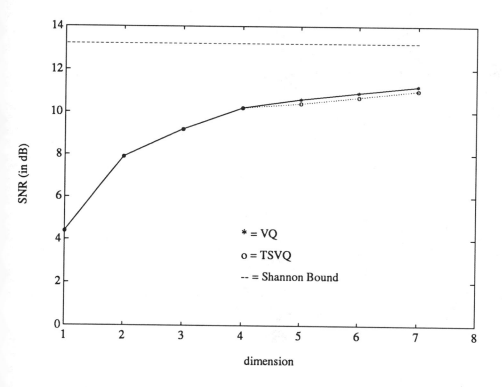

Figure 12.5: SNR: TSVQ for a Gauss Markov Source

In these examples, the performance of the suboptimal but far less complex TSVQ is always within 1 dB of the full (optimal) searched VQs. Further substantiation of this relatively small loss in performance may be found

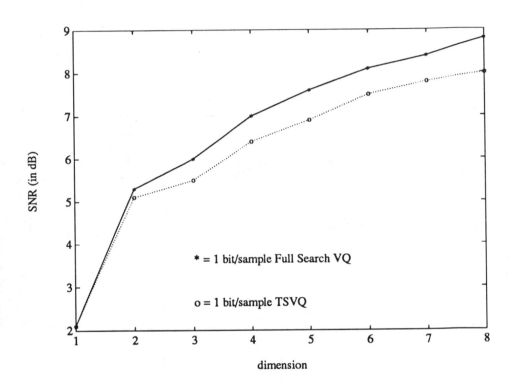

Figure 12.6: SNR: TSVQ for Sampled Speech

in [160] [229]. By using m-ary trees, performance indistinguishable from that of full search VQ can be obtained in some applications. For example, it was found by Wong et al. that for LPC VQ of the filter coefficients, a three layer tree with 4 bits, 4 bits, and 2 bits in the first, second, and third layer, respectively, gave essentially the same performance as a full search 10 bit tree [338].

12.5 Classified VQ

Classified VQ (CVQ) is similar to the use of a one stage tree-structured VQ where the criterion for selecting which of m branches to descend is based on an heuristic characteristic that the designer adopts to identify the mode of the particular input vector. Thus instead of a test codebook, an arbitrary classifier may be used to select a particular subset of the codebook to be searched. Varying size subsets are allowed and we can partition the codebook into unequal sized sub-codebooks. Figure 12.7 illustrates this scheme. The classifier generates an index, an integer from 1 to m, which

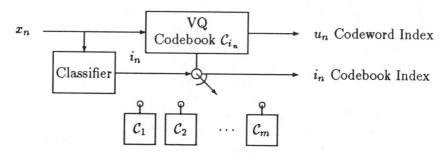

Figure 12.7: Classified VQ

identifies which sub-codebook C_i to search for a nearest neighbor. The codeword consists of the index i specifying which of the m codebooks is selected and $\log_2 \|C_i\|$ bits specifying the selected word in the codebook.

Many possibilities exist for the choice of classifier. It can be a simple VQ encoder that identified in which of m regions of the input space the current input vector lies. In this case, CVQ is very similar to a two stage TSVQ where the second stage nodes may each have different codebook sizes. In fact a TSVQ can be thought of as a special case of a tree structure where each node is a VQ encoder with a codebook whose size is not restricted. The classifier in a CVQ encoder can extract one or more features from the input vector, such as the mean (average amplitude of the samples), the energy (sum of the squares of the samples), the range (difference between the maximum and minimum amplitudes of the components), and other statistics.

The particular physical meaning of the vector may lead to many other features. For example, in a two-dimensional set of pixels representing an image block, an edge detector can be used to identify the presence or absence of edges (or high detail) and the direction and location of edges if present [259]. For a vector that represents a block of consecutive speech samples, the zero-crossing count (number of sign changes in proceeding from the first to last sample) can be a useful feature. Once such features are identified, the task remains to decide how many distinct modes should be used. A scalar or vector classifier must be chosen or designed to provide a complete classification operation. The classifier can be designed either heuristically or using the GLA with a training set of features that are extracted from a training set of input vectors.

Classified VQ Design

Once an m-class classifier has been designed, the sub-codebooks can be designed in one of two distinct ways. The first method is to pass the original training set T through the classifier and generate m subsets T_j of the training set. Each sub-codebook of designated size m_j is then designed using the GL algorithm with T_j as the training set. The concatenation of the sub-codebooks constitutes the complete codebook for CVQ. The designer may use heuristic considerations to select the sizes m_j with the constraint that they add up to the desired value of N. A bit allocation optimization algorithm such as those considered in Chapters 8, 16, and 17 can also be used. A specific design technique for bit allocation in classified VQ is considered, for example, in Riskin [268]. Alternatively, the sizes m_j can be chosen to be proportional to the fractional size of the training subsets T_j. In image coding, it has been found that certain classes are of greater perceptual significance than others so that it can be advantageous to make the sub-codebook size for such a class disproportionately large compared to the relative occurrence of such vectors in the training set [259].

12.6   Transform VQ

Instead of directly applying a signal vector $\mathbf{X}$ as input to a vector quantizer, an orthogonal linear transformation $\mathbf{T}$, as used in transform coding, can first be applied to the input vector. The transformed vector, $\mathbf{Y}$, can then be vector quantized, and the quantizer output $\hat{\mathbf{Y}}$ can be inverse transformed to yield the quantized approximation $\hat{\mathbf{X}}$ to the original input $\mathbf{X}$.

This configuration is shown in Figure 12.8. If the VQ dimension is the

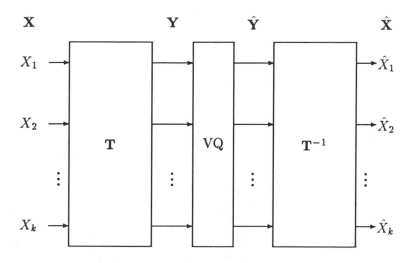

Figure 12.8: Transform VQ

same as the transform dimension, then it is easy to show that the optimal code vectors for this structure are simply the transforms **TX** of the optimal code vectors for the codebook that would be used for direct VQ of **X** without any transformation. Furthermore, the average distortion (squared error) will be exactly the same as in the case of direct VQ without using a transformation. This suggests that transformation offers no advantage but has the disadvantage of adding some extra complexity.

Nevertheless, there is a useful advantage to be gained from the use of transforms in conjunction with VQ. A key idea is that linear transforms often *compact* the information in the original vector into a subset of the vector components. Specifically, the discrete cosine transform in essence performs a mapping to the frequency domain, and often a signal vector has very little energy in the higher frequency regions of the spectral band. This property implies that a substantial fraction of the components of the transformed signal vector are likely to be very close to zero and may often be neglected entirely. This reduces the dimension of the vector to be coded in the transform domain and thereby the VQ complexity is reduced. Alternatively, the transform coefficients can be partitioned into groups and each group separately vector quantized. In particular, a vector representing the low energy components can be separately coded with a very low bit-rate (including fractional values of rate) rather than be neglected completely. Partitioning of the transform coefficients into groups makes it feasible to handle larger dimension vectors in contrast with direct VQ without transform coding.

Further advantages are similar to those that motivated traditional transform coding. One can often use larger block sizes for transforms than is practicable for VQ, thus effectively increasing the dimension of the code from that achievable by VQ alone, and human visual and auditory perception seem to be sensitive to quality in the transform domain which allows for easy adaptation of the bit allocation to transform vectors according to their perceptual importance.

The compaction reduces the dimensionality of the signal vector by transforming it and discarding the unneeded components (i.e., truncating the vector). The truncation, $\mathbf{H}$, takes a k-dimensional input vector, $\mathbf{Y}$ and maps it into a p-dimensional $(p < k)$ vector, $\mathbf{Z}$. Then, p-dimensional VQ is used to quantize the compacted vector, forming the quantized approximation $\hat{\mathbf{Z}}$. Reconstruction can be achieved by a padding operation, $\mathbf{H}^*$, that appends $k - p$ additional components with value zero to the vector (or possibly uses an optimal 0 resolution codeword), producing the k-dimensional vector $\hat{\mathbf{Y}}$. Finally, the inverse transform, $\mathbf{T}^{-1}$ is applied to $\hat{\mathbf{Y}}$ to obtain the final reproduction $\hat{\mathbf{X}}$. This scheme is indicated in Figure 12.9(a). Note that

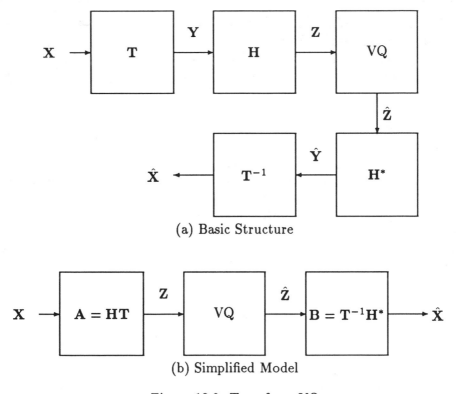

(a) Basic Structure

(b) Simplified Model

Figure 12.9: Transform VQ

the cascade of $\mathbf{T}$ and $\mathbf{H}$ is equivalent to performing a rectangular transformation $\mathbf{A} = \mathbf{HT}$ on $\mathbf{X}$ with a matrix of dimension $p \times k$. Similarly, the zero padding operation $\mathbf{H}^*$ followed by the inverse transform $\mathbf{T}^{-1}$ operation is equivalent to a rectangular matrix $\mathbf{B} = \mathbf{T}^{-1}\mathbf{H}^*$ of dimension $k \times p$ operating directly on $\hat{\mathbf{Z}}$ to produce $\hat{\mathbf{X}}$. This simplified structure for transform VQ is shown in Figure 12.9(b).

It can be seen that

$$\|\mathbf{X} - \hat{\mathbf{X}}\|^2 = \|\mathbf{Z} - \hat{\mathbf{Z}}\|^2 + \|\mathbf{Y} - \mathbf{H}^*\mathbf{H}\mathbf{Y}\|^2.$$

Hence, the average distortion in quantizing $\mathbf{X}$ is given by

$$D = E(\|\mathbf{Z} - \hat{\mathbf{Z}}\|^2) + \sum_{i=p+1}^{k} E[Y_i^2].$$

Thus the excess distortion added by coding $\mathbf{X}$ in this way is simply the truncated energy in the "high frequency" components of $\mathbf{Y}$ which have been discarded.

The codebook for quantizing $\mathbf{Z}$ can consist simply of the truncated version of the transformed code vectors $\mathbf{TY}_i$ that would be used without dimensionality reduction. It may be shown that although this is not the optimal codebook, the performance penalty is small when the truncated energy is low.

The complexity saving with this truncation technique can be quite substantial. Suppose that a resolution of r bits per vector component are allocated to code the k-dimensional vector $\mathbf{X}$ for a total bit allocation of rk bits. Without truncation the codebook size would be $N = 2^{rk}$. With truncation and using the same resolution r, the codebook size is reduced to $N = 2^{rp}$, which can be orders of magnitude smaller depending on the values of k and p. When k is fairly large, the savings in complexity can be very substantial while the performance degradation (due to the truncated energy) can be negligible. Transform VQ with truncation has been found to be a valuable complexity reducing technique for image coding [259]. Alternatively, for the same complexity, the resolution can be increased as the dimension is reduced so that increased performance is achievable.

Image transform VQ using the discrete cosine transform is treated in Aizawa et al. [8]. Early work on VQ in the transform domain may be found in King and Nasrabadi [205]. A similar application for speech compression can be found in Adlersberg and Cuperman [5].

Transform VQ with Vector Partitioning

An alternative approach to transform VQ is to not throw away any coefficients, but instead partition the transform coefficients into subsets and use

VQ separately on each subset of the transform coefficients. We shall treat the idea of vector partitioning more generally as a form of product code in Section 12.8, but the idea is useful in the present specific application. With this approach the transform coefficients are viewed as a collection of features that represent the signal and judiciously grouped subsets of features can form feature vectors that are separately and efficiently coded with VQ. Here there are two basic approaches. The first is to choose a fixed vector size, but use a bit allocation algorithm such as in Chapters 8, 16, and 17 to assign different resolutions to the different codebooks. Another method is to pick a fixed resolution for all of the codebooks (say the maximum feasible), but to adjust the vector dimensions in a reasonable way, e.g., by an iterative algorithm which increases the dimension (lowering the resolution per coefficient) for those vectors contributing very little to the overall average distortion and doing the opposite for vectors which contribute heavily. When used with a Fourier transform (FFT), the variable dimension approach provides noticeably better subjective performance and about 3 dB improvement in mean squared error over traditional (scalar quantized) transform coders with bit allocation using the high resolution approximation [59] [222].

In one approach to wide-band audio coding at a 32 kHz sampling rate, the transform (DCT) of a block of 512 samples is formed and partitioned into 127 coefficient vectors each of 4 adjacent coefficients. (The dc term is separately quantized and the 3 uppermost frequency coefficients are discarded.) A perceptual model of auditory masking effects is used to adaptively determine how much distortion is tolerable in each of the 127 vectors and an appropriate quantizer is then allocated to each vector [55].

Combined transform VQ and vector partitioning can also be used in image coding. Suppose we have an 8 by 8 block of pixels. A two-dimensional discrete-cosine transform can provide a compaction where only 21 of the 64 transform coefficients need be retained; but this dimension is still too high to encode with a single full search VQ. However it is known that certain subsets of transform coefficients, corresponding to specific regions of spatial frequencies, are relatively uncorrelated with other regions or frequency bands. Thus, the 21 coefficients can be partitioned into four subvectors. The first, of dimension one, is the "dc" coefficient corresponding to the block mean. The second is of dimension 5, the third of dimension 9 and the fourth of dimension 6. This is a two dimensional variation of the variable dimension transform VQ described above. In image coding, the "dc" coefficients from different blocks of an image in the same spatial region are highly correlated so that predictive coding or VQ can be used on these coefficients, one from each block. In this way, some of the spatial correlation that extends well beyond an individual block can be exploited.

Sub-band VQ

A closely related cousin to transform VQ with a Fourier transform is sub-band coding VQ where the "frequency domain" coefficients are obtained by using bandpass filters (often quadrature mirror filters) instead of using the FFT. Developments of VQ sub-band coding for speech may be found in [4] and [141] and for images in [330]. This is the two-dimensional extension of the sub-band coders described in Chapter 6.

Wavelet Transforms

A variation on both the transform coder and the sub-band coder that has excited considerable interest recently use one of the wavelet transforms. A collection of survey papers on the mathematics and applications of wavelet transforms may be found in [74] along with an extensive guide to the basic papers on the subject. We here limit ourselves to only a few comments and citations of some of the earliest applications of the idea to image compression.

In the abstract, a wavelet transformation can be viewed as simply an alternative transformation to the Fourier, cosine, and Karhunen-Loeve transforms already considered. The key idea is to form a decomposition of classes of signals into weighted sums of basis functions (complex exponentials for the Fourier transform and cosines for the cosine transform). Unlike traditional Fourier theory, the basis functions are formed by scaling and translating a single function and the mathematical properties of the decomposition are determined by the properties of the underlying function. The use of scaling permits one to construct approximations to the original signal at multiple resolutions, analogous to reconstructing an image at multiple resolutions by downsampling. Wavelets can provide decompositions of signals into orthogonal basis functions (as does the Fourier transform), but biorthogonal (orthogonal signals plus their negatives) and simply nonorthogonal transforms are also considered. Some wavelet transforms can be implemented by quadrature mirror filters such as are used in sub-band coding, but not all quadrature mirror filters perform a wavelet decomposition. The principal potential advantage of wavelet based image compression systems is that the regularity properties of the wavelet transform can provide good approximations to a diversity of signals with a small number of coefficients. It has also been claimed that wavelet decompositions offer better models of human auditory or visual processing. Antonini, Barlaud, Mathieu, and Daubechies [15] [14] reported good quality images using vector quantizer based wavelet transform coding combined with entropy coding at 1 bpp. The chose biorthogonal wavelets which could be implemented by relatively simple filters and used the generalized Lloyd algorithm to design codebooks for the various resolutions. High resolution quantizer theory was used to allocate bits among the codebooks and to choose the vector dimensions. Subsequently Antonini, Barlaud, and Mathieu [13] reported much improved performance using large dimensional lattice based vector quantizers on the wavelet transform coefficients. They reported peak signal-to-noise ratios in the range of 30 dB at only 0.08 bpp (with rate being measured by entropy).

In a different approach, Devore, Jawerth, and Lucier [97][98] developed a non-probabilistic approach to quantizer design based on limiting the class of input vectors to a suitably smooth class, finding the "optimal" performance achievable for this class in a certain sense, and then demonstrating that a simple scalar quantization scheme operating on the wavelet coefficients provided nearly optimal performance. The argued that magnitude error is more meaningful than squared error for use in nearest neighbor selection and they choose quantized coefficients to minimize a distortion of the form $||(c - \hat{c})\phi||$, where ϕ is a basis function and the norm is an L_1 distance, that is, an integrated absolute magnitude. They also measure rate by entropy (effectively assuming entropy coding) and report good performance at rates near 0.1 bpp, but their applications use piecewise constant wavelets (unlike the fairly smooth wavelets of Antonini et al.) and it is difficult to compare the two approaches directly. Both suggest, however, that wavelet transforms provide a potentially attractive form of transform or sub-band codes that can achieve good performance with simple lattice quantizers.

12.7 Product Code Techniques

An age-old technique for handling an unmanageably large task is to decompose it into smaller sub-tasks. This is, in essence, the idea of product codes. In VQ, the task that can often become too large is encoding. One way to reduce the task is to partition the vector into subvectors, each of lower dimensionality. Instead of a single VQ for the entire vector, each subvector can be separately encoded with its own codebook. By sending a set of indexes to the decoder, the decoder reconstructs the original vector by first decoding each subvector and then concatenating these vectors to regenerate the reconstructed approximation to the original large vector. This strategy is perhaps the simplest illustration of a product code technique.

As an example consider again a 512 by 512 pixel image. The image can be viewed as a single vector of dimension 262,144 which would be quite impossible to code directly using VQ. It can, however, be partitioned into 4 by 4 blocks, and each of these 16,384 blocks can be treated as a separate vector of dimension 16 and vector quantized. This is a natural and obvious way to break down an impossible problem into a large set of reasonable tasks. It hardly deserves the special name of "product code." Yet, there are less obvious techniques where useful savings can be achieved based on a slightly more general idea of product codes.

An obvious disadvantage of such a partitioning of the large vector into subvectors is that there typically exists statistical dependence between the different subvectors. This dependence can lead to greater coding efficiency if the large vector is directly coded as one entity. By separate coding of each subvector, we lose the possibility of exploiting that dependence. Rather than merely partition a vector into subvectors, we need some other way to decompose the original vector into a set of vectors that are individually easier to code. If we can find a way to partition the information into vectors that are approximately independent of one other, then the coding complexity can be greatly reduced without a substantial performance penalty. This is the general idea and the objective of product codes.

What precisely is a product code? Given a vector $\mathbf{X}$ of dimension $k > 1$, let $\mathbf{U}_1, \mathbf{U}_2, \cdots, \mathbf{U}_\nu$ be a set of vectors that are functions of $\mathbf{X}$ and jointly determine $\mathbf{X}$. Thus, there are functions f_i for $i = 1, 2, \cdots, \nu$ such that $\mathbf{X}$ can be decomposed or "analyzed" into component vectors $\mathbf{U}_i$ according to

$$\mathbf{U}_i = f_i(\mathbf{X}) \text{ for } i = 1, 2, \cdots, \nu. \tag{12.7.1}$$

Each vector $\mathbf{U}$ is sometimes called a *feature* vector because it is chosen to represent some characteristics or attributes of the original vector $\mathbf{X}$ while still only partially describing $\mathbf{X}$. Each feature vector should be easier to quantize than $\mathbf{X}$ because it takes on values in a more compact region of k-dimensional space or has a lower dimensionality. In particular, one or more of these component vectors can be scalars, i.e., vectors of dimension one. Furthermore, given the set of component vectors, there is a function g, such that the original vector $\mathbf{X}$ can be reconstituted or "synthesized" by

$$\mathbf{X} = g(\mathbf{U}_1, \mathbf{U}_2, \cdots, \mathbf{U}_\nu). \tag{12.7.2}$$

For each i, let $\mathcal{C}_i$ be a codebook of N_i reproduction code vectors that contain the reproduction values for $\mathbf{U}_i$. A *product code vector quantizer* is then a vector quantizer whose encoder observes $\mathbf{X}$ and generates a set of indexes $I_1, I_2, \cdots, I_\nu$ specifying the reproduction values $\hat{\mathbf{U}}_i \in \mathcal{C}_i$; $i = 1, \cdots, \nu$, which are in turn transmitted to the decoder. I_i specifies the selected code vector for reproducing the component vector $\mathbf{U}_i$ for each i. The code is called a product code because the requirement that $\hat{\mathbf{U}}_i \in \mathcal{C}_i$ for all i is equivalent to saying that the overall vector $(\hat{\mathbf{U}}_1, \cdots, \hat{\mathbf{U}}_\nu)$ is in the Cartesian product $\mathcal{C}$ of the ν codebooks, that is,

$$(\hat{\mathbf{U}}_1, \cdots, \hat{\mathbf{U}}_\nu) \in \mathcal{C} = \mathcal{C}_1 \times \mathcal{C}_2 \times \cdots \times \mathcal{C}_\nu.$$

The decoder then finds the reproduction vectors $\hat{\mathbf{U}}_1, \hat{\mathbf{U}}_2, \cdots, \hat{\mathbf{U}}_\nu$ and generates a quantized approximation to $\mathbf{X}$ given by

$$\hat{\mathbf{X}} = \mathbf{g}(\hat{\mathbf{U}}_1, \hat{\mathbf{U}}_2, \cdots, \hat{\mathbf{U}}_\nu). \tag{12.7.3}$$

The set of all possible reproduction vectors for $\mathbf{X}$ is therefore determined by the set of all possible combinations taking any code vector from each of the codebooks. There are

$$N = \prod_{j=1}^{\nu} N_j \tag{12.7.4}$$

reproduction vectors for $\mathbf{X}$. This set of vectors is, in effect, the codebook $\mathcal{C}$ with which $\mathbf{X}$ is quantized. The codebook is not optimal in general because its code vectors are constrained by the relationship (12.7.3) and the individual component codebooks $\mathcal{C}_i$. The encoding and storage complexities with this approach are each roughly equal to

$$\text{Complexity} = \sum_{j=1}^{\nu} N_j \tag{12.7.5}$$

since the encoding can be performed by separate encoding of each component vector, and each codebook is separately stored.

Figure 12.10 shows the general configuration of a product code VQ. The internal structure of the encoder is not indicated here since there are a variety of ways in which this can be done.

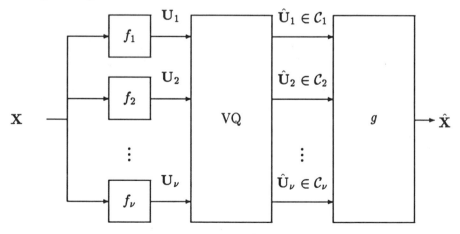

Figure 12.10: Product Code VQ: General Configuration

It is important to note that although each $\hat{\mathbf{U}}_i$ can take values only in $\mathcal{C}_i$, the selection of the codewords $\hat{\mathbf{U}}_i$ in the encoding process is in general interdependent; that is, it need not be true that the individual $\mathbf{U}_i$ are separately quantized to form the $\hat{\mathbf{U}}_i$. The reproduction values for the different vectors can depend on the choice of reproduction values for other vectors. In particular, since we have called the mapping a VQ, the implication is that the entire (concatenated) reproduction vector $(\hat{\mathbf{U}}_1, \cdots, \hat{\mathbf{U}}_\nu)$ is selected so as to be a minimum distortion reproduction in $\mathcal{C}$ to the entire input vector $(\mathbf{U}_1, \cdots, \mathbf{U}_\nu)$. In general this cannot be achieved by separately selecting the reproductions. It is occasionally desirable, however, to require that the separate $\mathbf{U}_i$ be quantized independently, in which case the system is called a *product code vector quantizer with independent quantization* and the structure becomes that of Figure 12.11. In this case there is no guarantee that the overall codeword in the product codebook will be a minimum distortion selection, but the encoding may be much simpler. Here, the

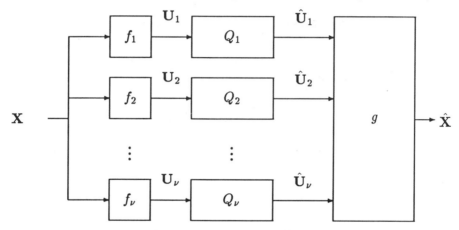

Figure 12.11: Product VQ with Independent Quantizers

analysis operation is performed by evaluating the functions f_i for a given input vector and using a separate nearest neighbor encoding of each component vector $\mathbf{U}_i$. As we shall later emphasize by example, *this structure is not necessarily the optimal encoding configuration for product code VQ!* We shall also see that in some cases properly constructed product encoders can be only slightly more complex than those with independent quantizers and yet still optimal, that is, find the nearest neighbor codeword in the product codebook. This requires using the codebooks in a coupled manner, but it can sometimes be easily accomplished by encoding the features in sequence, with the encoding for one code depending on previous selections. Specifically, a *sequential search product code* codes only one feature vector at a time, but generates this feature from the previously quantized feature vectors as well as from the input vector [54]. Thus, the kth feature vector is generated by

$$\mathbf{U}_1 = f_1(\mathbf{X}),$$

and

$$\mathbf{U}_i = f_i(\mathbf{X}, \hat{\mathbf{U}}_1, \cdots, \hat{\mathbf{U}}_{i-1})$$

for $i = 2, 3, \cdots, \nu$. In general, a sequential search product code can offer enhanced performance over the product code structure with independent quantizers. Keep in mind, however, that even though the encoder of a product code vector quantizer may be optimal for the given product code, in general the product code itself will not be optimal because of its structural constraint.

The next three sections treat a variety of special cases of product code books.

12.8   Partitioned VQ

The simplest and most direct way to reduce the search and storage complexity in coding a high dimensional vector is simply to partition the vector into two or more subvectors. Thus the input vector

$$\mathbf{X} = (X_1, X_2, \cdots, X_k)$$

can be partitioned into

$$\mathbf{X}_a = (X_1, X_2, \cdots, X_m)$$

an m-dimensional vector with $m < k$ and

$$\mathbf{X}_b = (X_{m+1}, X_{m+2}, \cdots, X_k)$$

a vector of dimension $k - m$.

Again using the squared error measure, we find

$$||\mathbf{X} - \hat{\mathbf{X}}||^2 = ||\mathbf{X}_a - \hat{\mathbf{X}}_a||^2 + ||\mathbf{X}_b - \hat{\mathbf{X}}_b||^2$$

from which it is obvious that the independent quantizer encoder structure of Figure 12.11 is optimal. The encoder must simply find the nearest code vector to $\mathbf{X}_a$ in the codebook $\mathcal{C}_a$ and as a separate and independent task, the nearest code vector to $\mathbf{X}_b$ in the codebook $\mathcal{C}_b$. Furthermore, the optimal codebook design procedure is to derive two separate training sets, $\mathcal{T}_a$ and $\mathcal{T}_b$ of m- and $(k - m)$-dimensional vectors, respectively, from the original training set $\mathcal{T}$ of k-dimensional vectors and separately find optimal codebooks of specified sizes for each training set.

Of course the partitioned VQ technique is directly generalized to a larger number than the two partitions considered above.

12.9 Mean-Removed VQ

Usually we deal with vectors that have zero statistical mean in the sense that the expected value of each component is zero. Nevertheless, the average amplitude of the components of a vector can differ widely from one vector to another and often this average value can be regarded as statistically independent of the variation of the vector, that is, of the way the components vary about this average. Furthermore, many vectors such as sampled image intensity rasters have only nonnegative components and hence have nonzero means. The local means over small blocks can vary quite widely over an image. The term *mean* of a vector will be used in this section specifically to refer to the *sample mean*, i.e., the average of the components of a vector. Thus the mean, m, of $\mathbf{X}$ is itself a scalar random variable given by

$$m = \frac{1}{k} \sum_{i=1}^{k} \mathbf{X}_i = \frac{1}{k} \mathbf{1}^t \mathbf{X} \qquad (12.9.1)$$

where $\mathbf{1} = (1, 1, \cdots, 1)^t$, the k-dimensional vector with all components equal to unity.

The mean-removed *residual*, $\mathbf{R}$, of the random variable $\mathbf{X}$ is defined as

$$\mathbf{R} = \mathbf{X} - \frac{1}{k}(\mathbf{1}^t \mathbf{X})\mathbf{1} = \mathbf{X} - m\mathbf{1}. \qquad (12.9.2)$$

Hence, $\mathbf{X}$ can be represented as the sum of a mean vector $m\mathbf{1}$ and the residual vector $\mathbf{R}$ according to:

$$\mathbf{X} = m\mathbf{1} + \mathbf{R}. \qquad (12.9.3)$$

The residual is the "mean-removed" version of the vector $\mathbf{X}$ and has zero mean. Thus we have a natural decomposition of the original vector into separate features, a mean (representing a general background level) and a residual (representing the shape of the vector about its mean). Quantizing these features using separate codebooks is referred to as MRVQ for "mean-removed VQ" or "mean-residual VQ."

Note that we have decomposed the vector $\mathbf{X}$ into two component feature vectors. The first is m which is actually a scalar. The second is $\mathbf{R}$ which has dimension k, the same dimension as $\mathbf{X}$ itself. Note, however, that $\mathbf{R}$ actually has only $k-1$ degrees of freedom since it lies in a $(k-1)$-dimensional subspace of $\mathcal{R}^k$. Why? This is easily seen by observing that $\mathbf{R}$ satisfies the constraint that

$$\mathbf{R}^t \mathbf{1} = \sum_{i=1}^{k} R_i = 0, \qquad (12.9.4)$$

where R_i is the ith component of the residual vector $\mathbf{R}$. This equation means that $\mathbf{R}$ lies on a hyperplane in $\mathcal{R}^k$ that passes through the origin.

This representation of $\mathbf{X}$ offers a simple, and often very valuable, product code. The reconstructed vector after quantization of the mean and the residual is given by

$$\hat{\mathbf{X}} = \hat{m}\mathbf{1} + \hat{\mathbf{R}}, \tag{12.9.5}$$

where $\hat{m}$ is a quantization level from a scalar codebook $\mathcal{C}_m$ of size N_m for the mean code levels, and $\hat{\mathbf{R}}$ is a code vector chosen from a codebook $\mathcal{C}_r$ of size N_r for the residual vectors. The equivalent codebook for $\mathbf{X}$ is the product codebook $\mathcal{C}_m \times \mathcal{C}_r$. Thus the decoder is as depicted in Figure 12.12.

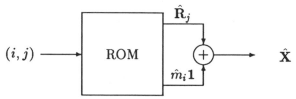

Figure 12.12: MRVQ Decoder

One immediate option for the encoder is to apply the independent quantizer encoder structure of Figure 12.11 to this product code. The corresponding structure for mean-removed VQ is shown in Figure 12.13(a) where the mean and residual are separated from the original input vector, each is quantized independently, and the indexes for the mean and residual are transmitted.

From Equations (12.9.3) and (12.9.5), we find the mean squared error distortion is

$$D = E[||\mathbf{X} - \hat{\mathbf{X}}||^2] = kE[(m - \hat{m})^2] + E[||\mathbf{R} - \hat{\mathbf{R}}||^2] \tag{12.9.6}$$

where we have assumed that the residual code vectors have zero mean, i.e., $\mathbf{1}^t\hat{\mathbf{R}} = 0$. From (12.9.6) we see that we can minimize the average distortion by finding the residual code vector with minimum Euclidean distance from the input residual vector and by separately finding the mean quantizing level nearest to the input mean. Thus in this case the product code encoder with independent quantizers is optimal in the nearest neighbor sense for the product codebook.

The codebook design procedure for mean-removed VQ using the encoding structure of Figure 12.13(a) and the squared error distortion is straightforward. A training set of means $\mathcal{T}_m$ is formed and used to design the mean codebook. Similarly, a training set of residuals $\mathcal{T}_r$ is formed and used to design the residual codebook. There is no interaction between the two designs. Note also that the centroid of a set of zero mean vectors will also

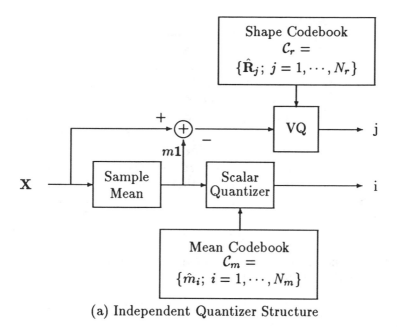

(a) Independent Quantizer Structure

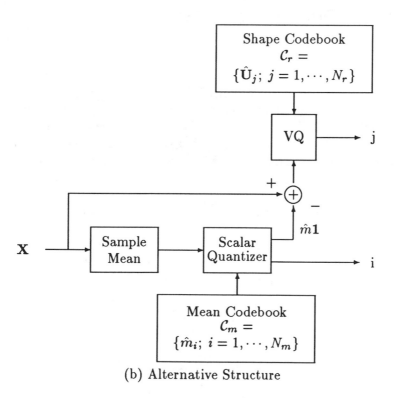

(b) Alternative Structure

Figure 12.13: Mean-Removed VQ

have zero mean. Hence, the residual codebook designed using the GLA will result in code vectors that have zero mean.

An alternative encoder structure for mean-removed VQ [24] is shown in Figure 12.13(b). The important distinction here is that the mean is first quantized and the *quantized mean* is then subtracted from **X**. Thus the residual is now computed with respect to the decoder's reproduction of the mean rather than with respect to the true mean. This new structure still has a product codebook, but the selection of the shape codeword is now influenced by that of the mean codeword and hence the system does not have the independent quantizer product code form but it is a sequential search product code. To distinguish this from the previous case, let **U** denote the residual obtained when the quantized mean is subtracted from **X**, that is,

$$\mathbf{U} = \mathbf{X} - \hat{m}\mathbf{1}. \tag{12.9.7}$$

Let $\hat{\mathbf{U}}$ denote the code vector used to quantize **U**, and let $\tilde{\mathbf{X}}$ denote the reproduced version of the input **X**. Then,

$$\tilde{\mathbf{X}} = \hat{m}\mathbf{1} + \hat{\mathbf{U}}. \tag{12.9.8}$$

Now, we obtain a new expression for the squared error distortion in quantizing **X**. From Equations (12.9.7) and (12.9.8), we get

$$D = E(\|\mathbf{X} - \tilde{\mathbf{X}}\|^2) = E(\|\mathbf{U} - \hat{\mathbf{U}}\|^2). \tag{12.9.9}$$

Note that, in this case, we do *not* make the assumption that the residual code vectors have zero mean. Superficially comparing (12.9.9) with (12.9.6) suggests that the alternative structure of Figure 12.13(b) will have better performance than the independent quantizer structure of Figure 12.13(a). This inference is based on the assumption that the same average distortion in quantizing the residuals **R** and **U** is achieved in the two cases. Is this true? Not necessarily.

If the residual codebook used in the alternative structure is the same as in the basic structure, we will be using a zero-mean code vector to approximate the residual **U** which does not have zero mean. It is easy to show that in this case the advantage of the alternate structure will be lost and precisely the same performance will be achieved as with the basic structure. To see this, observe that

$$\mathbf{U} = \mathbf{R} + (m - \hat{m})\mathbf{1} \tag{12.9.10}$$

so that

$$D = E(\|\mathbf{U} - \hat{\mathbf{U}}\|^2) = kE((m - \hat{m})^2) + E(\|\mathbf{R} - \hat{\mathbf{U}}\|^2). \tag{12.9.11}$$

Thus, the best code vector $\hat{\mathbf{U}}$ for quantizing $\mathbf{U}$ is equal to the best code vector used for quantizing $\mathbf{R}$ if the same residual codebook is used in both cases. This implies that $\hat{\mathbf{U}} = \hat{\mathbf{R}}$ and comparing (12.9.11) with (12.9.6) shows that there is no difference in performance between the two techniques.

On the other hand, the alternate structure is superior to the basic structure if an improved codebook design is used for it. Since the codebook is used to quantize the residuals $\mathbf{U}$ which do not have zero mean, the optimal codebook will not have zero mean. The codebook that was optimal for quantizing $\mathbf{R}$ will in general be suboptimal for quantizing $\mathbf{U}$ whose statistics are different than those of $\mathbf{R}$. Nevertheless, regardless of the codebook design, the alternate structure is not the optimal encoder for the product code. Superior performance for a given pair of codebooks is possible if the encoder performs a joint search for the best mean and shape.

The codebook design procedure for the alternate structure follows directly from the encoding structure. From the original training set T of input vectors extract the vector means to form the training set of means T_m and use the GLA to find the mean codebook as before. Generate a new residual training set T_u by removing the mean from each training vector, quantizing it using the mean codebook, and subtracting the quantized mean from (each component of) the training vector. The resulting set of residual vectors is the desired training set T_u. Finally, apply the GLA to this training set to obtain the desired codebook C_u. Note that the code vectors will not be constrained to have zero mean, although they will have very small mean values, particularly if the mean quantizer has high resolution. Thus the code vectors do not actually lie in a $(k-1)$-dimensional subspace of $\mathcal{R}^k$ but instead are found in a narrow slice of $\mathcal{R}^k$ characterized by a region $\mathcal{L}$ of the form:

$$\mathcal{L} = \{\mathbf{x} \in \mathcal{R}^k : |\sum_{i=1}^{k} x_i| < k\delta\}$$

where δ is the maximum quantizing error of the mean quantizer. We still have, however, a useful product code since the residual, although of the same dimensionality as the original input vector, is confined to a very restricted subset of the space $\mathcal{R}^k$.

As a design example, return to the VQ of the image example of the previous chapter. Figure 12.14 compares a closeup of Lenna's eye from the USC data base for the original image, the "vanilla" VQ of Chapter 11 at .5 bpp, and the MRVQ at the same bit rate. (The example is from [24] and was reprinted in [153].) The image still leaves much to be desired, but a simple modification to memoryless VQ yields a significant improvement in subjective quality, and the blockiness and unpleasant edge effects are much

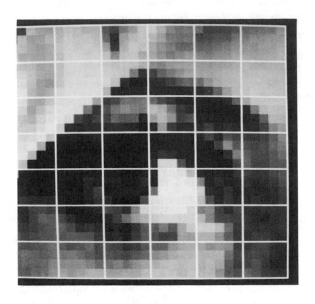

(a) Original Eye

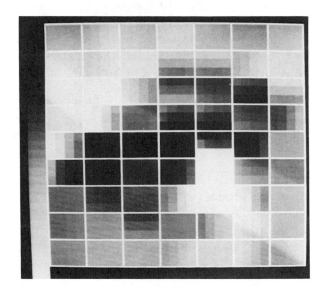

(b) Ordinary VQ Eye

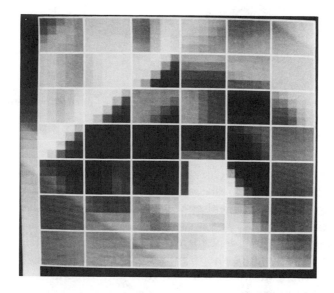

(c) Mean-Removed VQ Eye with DPCM Mean Coding

Figure 12.14: Mean-Removed VQ at .5 bpp: Eye

diminished.

12.10 Shape-Gain VQ

Another product code technique that decomposes the problem into that of coding a scalar and a vector is based on extracting the root mean-square value of the vector components. This quantity is called the *gain* and serves as a normalizing scale factor. The normalized input vector is called the *shape*. The idea of shape-gain VQ is that the same pattern of variation in a vector may recur with a wide variety of gain values. If this is true, it suggests that the probability distribution of the shape is approximately independent of the gain. We would then expect very little compromise in optimality with a product code structure. From another perspective, we can say that the code is permitted to handle the dynamic range of the vector separately from the shape of the vector. Shape-gain VQ was introduced in [42] and optimized in [278]. It is particularly useful in speech and audio coding where the waveform has a wide dynamic range of short term power levels.

The gain g of a vector $\mathbf{X}$ is a random variable given by

$$g = ||\mathbf{X}|| = \sqrt{\sum_{i=1}^{k} X_i^2}, \qquad (12.10.1)$$

the Euclidean norm of the vector. The *shape*, $\mathbf{S}$ of a vector $\mathbf{X}$ with nonzero gain is given by

$$\mathbf{S} = \frac{\mathbf{X}}{g} \qquad (12.10.2)$$

so that $||\mathbf{S}|| = 1$. All shape vectors in the shape codebook $\mathcal{C}_s$ have unit gain. In this product code decomposition, the shape vector lies on the surface of a hypersphere in k-dimensional space and is therefore easier to quantize than the original vector $\mathbf{X}$.

It turns out that the independent quantizer encoding structure of Figure 12.11 is *not* optimal for shape-gain encoding. That structure gives an encoding rule where we would find the best gain code level for quantizing the input gain, and, in a separate and independent search, choose the best shape code vector to code the input shape. This method might indeed give fairly reasonable performance, especially for high resolution, but as we shall see, it is not optimal.

In order to determine the optimal encoding structure, we begin by examining the performance objective. Again we assume the squared error distortion measure. Thus,

$$d(\mathbf{X}, \hat{g}\hat{\mathbf{S}}) = ||\mathbf{X} - \hat{g}\hat{\mathbf{S}}||^2 \qquad (12.10.3)$$

where $\hat{g}$ is the quantized version of g and $\hat{\mathbf{S}}$ is the quantized version of the shape $\mathbf{S}$. The gain and shape codebooks are denoted by $\mathcal{C}_g$ and $\mathcal{C}_s$ and have sizes N_g and N_s, respectively. Expanding this expression gives

$$d(\mathbf{X}, \hat{g}\hat{\mathbf{S}}) = ||\mathbf{X}||^2 + \hat{g}^2 - 2\hat{g}(\mathbf{X}^t\hat{\mathbf{S}}). \qquad (12.10.4)$$

This expression can be minimized over $\hat{g}$ and $\hat{\mathbf{S}}$ in two steps. First select the shape vector $\hat{\mathbf{S}}$ to minimize the third term, that is, pick the $\hat{\mathbf{S}}$ that maximizes $\mathbf{X}^t\hat{\mathbf{S}}$. Note that $\hat{g}$ is always positive and its value does not influence the choice of $\hat{\mathbf{S}}$. Note also the important fact that the input vector need not be itself gain normalized in order to choose the best shape vector. Such normalization would involve division by the input gain and would significantly increase the encoder complexity. The maximum correlation selection obviates any such computation. Once the shape codeword is chosen, select the $\hat{g}$ to minimize the resulting function of $\hat{g}$, i.e.,

$$||\mathbf{X}||^2 + \hat{g}^2 - 2\hat{g}(\mathbf{X}^t\hat{\mathbf{S}}) = ||\mathbf{X}||^2 + [\hat{g} - (\mathbf{X}^t\hat{\mathbf{S}})]^2 - (\mathbf{X}^t\hat{\mathbf{S}})^2, \qquad (12.10.5)$$

which is accomplished by chosing $\hat{g}$ to minimize

$$[\hat{g} - \mathbf{X}^t\hat{\mathbf{S}}]^2$$

for the previously chosen (unit norm) $\hat{\mathbf{S}}$. Note that the second step requires nothing more than a standard scalar quantization operation where the nearest gain codeword to the quantity $\mathbf{X}^t\hat{\mathbf{S}}$ in the gain codebook is selected.

Thus the optimal encoding rule is a two step procedure where the first step involves only one feature (the shape) and one codebook. The second step depends on the first step in its computation of the nearest neighbor for the second feature (the gain) in its codebook. In effect, the first step affects the distortion measure used to compute the second step. Observe that this is the reverse of the mean-removed VQ (alternate structure) where the scalar parameter was quantized first and the residual (corresponding to the shape) quantized second.

This procedure can also be seen to yield the optimal (nearest neighbor) product codeword by observing that

$$\min_{\hat{S},\hat{g}} \left((\hat{g})^2 - 2\hat{g}(\mathbf{X}^t\hat{S}) \right) = \min_{\hat{g}} \left((\hat{g})^2 - 2\hat{g}\max_{\hat{S}}(\mathbf{X}^t\hat{S}) \right).$$

The shape-gain quantizer including the encoder structure obtained from this derivation is depicted in Figure 12.15. The decoder is shown in Figure 12.16.

Codebook Design for Shape-Gain VQ

We now consider the task of codebook design for shape-gain VQ based on the encoding procedure of Figure 12.15. We begin with a training set $\mathcal{T}$ of input vectors, $\mathbf{x}_k$, that represent the random vector $\mathbf{X}$. The objective is to find the shape and gain codebooks that minimize the average distortion incurred in encoding the training vectors with the structure of Figure 12.15. There are several variations on the Lloyd algorithm by which this can be accomplished [278], but we shall consider the most basic one with demonstrable optimality properties. The variations use short cuts and approximations, but they yield very similar results.

A shape-gain VQ is completely described by three objects:

- The gain codebook $\mathcal{C}_g = \{\hat{g}_i; \ i = 1, 2, \cdots, N_g\}$,

- the shape codebook $\mathcal{C}_s = \{\hat{S}_j; \ j = 1, 2, \cdots, N_s\}$, and

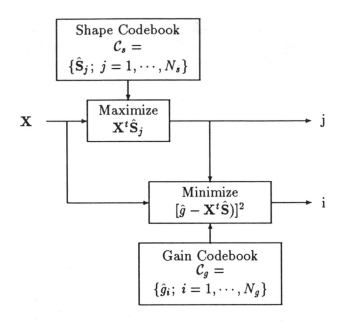

Figure 12.15: Shape-Gain VQ Encoder

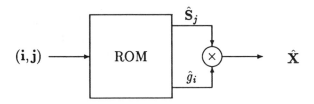

Figure 12.16: Shape-Gain Decoder

- a partition $\mathcal{R} = \{R_{i,j};\; i = 1, 2, \cdots, N_g;\; j = 1, 2, \cdots, N_s\}$ of $N_g \times N_s$ cells describing the encoder, that is, if $\mathbf{x} \in R_{i,j}$, then $\mathbf{x}$ is mapped into (i, j) and the resulting reproduction is formed from the shape-gain vector $(\hat{g}_i, \hat{S}_j)$. We express this as $g(\mathbf{x}) = \hat{g}_i$ and $S(\mathbf{x}) = \hat{S}_j$.

As in other design algorithms we must initialize these three components of the code and then iteratively improve them by optimizing each in turn for the other two. We now proceed to do this.

Define the gain partition

$$\mathcal{G} = \{G_i;\; i = 1, \cdots, N_g\}$$

by

$$G_i = \bigcup_{j=1}^{N_s} R_{i,j}.$$

G_i is the region of the input vector space that maps into the gain codeword $\hat{g}_i$. Similarly define the shape partition

$$\mathcal{A} = \{A_j;\; j = 1, \cdots, N_s\}$$

by

$$A_j = \bigcup_{i=1}^{N_g} R_{i,j},$$

the region mapping into the shape codeword S_j. Note that if we were using the independent quantizer structure of Figure 12.11, then indeed we would have $\mathcal{R} = \mathcal{G} \times \mathcal{A}$. This is not, however, true in general.

Denote the average distortion by

$$D(\mathcal{C}_g, \mathcal{C}_s, \mathcal{R}) = Ed(\mathbf{X}, g(\mathbf{X})S(\mathbf{X})).$$

Optimum partition for fixed gain and shape codebooks: For given $\mathcal{C}_s$ and $\mathcal{C}_g$, let $\mathcal{R}^*(\mathcal{C}_s, \mathcal{C}_g)$ denote the minimum distortion partition (with the usual arbitrary tie-breaking rule). Analogous to the Lloyd optimality conditions, for any encoder partition $\mathcal{R}$

$$D(\mathcal{C}_g, \mathcal{C}_s, \mathcal{R}) \geq D(\mathcal{C}_g, \mathcal{C}_s, \mathcal{R}^*(\mathcal{C}_s, \mathcal{C}_g)). \tag{12.10.6}$$

In other words, the minimum distortion selection of gain and shape together minimizes the average distortion for a fixed pair of codebooks.

Optimum gain codebook for fixed shape codebook and partition: Another necessary condition for optimality can be found by conditioning the input on the gain cells alone. We have that for any partition $\mathcal{R}$

$$
\begin{aligned}
D(\mathcal{C}_g, \mathcal{C}_s, \mathcal{R}) &= \sum_{i=1}^{N_g} E[d(\mathbf{X}, \hat{g}_i S(\mathbf{X}) | \mathbf{X} \in G_i)] \Pr(\mathbf{X} \in G_i) \\
&\geq \sum_{i=1}^{N_g} \inf_{g \geq 0} E[d(\mathbf{X}, g S(\mathbf{X}) | \mathbf{X} \in G_i] \Pr(\mathbf{X} \in G_i).
\end{aligned}
$$

The value of g which yields the infimum for a particular i and $\mathcal{R}$ (which determines the mapping $S(\mathbf{x})$) can be thought of as a *conditional gain centroid*, where the conditioning is on the partition $\mathcal{R}$ and hence on the shape encoding. Assume for the moment that these conditional centroids can be evaluated and denote the collection by

$$
\mathcal{G}^*(\mathcal{C}_s, \mathcal{R}) = \{\hat{g}_i;\ i = 1, 2, \cdots, N_g\},
$$

where the notation emphasizes that the conditional gain centroids depend both on the shape codebook and on the partition $\mathcal{R}$. Thus we have the condition

$$
D(\mathcal{C}_g, \mathcal{C}_s, \mathcal{R}) \geq D(\mathcal{G}^*(\mathcal{C}_s, \mathcal{R}), \mathcal{C}_s, \mathcal{R}). \tag{12.10.7}
$$

Optimum shape codebook for fixed gain codebook and partition: Similarly we can condition on the shape partition $\mathcal{A}$ and write

$$
\begin{aligned}
D(\mathcal{C}_g, \mathcal{C}_s, \mathcal{R}) &= \sum_{j=1}^{N_s} E[d(\mathbf{X}, g(\mathbf{X}) \hat{S}_j) | \mathbf{X} \in A_j] \Pr(\mathbf{X} \in A_j) \\
&\geq \sum_{j=1}^{N_s} \inf_{\mathbf{s}} E[d(\mathbf{X}, g(\mathbf{X}) \mathbf{s}) | \mathbf{X} \in A_j] \Pr(\mathbf{X} \in A_j),
\end{aligned}
$$

which defines a conditional shape centroid given the overall partition. If the collection of these conditional shape centroids is denoted by

$$
\mathcal{A}^*(\mathcal{C}_g, \mathcal{R}) = \{\hat{\mathbf{s}}_j;\ j = 1, 2, \cdots, N_s\},
$$

then we have the condition

$$
D(\mathcal{C}_g, \mathcal{C}_s, \mathcal{R}) \geq D(\mathcal{C}_g, \mathcal{A}^*(\mathcal{C}_g, \mathcal{R}), \mathcal{R}). \tag{12.10.8}
$$

Again note that the shape centroids depend on the gain codebook and the full partition.

All that remains to completely describe the algorithm is to compute the conditional centroids. These are quoted with the proofs left to the exercises as they are similar to the Lloyd proofs with more algebraic clutter. It can be shown that the shape centroid is given by

$$\hat{\mathbf{s}}_j = \frac{E[g(\mathbf{X})\mathbf{X}|\mathbf{X} \in A_j]}{\|E[g(\mathbf{X})\mathbf{X}|\mathbf{X} \in A_j]\|}, \qquad (12.10.9)$$

which for the case of a training set becomes

$$\hat{\mathbf{s}}_j = \frac{\sum_{i=1}^{N_g} \sum_{k:\mathbf{x}_k \in R_{i,j}} g(\mathbf{x}_k)\mathbf{x}_k}{\|\sum_{i=1}^{N_g} \sum_{k:\mathbf{x}_k \in R_{i,j}} g(\mathbf{x}_k)\mathbf{x}_k\|}.$$

Note that, as with the usual Euclidean centroid, we must compute a vector sample average over a partition cell. Here, however, we must normalize the average to have unit Euclidean distance.

The gain centroid is

$$\hat{g}_i = E(\mathbf{X}^t S(\mathbf{X})|\mathbf{X} \in G_i), \qquad (12.10.10)$$

which for a training set becomes

$$\hat{g}_i = \frac{1}{\|G_i\|} \sum_{j=1}^{N_s} \sum_{k:\mathbf{x}_k \in R_{i,j}} \hat{\mathbf{s}}_j^t \mathbf{x}_k,$$

which is again the sample average. Obviously the gain centroids should be set to 0 if the above averages produce a negative number.

Equations (12.10.7)–(12.10.8) and the centroids of (12.10.9)–(12.10.10) provide a set of necessary conditions analogous to the Lloyd conditions which can be used as the basis of an iterative code improvement algorithm. (In fact, several iterative algorithms with similar performance can be based on these conditions.) Perhaps the most natural is that of Table 12.4, which is known as the jointly optimized product code VQ design algorithm.

As in the Lloyd algorithm, each step can only improve or leave unchanged the average distortion. Unlike the Lloyd algorithm, there is no nice intuition of iteratively optimizing encoder for decoder and vice versa. Choosing the optimum partition indeed optimizes the encoder, but the two codebooks (corresponding to the decoder) must be alternately optimized for each other, with an optimal encoder in between.

A shape-gain VQ can be designed in an analogous manner for LPC-VQ with the residual energy playing the role of the gain. The details may be found in [278,312].

As an example, shape-gain quantizers were designed for the same speech data set considered previously for 1 and 2 bits per sample. For each resolution and dimension various sizes of shape and gain codebooks were tried.

Table 12.4: **Shape-Gain VQ Design**

Step 0. Initialization: Given $N_g, N_s, \epsilon > 0$, an initial product codebook $\mathcal{C}_g(0) \times \mathcal{C}_s(0)$, and a training set T, set $m = 0$ and $D_{-1} = \infty$. The initial gain and shape codebooks can be formed by separately running the Lloyd algorithm (or the splitting algorithm) on the gains and shapes of training vectors. Alternatively, a uniform quantizer can be used for the gain and a lattice quantizer for the shape.

Step 1. Find the optimum partition $\mathcal{R}^*(\mathcal{C}_g(m), \mathcal{C}_s(m))$ of T.

Step 2. Compute the average distortion

$$D_m = D(\mathcal{C}_g(m), \mathcal{C}_s(m), \mathcal{R}^*(\mathcal{C}_g(m), \mathcal{C}_s(m))).$$

If

$$(D_{m-1} - D_m)/D_m < \epsilon,$$

halt with $(\mathcal{C}_g(m), \mathcal{C}_s(m), \mathcal{R}^*(\mathcal{C}_g(m), \mathcal{C}_s(m)))$ describing the final quantizer. Else continue.

Step 3. Compute the optimum shape codebook using (12.10.9)

$$\mathcal{C}_s(m + 1) = \{\hat{s}_j\} = A^*(\mathcal{C}_g(m), \mathcal{R}^*(\mathcal{C}_g(m), \mathcal{C}_s(m))).$$

Step 4. Compute the optimum partition $\mathcal{R}^*(\mathcal{C}_g(m), \mathcal{C}_s(m + 1))$.

Step 5. Compute the optimum gain codebook using (12.10.10)

$$\mathcal{C}_g(m + 1) = \{g_i\} = G^*(\mathcal{C}_s(m + 1), \mathcal{R}^*(\mathcal{C}_g(m), \mathcal{C}_s(m + 1))).$$

Step 6. Set $m = m + 1$ and go to Step 1.

The initial codebooks were obtained using a variation of the splitting technique. Starting with 0 bit resolution codebooks, the higher resolution quantizers at each stage were obtained by splitting one or the other of the gain or shape codebooks. The detailed algorithm for designing the code from scratch is described in [278]. The results for the (experimentally) best choices of the gain and shape codebook sizes (giving the desired total bit rate) are summarized in Figures 12.17 and 12.18 for Gauss Markov and sampled speech, respectively.

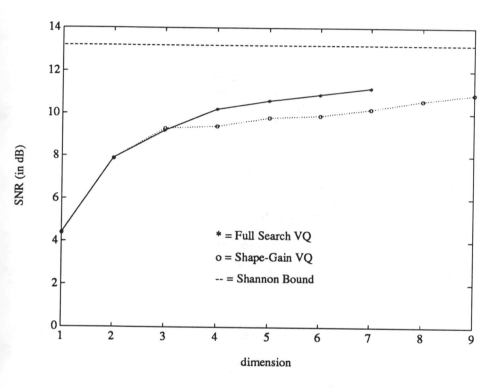

Figure 12.17: SNR for Shape-Gain VQ of Gauss Markov Source

As expected, for a common resolution and dimension, shape-gain VQ is inferior to full search VQ in terms of performance, but it has the reduced complexity, both computational and storage, of a product code. Because of this reduced complexity, shape-gain VQ is capable of being used at higher dimensions than full search VQ (since the shape codebook has fewer codewords to be stored and searched) and this can improve performance for a given bit-rate. This advantage is not realized for the Gauss Markov source, but it provides about a half dB more on speech.

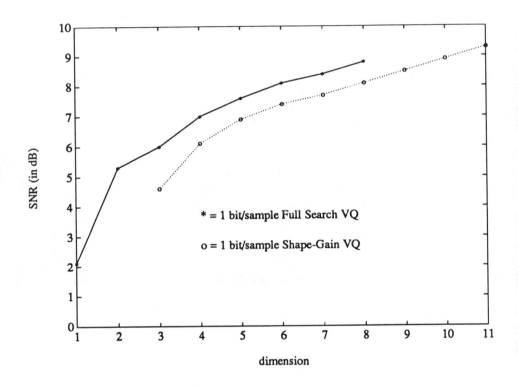

Figure 12.18: SNR for Shape-Gain VQ of Sampled Speech

Mean/Shape-Gain VQ

Often we can combine more than one product code technique to obtain a hybrid coding method that is still in the category of a product code. One such method is to combine mean-removal with shape-gain VQ. Given an input vector, we can first remove its mean and then normalize it by its gain to obtain a shape vector that is effectively normalized to have zero mean and unit gain. Codebooks designed for these shape vectors tend to be more robust since they are less dependent on an accurate statistical model of the source. A version of this technique has been used in image coding [240].

12.11 Multistage VQ

If a random vector $\mathbf{X}$ is such that it does not have a wide variation of gain or of mean values, then shape-gain or mean-removed VQ methods are not likely to be very helpful. If the dimension is quite large, partitioned VQ would certainly solve the complexity problem but might severely degrade performance when there is substantial statistical interdependence between different subvectors. If we are as much concerned with storage as with search complexity, then general tree-structured VQ and classified VQ are not helpful. Furthermore, transform VQ may be of very limited help if the degree of compaction achievable still results in too high a vector dimension. So what else is there?

One alternative technique that has proved valuable in a number of speech and image coding applications is *multistage* or *cascaded* VQ [198]. The technique is also sometimes referred to as *residual VQ*[27] [130]. The basic idea of multistage VQ (MSVQ) is to divide the encoding task into successive stages, where the first stage performs a relatively crude quantization of the input vector using a small codebook. Then, a second stage quantizer operates on the error vector between the original and quantized first stage output. The quantized error vector then provides a second approximation to the original input vector thereby leading to a refined or more accurate representation of the input. A third stage quantizer may then be used to quantize the second stage error vector to provide a further refinement and so on.

We consider first the special case of two-stage VQ as illustrated in Figure 12.19. The input vector $\mathbf{X}$ is quantized by the initial or first stage vector quantizer denoted by Q_1. The quantized approximation $\hat{\mathbf{X}}_1$ is then subtracted from $\mathbf{X}$ producing the error vector $\mathbf{E}_2$. This error vector is then applied to a second vector quantizer Q_2 yielding the quantized output $\hat{\mathbf{E}}_2$. The overall approximation $\hat{\mathbf{X}}$ to the input $\mathbf{X}$ is formed by summing the first and second approximations, $\hat{\mathbf{X}}_1$ and $\hat{\mathbf{E}}_2$. The encoder for this VQ scheme simply transmits a pair of indexes specifying the selected code vectors for each stage and the task of the decoder is to perform two table lookups to generate and then sum the two code vectors.

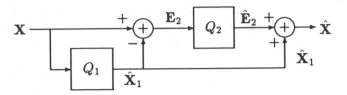

Figure 12.19: Two-Stage VQ

By inspection of the figure it may be seen that the input-output error is equal to the quantization error introduced by the second stage, i.e., $\mathbf{X} - \hat{\mathbf{X}} = \mathbf{E}_2 - \hat{\mathbf{E}}_2$. From this equation it can be readily seen that the signal to quantizing noise power ratio in dB (SNR) for the two stage quantizer is given by $\text{SNR} = \text{SNR}_1 + \text{SNR}_2$, where SNR_i is the signal-to-noise ratio in dB for the ith quantizer. In comparison with a single quantizer with the same total number of bits, a two-stage quantizer has the advantage that the codebook size of each stage is considerably reduced so that both the storage requirement and the total search complexity are substantially lowered. The price paid for this advantage is an inevitable reduction in the overall SNR achieved with two stages.

The two-stage VQ scheme can be viewed as a two-level TSVQ structure of the general form shown in Figure 12.1. To see this, let $\mathcal{K}_1 = \{\hat{\mathbf{X}}_{1,j}; \ j = 1, \cdots, N_1\}$ denote the codebook of quantizer Q_1 and $\mathcal{K}_2 = \{\hat{\mathbf{E}}_{2,i}; \ i = 1, \cdots, N_2\}$ the codebook of quantizer Q_2. Then, the first level TSVQ codebook coincides with the first stage codebook, i.e., $\mathcal{C}^* = \mathcal{K}_1$. The second level TSVQ codebooks correspond to shifted versions of the second stage codebook $\mathcal{K}_2$, i.e., $\mathcal{C}_j = \{\hat{\mathbf{X}}_{1,j} + \hat{\mathbf{E}}_{2,i}; \ i = 1, \cdots, N_2\}$. Thus, the 2-stage VQ structure is exactly equivalent to a specific two-level TSVQ. In this case, the MSVQ structure is more storage-efficient than the TSVQ because of the special structure of the second level codebooks.

The general multistage VQ method can be generated by induction from the two-stage scheme. By replacing the box labeled Q_2 in Fig. 12.19, with a two-stage VQ structure, we obtain 3-stage VQ. By replacing the last stage of an m-stage structure, we increase the number of stages to $m + 1$. The general configuration for an MSVQ encoder is easily inferred from Figure 12.20 which illustrates the 3 stage case. The corresponding decoder is shown in Figure 12.21. The vector quantizer at each stage operates on the quantization error vector from the previous stage. Each stage generates an index that is sent to the decoder.

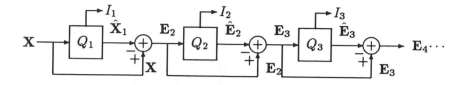

Figure 12.20: Multistage VQ Encoder

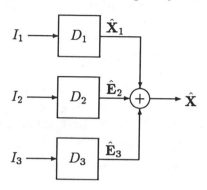

Figure 12.21: Multistage VQ Decoder

In the multistage approach, the input vector is represented as the *sum* of two or more vectors of the same dimension as the original, where each successive term of the sum can be considered as a refinement or *successive approximation* improvement of the previous terms. The reproduction vector is computed in the following form:

$$\hat{\mathbf{X}} = \hat{\mathbf{X}}_1 + \hat{\mathbf{E}}_2 + \cdots + \hat{\mathbf{E}}_m \qquad (12.11.1)$$

for the case of m stages. The first stage quantizer Q_1 approximates $\mathbf{X}$ with $\hat{\mathbf{X}}_1$ using a codebook $\mathcal{K}_1$ of size N_1. The second stage quantizer Q_2 is used to approximate the quantization error vector $\mathbf{E}_2 = \mathbf{X} - \hat{\mathbf{X}}_1$ using a codebook $\mathcal{K}_2$ so that

$$\hat{\mathbf{X}}_2 = \hat{\mathbf{X}}_1 + \hat{\mathbf{E}}_2$$

is an improved approximation to $\mathbf{X}$. Similarly, the third stage quantizer approximates the quantization error vector $\mathbf{E}_3 = \mathbf{X} - \hat{\mathbf{X}}_1 - \hat{\mathbf{E}}_2$ associated with the second stage using the codebook $\mathcal{K}_3$ to obtain

$$\hat{\mathbf{X}}_3 = \hat{\mathbf{X}}_1 + \hat{\mathbf{E}}_2 + \hat{\mathbf{E}}_3 = \hat{\mathbf{X}}_2 + \hat{\mathbf{E}}_3,$$

which provides a further refinement to $\mathbf{X}$; and so on, until the complete approximation is produced by adding the m vectors. There is a separate codebook $\mathcal{K}_i$ for each of the m vectors that form the reproduction vector. The overall codeword is the concatenation of codewords or indices chosen from each of these codebooks. The encoder transmits indexes $I_1, I_2, \cdots, I_m$ to the decoder, which then performs a table-lookup in the respective code-books and forms the sum given in (12.11.1). Thus this is a product code where the composition function g of the decoder is simply a summation of the reproductions from the different VQ decoders.

Note that the complexity is reduced from $N = \prod_{i=1}^{m} N_i$ to $\sum_{i=1}^{m} N_i$ and the equivalent product codebook is generated from the Cartesian product $\mathcal{K}_1 \times \mathcal{K}_2 \times \cdots \times \mathcal{K}_m$. Thus both the complexity and storage requirements can be greatly reduced using multistage VQ.

As usual, there is a performance penalty with this product code tech-nique. It is easy to show that the overall quantization error between input and output is equal to the quantization error introduced in the last stage; from this it can be shown that the signal to quantizing noise power ratio in dB (SNR) for the multistage quantizer is given by

$$\text{SNR} = \sum_{i=1}^{m} \text{SNR}_i, \qquad (12.11.2)$$

consistent with the two-stage result discussed earlier. If we make the ideal-ized assumption that the SNR of a quantizer grows linearly with the number of bits, this result would indicate that we can achieve the same performance by sharing a fixed quota of bits over several quantizers as by using all the bits in one quantizer while at the same time achieving a great complexity reduction. Alas, this is not so. The coding gain of a quantizer depends on the statistics of the input signal and the successive quantizers tend to have rapidly diminishing coding gains (i.e., the SNR per bit decreases) due to the vector components of successive stages tending to be less correlated. Gen-erally the quantization error vector has more randomness than the input to the quantizer since its components tend to be less statistically dependent than those of the input vector. Thus, in practice, multistage coders often have only two and occasionally three stages. As far as we know, there has been no report of a coding system using four or more stages.

Codebook design for multistage VQ is also performed in stages. First, the original training set $\mathcal{T}$ is used to generate the first stage codebook of the desired size in the conventional manner. Next, a new training set $\mathcal{T}_1$ is generated by applying the training set to the first stage VQ and generating the set of error vectors that represent the statistics of the vectors applied to the second stage. This training set is of the same size as the original and the vectors are of the same dimension. The process is repeated for successive stages. Note that the codebook design complexity is reduced here compared to the design for a single stage VQ of dimension N.

This codebook design procedure is not optimal in the sense that it does not find the best set of codebooks for sequentially encoding a vector stage-by-stage with the multistage structure. It is, however, *greedy* in the sense that it finds the codebook for the first stage that would be optimal if there were only one stage. Then it finds the best codebook for the second stage given the first-stage codebook and assuming there are only two stages. Similarly, each successive stage is optimal given all previous stage codebooks and assuming that it is the last stage. This design algorithm is consistent with the philosophy of progressive reconstruction as discussed in the context of TSVQ.

Recently, an improved design algorithm for MSVQ was proposed which yields slightly better results than the greedy algorithm usually used [57] for the usual sequential-search MSVQ encoder. Alternatively, if the sequential-search encoder is abandoned and exhaustive search is performed to find the best set of indexes $\{I_1, I_2, ..., I_m\}$ for given m stage MSVQ decoder, even better performance can be achieved for a given set of codebooks. The design of optimal codebooks for this purpose was considered by [27]. In this case the search complexity is essentially the same as with a single unstructured VQ codebook; only the storage requirement is reduced.

Figures 12.22–12.23 show the performance of multistage VQ for the Gauss Markov and sampled speech examples in comparison with TSVQ, of which it is a special case. As expected, the quality degrades from TSVQ, but the complexity and storage reduction may be worth the quality loss in some examples.

Multistage/Tree VQ

We have noted earlier that MSVQ can be regarded as a special case of TSVQ. In tree-structured VQ, the number of codebooks grows exponentially with the number of stages (or layers), whereas in multistage VQ the growth is linear with the number of stages. In fact, it is possible to have hybrid methods that give different compromises between the number of stages and the number of codebooks. The basic multistage technique described above is most efficient when used for only two or three stages. This number of stages may not be enough to reduce the encoding complexity of each stage to a desired level because of the large size codebooks that are needed at each stage.

One useful hybrid technique is to replace the full search VQ in each stage with a tree-structured (usually binary) VQ. This in effect divides each stage into many stages but with much smaller codebook sizes per stage. Each successive layer of the tree is another stage of coding and has a certain

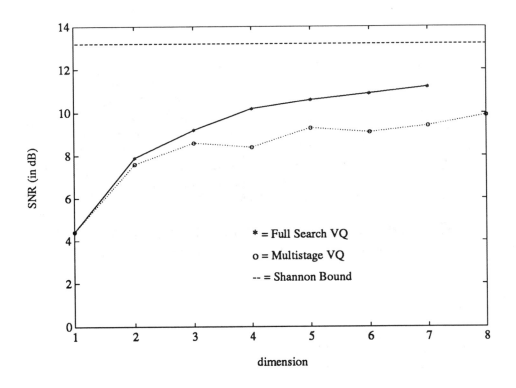

Figure 12.22: SNR for Multistage VQ for Gauss Markov Source

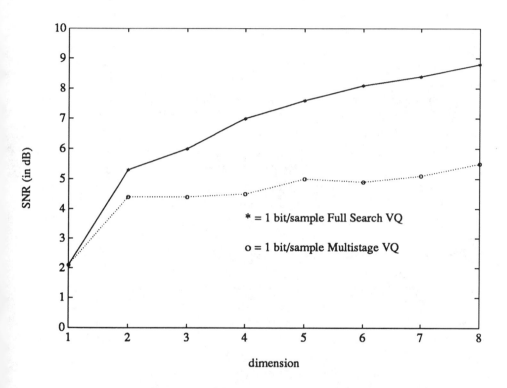

Figure 12.23: SNR for Multistage VQ for Sampled Speech

number of codebooks associated with it. At each stage of this structure one of several codebooks is selected for encoding the current vector. The number of codebooks at a given stage is called the *fanout* of that stage. As the encoding proceeds through the first tree the fanout grows until the end of that tree is reached. The transition to the next tree corresponds to an instant reduction in fanout of the structure and then the fanout increases again as the encoding proceeds through the second tree structure. A substantial complexity reduction is achieved with this combination and a moderate storage reduction is also achieved. For example, suppose the equivalent product codebook has size $N = 2^{18}$ and each of two stages is implemented as a binary tree structure of depth 9 with 2^9 terminal nodes. Then each input vector is quantized with 18 bits. The search complexity for each stage is 18 inner product computations (if implemented as a hyperplane decision test) and the storage for each tree is 1023 vectors (one $(k + 1)$-dimensional vector per hyperplane). This hybrid of multistage and tree-structured VQ is called *multistage tree VQ* (MSTVQ). One application where MSTVQ has been effective is in wideband audio coding [54].

In some applications, TSVQ, MSVQ, or MSTVQ are attractive choices for variable rate or embedded encoding. However, in such schemes the maximum rate for coding (when all stages of the structure are needed) can be very large and the storage requirement even with MSTVQ can become excessive or prohibitive. A method to extend the MSTVQ scheme to high rates while limiting storage has been proposed in [53], called constrained-storage MSTVQ (CS-MSTVQ). In this scheme, the standard MSVQ structure of Figure 12.20 is modified by allowing each stage to have a predetermined number of codebooks intermediate between the usual MSVQ case of one codebook per stage and the usual TSVQ case of m^{r-1} codebooks per stage for the rth stage of an m-ary tree structure. In this method, the first stage has one codebook, say of size m. Once an input vector is quantized with the first stage, both a quantized vector $\hat{X}$ and an index I are generated and both the index and the corresponding error vector are applied to the second stage. The index determines which of M codebooks (with $M \leq m$) is selected to encode the error vector from the first stage. Similarly, an error vector and an index from the second stage are passed to the third stage and the process is repeated. The index when applied to a given stage is mapped by a pointer function $\mu(I)$ to identify a particular codebook index from a set of M choices. Note that the fanout is controlled by the choice of M and each stage can have a different degree of fanout if desired. Figure 12.24 illustrates the general structure of this scheme which falls into the class of a sequential-search product code as described in [53]. In MSTVQ, the error vector passed from one stage to the next has one of N probability distributions depending on the value of the index that is also passed to the next stage. Codebook design for MSTVQ requires a method for finding: (a) M codebooks for a set of N vectors with different statistics and (b) a pointer function that assigns a particular codebook to each of N index values. When $N = M$, a separate codebook can be designed for each source and no "sharing" of codebooks is needed; for $M < N$ some or all of the codebooks must be shared for use by input error vectors with different statistics. A solution to the shared codebook design problem is described in the following section.

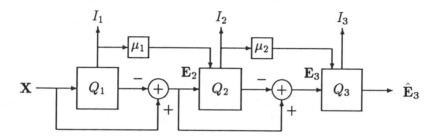

Figure 12.24: 3-Stage Constrained-Storage MSTVQ Structure

12.12 Constrained Storage VQ

Suppose we have a set of I random vectors $\mathbf{X}_i$ for $i = 1, 2, ..., I$, of dimension k each of which is to be vector quantized with the same resolution r in bits per vector component. In general, a separate codebook optimized for each source is needed to obtain the best performance for coding that source. To reduce the total storage requirement, codebook sharing can be introduced where a subset of sources is assigned to a particular codebook from a set of $M < I$ codebooks. Thus we need to find (a) M codebooks C_j for $m = 1, 2, ..., M$, each containing $L = 2^{rk}$ code vectors and (b) a pointer function $\mu(\cdot)$ that gives the index $m = \mu(i)$ identifying the particular codebook C_m or equivalently the corresponding quantizer Q_m for coding the ith source vector. A method that solves this problem is called constrained-storage VQ (CSVQ).

Assume that each vector quantizer Q_m performs a nearest neighbor encoding with a distortion measure $d(\mathbf{x}, \mathbf{y})$. The overall performance measure for this coding system is defined to be

$$D = \sum_{i=1}^{I} w_i \, E d(\mathbf{X}_i, Q_{\mu(i)}(\mathbf{X}_i))$$

where a suitably chosen set of nonnegative weight values w_n with $\sum_{i=1}^{I} w_i = 1$ are chosen to measure the relative importance of distortion in the different quantizers according to the needs of a particular application. This formulation can be applied to MSTVQ as well as a variety of other coding schemes where multiple vector sources must be quantized. This problem was formulated and solved in [55]. It was shown that the set of codebooks and the pointer function which are jointly optimal for minimizing the overall average distortion must satisfy two necessary conditions that are analogous to the necessary conditions for optimality of a single VQ. These conditions are:

(a) For each m, $\mu(i) = m$ if

$$Ed(\mathbf{X}_i, Q_m(\mathbf{X}_i)) \leq Ed(\mathbf{X}_i, Q_j(\mathbf{X}_i))$$

for all $j \in \{1, 2, ..., M\}$. (The smallest index m satisfying this condition can be chosen in the event of ties to avoid ambiguity.)

(b) Given a pointer function $\mu(\cdot)$, then, for each m, the optimal codebook $\mathcal{C}_m$ is that codebook which performs optimal VQ for the random vector whose pdf $g_m(\mathbf{x})$ is equal to the weighted average of the input pdf's assigned to this codebook. This pdf $g_m(\mathbf{x})$ is defined as

$$g_m(\mathbf{x}) = \frac{1}{q_m} \sum_{i \in S_m} w_i f_i(\mathbf{x})$$

where $q_m = \sum_{i \in S_m} w_i$ and $S_m = \{i : \mu(i) = m\}$ indicating the set of source vectors assigned to the mth codebook.

A codebook design procedure given in [55] is based on iteratively alternating between the two necessary conditions starting with a separate training set for each vector source. Specifically, an initial set of M codebooks is selected and the following steps are performed:

1. For each i, find the pointer value $m = \mu(i)$ according to

$$m = \min_m^{-1} Ed(\mathbf{X}_i, Q_m(\mathbf{X}_i)).$$

Then compute the overall average distortion D achieved with this pointer function and the current set of codebooks in encoding the training sets. If the fractional drop in D from the previous iteration is less than a chosen threshold value, terminate the algorithm.

2. Given the newest pointer function, find a new set of codebooks by designing for each m an optimal codebook $\mathcal{C}_m$ for the pdf $g_m(\mathbf{x})$.

The codebook design in the second step can be performed with the GLA by forming from the original training sets a composite training set that is statistically representative of the pdf g_m. A composite training set of size J for designing the mth codebook can be formed by simply collecting $\frac{w_i}{q_m}J$ training vectors from the ith training set for each $i \in S_m$.

The necessary condition for optimality of CSVQ are similar in spirit to the conditions for optimality of a single vector quantizer as described in Chapter 10. Condition (a) is very similar to a nearest neighbor clustering step, except that here we are clustering sources rather than training vectors [56]. Condition (b) is similar to the centroid condition except here we are forming a "centroid" of several probability distributions to get a new distribution. In fact, this similarity is more than an analogy. It can be seen that CSVQ is a generalization of VQ to the "quantization" of vector source distributions where the quantized approximation to a source distribution is a particular codebook from the set of CSVQ codebooks. Thus, a code vector generalizes to a codebook and a codebook generalizes to a set of codebooks. CSVQ has been applied successfully to wideband audio coding where tree structured codebooks previously solved the problem of encoding complexity and CSVQ solved the remaining problem of astronomic storage complexity [56].

12.13 Hierarchical and Multiresolution VQ

Frequently a signal has a very large number of contiguous samples that are statistically interdependent so that blocking it into vectors with very high dimensionality is most efficient for VQ. For example, speech samples can have significant correlation over an interval as much as hundreds of samples in length, corresponding to a phonetic unit of speech with duration as high as 50 to 100 ms. Also, image regions containing hundreds of pixels can have significant correlation corresponding to a particular object in the scene. Here we use the term "correlation" in a loose sense to include nonlinear statistical interdependence. The complexity barrier of course prohibits coding a high dimensional vector with other than very low rates. For example, with dimension 64 a rate of 0.2 bits per sample is a typical upper limit to avoid excessive complexity. Yet in many applications rates as high as 1 bit per sample or even higher are needed.

We now describe some alternatives to the constrained structure VQ methods described so far in this chapter. The alternatives are particularly suited for very high dimensional vectors. The general idea is to extract from the original vector which in this context we call a *supervector* a new feature vector of reduced dimensionality which describes attributes of the supervector that account for much of the correlation or interdependence among its samples. The feature vector is then coded by a suitable VQ technique and the quantized feature vector, either by itself or in conjunction with other information, can generate an approximate description of the supervector; alternatively, the feature vector can contribute to a decomposition of the supervector into lower dimensional subvectors. If the feature vector is partitioned into subvectors and the basic scheme is again applied to each subvector a *hierarchical* VQ structure can be generated. We shall see that these schemes include generalized versions of shape-gain VQ, mean-removed VQ, and multistage VQ, and they also lead to tree structures. When the feature vector is a subsampled version of the original vector, the hierarchical scheme falls into the category of a *multiresolution* coding technique since different stages of processing occur with different degrees of subsampling. In this case, the term *resolution* refers to time or space resolution (rather than quantizer or amplitude resolution) for vectors containing samples of a time waveform or samples of an image, respectively.

In a general approach to hierarchical VQ (HVQ), a supervector $\mathbf{X}$ of dimension $S = sr$ is applied to a feature extractor which partitions $\mathbf{X}$ into r subvectors $\mathbf{X}_i$ each of dimension s, extracts a scalar feature f_i from each subvector and then a feature vector $\mathbf{F}$ of dimension r is formed from these scalars [142]. The feature vector $\mathbf{F}$ is then vector quantized. The quantized feature vector $\hat{\mathbf{F}}$ is applied to the supervector to form a *reduced* supervector $\mathbf{Y}$ which typically has a reduced amount of statistical interdependence among its subvectors. To be specific, we shall first consider the case where the feature is the norm (gain) of the subvector as proposed in [142]. In this case, the supervector is reduced by normalizing each of its subvectors by the corresponding quantized gain obtained from the vector quantized feature vector. The reduced supervector $\mathbf{Y}$ is then partitioned into its subvectors $\mathbf{Y}_i$. This completes the first quantization stage of the HVQ coding method.

In the simplest version of HVQ there are only two stages; for the final (second) stage, the r subvectors $\mathbf{Y}_i$ are individually vector quantized. However, the total number of bits available to quantize the set of r subvectors can be assigned with a suitable bit allocation algorithm rather than equally distributed. The quantized feature vector $\hat{\mathbf{F}}$ provides the controlling information that determines the allocation.

For a 3-stage HVQ scheme, the second stage consists of applying each of the subvectors $\mathbf{Y}_i$ with dimension $s = mk$ to a feature extractor. The feature extractor partitions each $\mathbf{Y}_i$ into m sub-subvectors called *cells* each of dimension k. The gain f_{ij} of each cell is found and for each subvector a feature vector $\mathbf{F}_i$ of dimension m is formed from these gains. Each of these r feature vectors is vector quantized forming vectors $\hat{\mathbf{F}}_i$ and then used to reduce the corresponding subvector $\mathbf{Y}_i$ by normalizing each of its cells by the corresponding quantized gain value. Each reduced subvector $\mathbf{Z}_i$ so formed is then expected to have a reduced statistical interdependence among its cells. In the third and final stage the cells Z_{ij} are vector quantized with a bit allocation for each cell determined by the quantized gain associated with that cell. The gain-normalized cells will typically have very little statistical interdependence so that they may be individually coded with k dimensional quantizers. Figure 12.25 illustrates the key steps of a 3-stage HVQ structure. A box labeled FE is a feature extractor which partitions the incoming vector into subvectors, finds the feature (gain) of each subvector and then forms a vector of these features. Each box labeled P is a partitioning operation where a vector is split into subvectors. The boxes labeled R are reduction operations where the input vector and the associated vector of features are processed to obtain a reduced vector, in our case one whose subvectors have been normalized by quantized gain values. In [142] it was shown that optimal bit allocation among cell quantizers can be achieved by using the cell gains to control the allocation algorithm [142], [293].

In the original HVQ method introduced in [142] the reduction of each feature vector is performed in each step by normalizing with unquantized gain values. In that case the feature extraction at all levels is performed prior to any quantization step. On the other hand, reduction at each step based on the quantization results of the prior step has some intuitive appeal and gives the method some similarity to multistage VQ and is analogous to the second approach to mean-removed VQ where the mean is first quantized and then substracted from the input vector.

Since gain normalization is important for signals with wide dynamic range, the use of the gain as the feature in HVQ is most naturally suited for speech and audio signals. An alternate approach to hierarchical VQ that is more appropriate for image and video coding arises when the arithmetic mean of the subvector rather than the gain is used as a feature [180]. This scheme is a direct generalization of mean-removed VQ by adding a hierarchical coding structure. It may also be seen as a generalized version of multistage VQ. As before, we consider a supervector of dimension $S = sr$. We extract an r-dimensional feature vector from the subvector means and vector quantize this mean vector. In this case, the supervector is reduced by subtracting from each component of each of its subvectors the corresponding quantized mean of that subvector. In a two-stage version of this approach, the mean-removed subvectors are each vector quantized. This method can be viewed as a generalization of mean-removed VQ. In the 3-stage case, the second stage consists of partitioning each subvector into cells, extracting the mean of each cell, and forming a feature vector

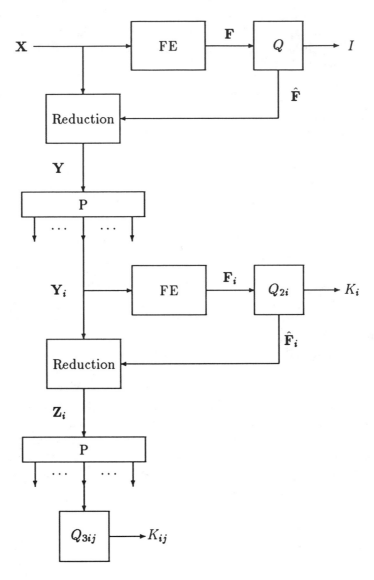

Figure 12.25: Hierarchical Vector Quantization

of means for each subvector. Each of these feature vectors is then vector quantized and reduced subvectors are formed by subtracting the quantized mean from each cell. In the third stage, the reduced cells are individually vector quantized.

Although a non-equal allocation of bits can be assigned to the reduced cells, the quantized means do not generally offer a useful guide for such an allocation. One heuristic way to allocate bits to the cells is to measure the gain or norm of each reduced cell. This quantity measures the rms error that would result from allocating zero bits to a cell. A bit allocation guided by cell gains could provide an effective allocation, however, the allocation must be transmitted to the decoder as side information since the cell gains are not known to the decoder. In [180], variable-rate coding was achieved by assigning either zero bits or a fixed nonzero bit allocation to each cell and sending to the decoder one bit of side information for each cell. Other variations of hierarchical coding were reported in [179], [181], [183], [182].

The process of removing the quantized means from the supervector may be viewed as a two step procedure. In the first step an approximation to the original supervector is formed by letting each subvector have all its components equal to the quantized mean of the original subvector. This operation can be recognized as an *interpolation* step, albeit a primitive type of interpolation that generates a piecewise constant approximation of the original. In the second step the interpolated supervector is subtracted from the original supervector. Similar comments apply to subsequent stages of feature extraction from subvectors. Thus, the HVQ method based on the mean as the feature involves the generation of multiple resolution approximations to the original vector and belongs in the general category of a multiresolution coding scheme. Of course, more sophisticated ways of performing interpolation can lead to better performance. Alternate perceptually based interpolation methods suited for images are considered in [179] and [182]. The general structure of subsampling followed by VQ followed by interpolation, called *interpolative VQ* or *IVQ* [174] and illustrated in Figure 12.26 is often a useful building block in image coding systems.

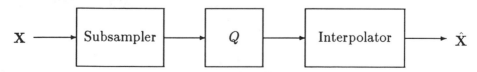

Figure 12.26: Interpolative VQ

In [182], a pyramidal approach to HVQ is described where three successive stages of subsampling of the supervector are performed prior to any quantization. The most coarsely subsampled version is coded first with VQ, then interpolation from the quantized vector gives an approximation that is subtracted from the unquantized intermediate resolution vector at the next level. This difference is then vector quantized and then interpolated to generate an approximation to the first level subsampled vector. The difference between this approximation and the original subsampled vector is then vector quantized and the result interpolated to generate a full resolution approximation to the original supervector. This method resembles the use of pyramid coding by Burt and Adelson [40] who introduced a multiresolution method for image coding based on scalar quantization called "pyramid" coding because one first constructs a pyramid-like data structure where the bottom level contains all of the samples of the image, the next level contains a downsampled version of the image, the next layer contains a further downsampled version, an so on. A difference pyramid can then be formed by taking the difference between a given level and the approximation to that level formed by interpolation from the adjacent level with fewer samples. The interpolation can be considered as a prediction of the higher resolution (in space) image from the lower. By quantizing these differences, the image can be digitized. An important difference between the between the multiresolution technique described here and the Burt and Adelson pyramid is that the structure here performs the quantization in a "closed-loop" form, doing the interpolation based on a quantized version of the lower spatial resolution image, while most traditional pyramid codes quantize "open loop" in the sense of quantizing an ideal difference image.

12.14   Nonlinear Interpolative VQ

Interpolative VQ, described in the previous section, can be viewed as a special case of the more general coding paradigm shown in Figure 12.27 where a k-dimensional feature vector $\mathbf{U}$ is extracted from an n-dimensional signal vector $\mathbf{X}$ and $k < n$. The VQ encoder $\mathcal{E}$ and decoder $\mathcal{D}$ each have a copy of the codebook $\mathcal{C}$ containing N code vectors of dimension k. The quantized feature vector $\hat{\mathbf{U}}$ is then used to generate an estimate $\hat{\mathbf{X}} = g(\mathbf{U})$ of the signal vector by a predetermined linear or nonlinear "interpolation" operation. This operation can be viewed as a generalization of the usual case of interpolation since it maps a reduced dimensionality characterization of the vector to an approximation to the original vector. If a nonlinear interpolator is used, the structure is appropriately called *nonlinear interpolative VQ* (NLIVQ). This structure is often a useful complexity reducing method for quantizing a very high dimensional vector. For examples of several types of feature extractors see [145].

Since the input to the interpolator in Figure 12.27 is one of a finite

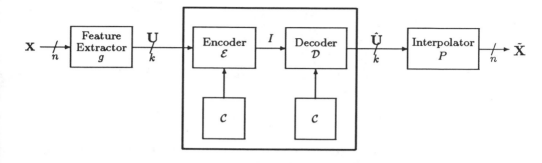

Figure 12.27: Nonlinear Interpolative VQ

number, N, of code vectors, the interpolation operation on each code vector can be computed in the design stage to yield N interpolated vectors of dimension n. These can be stored in a codebook C^* and the decoding and interpolation operations combined into a single operation called *interpolative decoding*. This yields the alternative structure of Figure 12.28.

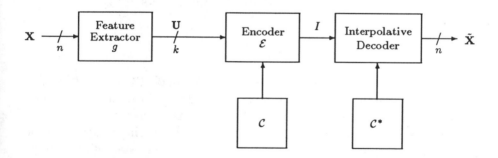

Figure 12.28: Nonlinear Interpolative VQ: Alternate Form

Note that the complexity associated with real-time computation of the interpolation is eliminated. Instead we have a very interesting new paradigm for VQ where the encoder and decoder have distinct codebooks with different dimensionalities.

There are a variety of choices for the interpolator in IVQ. It can be based on heuristic methods leading to either linear or nonlinear schemes. Alternatively, it is possible in principle to find the *optimal* nonlinear interpolator. In fact, this is a standard nonlinear estimation problem, discussed in a general context in Chapter 4. In particular, from Theorem 4.2.1, the

MMSE estimate of $\mathbf{X}$ given $\hat{\mathbf{U}}$ is given by

$$\hat{\mathbf{X}} = E[\mathbf{X}|\hat{\mathbf{U}}]$$

Since $\hat{\mathbf{U}}$ is determined by the transmitted index I, the ith interpolated vector $\mathbf{c}_i$ is thus given by

$$\mathbf{c}_i = E[\mathbf{X}|\mathcal{E}(g(\mathbf{X})) = i]$$

Thus, once the feature extractor and VQ codebook $\mathcal{C}$ are known, the design of the interpolative decoder codebook $\mathcal{C}^*$ can be determined in this way. In practice, however, the probability distribution of the signal vector is either not known or if known leads to an intractable computation. As in conventional VQ design, we revert to the use of a training set to design the codebook $\mathcal{C}^*$.

Let $\mathcal{T}$ denote a training set of M signal vectors $\mathbf{x}_j$ which is statistically representative of the signal vector $\mathbf{X}$ and assume the training ratio $M/N \gg 1$. The encoder $\mathcal{E}$ induces a partition of the training set into regions given by

$$R_i = \{\mathbf{x}_j : \mathcal{E}(g(\mathbf{x}_j)) = i\}$$

for $i = 1, 2, \ldots, N$. It then follows that the code vectors are given by

$$\mathbf{c}_i = \frac{1}{N_i} \sum_{j : \mathbf{x}_j \in R_i} \mathbf{x}_j .$$

In this way, we obtain the interpolative codebook $\mathcal{C}^*$ for representing the high dimensional vector $\mathbf{X}$ from the quantized reduced dimensionality feature vector. Note that no additional complexity is involved in performing the optimal nonlinear interpolation for NLIVQ with the structure of Figure 12.28 compared to the use of suboptimal or heuristic interpolation. Of course, with a finite training set, the code vectors obtained are only approximations to the optimal ones, but the approximation can be made arbitrarily close to optimal by using a sufficiently large training set. The design of the codebook $\mathcal{C}^*$ is relatively simple since it does not require an iterative algorithm as does conventional VQ design. Furthermore, no issue of local rather than global optimality arises here as it does in ordinary VQ design. NLIVQ with the simplified structure considered here offers a versatile and effective building block for hierarchical, multiresolution, and other coding systems.

NLIVQ has an alternative, but equivalent, interpretation in terms of the Lloyd optimality properties. We shall see that it can be viewed as being a constrained encoder for the original (not downsampled) vector $\mathbf{X}$ in combination with an optimal decoder. Suppose, for example, that we wish to

OXOXOXOX

XOXOXOXO

OXOXOXOX

XOXOXOXO

Figure 12.29: 2:1 Downsampling pattern

code a rectangular image block of $4 \times 8 = 32$ pixels by first downsampling by 2:1, using a vector quantizer on the resulting 16-dimensional vector, and then using optimal nonlinear interpolation to provide a final reproduction of the original 32-dimensional vector. This might be done using the pattern of Figure 12.29 where the xs and os together denote the original 32 pixels in the rectangular block and the xs alone represent a downsampled subset of the pixels constituting a $4 \times 4 = 16$-dimensional square block. Using NLIVQ a VQ would be designed for the xs alone, that is, the Lloyd algorithm would be run on these 16-dimensional vectors formed by downsampling successive nonoverlapping 4×8 blocks. Once the final codebook was obtained, the missing pixels (the os) would be interpolated from the 16-dimensional codewords via the conditional expectation of the indices produced by the encoder. The alternative view is to instead consider a VQ that operates on the full 32-dimensional vector: the codebook will contain 32-dimensional reproduction vectors. The encoder, however, will be constrained to depend only on the xs when selecting the index. The Lloyd optimality conditions then have two implications:

- The encoder should minimize the distortion (subject to the imposed constraint). In order to minimize the average distortion between the 32-dimensional input block and a 32-dimensional reproduction vector, the encoder can do no better than to pick the reproduction vector minimizing the distortion between the x pixels in the reproduction vector and the x pixels in the input block (assuming the distortion measure is, like the squared error, additive). Thus the encoder produces the same index as a 16-dimensional VQ operating only on the x pixels.

- The decoder should be optimum for the encoder, that is, corresponding to each index produced by the encoder is a 32-dimensional codeword that is the conditional expectation of the full 32-dimensional input vector given the encoder index.

Clearly the encoder is exactly as before, it is simply a nearest neighbor selection using the downsampled vector. The decoder might at first glance appear different, since it uses an optimal 32-dimensional decoder for the given encoder viewed as a mapping of 32-dimensional input vectors into indices. The conditional expectation of the full 32-dimensional input vector given the encoder index can, however, be decomposed into two pieces: the 16-dimensional conditional expectation of the x pixels, which forms the centroid or Lloyd optimal codeword for the 16-dimensional code, and the conditional expectation of the o pixels, which forms the optimal nonlinear interpolation of the pixels not included in the 16-dimensional code based on the 16-dimensional codeword index. In other words, the downsampled code combined with the optimal nonlinear interpolator can be viewed as simply a constrained encoder combined with an optimal decoder for the full block. The combination of downsampled optimal decoder and nonlinear interpolator is equivalent to a non-downsampled optimal decoder.

12.15   Lattice Codebook VQ

We have seen in Chapter 10 that multidimensional lattices provide a highly structured VQ with efficient encoding algorithms. A lattice codebook VQ is a VQ whose codebook consists of a subset of a lattice or a translation of a subset of a lattice, the most common example being a cubic lattice. In Chapter 10, little care was taken to describe how a suitable subset of a lattice should be chosen, but implicitly one used a collection of lattice points covering the region of space containing most of the source probability. If the source probability is confined to a bounded region of space, then it is natural to confine the lattice codebook to the same space. Given a lattice and such a region, a key parameter is the density of the lattice, that is, how many lattice points per unit volume of space are contained in the codebook. The higher the density, the higher the bit rate and the smaller the average distortion.

In the important special case of a cubic lattice, forming the lattice codebook by confining the cubic lattice to a k-dimensional cube centered at the origin yields a VQ that is equivalent to using a uniform scalar quantizer k successive times, that is, each component of the input vector is applied to the same scalar quantizer. Suppose for example that the scalar codebook has an odd number N of levels separated by Δ with a level at 0 and the remainder symmetric about 0. In this case the resulting k-dimensional VQ codebook can be viewed as a subset of a cubic lattice with minimum point separation Δ, that is, of the lattice

$$\Lambda = \{\Delta \sum_{i=1}^{k} l_i \mathbf{u_i}; \text{ all integers } l_i; \ i = 1, \cdots, k\},$$

where the $\mathbf{u}_i$ are the orthogonal unit vectors in each dimension. The subset is formed by taking all of these points within the *support set* defined as the k-dimensional cube

$$C = \{\mathbf{x} : |x_i| \leq \frac{N}{2}\Delta;\ i = 1, 2, \cdots, k\}.$$

Thus the lattice codebook corresponding to ordinary uniform quantization used k times to form a k-dimensional VQ is $C \bigcap \Lambda$. We saw in Chapter 9 that this particular quantizer can perform within 0.255 bits of the Shannon optimum as given by the rate distortion function if

- the source is memoryless with a smooth density function,

- the resolution is large,

- the overload noise is negligible in comparison with the granular noise,

- and rate is measured by entropy, that is, variable length noiseless codes are permitted to reduce the number of bits per symbol to the entropy of the quantized source.

Thus in some applications a lattice codebook VQ combined with entropy coding can provide a nearly optimal code with reasonable implementation complexity. Better performance (nearer the optimum) can be obtained by using a better lattice.

Unfortunately, however, lattice quantizers have two inherent shortcomings. The first is that their structure can prevent them from being optimal in the sense of minimizing average distortion for a given rate except for sources uniformly distributed over some region of space, e.g., memoryless sources with uniform distributions. The uniform distribution of reproduction vectors is well matched to uniformly distributed source vectors, but it is suboptimum for nonuniform distributions. One can show using high resolution quantization theory that for an arbitrary input pdf $f_{\mathbf{X}}(\mathbf{x})$, the optimal density for VQ reproduction vectors is proportional to

$$f_{\mathbf{X}}(\mathbf{x})^{\frac{k}{k+2}}$$

for the case of the mean squared error distortion [349][137][159]. We shall see that for large dimensions the Shannon-McMillan-Breiman theorem ensures that the vector pdf will be approximately uniform over a particular region of vector space and in this case a subset of a lattice restricted to this region of approximately uniform probability density can give good performance.

The second problem with general lattice codebook VQs is the implicit assumption of very large resolution; the codes remain primarily suited for large bit rate and small distortion applications.

A third problem which should be kept in mind is that for an ordinary lattice code to provide near optimal performance, it must be used in conjunction with a noiseless code.

In this section we sketch techniques by which the first problem can be ameliorated for large dimensional codes. The second problem can be addressed, but lattice codebook VQs are primarily useful for large to moderate bit rates.

The basic idea is to choose a good region or subset (support set) of the lattice to provide a codebook well matched to a particular source. In particular we would like to do better than the simple multidimensional cube usually used. If properly done this should cost little or no more in terms of encoding complexity while providing a significant gain in the efficiency of the use of reproduction levels. As in the discussion of general lattices, we will not go into much detail of actual encoding algorithms. The goal will be to show how a subset of a lattice can be chosen to provide a good code. The encoding problem in fact consists of two aspects: the selection of a nearest neighbor (or nearly so) in the lattice codebook to the input vector and the assigning of a binary channel codeword or index to the lattice point selected. It may be surprising that indexing can pose a difficulty since for the small dimensional, small bit rate codes emphasized in this book the assigning of a binary index to VQ codewords is easily accomplished by table lookup. When the dimension and bit rate are large, however, table lookups must be replaced by efficient algorithms for computing the indices. Such algorithms have been studied for both the channel coding and vector quantization applications. The selection of nearest neighbors in a lattice to a given input vector is considered in depth in [76] [78] [79] [6] [113] [124]. The indexing problem is treated in [113] [77] [125].

To describe the first basic idea, some additional Shannon information theory ideas are needed which will build on those described in Chapter 9. Our discussion will not attempt to be mathematically rigorous, but the ideas can all be made so. The reader interested in pursuing the details of the first portion of this section can find deeper treatments in Sakrison [281] and Fischer [113] [114]. For simplicity we focus on the coding of memoryless sources as in [113]. For this case the particular Shannon limit theorem needed follows from the weak law of large numbers.

Suppose that $\{X_n\}$ is an iid source described by a marginal probability density function $f_X(x)$. The joint density function for a k-dimensional vector $\mathbf{X}$ will therefore be given by

$$f_{\mathbf{X}}(\mathbf{x}) = \prod_{i=1}^{k} f_X(x_i).$$

The weak law of large numbers implies the following version of the Shannon-McMillan-Breiman theorem for densities (see Barron [28] or Gray [157] for general theorems of this type):

$$-\frac{1}{k} \sum_{i=1}^{k} \log f_X(X_i) \quad \underset{k \to \infty}{\to} \quad -E[\log f_X(X)]$$

$$= -\int_{-\infty}^{\infty} f_X(x) \log f_X(x)\, dx$$

$$\equiv h(X), \qquad\qquad (12.15.1)$$

where the limit is in the sense of convergence in probability and $h(X)$ is the differential entropy of the random variable X_0 as defined in Chapter 9. This result is sometimes referred to as a "continuous AEP" since the AEP or asymptotic equipartition property is a common name for the discrete alphabet version of this result.

Convergence in probability means that for any $\epsilon > 0$

$$\lim_{k \to \infty} \Pr(|-\frac{1}{k} \sum_{i=1}^{k} \log f_X(X_i) - h(X)| > \epsilon) = 0. \qquad (12.15.2)$$

Since the left-hand side of (12.15.1) can be written as

$$-\frac{1}{k} \log f_{\mathbf{X}}(\mathbf{X})$$

we can interpret (12.15.1) as saying that if the dimension k is large, then almost all of the probability is concentrated on those vectors $\mathbf{x}$ for which

$$f_{\mathbf{X}}(\mathbf{x}) \approx 2^{-kh(X)}. \qquad\qquad (12.15.3)$$

In other words, memoryless sources produce large dimensional vectors which with high probability have an approximately constant pdf and hence are approximately uniformly distributed over the region of vector space where the pdf $f_{\mathbf{X}}(\mathbf{x})$ equals $2^{-kh(X)}$. This suggests that a lattice quantizer confined to this region might perform quite well. We will not be wasting lattice points where there is little input probability, yet we will be taking advantage of the uniform distribution of lattice points by constraining reproduction vectors to lie in a region of approximately uniform input probability.

A slightly different strategy would be to not only consider lattice points on or near the constant probability surface in k-dimensional space

$$B = \{\mathbf{x} : f_{\mathbf{X}}(\mathbf{x}) = 2^{-kh(X)}\}, \qquad (12.15.4)$$

but to instead consider lattice points for all vectors contained within B, that is, within the volume enclosed by the surface B. This might be a good idea in some applications because it can lead to a simpler code and a simpler analysis at the expense of some possibly wasted reproduction vectors. It follows from the Shannon-McMillan-Breiman theorem, however, that the volume of the little used reproduction vectors is quite small when the dimension is large.

This coding strategy of confining the lattice to a specific contour of constant probability (or to a superset of this contour) will be feasible if the support region has an easily described structure. We consider two special cases where such simple structure is evident.

First suppose that the source is an iid Gaussian source with marginal pdf

$$f_X(x) = \frac{1}{\sqrt{2\pi\sigma^2}} e^{-\frac{1}{2\sigma^2}x^2}.$$

The differential entropy of this source is

$$h(X) = \log\sqrt{2\pi\sigma^2 e}$$

and hence from (12.15.3) almost all of the source probability will be concentrated on the set of points $\mathbf{x}$ for which

$$\left(\frac{1}{2\pi\sigma^2}\right)^{\frac{k}{2}} e^{-\frac{1}{2\sigma^2}\sum_{i=1}^{k} x_i^2} \approx 2^{-\frac{k}{2}\log 2\pi\sigma^2 e} = \left(\frac{1}{2\pi\sigma^2 e}\right)^{\frac{k}{2}}$$

and hence

$$e^{-\frac{1}{2\sigma^2}\sum_{i=1}^{k} x_i^2} \approx e^{-k/2}$$

or

$$\frac{1}{k}\sum_{i=1}^{k} x_i^2 \approx \sigma^2.$$

The collection of vectors $\mathbf{x}$ satisfying this expression forms a shell including the surface of a k-dimensional sphere of radius σ. A lattice codebook VQ for this source with k large could be constructed by populating the codebook by lattice points constrained to lie on (or near) the surface of this k-dimensional sphere. Alternatively, all lattice points within the sphere could be used. Note that for the given density the pdf will be higher at internal points than on the shell. If the codebook is to only include the surface of the sphere or a small spherical shell including the surface, only lattice points having a Euclidean norm near σ would be included in the code. If the lattice chosen were a cubic lattice, a possible encoding technique would be to first project the input vector onto the sphere by normalizing it to have norm σ, and then to scalar quantize each separate coordinate to produce a point in the cubic lattice close to the sphere surface. If the resulting lattice vector is in the codebook, the nearest-neighbor selection is complete. Otherwise a further operation must be performed to produce a valid codeword. A variety of encoding algorithms including the binary indexing portions are described in [113]. Other lattices might perform better, but at the expense of a more complicated encoding. The scaling operation inherent in such encoding algorithms is itself not trivial and adds to the complexity of the code.

In the case of a Laplacian source emphasized by Fischer [113],

$$f_X(x) = \frac{\lambda}{2} e^{-\lambda|x|}$$

and

$$h(X) = \log(\frac{2e}{\lambda}). \tag{12.15.5}$$

Here (12.15.3) implies that most of the probability will be concentrated on **x** for which

$$(\frac{\lambda}{2}) e^{-\lambda \sum_{i=1}^{k} |x_i|} \approx 2^{-h(X)}$$

or

$$\sum_{i=1}^{k} |x_i| \approx \frac{k}{\lambda}. \tag{12.15.6}$$

If one takes the approximation as exact, this defines a multidimensional pyramid (or a sphere with respect to the l_1 norm). Thus in this case one restricts the lattice to the surface of a multidimensional pyramid, which led to the name "pyramid VQ" in [113]. (This nomenclature should not be confused with the "pyramid codes" constructed using multiresolution techniques as in Burt and Adelson [40].) As with the Gaussian source, the Laplacian source can be encoded in two stages in the case of a cubic lattice. The input vector can first be mapped into the closest vector satisfying (12.15.6), i.e., the closest vector in the face of the pyramid. This can then be scalar quantized to produce a nearby lattice point. If all lattice points close to the pyramid faces are in the codebook, this completes the encoding. If some are not (e.g., only lattice points actually in the face might be used), then a further stage of encoding may be necessary.

Applying high resolution quantizer arguments similar to those used in Chapters 5, 6, and 9, Fischer demonstrated that quantizers constructed in the above manner provide performance roughly $(1/2)\log(\pi e/6)$ worse than the Shannon rate-distortion optimum, that is, roughly the same as that promised by uniform quantization and entropy coding. The lattice codebook techniques, however, do not require variable-rate coding and the associated circuitry. An advantage over VQs designed by clustering is the ability to use large dimensions and hence achieve greater quality at a fixed rate.

The lattice codebook VQs described to this point retain the implicit assumption of high resolution, that is, many bits. They can be used at moderate bit rates by modifying them. For example, one can use a gain/shape code to separately code a gain term describing the size of the sphere or pyramid and then use one of a set of spheres or pyramids to encode the shape [113].

Overload Distortion in Lattice Codebooks

Eyuboğlu and Forney [108] have recently introduced an alternative approach to the design of lattice codebooks that shows great promise for providing a design technique that explicitly considers overload distortion (unlike the usual high resolution quantization approximations which permit only granular noise) and balances granular quantization noise with overload quantization noise at practical bit rates and dimensions. Their technique is a source coding dual of Ungerboeck's lattice and trellis coded modulation techniques [314] and takes further advantage of the structure of lattices.

Consider the Gaussian iid source as a specific example. We have seen above that with high probability, input vectors will fall on the surface of a k-dimensional sphere. We now focus on the sphere itself as the support set and not just on the surface. The lattice points within the sphere will do a good job of quantizing those input vectors that lie within the sphere. Those vectors outside, albeit improbable, will cause larger distortions. We consider the support region to be the granular region and the region outside to be the overload region. The traditional high resolution theory treats only the granular region, assuming that the overload noise is negligible. Eyuboğlu and Forney focused on two distinct effects: the collection of lattice points in the granular region and on the probability of overload.

The lattice providing the basic collection of candidate VQ reproduction vectors determines the average distortion in the granular region. For this reason they refer to this lattice as the granular lattice. By choosing better lattices in higher dimensions one can improve performance by lowering the average distortion. We have seen that the simple cubic lattice performs within 0.255 bits (in entropy) of the optimum performance as given by the rate distortion function, and hence it would appear that at most these 0.255

bits could be achieved by using a better, but more complicated, lattice. This conclusion is misleading, however, because the entire development considers only granular noise, that is, the quantization error due to input vectors outside of the support region are not taken into consideration. An additional separable performance gain can be achieved by proper design of the support region.

High resolution quantization arguments show that the average distortion due to granular noise depends primarily on the distortion measure and on the lattice structure. In particular, it depends on the shape of the quantizer cell or Voronoi region centered at the origin (and replicated by all granular lattice points). The average distortion due to granular noise is fairly independent of the specific probability density of the input vector. For a support region of a given volume, the average distortion due to granular noise does not depend strongly on the specific shape of the support region or on its probability. This fact is common to all previous theory and vector quantizer design algorithms based on high rate arguments. The overload distortion, however, is strongly influenced by the shape of the input pdf.

As an example of the effect of the shape of the support lattice on the overall distortion, consider the effect of using a cubic lattice with a fixed density of points and two possible support regions. The first is the usual Euclidean cube resulting in a uniform quantizer. The second is a multidimensional sphere having the same volume. The two lattice codebooks will have the same size or bit rate, but the spherical support region will result in a smaller overload probability because of the spherical symmetry of the Gaussian source. This smaller overload probability intuitively should result in smaller overall average distortion. Equivalently, if one begins with the two choices and the requirement that the cubic and spherical region be chosen so as to result in the same probability of overload, then the spherical region will have the smaller volume. For a fixed lattice density, this will imply a smaller codebook size and hence a smaller bit rate. As the lattice density determines the granular distortion, both choices yield the same granular distortion. The spherical support region, however, yields either a lower probability of overload for a given support volume or bit rate, or it yields a lower support volume or bit rate for a given probability overload. Note in the latter case of fixed overload probability and reduced volume, we could return to the original bit rate by using a higher lattice density, which would result in lower granular noise.

Another way of interpreting the effect of the choice of the support region is to consider "limiting" the input prior to quantization. Standard practice for preventing overload in a scalar quantizer is to limit the input by mapping all values larger than the top of the granular region into the top value and all values smaller than the bottom of the granular region into the bottom value, thus forcing the limited source to place all its probability on the granular region. All of the overload noise is contained in the limiting operation itself. Limiting each scalar input separately, however, results in a cubic support region for the corresponding VQ. Yet we know from earlier arguments that it is desirable to choose a support region so that the bit rate can be reduced for a fixed probability of overload (or the probability of overload can be reduced for a fixed bit rate and granular distortion). Selecting the support region, which we saw was spherical in the case of a memoryless Gaussian source, can be considered as a form of multidimensional limiting of the input signal.

Unfortunately, spherical support regions still yield encoding algorithms that can be complex for large dimensions, especially with the assignment of binary indices to codebook members in an algorithmic manner. This led Eyuboğlu and Forney to the second key step, which paralleled the use of lattices in channel coding applications. Instead of using a spherical region, the support region is chosen to have the shape of a convex polytope; specifically, the region is the Voronoi cell corresponding to the origin of a "large" lattice. (They also consider the use of the Voronoi region of a trellis code.) The amplitude scale of the lattice is adjusted so that this one cell is appropriately large to serve as the region of support. This large lattice can be a sublattice of the granular lattice. Suitably chosen lattices in sufficiently high dimensions can yield Voronoi regions that closely approximate a sphere. The lattice VQ of Eyuboğlu and Forney [108] is the source coding dual of the channel codes of [77][125] (the gain due to the support region selection corresponds to the "shaping gain" of the channel codes).

The choice of a Voronoi region of a large lattice as a support region for the granular lattice has several benefits. One is that one can obtain high rate approximations to the resulting relations of rate, granular distortion and overload probability. These results show how gains exceeding those promised by the purely granular results can be achieved. A second benefit is that it leads to simple algorithms for indexing the codewords so that the need to store indices is greatly reduced. A potential additional benefit is that the structure of the support region suggests alternative possible "limiting" techniques for encoding input vectors that overload the quantizer, that is, fall outside the support region. As an example, one could use efficient lattice encoding techniques on the larger lattice, called the *boundary lattice*, to encode the input vector. The resulting error vector will be the original vector (except for a sign change) if the original vector is within the support region. The error vector will always lie in the support region, however, even if the original input does not. This can be viewed as a multidimensional analogy to modular arithmetic. For example, consider a quantizer mapping every real number r into $q(r) = \lfloor r \rfloor$, the greatest integer less than or equal to r. The resulting error $r - q(r)$ is then simply r mod 1, the fractional part of r. Forming this quantization error thus "folds" all real numbers into the unit interval and is a variation on the idea of modulo-PCM where an input scalar is limited in modular fashion rather than by ordinary limiting. Forming the multidimensional error by using a lattice quantizer similarly maps every input vector into the Voronoi region of the origin and can be considered as a form of modulo-VQ. As an alternative, if the boundary lattice encoding yields anything other than 0, some other scheme can be applied to convert the overload input vector into a new vector in the granular region.

This technique combines a granular lattice to do the actual quantization with a large sublattice to form a multidimensional limiting of the input vector to a suitably chosen support region which minimizes the volume of the granular region while keeping the probability of overload constant (or, alternatively, minimizes the volume of the granular region and hence lowers the rate while keeping the probability of overload constant). It is shown in [108] that this combination can provide codes arbitrarily close to the Shannon optimum and that good performance can be achieved at practical rates. Future papers by Eyuboğlu and Forney promise to provide practical implementation algorithms and extensions to sources with memory.

12.16   Fast Nearest Neighbor Encoding

While many of the constrained VQ methods described in this chapter offer reduced encoding complexity, several methods have been developed for reduced complexity (i.e., fast) encoding while retaining a given unstructured codebook. The objective is to find the nearest neighbor code vector to a given input vector without incurring the complexity of an exhaustive search through the codebook. In some methods, the encoder does not always find the nearest neighbor but rather a code vector that is one of the nearest neighbors. Here we briefly review a few of these techniques. It should be noted that the relative advantage of a particular technique is generally implementation dependent. In particular, a programmable signal processor chip may require an entirely different algorithm than an application-specific integrated circuit. Furthermore, unlike exhaustive search which has a fixed search time, many techniques have a random search time. In some applications, such as image compression for storage, the average search time is a more important indicator of performance than the worst-case search time. On the other hand, the reverse is true in some other applications such as speech coding for telecommunications.

One of the simplest methods for fast search, called the *partial distance* method, can typically reduce the average search time by a factor of 4 and requires only a small modification of the exhaustive search nearest neighbor encoding rule given in Chapter 10 and is based on the squared Euclidean distortion measure [66],[29]. For a given input vector $\mathbf{x}$ in k dimensions, the search proceeds sequentially through the codebook and after the mth code vector has been tested, the minimum distortion so far, D, and the best index so far, I are retained. Then to test the $m + 1$th code vector $\mathbf{c}$, a counter r is initialized with value 1. and the following operation is performed for $r = 1, 2, \ldots, k$.

Test: Compute the partial distance $D_r = \sum_{i=1}^{r}(x_i - c_i)^2$. If $D_r > D$ then reject this code vector and go to the next code vector. If $D_r < D$ and $r < k$, increment the counter (i.e., r is replaced with $r + 1$) and go to Test. If $D_r > D$ and $r = k$, replace D with D_r and I with $m + 1$ and go to the next code vector.

The validity of this algorithm is self-evident. Also, it should be noted that the algorithm always results in locating the true nearest neighbor since no approximations are made. The worst-case search time is no better than exhaustive full search although it would be virtually impossible for the worst-case to occur. The utility of this algorithm is limited to applications where the time incurred in a comparison operation is relatively small compared to a multiplication operation.

One general approach to reducing the encoder complexity is based on identifying the geometric boundaries of the Voronoi cells and storing a suitable data structure to ease the real-time search process [66], [65]. Much of the encoding complexity is transferred from the real-time encoding operation to the off-line design of the data structure.

An example of such a technique is the *projection* method, where the data structure is obtained by finding for each Voronoi cell the boundaries of the smallest hyper-rectangular region in k-dimensional space that contains the nearest neighbor cell [66]. Specifically, the minimum and maximum values of the jth coordinate of points in cell R_i are found by projecting the cell onto the jth coordinate axis. This gives for each cell a set of $2N$ hyperplanes (each orthogonal to one of the coordinate axes) that specify the faces of the bounding hyper-rectangle. This provides a partition of each coordinate axis into $2N - 1$ intervals. A table is then generated for each coordinate indicating which Voronoi cells have a projection that lies in each interval. Various ways of organizing the table are possible. The search procedure then scalar quantizes one component at a time of the input vector and identifies the set of candidate cells. As successive components are quantized, the intersection of the prior set of candidates with those associated with the current components are formed until until all components have been examined. Then, an exhaustive search is performed over the final set of candidates, which is much smaller in number than the original codebook size N. In principal this method will always yield the nearest neighbor, however, the task of finding the exact projections is extremely demanding and appears to be computationally infeasible. In practice, the projections can be found by encoding a large training set, storing the clusters so obtained and allowing the projection of the ith cluster to substitute for the projection of the cell itself.

Perhaps the most widely studied fast search technique is based on the use of *K-d trees* [31], [128], [223], [260], [107]. A K-d tree is a binary tree with a hyperplane decision test at each node where each hyperplane is orthogonal to one of the coordinate axes of the k-dimensional space and a set of *buckets* or terminal nodes. The search process is simplified since each node requires the examination of only one component of the input vector to see if it lies above or below a threshold. Each bucket consists of a small set of candidate code vectors that remain eligible after the search process followed the path leading to this bucket. The K-d tree can be very efficient for codebook search but it requires an effective design procedure based on training data. Note that the projection method described earlier is actually a special case of a K-d tree.

Another technique that has been frequently used by VQ researchers is based on the triangle inequality for distortion measures that are monotonic functions of a metric. In particular, it is applicable to the squared Euclidean distance. The idea is to use some *anchor points* as reference points from which the distance to each code vector is precomputed and stored. The encoder then computes the distance between the input vector and each anchor point after which some simple comparisons using the precomputed data eliminate a large number of code vectors as candidates for the nearest neighbor. To illustrate the idea in rudimentary form, suppose we have one anchor point $\mathbf{a}$. Let $\mathbf{c}_1$ be a code vector that is a currently eligible candidate for the nearest neighbor. Assume we know the value of $d_i = \|\mathbf{c}_i - \mathbf{a}\|$ for $i = 1, 2, \ldots, N$, where N is the codebook size. The triangle inequality gives

$$\|\mathbf{x} - \mathbf{c}_j\| > d_j - \|\mathbf{x} - \mathbf{a}\|.$$

Therefore, if we have $d_j > \|\mathbf{x} - \mathbf{c}_1\| + \|\mathbf{x} - \mathbf{a}\|$, it follows that $\|\mathbf{x} - \mathbf{c}_j\| > \|\mathbf{x} - \mathbf{c}_1\|$, which eliminates $\mathbf{c}_j$ as a candidate. A variety of refinements have emerged to make this approach an efficient method for codebook searching. In general, there is a trade-off between the search speed and the size of the precomputed data table required. (In the above illustration only one anchor point is used so that the table consists of only the N values of d_i.) For further details of search methods based on the triangle inequality, see [322], [299], [238], [239], [261], [246].

12.17 Problems

12.1. The splitting TSVQ design algorithm designs a tree-structured code from the root node out. Suppose that instead you have a VQ codebook with 2^R entries (R an integer) and wish to design a binary tree to search it, that is, to construct a TSVQ having the prespecified labels for all the terminal nodes. Describe an algorithm to accomplish the design and discuss its potential advantages and disadvantages over full search VQ with the same codebook and a TSVQ designed by the splitting technique.

12.2. Show that if a random vector $\mathbf{X}$ has an optimal codebook $\mathcal{C} = \{\mathbf{y}_i\}$, then an optimal codebook for the transformed vector $\mathbf{TX}$, where $\mathbf{T}$ is a unitary transformation, consists of the transformed codebook $\mathbf{T}\mathcal{C} = \{\mathbf{Ty}_i\}$.

12.3. Prove the conditional centroids formulas (12.10.9)–(12.10.10) for the product code design.

12.4. In searching through a binary TSVQ codebook a binary decision is needed at each node. With the usual MSE criterion this node decision test can be performed using an inner product method with only k multiplies (and an addition and a comparison operation). However, for an m-ary TSVQ the node decision test is often said to require mk multiplies, with $k = 2$ being an exception to this rule. Show that in fact it is always possible to do an m-ary node decision test with an average number of multiplications less than $3k$ for a ternary TSVQ codebook ($m = 3$).

We know that, with a mean squared error criterion, searching a balanced m-ary TSVQ codebook requires $m \log_m N$ vector-compare operations, instead of mN operations in full search VQ, where N is the number of codewords. In a similar fashion, we would like to compare the computational complexity of codebook design between the two schemes. Assume that the number of Lloyd iterations to codebook convergence is the same everywhere. Also, consider only the number of vector-compare operations in the partitioning of training vectors in the Lloyd design loop. The training set size is M.

12.5. In the process of designing a standard binary tree for tree-structured VQ, it is possible to define a value of the distortion associated with each node of the tree in the following way. Each node of the tree (except the root node) is associated with a node vector (either a test vector or a code vector) that was designed by taking the average of all vectors in a particular cluster of training vectors. Define the distortion of that node as the average distortion between the node vector and each vector in the cluster used to obtain that vector. It is also possible to associate with each node a probability value which represents (approximately) the probability that when an input vector is encoded, the path taken through the tree will reach this node. A balanced tree can be "pruned" (leading to an unbalanced tree) by removing a pair of terminal branches ("leaves") that come from a particular parent node. By repeated use of this pruning operation on the tree, many possible unbalanced subtrees can be generated from a given balanced tree.

(a) Explain how the node probabilities can be computed using the training set as part of the tree design process.

(b) Explain how an unbalanced tree can be used to encode an input vector with a variable length binary word (so that different input vectors may lead to a different number of bits being used to identify the quantized value).

(c) Explain the effect of pruning on both the average distortion and on the average bit-rate in encoding an input vector—in comparison with the use of the original balanced (unpruned) tree.

(d) Suggest an algorithm for pruning a tree to achieve a 'good' trade-off between average rate and average distortion.

12.6. A stationary scalar-valued random sequence x_k is encoded with 8 bits per sample. The samples arrive at a rate of 1000 b/s (bits per second). In order to have variable-rate capability, the encoder data stream must be split into two channels, $S1$ and $S2$, with rates of 6 kb/s and 2 kb/s, respectively. Occasionally $S2$ will be dropped (lost or abandoned) and the receiver will be forced to recover the signal using only $S1$. The encoder does not know when $S2$ is lost and must always use a fixed method of generating the bits for each channel. (There is no feedback channel.) Find encoding schemes for allocating bits to $S1$ and $S2$ so that when both channels are available, a performance as close as possible to that achievable with an encoder designed for a single 8 kb/s fixed rate channel and when $S1$ is available, a performance as close as possible to that achievable with an encoder designed for a single 6 kb/s fixed rate channel. Describe separately, your proposed solution to this problem for each case specified below and explain the rationale of your method. Do not worry about complexity. You won't be required to build the coders you propose.

Explain the rationale for your method and why it achieves the desired objectives.

Case (a) The source is scalar quantized sample by sample without exploiting correlation that might exist in the sequence. (The encoder has no memory and sends out 8 bits to $S1$ and $S2$ after each observation of an input sample.)

Case (b) The source is blocked into vectors corresponding to successive pairs of samples and for each pair, the encoder generates 12 and 4 bits for $S1$ and $S2$ respectively after observing the current pair of samples. The coder can only exploit correlation between the two samples in a pair but not from one vector to another.

Case (c) The encoder has a DPCM structure, with little or no modifications from a conventional DPCM encoder that would be used to generate 8 kb/s for a fixed rate application.

12.7. To reduce complexity in coding a large dimensional vector, it is proposed to find $\mathbf{U}$, an $m \times k$ matrix, which maps an input vector $\mathbf{x}$ of dimension k into a reduced dimensionality vector $\mathbf{y}$ of dimension m given by $\mathbf{y} = \mathbf{U}\mathbf{x}$ where $m < k$. Then, $\mathbf{y}$ is vector quantized to $\hat{\mathbf{y}}$ and an approximation to $\mathbf{x}$ is obtained according to $\hat{\mathbf{x}} = \mathbf{W}\hat{\mathbf{y}}$, where $\mathbf{W}$ is an $k \times m$ matrix, in such a way that $\mathbf{x}$ is approximately recoverable from $\mathbf{y}$. Given the statistics of $\mathbf{x}$, find a method for designing the matrices $\mathbf{U}$ and $\mathbf{W}$ for given values of k and m, so that

(i) $E(||\mathbf{x} - \mathbf{W}\mathbf{U}\mathbf{x}||)^2$ is reasonably small (or as small as possible if you can find such a solution) and so that

(ii) $E(||\mathbf{x} - \hat{\mathbf{x}}||^2)$ is approximately equal to $E(||\mathbf{y} - \hat{\mathbf{y}}||)^2$. Note that condition (ii) means that the vector quantization of $\mathbf{y}$ can be based on approximating $\mathbf{y}$ by $\hat{\mathbf{y}}$ as close as possible. Explain why your method achieves the desired objectives.

12.8. (a) Find an efficient or, if you can, an optimal way to encode a vector $\mathbf{x}$ into three indices, I, J, and K, which are transmitted to the decoder and then applied to mean (scalar), gain (scalar), and shape (vector) decoders respectively, producing m_i, g_j, and s_k as the respective outputs of the three decoders. The reconstructed output is the vector $\hat{\mathbf{x}} = m_i + g_j s_k$. Assume the decoders are given in designing the encoding algorithm.

(b) For the encoding method you have proposed, indicate how the codebooks should be designed for a given training set of input vectors.

12.9. For a multi-stage vector quantizer, for coding k-dimensional vectors at a rate of r bits per sample with each stage having the same size codebook and full search through each codebook, how many multiplications are needed per input vector? How much storage is needed? If the ratio k/s is kept constant and the dimension is increased, how does the encoding complexity increase with dimension? As always show how you derive your answers.

12.10. The first stage quantization error $\mathbf{E}_1$ of a two stage vector quantizer will have a large magnitude on the average when certain code vectors are selected for representing the input $\mathbf{X}$ and will have a small magnitude when certain other indexes are selected for representing the input $\mathbf{X}$. Specifically,

$$v_i^2 = E[||\mathbf{E}_1||^2 | \mathbf{X} \in R_i]$$

is the conditional squared gain of the error vector given that a particular index i was selected to encode $\mathbf{X}$. The set R_i denotes a partition cell of the first stage quantizer. Yet, the second stage quantizer uses one fixed codebook for quantizing the first stage error regardless of this fact. We should be able to do better by making use of our knowledge of the result of the first stage encoding to improve the quantization of the first stage error without increasing the bit rate. Note that v_i for each i can be computed once the first stage codebook is designed.

(a) How can v_i be computed for a particular i from a training set of input vectors: $\mathcal{T} = \{\mathbf{x}_1, \mathbf{x}_2, ..., \mathbf{x}_M\}$.

(b) Suppose we have a table of values for v_i and a table lookup operation $H(i) = v_i$ for each i in the range of index values $1, 2, \cdots, N$ where N is the first-stage codebook size. How can we make use of this information to design a better second stage quantization scheme for the same bit rate? Describe your proposed coding scheme clearly and illustrate with a block diagram (including a block for the operation H that maps i into a gain value).

Chapter 13

Predictive Vector Quantization

13.1   Introduction

The coverage of VQ has focused thus far on the coding of a single vector extracted from a signal, that is, on *memoryless* VQ where each input vector is coded in a manner that does not depend on past (or future) actions of the encoder or decoder. This vector is typically a set of parameters extracted from a finite segment of a signal or a set of adjacent samples of a signal. The segments are themselves usually blocks of consecutive samples of speech or two-dimensional blocks of pixels in an image. Usually we need to quantize a sequence of vectors where each vector can be assumed to have the same pdf, but the successive vectors may be statistically dependent. Separate use of VQ for each vector does not take this dependence into consideration. Vector coders possessing memory may be more efficient in the sense of providing better performance at given bit rates and complexity by taking advantage of the inter-vector or inter-block dependence or correlation.

One way of incorporating memory or context into VQ is to use *predictive vector quantization* (PVQ), a straightforward vector extension of traditional scalar predictive quantization or DPCM as treated in Chapter 7. By replacing both the scalar quantizer and the scalar predictor by their vector counterparts, we obtain the general scheme of Figure 13.1. As in the scalar case, the past codewords $\hat{\mathbf{e}}_n$ or, equivalently, the reproductions $\hat{\mathbf{X}}_n$ are used to predict the new input $\mathbf{X}_n$, which is now a vector. Hence the predictor must predict several samples or symbols or pixels based on the past. The difference between the input vector and the prediction is formed and vector quantized. The technique of PVQ was introduced in [86] and

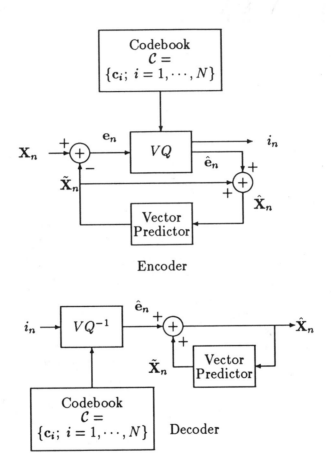

Figure 13.1: Predictive Vector Quantizer

[117] in 1982. More extensive studies of PVQ followed in [85], [87] and [58]. Studies of PVQ for quantized control applications appear in [118] and [112]

The sequence of vectors $\{\mathbf{X}_n\}$ to be coded by PVQ can be generated from a scalar process $x(n)$ by forming successive blocks of k consecutive samples, resulting in a *blocked scalar process*. By converting the serially arriving samples in each block into a parallel set of samples, a sequence of k-dimensional vectors is generated. Alternatively, the sequence of vectors could be an intrinsically vector process where each vector represents a set of features or samples simultaneously extracted from a signal rather than a blocked scalar process. Both cases are amenable to the use of PVQ. Examples of intrinsically vector processes are:

1. *Microphone Array.* Each vector represents the set of amplitude samples observed at one instant of time in an array of k microphones in an audio conferencing system. Successive vectors correspond to successive sampling instants in time. Successive vectors will generally be highly correlated.

2. *Video.* Each vector represents a particularly located segment or block of $m \times n = k$ pixels in a sampled image and successive vectors correspond to successive blocks at the same spatial location of a time varying image as seen by a video camera. Successive vectors describe the same spatial location at successive time instants. The sequence of vectors is quite highly correlated, particularly during periods when there is little motion in the scene under observation by the camera. See for example [26] [25].

3. *LPC Parameter Sequence.* In many schemes for low bit-rate coding of speech waveforms, the waveform is initially partitioned into segments called *frames*, each frame is stored in a buffer, and an analysis procedure computes a set of k linear prediction parameters (such as reflection coefficients) directly from the samples in the buffer. (The computational methods are discused in Chapter 4.) Each vector characterizes the spectral envelope of one frame. With a typical frame size of $20 - 25$ ms, successive vectors have significant correlation because of the relatively slow variation of the short term spectrum of the speech waveform. See, in particular, [346].

4. *Blocked Speech Signal.* A speech waveform $x(n)$ is partitioned into blocks of k samples to form a blocked scalar process, leading to a vector sequence. In this case each vector has the form

$$\mathbf{X}_m = [x(mk), x(mk + 1), \cdots, x(mk + k - 1)]^t.$$

It is evident that successive blocks will have substantial correlation since nearby speech samples are highly correlated. See, [3], [87].

5. *Image Block Sequence.* An image is partitioned into $m \times n = k$ blocks where k is typically 16 and each block is formed into a vector. This is a two-dimensional example of a blocked scalar process. By block raster scanning across successive rows of blocks, a sequence of vectors is formed. Vectors corresponding to spatially adjoining or nearby blocks will generally be highly correlated.

The first three examples describe intrinsically vector sequences whereas the last two examples are vector sequences obtained by blocking scalar processes. In the case of intrinsically vector sequences, the vector dimension may be constrained by the application. For example, for the linear predictor parameter set, often there is very little advantage to increasing the number of parameters beyond 10. On the other hand, in the case of blocked scalar processes, the vector dimension k is a design parameter that may have an important role in the overall performance of a PVQ system.

Although quantizing larger dimensional vectors with VQ almost always improves performance, the same is not true for vector predictors of blocked scalar processes. By predicting an entire future vector based on the past rather than predicting only a single future scalar, one is effectively going farther out on the proverbial limb by trying to predict more events farther into the future given the known past. The more remote components of the vector will be much less predictable than the components that are closer in time to the observed past. The farther into the future we try to predict (the larger the vector dimension), the more stale the given information becomes and the accuracy of the resulting prediction deteriorates. Nevertheless, this loss of predictor performance is more than compensated for by the gain in vector quantizer performance resulting from the removal of correlation from the vector sequence and the shaping of the vector pdf when the residual is formed. There is no theorem proving this, but in all cases known to us the use of prediction in a sensible manner and with a reasonable choice of vector dimension improves performance in PVQ, sometimes significantly.

In this chapter we describe the fundamentals of PVQ, which are natural extensions of the corresponding scalar results, and develop design algorithms for PVQ systems. Given a predictor, the VQ can be designed by running the Lloyd algorithm in a closed-loop form. As in the scalar predictive quantizer, we can design the predictor using ideas from linear prediction theory by assuming that the decoded reproduction is a good approximation to the original input. We shall also refer to another design approach which attempts to jointly optimize the VQ and the predictor for each other. Unfortunately this mathematically appealing approach gives a negligible performance improvement over the more simple-minded approach and hence the increased design complexity is not justified in general. It is worth knowing, however, that the simple techniques perform as well as they do. Because the simplest and most useful approach to predictor design is based on linear prediction, a substantial portion of this chapter is devoted to developing the general vector linear prediction theory. Although the full power of the theory is not required in some of the applications (e.g., image PVQ can do quite well with the vector prediction ideas of Chapter 4), the theory has several important areas of application in signal compression.

After developing the basic ideas of PVQ, we introduce the basic theory of finite memory vector linear prediction, where a vector is predicted from observed values of a finite number of preceding vectors. The basic theory of optimal vector prediction for a stationary vector sequence is outlined and the special case of blocked scalar processes is examined.

13.2 Predictive Vector Quantization

The basic predictive vector quantizer structure is depicted in Figure 13.1 and, as previously noted, is a direct generalization of the scalar predictive quantizer. An estimate (or prediction) $\tilde{\mathbf{X}}_n$ of the input vector $\mathbf{X}_n$, is made from past observations of reconstructed vectors $\{\hat{\mathbf{X}}_k; \ k < n\}$. Typically the predictor is assumed to be of finite order m in the sense that it depends only on m past values, that is, the predictor has the form

$$\tilde{\mathbf{X}}_n = P(\hat{\mathbf{X}}_{n-1}, \hat{\mathbf{X}}_{n-2}, \cdots, \hat{\mathbf{X}}_{n-m}).$$

This is a *closed-loop* prediction of $\mathbf{X}_n$ and must be distinguished from the open-loop case developed for scalars in Chapter 4 and to be developed for vectors later in this chapter. In the open-loop problem, the observables are past values of the actual input sequence rather than digital reproductions.

The prediction is subtracted from the input vector forming the difference vector $\mathbf{e}_n = \mathbf{X}_n - \tilde{\mathbf{X}}_n$ which is then vector quantized to form $\hat{\mathbf{e}}_n = Q(\mathbf{e}_n)$. The reconstructed approximation $\hat{\mathbf{X}}_n$ to the original input vector $\mathbf{X}_n$ is then formed by adding the prediction to the quantized difference vector:

$$\hat{\mathbf{X}}_n = \hat{\mathbf{e}}_n + \tilde{\mathbf{X}}_n.$$

As in the case of scalar predictive quantization, we have the obvious but very important result:

Theorem 13.2.1 Fundamental Theorem of Predictive Vector Quantization: *The overall reproduction error in predictive vector quantization is equal to the error in quantizing the difference signal presented to the vector quantizer.*

Although the result is obvious, it is worthwhile to emphasize that it follows from the simple fact that

$$\mathbf{X}_n - \hat{\mathbf{X}}_n = (\mathbf{X}_n - \tilde{\mathbf{X}}_n) - (\hat{\mathbf{X}}_n - \tilde{\mathbf{X}}_n) = e_n - \hat{e}_n. \qquad (13.2.1)$$

This result holds regardless of the specific nature of the predictor, which can be linear or nonlinear. Predictors are usually designed to depend on only a finite number of past vectors, often only one. Note that if the predictor is trivial, that is, if it just produces a vector $\mathbf{0} = (0, 0, \cdots, 0)^t$ for any input, then the system reduces to ordinary memoryless VQ. Thus memoryless VQ can be seen as a special case of predictive VQ where the predictor has zero order.

As in the scalar case, separate gains for prediction and quantization can be defined. The *closed-loop prediction gain ratio* is defined by

$$G_{\text{clp}} = \frac{E[||\mathbf{X}_n||^2]}{E[||e_n||^2]} \qquad (13.2.2)$$

and the coding gain ratio (or SNR power ratio) of the vector quantizer is defined as

$$G_{\mathbf{Q}} = \frac{E[||e_n||^2]}{E[||e_n - \hat{e}_n||^2]}. \qquad (13.2.3)$$

Thus, from (13.2.1), (13.2.2), and (13.2.3), the overall signal-to-noise power ratio is given by

$$\frac{E(||\mathbf{X}_n||^2)}{E(||\mathbf{X}_n - \hat{\mathbf{X}}_n||^2)} = G_{\text{clp}} G_{\mathbf{Q}}$$

or in dB units,

$$\text{SNR}_{\text{sys}} = 10 \log_{10} G_{\text{clp}} + \text{SNR}_{\mathbf{Q}}, \qquad (13.2.4)$$

where SNR_{sys} is the overall signal-to-noise ratio and $\text{SNR}_{\mathbf{Q}} = 10 \log_{10} G_{\mathbf{Q}}$ is the signal-to-noise ratio of the quantizer.

In general, the signal-to-noise ratio achieved for a vector quantizer operating on the difference signal e_n for a particular codebook size N will be less than the signal-to-noise ratio achieved for the same size vector quantizer operating on the input signal $\mathbf{X}_n$. This is largely due to the fact that the effect of vector prediction is to produce an error vector e_n that has components which are much less correlated than those of $\mathbf{X}_n$. However, the correlation between successive input vectors is being efficiently exploited by the vector predictor. The prediction gain more than compensates for the slight drop in coding gain of the quantizer in the closed-loop environment and PVQ results in increased performance over memoryless VQ.

Predictive Vector Quantizer Design

The design of a predictive vector quantization scheme requires both a predictor and a VQ codebook design. The simplest approach to the design problem is to use the *open-loop* design methodology. With this approach, the predictor is designed based on the statistics of $\mathbf{X}_n$, generally by using empirical observations (a training sequence) of $\mathbf{X}_n$. The actual procedure to design a vector predictor from a given training sequence will be considered later. Once the predictor has been designed, the VQ codebook can also be designed using an open-loop approach. The codebook design is based on the ideal prediction residual

$$\mathbf{W}_n = \mathbf{X}_n - P(\mathbf{X}_{n-1}, \mathbf{X}_{n-2}, \cdots, \mathbf{X}_{n-m}),$$

where P, the predictor already designed, is indicated here as a finite memory predictor. An input training sequence $\{\mathbf{x}_n; n = -m, \cdots, 1, 2, \cdots, L\}$ can then be used to generate a residual training set i.e., a training set of open-loop prediction error vectors, $\{\mathbf{w}_n; n = 1, 2, \cdots, L\}$ which in turn can be used to design a VQ using the generalized Lloyd algorithm.

Combining the predictor with the quantizer then yields a complete PVQ system. The open-loop approach is the simplest design methodology since it permits one to design the predictor first without taking the quantizer into account. Then the quantizer is separately designed making use of this predictor only for the formation of the VQ training sequence.

Once the predictor and VQ codebook have been designed, the predictor is then used in the PVQ structure of Fig. 13.1, operating on the reconstructed signal $\hat{\mathbf{X}}_n$ to produce the prediction $P(\hat{\mathbf{X}}_{n-1}, \hat{\mathbf{X}}_{n-2}, \cdots, \hat{\mathbf{X}}_{n-m})$ of the next input, thereby "closing the loop." This predictor is not necessarily optimal for the given signal actually being used (the reproduction instead of the original), but it should provide performance close to the optimal if the reproduction is of sufficiently high quality. Also, the quantizer now operates on the closed-loop residual vectors whose statistics may differ from the open-loop residual for which it was designed. Hence the quantizer is not necessarily optimal. Again, if the reproduction sequence is of sufficiently high quality (i.e., if the quantizer resolution is sufficiently high) then the quantizer will be approximately optimal.

Two approaches for designing the vector quantizer in a closed-loop fashion for a given fixed predictor will be described here. Both either use or mimic the Lloyd algorithm, but neither approach inherits the optimality properties of the Lloyd iteration because of the feedback loop in the PVQ structure [86], [87], [58]. The first approach to VQ codebook design, called the *closed-loop design*, for a given predictor was proposed in [139]. The method coincides with the vector extension of a scalar predictive quantizer design method due to Y. Linde in unpublished work (cited in [58]). In this method, the open-loop predictor is first designed from the training data and this predictor henceforth remains unchanged. The codebook design is an iterative procedure that starts with the selection of an initial codebook. Then the PVQ system with a given "old" codebook encodes the training sequence. Next, a cluster of those residual vectors (formed in the closed-loop operation of the PVQ encoder) which were assigned to a particular code vector is generated; for each such cluster the centroid is computed and a new codebook is formed from these centroids. The new codebook is then substituted into the PVQ structure and the process continues with another encoding of the training set. This process is repeated until a stopping criterion is satisfied.

The basic algorithm for a given initial codebook, a given vector predictor P, and a given quantizer distortion measure $d(\cdot, \cdot)$ is summarized in Table 13.1.

Note that the given training sequence $\{x_n\}$ is used in each iteration to generate a residual sequence $\{e_n\}$, which depends on the previous codebook and itself functions as the effective training sequence for the design of the next codebook. The algorithm is, in effect, trying to design a codebook from a residual training sequence that depends on the codebook and the training sequence of residuals changes with each iteration. Thus, Step 1 is not in general optimal and the distortion need not decrease monotonically. In practice, however, it has been found to be an effective algorithm and a substantial coding gain can be achieved over the open-loop design approach described earlier.

The initial codebook for this algorithm can be designed using the open-loop approach, i.e., by applying the Lloyd algorithm on a training set of open-loop residual vectors. Alternatively, the codebook can be designed from scratch using the natural extension of the splitting method applied to the closed-loop algorithm.

Another closed-loop approach to designing the VQ is to use the fixed predictor P and an old codebook to generate a residual sequence (as in Step 1 above) by encoding the original sequence of input vectors $\{x_n\}$. The residual sequence so generated is then retained for subsequent use as the training sequence for the VQ design. Then the Lloyd algorithm is run on this sequence to produce a new codebook. This alternate method is simpler to implement than the closed-loop method since multiple PVQ encodings are avoided. Note that this approach can be viewed as somewhat less of a "closed-loop" design since only the training sequence is generated closed-loop and the codebook design is then run open-loop on the fixed training sequence. Hence we will refer to this method as the *semi-closed-loop* approach.

Table 13.1: **Closed-Loop VQ Design**

Step 0: Given a training sequence $\{\mathbf{x}_n; n = -m, \cdots, 1, 2, \cdots, L\}$, an initial codebook $\mathcal{C}_1$, and ϵ, set $D_0 = \infty$, $k = 1$.

Step 1: Encode the training sequence using the PVQ described by the vector predictor P and the codebook $\mathcal{C}_k = \{\mathbf{c}_l; l = 1, \cdots, N\}$. This results in the sequence of indices

$$u_n = \min_l {}^{-1} d(\mathbf{e}_n, \mathbf{c}_l)$$

where $\mathbf{e}_n = \mathbf{x}_n - \tilde{\mathbf{x}}_n$, and the corresponding reproductions are

$$\hat{\mathbf{x}}_n = \tilde{\mathbf{x}}_n + \mathbf{c}_{u_n},$$

where

$$\tilde{\mathbf{x}}_n = P(\hat{x}_{n-1}, .., \hat{x}_{n-m}).$$

For the average distortion D_k between the input and output of the PVQ system resulting from the encoding, test if $(D_{k-1} - D_k)/D_{k-1} < \epsilon$. If yes, stop. If no, continue to Step 2.

Step 2: Modify the codebook by replacing the old code vectors by the centroids of all residual vectors that map into them in Step 1, that is, replace the old $\mathbf{c}_l$ by

$$\mathbf{c}_l' = \min_{\mathbf{c}} {}^{-1} \sum_{n:u_n = l} d(\mathbf{e}_n, \mathbf{c}),$$

where the $\mathbf{e}_n$ are the same as in Step 1, that is, the $\mathbf{e}_n$ were formed by the PVQ operating with the old codebook. The collection of centroids constitutes $\mathcal{C}_{k+1}$. Set $k+1 \rightarrow k$. Go to Step 1 with the new codebook $\mathcal{C}_k$ becoming the old codebook.

The semi-closed-loop approach is in essence a natural modification of the open-loop approach. Instead of using an idealized residual formed by running the predictor on the input and forming a residual without any quantization, we form a more genuine residual by running the predictor on the quantized input to generate a training sequence. The algorithm is not truly closed-loop because this residual is then used for the improvement of the VQ residual codebook and experimental results confirm that the closed-loop design gives superior performance.

All of the above methods yield a VQ that is at least intuitively well-matched to the original predictor. Once completed, one can in principle attempt to improve the predictor for the given VQ. One way to accomplish this is to mimic the semi-closed-loop design approach used for the VQ: Use the old predictor and VQ to produce a residual sequence. This residual sequence can then be used as an empirical distribution for estimating an optimal predictor. One can then iterate these two operations, trying to find a good VQ for the predictor and vice-versa. A more theoretically sound approach to such joint optimization investigated in [58] is to define the general optimization problem in terms of the VQ partition and reproduction vectors and linear predictor coefficients and then apply optimization techniques such as the stochastic gradient and coordinate descent methods of adaptive signal processing [334]. We do not pursue these more complicated approaches because in several applications they have not proved to be worth the effort: The simple open-loop predictor design followed by a closed-loop VQ design yields coders that have nearly identical performance to those produced by the jointly optimized design [58]. Note that the closed-loop design method of Table 13.1 can be viewed as a special case of the jointly optimized design of [58] where the predictor improvement step is omitted. In general, the closed-loop design has been found to yield a useful improvement over the open-loop method.

Finally we return to the issue of what form of predictor to use in the open-loop design. The simplest form is a finite memory linear predictor, the vector extension of the finite order scalar linear predictors of Chapter 4. Most of the remainder of this chapter is devoted to the theory and design of (open-loop) vector linear predictors. Then we briefly present a VQ approach to the design of nonlinear predictors. The final section presents some design examples.

13.3   Vector Linear Prediction

Let $\{\mathbf{X}_n\}$ denote a stationary random sequence of k-dimensional vectors with finite second moment or "average energy," i.e., $E(||\mathbf{X}_n||^2) < \infty$.

A finite-memory vector linear predictor with memory size m is defined by a set of m linear predictor coefficient matrices, $\mathbf{A}_j$, (also called *coefficient matrices* for short) for $j = 1, 2, 3, \cdots, m$. The predictor, also called an *mth order predictor*, forms an estimate $\tilde{\mathbf{X}}_n$ of $\mathbf{X}_n$ from the m prior observations $\mathbf{X}_{n-1}, \mathbf{X}_{n-2}, \cdots, \mathbf{X}_{n-m}$, according to

$$\tilde{\mathbf{X}}_n = -\sum_{j=1}^m \mathbf{A}_j \mathbf{X}_{n-j} \qquad (13.3.1)$$

where each $\mathbf{A}_j$ is a $k \times k$ matrix. Note that this predictor is using every component of each of the m prior vectors of the process to estimate any one component of the current vector.

As usual, a measure of performance of the predictor is needed in order to define and determine an optimal predictor. The *prediction residual vector* or *error vector* is defined as

$$\mathbf{e}_n = \mathbf{X}_n - \tilde{\mathbf{X}}_n \qquad (13.3.2)$$

and we adopt the usual mean squared error measure given by

$$D_m = E(\|\mathbf{e}_n\|^2). \qquad (13.3.3)$$

A vector predictor for a given memory size m will be said to be *optimal* if it minimizes this measure over all possible sets of m coefficient matrices. Later, in discussing the design of predictors from empirical data, we shall assume ergodicity and make use of a time average of the squared error norm as our performance measure.

As in the scalar case, it is convenient to have a measure of performance that relates the mean square error to the energy of the vector to be predicted. We define the *prediction gain ratio*, also called the *open-loop prediction gain ratio* to distinguish from the closed-loop case to be described later, by

$$G_m = \frac{E(\|\mathbf{X}_n\|^2)}{D_m} \qquad (13.3.4)$$

where m indicates the memory size, and the *prediction gain* or *open-loop prediction gain* by

$$10 \log_{10} G_m,$$

measured in dB units. With this background we can now consider the theory of optimal vector prediction for stationary random sequences of vectors.

To describe the statistical character of the random vector sequence, we need to specify the second-order statistics of the process $\{\mathbf{X}_n\}$. Define the $k \times k$ correlation matrix, $\mathbf{R}_{ij}$ by

$$\mathbf{R}_{ij} = E[\mathbf{X}_{n-i} \mathbf{X}_{n-j}^t]. \qquad (13.3.5)$$

Note that $\mathbf{R}_{ij} = \mathbf{R}_{ji}^t$. Also, by stationarity, $\mathbf{R}_{ij}$ will depend only on the difference, $i - j$ and when $i = j$ it reduces to $\mathbf{R}_k$, the usual kth order correlation matrix of $\mathbf{X}_n$, defined in Chapter 4. Frequently, it can be assumed that the mean of the process is zero, i.e., $E\mathbf{X}_n = \mathbf{0}$ where $\mathbf{0}$ is the k-dimensional vector whose components are all zero. In that case, the correlation matrices become covariance matrices. Later we consider the case of nonzero mean.

Nondeterministic Vector Processes

In solving for the optimal vector predictor, we shall make the assumption that the sequence $\{\mathbf{X}_n\}$ has the property of being nondeterministic. This property was defined previously for scalar sequences and here we extend the definition to vector sequences.

A random sequence of vectors is said to be *deterministic* if for any integer m, it is possible to find a set of m vectors $\mathbf{a}_1, \mathbf{a}_2, \cdots, \mathbf{a}_m$ that are not all zero (not all equal to the zero vector) for which

$$\mathbf{X}_1^t \mathbf{a}_1 + \mathbf{X}_2^t \mathbf{a}_2 + \cdots + \mathbf{X}_m^t \mathbf{a}_m = 0. \tag{13.3.6}$$

The random sequence is said to be *nondeterministic* if it is not deterministic. Thus, for a nondeterministic process it is impossible to exactly specify one component of one vector by a linear combination of the remaining components of that vector and the components of a finite number of adjacent vectors. It follows that for a nondeterministic process if

$$\sum_{i=1}^{m} \mathbf{A}_i \mathbf{X}_i = \mathbf{0} \tag{13.3.7}$$

(with probability one), then each $\mathbf{A}_i = \mathbf{0}$, where $\mathbf{0}$ is the $k \times k$ matrix of all zeros, for otherwise at least one row of this matrix equation will violate the condition (13.3.6).

Optimal First Order Vector Linear Predictor

It is worth pointing out at this time that if one has a stationary vector process and if one is interested only in the first order linear predictor (i.e., the prior vector in the sequence is used to estimate the current vector), then the optimal solution follows immediately from Theorem 4.2.2 by replacing $\mathbf{Y}$ with $\mathbf{X}_n$, $\mathbf{X}$ with $\mathbf{X}_{n-1}$ and $\mathbf{A}$ with $-\mathbf{A}_1$ to obtain

$$\mathbf{A}_1 = -E[\mathbf{X}_n \mathbf{X}_{n-1}^t] \mathbf{R}_{11}^{-1} = -\mathbf{R}_{01} \mathbf{R}_{11}^{-1}. \tag{13.3.8}$$

The correlation matrices can be estimated from the same training sequence used to design the codebook in a PVQ system. This simple vector predictor design is sufficient in many applications. A more sophisticated understanding of optimal vector linear prediction is useful in other applications such as the predictive coding of LPC parameter vectors.

Optimal Vector Linear Predictor

To find the optimal vector predictor for $m \geq 1$ one could extend the above result or use the calculus approach and examine how the mean squared error D_m depends on each component of each coefficient matrix, then differentiate D_m with respect to each such coefficient, set the derivative equal to zero, and thereby obtain a set of mk^2 linear equations in the mk^2 unknown components of the m coefficient matrices.

A much simpler alternative is to use the orthogonality principle of Theorem 4.2.4 for each component of the estimate $\tilde{X}_n$. Note that the ith component $\tilde{X}_{ni}$ of $\tilde{X}_n$ depends only on the ith row of each coefficient matrix. Hence the estimate $\tilde{X}_{ni}$ of the ith component X_{ni} of X, can be optimized separately for each i. The orthogonality principle says that the ith component e_{ni} of the prediction error vector e_n must be orthogonal to each observable random variable. The observable random variables in this case are all the components of each of the m observed random vectors. Thus, we can express the orthogonality condition as

$$E[e_{ni}X_{n-j}] = 0 \text{ for } j = 1, 2, \cdots, m; \ i = 1, 2, \cdots, k, \qquad (13.3.9)$$

or, more compactly,

$$E[e_n X_{n-j}^t] = 0 \text{ for } j = 1, 2, \cdots, m. \qquad (13.3.10)$$

Thus, by substituting (13.3.2) and (13.3.1) into (13.3.10), we obtain

$$E([X_n + \sum_{\nu=1}^{m} A_\nu X_{n-\nu}] X_{n-j}^t) = 0 \qquad (13.3.11)$$

or

$$R_{0j} = -\sum_{\nu=1}^{m} A_\nu R_{\nu j} \text{ for } j = 1, 2, \cdots, m. \qquad (13.3.12)$$

Note that (13.3.12) is the matrix analog of the Wiener-Hopf equation or normal equation introduced in Chapter 4. In particular, in the simplest case of vector prediction with memory size one, that is, if X_n is to be predicted from the previous input X_{n-1}, the equations reduce to

$$A_1 R_{11} = -R_{01}. \qquad (13.3.13)$$

Since the process is nondeterministic, the components of the vector $\mathbf{X}_n$ are linearly independent so that from Chapter 4, $\mathbf{R}_{11}$, the correlation matrix previously denoted as $\mathbf{R}_k$, is positive definite and the optimal predictor coefficient matrix is given explicitly by

$$\mathbf{A}_1 = -\mathbf{R}_{01}\mathbf{R}_{11}^{-1}, \tag{13.3.14}$$

which agrees with (13.3.8) as derived from Theorem 4.2.2.

In principle, the general case of memory size m can be solved by a single matrix inversion that is exactly equivalent to solving the system of equations (13.3.12). Let

$$\mathbf{Y}_n^t = [\mathbf{X}_n^t \ \mathbf{X}_{n-1}^t \cdots \mathbf{X}_{n-m+1}^t], \tag{13.3.15}$$

and consider the problem of finding an optimal first order predictor for $\mathbf{Y}_n$ from observation of $\mathbf{Y}_{n-1}$. The optimal linear predictor then has the form $\hat{\mathbf{Y}}_n = -\mathbf{B}\mathbf{Y}_n$. It is easy to see from the orthogonality principle (or from the result of Theorem 4.2.2) that the optimal $mk \times mk$ matrix $\mathbf{B}$ is given by

$$\mathbf{B} = \begin{bmatrix} \mathbf{A}_1 & \mathbf{A}_2 & \cdots & \cdots & \mathbf{A}_m \\ -\mathbf{I} & 0 & \cdots & \cdots & 0 \\ 0 & -\mathbf{I} & \cdots & 0 & 0 \\ \cdots & \cdots & \ddots & \vdots & \vdots \\ 0 & 0 & \cdots & -\mathbf{I} & 0 \end{bmatrix} \tag{13.3.16}$$

where $\mathbf{I}$ is the $k \times k$ identity matrix and $\mathbf{A}_i$ are the optimal coefficient matrices given by (13.3.12).

Note that (13.3.12) can be expressed as a single matrix equation:

$$\begin{bmatrix} \mathbf{R}_{11} & \mathbf{R}_{12} & \cdots & \mathbf{R}_{1m} \\ \mathbf{R}_{21} & \mathbf{R}_{22} & \cdots & \mathbf{R}_{2m} \\ \vdots & \vdots & \ddots & \vdots \\ \mathbf{R}_{m1} & \mathbf{R}_{m2} & \cdots & \mathbf{R}_{mm} \end{bmatrix} \begin{bmatrix} \mathbf{A}_1^t \\ \mathbf{A}_2^t \\ \vdots \\ \mathbf{A}_m^t \end{bmatrix} = - \begin{bmatrix} \mathbf{R}_{10} \\ \mathbf{R}_{20} \\ \vdots \\ \mathbf{R}_{m0} \end{bmatrix}, \tag{13.3.17}$$

or more compactly, $\mathcal{R}\mathbf{A}^t = -\mathbf{V}$, where $\mathcal{R}$, $\mathbf{A}$, and $\mathbf{V}$ are defined to correspond to (13.3.17).

The matrix $\mathcal{R}$ is a square $km \times km$ "supermatrix" and it is not in general Toeplitz. Note also that $\mathcal{R}$ is the correlation matrix of the mk-dimensional random vector $\mathbf{Y}_n$. It is therefore a positive semi-definite matrix and will be positive definite and hence invertible if the random vector sequence $\mathbf{X}_n$ is nondeterministic. (See Problem 13.3.) In this case, we can solve (13.3.17) explicitly for the coefficient matrices by inverting this matrix. However, this is typically a very high complexity operation and direct inversion could suffer from numerical instability. Fortunately, a generalized version of the Levinson-Durbin algorithm exists which is computationally much more efficient for finding the coefficient matrices for vector linear predictors in the general case of mth order stationary k dimensional vector sequences [335], [331]. The algorithm can be derived from the concept of backward prediction of a vector sequence. The ideas are very similar to the scalar case and directly generalize the backward linear prediction theory described in Chapter 4 to the case of a vector random sequence. The distortion, and hence the prediction gain, can also be computed at each iteration, providing a simple way to determine how the performance improves as the memory size increases.

Once an optimal predictor is determined, it is useful to have a formula for the mean squared prediction error, D_m. From the definition (13.3.3) and from the orthogonality condition (13.3.11), we can express D_m in the form

$$D_m = E[e_n^t \mathbf{X}_n] = E[||\mathbf{X}_n||^2] + \sum_{i=1}^{m} E[\mathbf{X}_n^t \mathbf{A}_j \mathbf{X}_{n-j}]$$

$$= \mathbf{R}_{00} + \mathrm{Tr}(\sum_{i=1}^{m} \mathbf{A}_j \mathbf{R}_{j0}), \qquad (13.3.18)$$

where as usual Tr denotes the trace of a matrix and we have used the property that for a $k \times k$ matrix $\mathbf{B}$ and two k-dimensional vectors $\mathbf{u}$ and $\mathbf{v}$,

$$\mathbf{u}^t \mathbf{B} \mathbf{v} = \mathrm{Tr}(\mathbf{B}\mathbf{v}\mathbf{u}^t).$$

very nice!

Alternatively, we can express the mean squared prediction residual D_m as $D_m = \mathrm{Tr}(\mathbf{R}_e^+)$ where $\mathbf{R}_e^+$ is the correlation matrix of the prediction error vector for an optimal predictor, i.e.,

$$\mathbf{R}_e^+ = E[e_n e_n^t] = E[e_n \mathbf{X}_n^t]$$

$$= \mathbf{R}_{00} + \sum_{i=1}^{m} \mathbf{A}_j \mathbf{R}_{j0}$$

$$= \sum_{i=0}^{m} \mathbf{A}_j \mathbf{R}_{j0} \qquad (13.3.19)$$

where we have used the orthogonality condition (13.3.10) and for convenience, we define the matrix $\mathbf{A}_0 \equiv \mathbf{I}$. It then follows that

$$D_m = \mathrm{Tr}(\mathbf{R}_e^+) = \mathrm{Tr}(\sum_{i=0}^{m} \mathbf{A}_j \mathbf{R}_{j0}).$$

Minimum Delay Property

The vector predictor can be regarded as a multichannel filter operating on the input vector process. Thus the mth order predictor has transfer function

$$\mathbf{P}(z) = -\sum_{j=1}^{m} \mathbf{A}_j z^{-j}$$

and the corresponding prediction error filter is

$$\mathbf{H}(z) = \mathbf{I} + \sum_{j=1}^{m} \mathbf{A}_j z^{-j}.$$

The filter $\mathbf{H}(z)$ is said to be *minimum phase* or *minimum delay* if all the roots of $\det \mathbf{H}(z) = 0$ lie inside the unit circle in the z plane. This condition insures that the inverse filter $\mathbf{H}^{-1}(z)$ will be stable. Indeed this condition is needed for stability of the feedback loop in the PVQ system. As in the scalar case, it can be shown that the optimal vector predictor of order m has a minimum phase prediction error filter if the process is stationary and nondeterministic. We now prove this for the special case of a first order predictor. Then we show that the general result for mth order can be deduced from the first order result.

In the first order predictor, the prediction error filter is simply $\mathbf{H}(z) = \mathbf{I} + \mathbf{A}_1 z^{-1}$. Hence it suffices to show that the eigenvalues of the coefficient matrix $\mathbf{A}_1$ are less than unity in magnitude.

Theorem 13.3.1 *If $\mathbf{A}_1$ is the optimal coefficent matrix for first order prediction of a stationary nondeterministic process $\mathbf{X}_n$, then the eigenvalues of $\mathbf{A}_1$ are less than unity in magnitude.*

Proof: Let $\mathbf{b}$ be an eigenvector of $\mathbf{A}^t$ with corresponding eigenvalue λ. From the orthogonality principle, the optimality of $\mathbf{A}_1$ implies that the prediction error $\mathbf{e}_n$ is uncorrelated with $\mathbf{X}_{n-1}$. Hence, from $\mathbf{X}_n = \mathbf{e}_n - \mathbf{A}_1 \mathbf{X}_{n-1}$, we have

$$E[(\mathbf{b}^t \mathbf{X}_n)^2] = E[(\mathbf{b}^t \mathbf{e}_n)^2] + E[(\mathbf{b}^t \mathbf{A}_1 \mathbf{X}_{n-1})^2] > \lambda^2 E[(\mathbf{b}^t \mathbf{X}_{n-1})^2]$$

where the inequality follows from the fact that the process is nondeterministic which implies that $E[(\mathbf{b}^t \mathbf{e}_n)^2]$ is strictly positive. This gives the inequality

$$(1 - \lambda^2) E(\mathbf{b}^t \mathbf{X}_n)^2 > 0$$

which implies that $|\lambda| < 1$.   □

This simple proof is due to E. Yair.

The minimum phase property of the mth order optimal linear vector predictor can be established by applying the above result to the first-order prediction case for the "supervector" $\mathbf{Y}_n$ defined by (13.3.15). From the above theorem, the optimal predictor coefficient matrix $\mathbf{B}$, as given in (13.3.16), must have its eigenvalues strictly inside the unit circle. By examining the matrix $\mathbf{B} - z\mathbf{I}_{mk}$, where $\mathbf{I}_{mk}$ is the $mk \times mk$ identity matrix, it can be shown that

$$\det[\mathbf{B} - z\mathbf{I}_{mk}] = z^{mk} \det[-\mathbf{H}(z)] \qquad (13.3.20)$$

from which it follows that $\mathbf{H}(z)$ is a minimum phase transfer function. (See Problem 4.2.)

Blocked Scalar Processes

An important special class of vector prediction problems arises when a scalar valued random sequence is partitioned into blocks of k samples with each block defining a k-dimensional vector. Let $\{s(n)\}$ denote a scalar-valued stationary random sequence. The corresponding *blocked scalar sequence* $\{\mathbf{X}_n\}$, a sequence of k-dimensional vectors, is defined by

$$\mathbf{X}_n = [s(nk - k + 1), s(nk - k + 2), \cdots, s(nk)]^t. \qquad (13.3.21)$$

In this case, the elements of each correlation matrix $\mathbf{R}_{ij}$ are particular autocorrelation values of the scalar process $\{s(n)\}$. Thus, denoting $r_i = E[s(n)s(n - i)]$, we find in particular that,

$$\mathbf{R}_{01} = \begin{bmatrix} r_k & r_{k-1} & \cdots & r_1 \\ r_{k+1} & r_k & \cdots & r_2 \\ \vdots & \vdots & \ddots & \vdots \\ r_{2k-1} & r_{2k} & \cdots & r_k \end{bmatrix}. \qquad (13.3.22)$$

It can be observed that the matrix $\mathcal{R}$ whose structure is shown in (13.3.17) is now a block Toeplitz matrix.

The equations for the optimal predictor for a blocked scalar process have some special properties not possessed in the general case of vector processes. This results in a simplification of the vector Levinson-Durbin algorithm since the backward coefficient matrices are simply determined from the forward coefficient matrices in this case.

Vector prediction of a blocked scalar process has some interesting features. Consider for simplicity the case of first order vector prediction. In this case, we are simultaneous predicting the next k samples of the scalar process from observation of the previous k samples. Clearly the nearby future samples can be predicted better than the more remote future samples.

Increasing the vector dimension k has two conflicting effects: it allows more past samples to be used to help improve the prediction quality but it imposes a harder burden on the predictor by having to predict further into the future. Experimental results show that the prediction gain first increases with k and then gradually decreases as k continues to grow. In general, scalar prediction will achieve a better prediction gain when its memory size m is equal to the vector dimension k used in first order vector prediction for the same underlying process. This is not surprising since scalar prediction can only benefit from an increase in prediction order and it does not suffer the disadvantage of having to predict more than one step into the future. Nevertheless, vector prediction of a blocked scalar process can be beneficial in PVQ if the vector dimension is reasonably chosen. The reason follows from (13.2.4); the coding gain offered by VQ over scalar quantization compensates for the reduced prediction gain offered by vector prediction and the overall result is still quite beneficial for moderate vector dimensions. Some specific results on speech coding with PVQ is reported in [87].

13.4 Predictor Design from Empirical Data

In practice it is usually necessary to use empirical data to describe the statistical character of a sequence of random vectors since adequate theoretical models are either lacking or mathematically intractable for use in predictor design. Suppose we have a finite set of consecutive observations $\mathbf{X}_n$ for $n \in \mathcal{M} = \{0, 1, 2, \cdots, L\}$ of a k-dimensional stationary vector process, and we wish to find an mth order vector linear predictor that will be nearly optimal for the process.

As in the scalar case of linear predictor design, described in Chapter 4, we choose our performance objective to be based on the minimization of the error incurred in predicting values of the observed vectors from m prior vectors. Specifically, we wish to find a predictor that minimizes

$$D = \sum_{\mathcal{N}} E(\|\mathbf{e}_n\|^2) = \sum_{n \in \mathcal{N}} E(\|\mathbf{X}_n + \sum_{j=1}^{m} \mathbf{A}_j \mathbf{X}_{n-j}\|^2) \qquad (13.4.1)$$

where $\mathcal{N}$ is a finite interval to be specified later. Note that the summation over this interval replaces the expectation operation used in the previous sections.

Applying the orthogonality principle to this performance measure, we obtain the necessary condition for optimality of the coefficient matrices:

$$\sum_{n \in \mathcal{N}} \{\mathbf{e}_n \mathbf{X}_{n-j}^t\} = 0 \text{ for } j = 1, 2, \cdots, m \qquad (13.4.2)$$

paralleling the result (13.3.10). This simplifies to

$$\tilde{\mathbf{R}}_{0j} = -\sum_{\nu=1}^{m} \mathbf{A}_\nu \tilde{\mathbf{R}}_{\nu j} \text{ for } j = 1, 2, \cdots, m, \qquad (13.4.3)$$

where

$$\tilde{\mathbf{R}}_{ij} = \sum_{\mathcal{N}} \mathbf{X}_{n-i} \mathbf{X}_{n-j}^t \qquad (13.4.4)$$

are the empirical correlation matrices. Assuming the process $\mathbf{X}_n$ is stationary and ergodic, for sufficiently large observation intervals, $\mathcal{N}$, $\tilde{\mathbf{R}}_{ij}$ normalized by the number of observations $\|\mathcal{N}\|$, will be very close to the correlation matrix $\mathbf{R}_{ij}$.

We now consider two different computational methods based on two distinct choices of the set $\mathcal{N}$ used in the error measure (13.4.1). For a more detailed discussion of this topic, see [62].

Autocorrelation Method

In this method, we use a rectangular window on the observed process and replace the true value of $\mathbf{X}_n$ by zero for all values of n outside of the observation interval, $\mathcal{M}$:

$$\mathbf{X}_n = 0 \text{ for } n \notin \mathcal{M}. \qquad (13.4.5)$$

Thus, the error measure (13.4.1) is summed over the set

$$\mathcal{N} = \mathcal{M} = \{0, 1, 2, \cdots, L\}. \qquad (13.4.6)$$

Note that the first $m-1$ terms in the error sum (13.4.1) are not producing correct prediction errors for the true process since they depend on the values of $\mathbf{X}_n$ for n outside of the observable window and such values of $\mathbf{X}_n$ are assigned the value zero. With this definition, (13.4.3) becomes

$$\tilde{\mathbf{R}}_{ij} = \sum_{n=-\infty}^{\infty} \mathbf{X}_{n-i} \mathbf{X}_{n-j}^t \qquad (13.4.7)$$

and it can be seen that $\tilde{\mathbf{R}}_{ij}$ is in this case a function of the difference $(i-j)$ and satisfies the symmetry property that $\tilde{\mathbf{R}}_{ij}^t = \tilde{\mathbf{R}}_{ji}$. In particular for $i = j$, we find that $\tilde{\mathbf{R}}_{ii}$ is a Toeplitz matrix. Furthermore, the supermatrix $\tilde{\mathcal{R}}$ corresponding to the empirical estimate of $\mathcal{R}$, having ijth submatrix $\tilde{\mathbf{R}}_{ij}$, is a block-Toeplitz matrix.

The coefficient matrices $\mathbf{A}_i$ can be determined by solving (13.4.3) and because of the properties of $\tilde{\mathbf{R}}_{ij}$ resulting from the autocorrelation method, the vector Levinson-Durbin algorithm can be used to efficiently compute the optimal predictor.

Covariance Method

In this method, the range of values used for the error measure is

$$\mathcal{N} = \{m, m+1, m+2, \cdots, L\} \tag{13.4.8}$$

so that each term in the error sum (13.4.1) depends only on the observed values of the process $\mathbf{X}_n$. This is a more accurate indicator of the predictor performance since it avoids the need to assume zero values for the observables outside of the observable region. In this case the correlation matrices $\tilde{\mathbf{R}}_{ij}$ are given by

$$\tilde{\mathbf{R}}_{ij} = \sum_{n=m}^{L} \mathbf{X}_{n-i} \mathbf{X}_{n-j}^t, \tag{13.4.9}$$

and unfortunately they do not have the symmetry found in the autocorrelation method. Consequently, the normal equations (13.4.3) cannot be solved by the vector Levinson-Durbin algorithm and must be solved by directly using

$$\tilde{\mathbf{R}}\mathbf{A}^t = -\tilde{\mathbf{V}}. \tag{13.4.10}$$

Also $\tilde{\mathbf{R}}$ is symmetric and under certain conditions it can be shown to be nonnegative definite with probability one. Hence, it can be solved in a numerically stable way using the Cholesky decomposition.

13.5   Nonlinear Vector Prediction

Optimal nonlinear predictors are generally very complicated and they cannot be easily designed from empirical data, except when the data is known to be Gaussian and the optimal nonlinear predictor is therefore linear. Some researchers have attempted to design nonlinear predictors by selecting *ad hoc* parametric models for the nonlinear operation and using training data to estimate the parameters. These methods are occasionally useful, but they are generally suboptimal due to the limited capabilities of the model. However, there is one approach that is not based on a parametric model of the nonlinearity yet it allows design based on training data. This method is based on the use of VQ and is therefore particularly appropriate for inclusion in this book. (See the closely related discussion of NLIVQ in Chapter 12.)

Suppose we wish to predict a k-dimensional vector $\mathbf{Y}$ from an n-dimensional vector $\mathbf{X}$. In Chapter 4 we considered this problem and found that the solution can be applied to solving a large variety of estimation or prediction problems. For the general nonlinear case Theorem 4.2.1 showed that the best MMSE prediction (or estimate) of a random vector $\mathbf{Y}$ from observation of $\mathbf{X}$ is $\hat{\mathbf{Y}} = E(\mathbf{Y}|\mathbf{X})$. In other words, the optimal prediction of $\mathbf{Y}$ when $\mathbf{X} = \mathbf{x}$ is given by $\hat{\mathbf{Y}} = g(\mathbf{x})$, where the function $g(\cdot)$ is given by $g(\mathbf{x}) = E(\mathbf{Y}|\mathbf{X} = \mathbf{x})$. Unfortunately, the evaluation of this function is generally intractable and in most applications the joint statistics of $\mathbf{X}, \mathbf{Y}$ are not known analytically. This difficulty is one of the major reasons that most applications use linear prediction methods and give little or no attention to nonlinear prediction, even when the statistics are clearly not Gaussian. In this section, we describe a mechanism for designing nearly optimal nonlinear predictors from training data.

The main idea of this approach is to assume we can adequately estimate $\mathbf{Y}$ from *quantized* observations of $\mathbf{X}$ once we have a vector quantizer with sufficiently high resolution that is optimized for quantization of $\mathbf{X}$. Suppose that we have already designed a vector quantizer Q with encoder $\mathcal{E}$, decoder $\mathcal{D}$, and codebook $\mathcal{C}$ of size N for the n-dimensional vector $\mathbf{X}$. It then follows from Theorem 4.2.1, that the optimal prediction $\hat{\mathbf{Y}}$ of $\mathbf{Y}$ from the quantized observation vector $Q(\mathbf{X})$, itself a random vector of dimension n, is given by

$$\hat{\mathbf{Y}} = E[\mathbf{Y}|Q(\mathbf{X})], \tag{13.5.1}$$

which takes on the values

$$\mathbf{y}_i = E[\mathbf{Y}|\mathcal{E}(\mathbf{X}) = i]. \tag{13.5.2}$$

Since there is only a finite number N of possible quantized observations, we can precompute and store all possible *prediction vectors* $\mathbf{y}_i$ in a *predictor codebook*. The resulting predictor codebook contains N vectors $\mathbf{y}_i$ of dimension k.

The operation of the nonlinear predictor is illustrated in Figure 13.2. The vector $\mathbf{X}$ is applied to the VQ encoder $\mathcal{E}$ producing the index I as a result of a nearest-neighbor search through the VQ codebook $\mathcal{C}$ of dimension n. The index is then applied to a predictive decoder that is simply a VQ decoder which performs a table-lookup to find the k-dimensional prediction vector at the Ith location of the codebook producing the final output $\hat{\mathbf{Y}} = \mathbf{y}_I$.

Although the structure of Fig. 13.2 appears very similar to a standard VQ system, the encoder and decoder codebooks differ and in particular they contain vectors with different dimensions. Furthermore, the transfer of the index I from encoder to the predictive decoder need not represent the transmission of information over a communication channel as it usually does in ordinary VQ. The entire structure can be viewed as a single black box that operates on the n-dimensional input vector to generate a k-dimensional

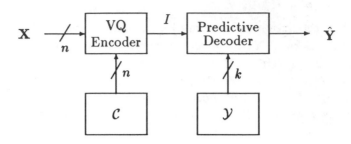

Figure 13.2: Nonlinear Predictor

prediction vector. This operation can be used in PVQ in place of a linear predictor.

The design of the nonlinear predictor from a training set is particularly simple. Assume that the VQ codebook has already been designed. Let $\{\mathbf{x}^{(j)}, \mathbf{y}^{(j)};\ j = 1, \cdots, M\}$ be a training set of M pairs of vectors that represents the statistics of the source. Let

$$\mathcal{R}_i = \{\mathbf{x}^{(j)} :\ \mathcal{E}(\mathbf{x}^{(j)}) = i\}. \tag{13.5.3}$$

The ith prediction vector, $\mathbf{y}_i$, is given by (13.5.2) where the expectation is now evaluated over the training set, so that:

$$\mathbf{y}_i = \frac{1}{|\mathcal{R}_i|} \sum_{j\,:\, \mathbf{x}^{(j)} \in \mathcal{R}_i} \mathbf{y}^{(j)} \tag{13.5.4}$$

where $|\mathcal{R}_i|$ is the size of the set $\mathcal{R}_i$. Thus, the prediction codebook design does not require an iterative algorithm, unlike the generalized Lloyd algorithm, and is a trivial computational task.

This approach to nonlinear predictor design leads to suboptimal VQ designs, but the degree of suboptimality is under the designer's control. There are two sources of suboptimality: the finite size N of the VQ codebook used to represent the observation vector $\mathbf{X}$, and the finite size M of the training set used to design the prediction codebook. It is intuitively evident that as N and M both approach infinity the predictor will approach the optimal nonlinear predictor.

One special case of nonlinear prediction of a vector from another vector is scalar mth order prediction, where the input vector $\mathbf{X}$ consists of m past samples of the speech waveform and the prediction $\mathbf{Y}$ has dimension one, being the next waveform sample to be predicted. In nonlinear scalar prediction of speech, this scheme was found to be very effective for whitening the spectrum of the residual far beyond what is possible with the same order linear predictor [324]. This study validated the method by testing its performance on a non-Gaussian moving average process for which the optimal nonlinear predictor is known. The method was also shown to be effective in enhancing the performance of DPCM. The basic concept of this VQ approach to nonlinear prediction was described in a more general context in Chapter 12. See also [145].

13.6 Design Examples

Figures 13.3 and 13.4 demonstrate the SNR performance of PVQ in comparison with memoryless VQ for the same Gauss-Markov and sampled speech sources described in Chapter 11. The Gauss-Markov example (with a correlation coefficient of 0.9) was run using the jointly optimized stochastic gradient technique of [58], but the closed-loop VQ design (with open-loop predictor design) of [86] [87] yields similar results. It should be noted that the optimal performance for the rate of 1 bit per sample as given by Shannon's rate-distortion function is 13.2 dB.

The results for sampled speech are shown for both the closed-loop VQ design (Table 13.1 of [86] [87]), denoted PVQ2 in the figure, and the jointly optimized stochastic gradient algorithm of [58], denoted PVQ1 in the figure. Results are presented for both first order and second order predictors for the speech source. Note that the performance of both design algorithms is quite close, which favors the simpler open-loop predictor design approach. The intuition is that a good quantizer can make up for any inaccuracies of the predictor. PVQ provides a performance improvement of about 0.7 dB for first order predictors to about 1 dB or more for second order predictors in this example.

Another interesting application of PVQ is for the interframe coding of linear prediction coefficients in LPC based speech coders where every frame (20 to 30 ms) requires the computation and transmission of a vector of LPC parameters. PVQ was successfully applied to this application by designing a one-step vector predictor for each of several statistically defined classes and adaptively selecting one of the predictors in each frame. The prediction residual vector was quantized using a weighted MSE distortion measure suited to the particular parameter set [346]. More recently enhanced performance for higher resolution applications was reported with this method by quantizing the residual vector with "split VQ" where the first 4 components and last 6 components of the 10-dimensional residual vector are treated as separate subvectors [325]. The specific LPC parameters used in these studies are line spectral frequencies (LSFs) also known as *line spectral pairs* (LSPs). They are determined from the reflection coeffi-

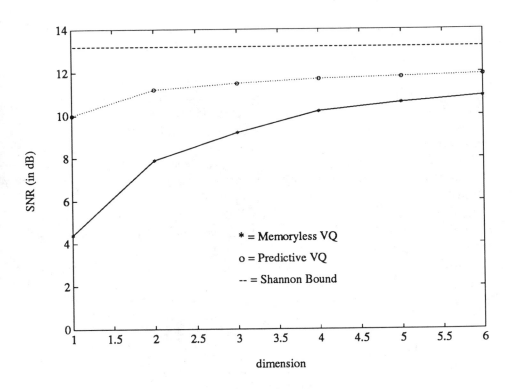

Figure 13.3: PVQ vs. VQ for a Gauss-Markov Source

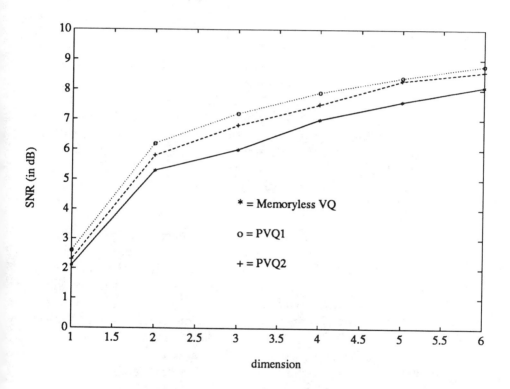

Figure 13.4: PVQ vs. VQ for Sampled Speech

cients and provide an equivalent but efficient representation of the optimal linear predictor for a given frame of speech.

One of the earliest PVQ schemes was developed for images by Hang and Woods [175]. The following examples provide a particularly simple application of PVQ to the USC and MRI images considered previously. The vector predictors use only a small number of selected pixels to form a linear prediction for each pixel in the block to be coded. The linear predictors are designed using a simple vector Wiener-Hopf equation based on the sample autocorrelation of the original pixels in the training sequence. Thus the PVQ design is open-loop for the predictor and closed-loop for the vector quantizer. The linear predictors were chosen here to be extremely simple. Each pixel in the block to be encoded is predicted using only a few nearby pixels in the adjacent blocks. The pixels being used for the prediction are indicated in Figures 13.5 and 13.6 for the USC and MRI images, respectively. For the USC images, pixel Y is predicted using pixels X_1 through X_5 from adjacent subblocks. The prediction is based on the two pixels in the same row in the subblock to the left, the two pixels in the same column in the subblock above, and the single pixel in the lower right-hand corner of the block to the upper left. The high correlation of the images resulted in negligible improvement if more pixels were used in the prediction. For the MRI images a smaller base for prediction was found to suffice. Here pixel Y is predicted using pixels X_1, X_2, and X_3 from adjacent subblocks. In these applications, PVQ is not being used in the general way it has been described in this chapter. Rather than predict the current block by a linear combination of observable blocks, only selected pixels from the neighboring blocks are chosen as observables according to heuristic knowledge of which ones are important for this purpose. In some cases, scalar predictors are designed for each pixel to be predicted. This is of course a special case of the general form of PVQ which reduces complexity by eliminating the use of unneeded observables (e.g., remote pixels in adjacent blocks).

The simple vector Wiener-Hopf equations of Theorem 4.2.2 were used by computing the required cross and autocorrelations as sample correlations from the training sequence. The use of prediction provides approximately 2 dB improvement in PSNR for the USC image and approximately 4 dB improvement for the MR image. A natural observation is that prediction can help by varying degrees depending on the nature of the image.

Figures 13.7 and 13.8 shows the relative SNR performance of memoryless VQ and PVQ designed using the simple open-loop predictor and VQ design technique for the USC data base test image and the magnetic resonance test image of Chapter 11, respectively. Figure 13.9 shows the corresponding images for Lenna at .3125 bpp.

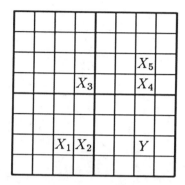

Figure 13.5: Prediction pixels for USC images

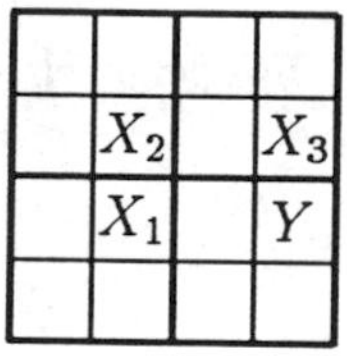

Figure 13.6: Prediction pixels for MRI images

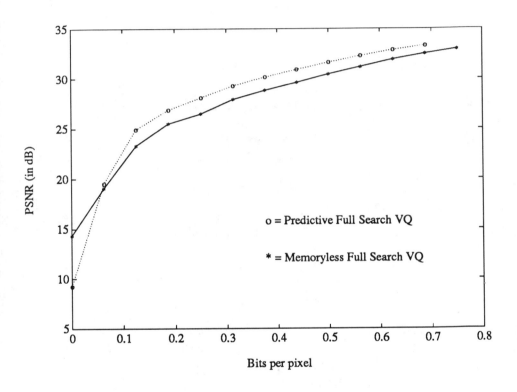

Figure 13.7: SNR of PVQ vs. VQ: Lenna

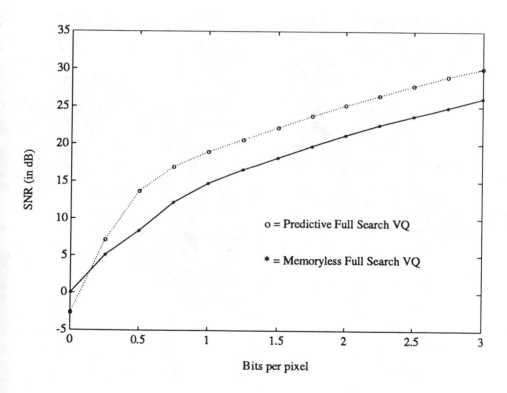

Figure 13.8: SNR of PVQ vs. VQ: MRI

Figure 13.9: PVQ vs. VQ: Lenna at .3125 bpp

Another application of PVQ to image coding was reported in [248]. In this study different design methods for the predictor coefficient matrices were compared and the performance for different prediction orders and vector dimensions was examined.

13.7 Problems

13.1. Let $x(n)$ be a stationary first order autoregressive Gauss-Markov process with zero mean, unit variance and adjacent sample correlation coefficient of 0.9. Suppose the process is formed into a blocked scalar process with a typical block represented as a vector of the form $(x(k), x(k-1))^t$. Find the optimal first order linear predictor for the resulting vector process.

13.2. By making use of Gauss type block row or column operations on the matrix $\mathbf{B} - z\mathbf{I}_{mk}$ discussed in Section 13.3, prove the validity of Equation (13.3.20).

13.3. Prove that the autocorrelation matrix $\mathcal{C}$ of the random vector sequence $\mathbf{Y}_n$ defined by

$$\mathbf{Y}_n^t = [\mathbf{X}_n^t \ \ \mathbf{X}_{n-1}^t \cdots \mathbf{X}_{n-m+1}^t]$$

is positive definite if the underlying vector sequence $\mathbf{X}_n$ is nondeterministic.

13.4. Describe variations on the open-loop, closed-loop, and semi-closed-loop predictive vector quantizer design techniques for tree-structured vector quantizers.

13.5. Suppose that the vector predictor in a PVQ system is constrained to produce only two possible distinct outputs (which precludes using a linear predictor). Describe how you might select these values and the corresponding predictor function.

Chapter 14

Finite–State Vector Quantization

14.1   Recursive Vector Quantizers

Like a predictive vector quantizer, a finite-state vector quantizer is a vector quantizer with *memory*. In general, a coding system with memory means that the index or channel symbol generated by the encoder depends not only on the current input vector but also on prior input vectors; correspondingly, the current reproduction produced by the decoder depends not only on the current channel symbol but also on the past history of the channel symbols received by the decoder. Both finite-state and predictive vector quantization can be considered as special cases of a more general class of VQ systems with memory called *recursive vector quantization* or *feedback vector quantization* [201], [202]. Recursive vector quantization is the generalization to vectors of the scalar recursive quantizers considered in [111], [236], [133]. It is instructive to begin with this more general class and then specialize it to the finite-state case.

A recursive or feedback vector quantizer is depicted in Figures 14.1–14.2. Given an input sequence of random vectors $\mathbf{X}_n$, $n = 0, 1, \cdots$, the encoder produces both a sequence of channel symbols u_n, $n = 0, 1, \cdots$, and a sequence of *states* S_n, $n = 0, 1, \cdots$, which effectively describe the encoder's behavior in response to the sequence of input vectors. As in the usual usage in control system theory, the state summarizes the influence of the past on the current operation of the encoder. Assume that the source vectors $\mathbf{X}_n$ are as usual k-dimensional real vectors, that is, $\mathbf{X}_n \in \mathcal{R}^k$. The channel symbols u_n take values in a channel alphabet $\mathcal{N}$, typically the set of all binary R-tuples or, equivalently, the set of integers from 1 to $N = 2^R$

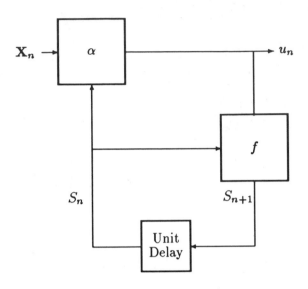

Figure 14.1: Recursive VQ: Encoder

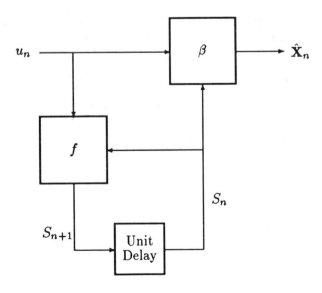

Figure 14.2: Recursive VQ: Decoder

when the channel symbol is represented by an integer index value, as in prior chapters. The state variable S_n takes values in a set S which we shall call the *state space*. As usual, R is the rate of the coding scheme in bits per input vector and hence R/k is the resolution or rate in bits per input sample (or bits per input vector component).

In order to ensure that the decoder can track the encoder state from the sequence of channel symbols without additional side information, we require that the states be determinable from knowledge of an initial state $S_0 = s_0$, together with the channel symbols. More explicitly, we assume that the next state S_{n+1} is determined by the current state S_n together with the previous channel symbol u_n by some mapping f:

$$S_{n+1} = f(u_n, S_n), \quad n = 1, 2, \cdots, \tag{14.1.1}$$

where f is called the *next-state function* or *state transition function*. Given the state S_n, the encoder mapping depends on the current state according to:

$$u_n = \alpha(\mathbf{X}_n, S_n). \tag{14.1.2}$$

The decoder also uses the current state to generate the reproduction vector from the channel symbol,

$$\hat{\mathbf{X}}_n = \beta(u_n, S_n). \tag{14.1.3}$$

For an initial state $S_0 = s_0$ that is specified for both the encoder and decoder, (14.1.1) and (14.1.2) completely describe the action of the encoder and (14.1.1) and (14.1.3) completely describe the operation of the decoder for time $n = 0, 1, \cdots$ (provided, of course, that the functions f, α, and β are described). These equations, however, are still more general than is desired, but we shall call a coding system having this general recursive structure a *recursive coding system* or *feedback source coding system*. By a *source coding system*, we mean a particular combination of encoder and decoder that represents an input source with a finite rate digital signal and reproduces an approximation to that source from the digital signal. We do not yet use the word "quantizer" but reserve this word for a more specific class of coding system.

Analogous to the basic operation of memoryless vector quantizers, we now make the added assumption that the encoder operates in a nearest neighbor or minimum distortion fashion. To make this quantitative, suppose that the encoder and decoder are in state s. The decoder has available a specific state-dependent collection of possible outputs,

$$C_s = \{\beta(u, s); \text{ all } u \in \mathcal{N}\},$$

that is, the collection of all possible reproduction vectors obtained by trying all possible channel vectors in the decoder function for a fixed state s. We shall call this collection the *state codebook* for the state s. (Note that the state codebook refers to the reproduction codebook for a fixed state, not a codebook containing states.) We shall call a coding system a *recursive vector quantizer* or a *feedback vector quantizer* if it has the form of (14.1.1)–(14.1.3) and in addition has the minimum distortion property

$$\alpha(\mathbf{x}, s) = \min_{u}^{-1} d(\mathbf{x}, \beta(u, s)), \qquad (14.1.4)$$

or, in other words, if the code has the property that

$$d(\mathbf{x}, \beta(\alpha(\mathbf{x}, s), s)) = \min_{\hat{\mathbf{x}} \in \mathcal{C}_s} d(\mathbf{x}, \hat{\mathbf{x}}). \qquad (14.1.5)$$

Thus a recursive vector quantizer always chooses the channel index which will produce the minimum possible distortion reproduction at the decoder given that both encoder and decoder are in a common state. Note that a recursive vector quantizer is fully specified by the next-state function f, the decoder mapping β, the state space $\mathcal{S}$, and the initial state s_0. The encoder mapping α is determined by the minimum distortion property (14.1.4). Consequently, at at any particular time instant n, the encoder and decoder perform exactly as a memoryless nearest-neighbor VQ with codebook $\mathcal{C}_{S_n}$. The encoder searches for the address of the best matching code vector $\mathbf{x}_i$ to the current input and sends the index i; the decoder receives this index and extracts the code vector $\mathbf{x}_i$ from its copy of this codebook. The only difference from a memoryless VQ is that at each successive time instant both coder and decoder switch to a new codebook. In a sense, recursive VQ is like memoryless VQ where the codebook common to both encoder and decoder varies with time. Of course, the mechanism for this variation is not arbitrary but rather is determined by the state transition function in response to the sequence of channel indexes.

An ordinary memoryless vector quantizer is a recursive vector quantizer with only one state. A predictive vector quantizer with memory size m can be viewed as a particular type of recursive vector quantizer wherein the current state is the current set of observables on which the prediction is based, i.e., the last m decoded reproduction vectors. A given state determines the prediction $\tilde{\mathbf{X}}_n$ and the state codebook for this state is formed by combining a single codebook for error vectors with the current prediction vector $\tilde{\mathbf{X}}_n$. More explicitly, if $\mathcal{C}_e$ is the residual VQ codebook, and the predictor operation is represented as $\tilde{\mathbf{X}}_n = g(S_n)$, then

$$\mathcal{C}_{s_n} = \{\hat{\mathbf{x}} : \hat{\mathbf{x}} = g(s_n) + \hat{\mathbf{e}}; \ \hat{\mathbf{e}} \in \mathcal{C}_e\}.$$

(See Problem 14.2.) In this chapter we shall explore another class of recursive vector quantizers, where the number of states is finite. First, however, several observations on the more general structure are in order.

The basic design goal for recursive VQ is the same as for memoryless VQ: We wish to design the coding system to minimize the long term average distortion. This can be done by basing the design on a long training sequence $\{x_i; i = 0, 1, \cdots, L\}$ of data and minimizing

$$\Delta = \frac{1}{L} \sum_{i=1}^{L} d(x_i, \hat{x}_i).$$

Unlike the memoryless VQ case, we cannot easily relate this to an expected average distortion since recursive quantizers start with an initial condition and are in general nonstationary. While they do possess ergodic properties in the sense that long term sample averages will converge, this approach is beyond the level of this book. The interested reader is referred to material on the theory of asymptotically mean stationary sources [201] [203] [202] [170] [155]. Since we have assumed an encoder structure, the design problem is to find the state codebooks and the next-state function. There does not yet exist any general approach to designing such coding systems and the existing algorithms focus on special cases with sufficient additional structure to suggest good design algorithms, e.g., as in the previously considered case of predictive vector quantizers.

Although recursive VQ is more general than memoryless VQ, the performance bounds of information theory show that the optimal achievable performance in terms of minimizing the average distortion for a given rate is actually the same for the two classes [159]. In other words, any rate/performance pair achievable by one structure is also achievable with the other, at least for well-behaved sources. These theoretical results are asymptotic, however, in that the potential performance of memoryless and recursive VQ is the same only when arbitrarily large dimensional vectors are allowed. Recursive VQ can and often does provide advantages for coders of fixed reasonable dimensions as we have seen in the case of predictive VQ. A recursive VQ can yield a coding system of significantly lower complexity than a memoryless VQ with the same rate and average distortion because it can take advantage of many small state codebooks which together provide a large repertoire of codewords. Provided the state transition function works well, the search complexity at each time is small, yet the overall collection of codewords is large. If the bit rate of a recursive VQ and a memoryless VQ are the same, the recursive VQ can use smaller dimensional vectors and hence smaller codebooks while achieving comparable performance. Alternatively, instead of achieving a fixed rate/distortion performance at a lower complexity, recursive VQ can often be designed to achieve either a given average distortion at a lower bit rate or a given bit rate with a lower average distortion than can a memoryless VQ of similar complexity.

The use of the nearest-neighbor encoder may at first seem natural, but it merits some reflection. Recall that the minimum distortion encoder is optimum for a memoryless VQ and hence one might hope that a similar property held for the more general case of recursive VQ. Unfortunately this is not the case. Because of the memory of the system, it is possible in theory that good short term decisions can lead to bad long term behavior, e.g., a codeword with very small distortion could lead to a state with a very bad codebook for the next vector. In the language of optimization algorithms, the minimum distortion algorithm is "greedy" and acts only to optimize the short term performance. Nonetheless, the minimum distortion encoder is intuitively satisfying and no better general encoder structures of comparable complexity for encoding a single input vector have yet been found. Hence we will consider only such encoders (in fact, we *defined* a recursive VQ to have such an encoder) and a goal of good design will be to reconcile this encoder with next-state functions as well as possible. The encoder can be improved by permitting the encoder to look ahead for several vectors, thereby permitting a better long-term minimum distortion fit at the expense of increased complexity and delay. This type of encoder will be considered in the chapter on trellis encoding systems.

14.2   Finite-State Vector Quantizers

A finite-state vector quantizer (FSVQ) is simply a recursive VQ with only a finite number of states; that is, in *finite-state vector quantization* (FSVQ), the coding system has the form of (14.1.1)–(14.1.5) with a finite set $S = \{\sigma_1, \cdots, \sigma_K\}$ such that the state S_n can only take on values in S. With no loss of generality, we can assume that $S = \{1, 2, \cdots, K\}$; that is, just call the states by integer names. Again note that ordinary memoryless VQ fits in this framework since it is simply FSVQ with only one state. For convenience, we use the abbreviation FSVQ to denote either finite-state vector quantization or finite-state vector quantizer as determined by context.

The state of an FSVQ encoder can be considered as a form of prediction of the next input vector based on the past encoded vectors or, equivalently, on the past reproductions available to the decoder. Each state determines a codebook and clearly choosing a good codebook is effectively equivalent to predicting the behavior of the next input vector given the current state. This is similar to the idea of a predictive VQ which by coding residuals of a prediction based on past reproductions effectively produces a separate codebook for each "state" described by the previous reproduction: the state codebook is the collection of all words of the form $\tilde{\mathbf{X}}_n + \hat{\mathbf{e}}$ for all words $\hat{\mathbf{e}}$ in the PVQ residual codebook, where $\tilde{\mathbf{X}}_n$ is the current prediction for the current vector. For a difference distortion measure such as the squared error, choosing the best reproduction

$$\hat{\mathbf{X}}_n = \tilde{\mathbf{X}}_n + \hat{\mathbf{e}}$$

over all possible codewords $\hat{\mathbf{e}}$ in the residual codebook is exactly equivalent to first forming the prediction residual

$$\mathbf{e}_n = \hat{\mathbf{X}}_n - \tilde{\mathbf{X}}_n$$

and then finding the minimum distortion residual codeword $\hat{\mathbf{e}}$ to $\mathbf{e}_n$ and then defining the new reproduction to be

$$\hat{\mathbf{X}}_n = \tilde{\mathbf{X}}_n + \hat{\mathbf{e}}.$$

Thus a PVQ, like an FSVQ, can be thought of as having a succession of codebooks determined by a state computable from past encoder outputs. In general, however, ordinary PVQ does not in principle have a finite number of states since there is no bound to the number of possible reproduction vectors. In theory, even a scalar Delta modulator can produce an infinity of possible reproductions. In practice this number of possible reproductions is usually truncated, but the number of possible state codebooks remains enormous for a general PVQ. An FSVQ can be thought of as a PVQ where the number of possible predictions of the future is constrained to be some reasonably small finite number, but a general FSVQ need not have the particular form of a PVQ of forming each state codebook by adding the codewords in a fixed residual codebook to a predicted vector. Hence the design techniques for FSVQ will differ from those of PVQ both in the encoder rules used and in the lack of linear predictors.

The restriction to a finite number of states permits a more concrete description of FSVQ. An FSVQ can be viewed as a collection of K separate memoryless vector quantizers together with a selection rule that determines which of K VQ codebooks is used to encode the current input vector into a channel index. Specifically, the current state specified which codebook is used and the resulting channel index combined with the current state determine the next state and hence the next VQ codebook. Since the decoder is able to track the sequence of states, it knows which codebook should be used at each time to decode the current index. Thus an FSVQ is an example of a *switched vector quantizer*. An FSVQ encoder can be depicted

by specializing Figure 14.1 to the finite-state case as in Figure 14.3 which
shows the encoder switching among possible codebooks as determined by
the state. The encoder sends the index of the minimum distortion codeword in the current state codebook. Let the index produced by minimum
distortion encoding of the input $\mathbf{X}$ using the codebook $\mathcal{C}_s$ corresponding
to state s be denoted $\mathcal{E}_s(\mathbf{X})$ and the decoded reproduction for index i and
state s denoted by $\mathcal{D}_s(i)$. Then

$$u_n = \mathcal{E}_{S_n}(\mathbf{X}_n), \tag{14.2.1}$$

and the decoded reproduction is given by the word with index u_n in the
codebook $\mathcal{C}_{S_n}$ determined by the current state S_n, which

$$\hat{\mathbf{X}}_n = \mathcal{D}_{S_n}^{-1}(u_n). \tag{14.2.2}$$

The corresponding decoder is shown in Figure 14.4.

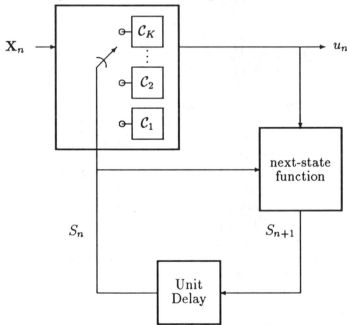

Figure 14.3: FSVQ Encoder

Ideally, the separate codebooks should be designed to suit different types
of behavior typical of the source while the state selection procedure should
be a good predictor of which behavior mode of the source will next follow. From this viewpoint, an FSVQ can be thought of as a simple form

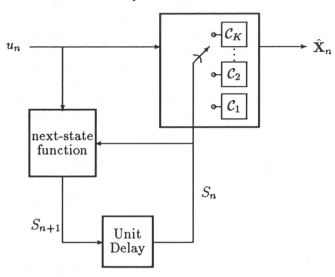

Figure 14.4: FSVQ Decoder

of adaptive vector quantization with "backward adaptation" to be studied in Chapter 16. If instead of forcing the decoder to determine the state sequence from the channel sequence, we were permitted an arbitrary choice of state (e.g., the best possible available codebook) for each input vector and then allowed to inform the decoder of the state via a separate side information channel (requiring additional bits), then the system would be adaptive with "forward adaptation" as in Chapter 16 or [195]. Such a system would also include as a special case a form of universal source code [245], but this case is not included in the FSVQ structure currently under consideration. We do not adopt the terminology of "adaptive coding system" for the FSVQ structure since the term is more commonly applied to a system which changes structure at a slow rate relative to the rate of arrival of incoming vectors in order to adapt to slowly varying statistical characteristics of the source.

Observe that FSVQ can be considered as a form of classified VQ with each state implicitly representing a class. Unlike CVQ, however, FSVQ is not permitted to send the additional bits or side information usually used to specify the class; the decoder must determine the class from the previous state (class) and the channel codeword. This connection leads to a powerful design technique for FSVQ similar to that used for predictive VQ: Open-loop CVQ will be seen to lead to FSVQ by simply "closing the loop," that is, by classifying the previous reproduction instead of the current input vector. This FSVQ can then be improved by an iterative recursion similar to the Lloyd algorithm and the closed-loop PVQ design.

If in (14.1.1)–(14.1.3) we do not require that the encoder be a nearest neighbor mapping, but we do require that the state have only a finite number of possible values, then the coding scheme is an example of a *finite-state source code*. Such coding systems date back to Shannon [289] for their definition and their theoretical properties have been studied by Gaarder and Slepian [132] and Ziv [352]. The coding systems considered here do not have the most general possible form of a finite-state coding system because of the assumption that the decoder and encoder know each other's state; this class of coding methods is dubbed *tracking finite-state source codes* by Gaarder and Slepian. Observe also that if PVQ is performed on an already digital source (such as a high rate scalar quantized source), then it, too, is an FSVQ. See Problem 14.4.

In summary, an FSVQ is specified by a state space $\mathcal{S} = \{1, 2, \cdots, K\}$, an initial state s_0, a decoder mapping $\beta(u, s)$, and a next-state function $f(u, s)$ defined for all $s \in \mathcal{S}$ and $u \in \mathcal{N}$ (recall that these imply the encoder mapping α via (14.1.4)). We shall denote an FSVQ by $\{\mathcal{S}, s_0, \beta, f\}$. We denote the state codebooks as before by

$$\mathcal{C}_s = \{\beta(u, s); \text{ all } u\}, \tag{14.2.3}$$

and we define the *super codebook* $\mathcal{C}$ as the union of all the state codebooks, that is, the collection of all the reproduction vectors in all of the state codebooks:

$$\mathcal{C} = \bigcup_{s \in \mathcal{S}} \mathcal{C}_s = \{\beta(u, s); u \in \mathcal{N}, s \in \mathcal{S}\}. \tag{14.2.4}$$

Note that if there are N code vectors in each state codebook and if there are K states, then there are no more than KN vectors in the super codebook. There may be many fewer because of duplication of words among the various state codebooks. Lastly, we define for each state s the *available next state set* $\mathcal{S}_s$ as the collection of possible successor states to s, that is, the collection of all σ for which $\sigma = f(u, s)$ for some channel vector $u \in \mathcal{N}$. For each s, $\mathcal{S}_s$ can have no more than N entries since given the current state the next state is determined by the current channel vector u.

14.3   Labeled-States and Labeled-Transitions

In FSVQ, there are two different viewpoints of the relationship between the state sequence and the reproduction sequence that have proven useful for the development of FSVQ design algorithms. We shall see that one can think of each reproduction vector as a *label* associated either with a state or with a transition between two consecutive states of the FSVQ. These two types of FSVQ, *labeled-state* and *labeled-transition*, are special cases of two classes of finite-state machines, the Moore machines (labeled-state finite-state machines) and Mealy machines (labeled-transition finite-state machines)[184].

Briefly, in a labeled-state FSVQ the states can be considered to be labeled by the reproduction vectors that will be produced if the encoder and decoder are driven into that state. The state codebook C_s of a state s, which contains all the codewords available in s, simply contains the labels of all states that can be reached from s. For a labeled-state FSVQ there are as many reproduction vectors in the super codebook as there are states. If the state codebooks have N words and there are K states, clearly $N \leq K$. Furthermore, since the encoder must act in a nearest-neighbor fashion, the selection of the nearest-neighbor reproduction also implies the choice of the next state. Thus the next-state function in a labeled-state FSVQ is determined by the nearest-neighbor choice.

Alternatively, a labeled-transition FSVQ associates a reproduction vector with every transition between states, that is, with every pair of states that are connected by the next-state rule. If there are K states and N words in each state codebook, the super codebook can have up to KN possible reproduction vectors, possibly many times the number of states.

We now consider in somewhat more detail the descriptions and properties of the two types of FSVQs. The main result of this section is that all FSVQs can be represented in either form, that is, the two classes are equivalent, although the specific details may differ between the labeled-state and labeled-transition representation of a given coding system.

Labeled-State FSVQ

Given an FSVQ $\{S, s_0, \beta, f\}$, let $\mathbf{X}_n$, u_n, S_n, and $\hat{\mathbf{X}}_n$, $n = 0, 1, \cdots$, denote the input sequence, channel sequence, state sequence, and reproduction sequence, respectively. We shall say that an FSVQ is a *labeled-state FSVQ* if the decoder mapping β which produces the decoder output depends on the current state and channel symbol only through the induced next state, that is, if there is a mapping $\gamma(s)$ such that

$$\beta(u, s) = \gamma(f(u, s)). \qquad (14.3.1)$$

In other words, the current reproduction $\hat{\mathbf{X}}_n$ is determined by the next state S_{n+1},

$$\hat{\mathbf{X}}_n = \gamma(S_{n+1}).$$

In this apparently very special case we can view the decoder outputs as *labels* assigned to the states so that the operation of the decoder is simply to view the current state S_n and current channel symbol u_n, produce the appropriate next state S_{n+1} using the next-state function, and advance to the new state and produce its label as output.

As with any recursive quantizer, the labeled-state FSVQ must satisfy the nearest neighbor condition (14.1.4), which constrains the encoder mapping α to have the form:

$$\alpha(\mathbf{x}, s) = \min_u{}^{-1} d(\mathbf{x}, \gamma(f(u, s))).$$

This implies that the minimum distortion satisfies

$$d(\mathbf{x}, \gamma(f(u, s))) = \min_{s' \in \mathcal{S}_s} d(\mathbf{x}, \gamma(s')),$$

where the minimum is over all possible next states reachable from the current state.

At first it might appear that a labeled-state FSVQ is a very special case, but in fact it is not special at all. Any FSVQ is equivalent to a labeled-state FSVQ since, as we shall see, given any FSVQ we can find a labeled-state FSVQ, the "revised" system, such that both FSVQs will produce the same reproduction sequence for a given input sequence. However, the state space, next-state function, decoder, and state sequence may be different. This fact can be demonstrated without recourse to formal theorem statements and proof by constructing the labeled-state FSVQ corresponding to an arbitrary FSVQ.

Suppose that we have an arbitrary FSVQ $\{\mathcal{S}, s_0, \beta, f\}$. We will construct a labeled-state FSVQ $\{\mathcal{S}', \sigma_0, \psi, g\}$ which produces the same output sequence for a given input. To do this we define a new or revised state space $\mathcal{S}'$ consisting of all pairs (w, s) with $w \in \mathcal{N}$, the old channel symbol alphabet, and $s \in \{1, 2, \cdots, K\}$, the old state space. Thus each revised state σ looks like an ordered pair (u, s) of a channel index u and an element s of the original state space. The number of possible states has grown from K to NK, but it is still finite. The revised state is initialized at $n = 1$ with the value $\sigma_1 = (u_0, s_0)$, where $u_0 = \alpha(\mathbf{x}_0, s_0)$ is the initial index value and s_0 is the initial state for the original system.

To define the new FSVQ, we first specify its next-state function g and decoder mapping ψ according to

$$g(w, \sigma) = (w, f(\sigma)), \qquad \psi(w, \sigma) = \beta(w, f(\sigma)).$$

Now suppose that the sequence of channel symbols to the new FSVQ decoder is identical to the sequence sent by the old encoder, so that $w_n = u_n$. We then show that the same reproduction sequence is produced as with the old FSVQ decoder. Suppose that for some n

$$\sigma_n = (u_{n-1}, s_{n-1}), \tag{14.3.2}$$

that is, the revised state at time n is the combination of the original state one time unit earlier combined with the channel vector produced in that state. Then, we have

$$
\begin{aligned}
\sigma_{n+1} &= g(w_n, \sigma_n) = (u_n, f(\sigma_n)) \\
&= (u_n, f(u_{n-1}, s_{n-1})) = (u_n, s_n).
\end{aligned}
$$

Hence, since (14.3.2) holds for $n = 1$, by induction it holds for all n. Also, the reproduction $\hat{\mathbf{X}}_n'$ produced by the revised decoder is given by

$$
\begin{aligned}
\hat{\mathbf{X}}_n' &= \psi(w_n, \sigma_n) = \beta(u_n, f(\sigma_n)) \\
&= \beta(u_n, s_n) = \hat{\mathbf{X}}_n.
\end{aligned}
$$

Observe that the new decoder is still given by β, that is, $\hat{\mathbf{X}}_n' = \beta(\sigma_n)$ so that the labeled-state condition holds. The only difference from the old system is that what used to be considered a channel vector and state as argument of β is now just the revised state. Thus, the revised decoder generates the same sequence of reproductions as the original decoder, provided the channel symbols are the same. The latter condition follows from the minimum distortion condition, which determines the best selection of channel index based on the input vector and the current state. Hence, the same sequence of channel symbols is produced in the two FSVQ systems.

We have thus shown that any FSVQ has an equivalent labeled-state FSVQ and hence there is no loss in generality assuming this structure.

Labeled-Transition FSVQ

Instead of assuming that the reproduction vectors are labels of states, we can consider them as labels of state transitions, that is, they are produced as we travel from one state to the next rather than on arrival in the next state. We shall say that an FSVQ is a *labeled-transition FSVQ* if there is a mapping $\Gamma(s, s')$ such that

$$
\beta(u, s) = \Gamma(s, f(u, s)), \tag{14.3.3}
$$

that is, the decoder outputs can be considered as labels on the transitions between states. In other words, the current reproduction $\hat{\mathbf{X}}_n$ is determined by both the current state s_n and next state s_{n+1},

$$
\hat{\mathbf{X}}_n = \Gamma(s_n, s_{n+1}) = \Gamma(s_n, f(u_n, s_n)).
$$

Note that this is clearly a generalization of a labeled-state FSVQ since we could form a labeled-state FSVQ by having Γ depend only on its second argument. Also, by its definition, the class of all labeled-transition FSVQs is contained in the class of all FSVQs. This leads to the conclusion that the set of all FSVQs is bigger than the set of labeled-transition FSVQs which in turn is bigger than the set of labeled-state FSVQs (where "bigger" and "smaller" include the possibility of equality). But we have seen that the set of all FSVQs and the set of labeled-state FSVQs are equivalent and hence all three classes are equivalent.

We have now argued that there is no loss in generality in restricting ourselves to FSVQs that are either labeled-state or labeled-transition since any FSVQ is equivalent to another FSVQ of these forms. This equivalence is in fact a special case of a well known result of the theory of finite state machines to the effect that Mealy machines are equivalent to Moore machines [184]). Why then consider these two equivalent structures? Primarily because different design algorithms will be better (or uniquely) suited for one structure or the other. In addition, running iterative improvement algorithms on equivalent coders may not yield equivalent coders; hence, given a coder of one form it may be useful to try iterative improvement algorithms both on the original coder and on an equivalent coder of the other form. Lastly, the two structures do have different characteristics which may be better matched to a particular application, as we now consider.

As an example, suppose that we have a labeled-transition FSVQ with K states and state codebooks of size $N = 2^R$ for each state and hence a super codebook of size no greater than KN. As previously described for a general FSVQ, we can construct an equivalent labeled-state FSVQ with the same behavior by defining a new state space where each state consists of a channel index and an original state and channel symbols. This new FSVQ has a much increased state space and hence has greater codebook storage requirements, but it also possesses a certain additional freedom for the designer that may more than compensate for this apparent disadvantage. By redesigning the next-state function for the given collection of states, one can now cause a given reproduction codeword to have as successors any N of the super codewords. On the other hand, redesigning the next-state function of a labeled-transition FSVQ only permits one to choose which state codebook, that is, which of a fixed collection of codewords, can succeed the given codeword. Quantitatively, any FSVQ and in particular a labeled-transition FSVQ has K^{NK} ways to choose a next-state function $f(u,s)$, since for each channel index u and current state s there are K choices for the value of f and there are NK pairs (u,s) for which f must be assigned a value. There may be multiple transitions between states in the sense that more than one input can drive a given state into another given state. Such multiple transitions can of course bear different labels. The equivalent labeled-state system that can be generated from the given labeled-transition FSVQ has KN states, but now multiple transitions are not repeated as they are wasted. Here there are

$$\binom{KN}{N}^{KN}$$

possible choices of the next-state function since for each of the KN states we can connect N successor states in $\binom{KN}{N}$ possible ways. This additional freedom can lead to quite different codebooks when the initial code is altered by an iterative improvement technique.

Intuitively, in both forms of FSVQ the state itself can be considered to correspond to a coarse approximation of the last source vector and hence a form of prediction of the next source vector to be encoded. (In fact, in one of the first developments of FSVQ design techniques, Haoui and Messer-schmitt [176] referred to their FSVQ system as a "predictive vector quantizer," a name we have used for a quite different non-finite-state system.) In the labeled-state case the approximation is simply the last reproduction codeword chosen, the label of the current state. In the labeled-transition case, the approximation is given not by a specific codeword, but by a specific codebook–the state codebook consisting of the labels of the transitions leaving the state. Thus the labeled-state code can perhaps use the extra states to produce a better coarse approximation of the source through the state sequence than can a labeled-transition code.

These observations suggest that for a fixed rate, the labeled-state FSVQ may be capable of taking advantage of its additional storage to provide better performance than can an equivalent labeled-transition FSVQ. To date, however, labeled-transition FSVQ algorithms have been better understood and have provided better performance.

14.4 Encoder/Decoder Design

As in the memoryless VQ design, we will attempt to optimize various parts of the coding system while holding fixed the remaining parts. Unlike the memoryless VQ design, we will be unable to find genuine optimal procedures and will have to content ourselves with finding good ones. For purposes of design, we will drop the requirement of a minimum distortion encoder in the same way that we did when designing a memoryless VQ. Permitting a more general encoder will allow algorithms to attempt to optimize decoders to encoders.

In this section we consider the portion of the design that most resembles memoryless quantizer design. We seek a good encoder α and decoder β for a given next-state function f. This leads to an iteration in which the encoder and decoder are alternately updated in a manner that is very similar to the memoryless VQ case. Suppose that we have a state space $\mathcal{S}$, an initial state s_0, and a next-state function f. What is the best decoder $\beta(u, s)$? Producing the best decoder for the given next-state function fixed is exactly equivalent to designing the state codebooks $\mathcal{C}_s = \{\beta(u, s); \ u \in \mathcal{N}\}$ for all $s \in \mathcal{S}$. Hence we also refer to decoder design as the *state codebook improvement algorithm*. The solution is only a slight variation of the memoryless VQ case and can be viewed as a simple extension of the decoder improvement algorithm developed by Stewart et al. for trellis encoders (which we treat later) [302]. Here we consider the general FSVQ form as there is no benefit to confining interest to labeled-state or labeled-transition VQs.

We first consider the task of finding the best decoder when the encoder α is also given. Suppose that $\{\mathbf{x}_l; \ l = 1, 2, \cdots, L\}$ is a training sequence. Encode the training sequence using the initial state s_0, the encoder α, and the next-state function f, that is, for $l = 0, 1, \cdots, L - 1$ form

$$
\begin{aligned}
u_l &= \alpha(\mathbf{x}_l, s_l) \\
s_{l+1} &= f(u_l, s_l).
\end{aligned}
$$

This produces the sequence $\{u_l, s_l\}$ so that we have available for the design the training sequence of triples $\{\mathbf{x}_l, u_l, s_l\}$.

We wish to find the mapping β that minimizes

$$
\Delta = \frac{1}{L} \sum_{l=1}^{L} d(\mathbf{x}_l, \beta(u_l, s_l)).
$$

As in the case of memoryless VQ, this sum can be rearranged to show that the optimal decoder code vectors consist of centroids. Here, however, we must condition both on the channel index *and on the state*. Define $\mathcal{M}(u, s)$ as the collection of all time indices n for which $u_n = u$ and $s_n = s$ and let $L(u, s)$ be the number of such time indices. Then

$$
\begin{aligned}
\Delta &= \sum_{u,s} \frac{L(u, s)}{L} \frac{1}{L(u, s)} \sum_{n \in \mathcal{M}(u,s)} d(\mathbf{x}_n, \beta(u, s)) \\
&= \sum_{u,s} p(u, s) E(d(\mathbf{X}, \beta(u, s)) | u, s),
\end{aligned}
$$

where we have defined the relative frequency

$$
p(u, s) = \frac{L(u, s)}{L}
$$

and the sample average conditional expectation

$$E(d(\mathbf{X}, \beta(u, s))|u, s) = \frac{1}{L(u, s)} \sum_{n \in \mathcal{M}(u, s)} d(\mathbf{x}_n, \beta(u, s)).$$

Writing it in this way shows that the required computation is almost exactly that required in the memoryless case: $\beta(u, s)$ should be a generalized centroid minimizing the conditional expectation given the channel vector and the state, that is,

$$\beta(u, s) = \min_{\mathbf{y}}^{-1} E(d(\mathbf{X}, \mathbf{y})|u, s) = \min_{\mathbf{y}}^{-1} \frac{1}{L(u, s)} \sum_{n \in M(u, s)} d(\mathbf{x}_n, \mathbf{y}).$$

We shall denote the above inverse minimum, or centroid, as $\mathrm{cent}(u, s)$. Thus, for example, if the distortion measure is the mean squared error, then

$$\beta(u, s) = \frac{1}{L(u, s)} \sum_{n \in \mathcal{M}(u, s)} \mathbf{x}_n.$$

The above development clearly provides the optimum decoder for the given encoder and next-state function. But the given encoder need not be optimal for the resulting decoder. Thus, some further effort is needed to find an effective (if not optimal) coder/decoder pair for the given next-state function.

Having now optimized the state codebooks for the encoder and next-state function, we would like to find the best encoder for the given encoder and next-state function. We could now mimic the memoryless VQ design by replacing the encoder α by the minimum distortion encoder for the current decoder and given next-state function. Unlike the memoryless case, however, this may not be the optimum encoder for the given decoder and next-state function; that is, the greedy encoding algorithm could lead to bad states. Nonetheless, it is what is usually done and rarely does it make things worse. This technique for "closing the loop" is essentially the same as that used for predictive VQ in Chapter 13. Even if the choice of encoder increases distortion, one can continue the algorithm by again finding the optimum decoder, then using a minimum distortion encoder, then finding the optimum decoder, and so on. In all simulations we know of, the algorithm converges to a stable code and overcomes any transient worsening of performance. There is no proof, however, that this should be the case. We could simply stop iterating if the distortion increases, but in fact we have found in all experiments that the iteration does better if it is allowed to continue to convergence.

To summarize we now have the algorithm of Table 14.1. The state code-

Table 14.1: **State Codebook Improvement Algorithm**

Step 0. Given a state space $\mathcal{S}$, an initial state s_0, an encoder α_0, a next-state function f, and a training sequence $\{x_l; \ l = 1, 2, \cdots, L\}$. Set $\epsilon > 0$, $m = 1$, $D_0 = \infty$.

Step 1. Encode the training sequence using α_{m-1} to obtain $\{u_l, s_l\}; \ l = 1, 2, \cdots, L$. For this encoder define the decoder (or state codebooks) by

$$\beta_m(u, s) = \operatorname{cent}(u, s).$$

Step 2. Replace the encoder α_{m-1} by the minimum distortion encoder for β_m, call this encoder α_m.

Compute

$$D_m = \frac{1}{L} \sum_{l=1}^{L} d(x_l, \beta_m(\alpha_m(x_n, s_n), s_n)).$$

If $|D_m - D_{m-1}|/D_{m-1} < \epsilon$, then quit. Else set $m \leftarrow m + 1$ and go to Step 1.

book improvement algorithm provides a means of optimizing the codebook for a given finite state code and iteratively improving the encoder/decoder combination for a given next-state function. We next turn to the more difficult task of finding a next-state function and thereby finding a complete FSVQ.

14.5 Next-State Function Design

Ideally, the design of next-state functions should parallel the methods used in the design of memoryless quantizers: Fix everything but the next-state function (i.e., the encoder, decoder, and number of states) and then find the best next-state function for the given part of the system. Unfortunately, no methods have yet been found for doing this. Instead, we take a somewhat different tack and break the design of an FSVQ into three separate stages, the last of which we have already seen.

1. Initial classifier design.

2. State space, next-state function, and state codebook design.

3. Iterative state codebook improvement.

The first step is intended to produce a means of classifying input vectors into states. Although we shall see other classification rules, one simple technique is to design a memoryless VQ for the input vectors and to consider the VQ outputs as possible states of the input for an FSVQ to track. This initial classifier is then used construct the items obtained in the second step: the state space, next-state function, and corresponding state codebooks. Once obtained, the codebook improvement algorithm of the preceding section can be run to fine tune the codebooks. In this section we focus on the interrelated first and second steps.

The three-step procedure lacks the iterative optimization character of the memoryless VQ design. For example, once the next-state function is found and the codebooks optimized for it, we do not then try to better match the next-state function to the codebooks. We will later describe an alternative approach that does try to mutually optimize the next-state function and codebooks, but here we consider only the single pass on finding the next-state function. This approach is clearly *ad hoc* and no theoretical results suggest that the resulting code is even locally optimal, but the approach does have the advantage of being simple and reasonably intuitive and permitting direct application of memoryless VQ design techniques to the first and third steps. The more complicated algorithms that attempt to jointly optimize all facets of the system have not to date provided significantly better performance. The best FSVQs so far have been variations on the above three-step design procedure that use smart classification techniques in the next-state function design.

The initial classifier is a mapping from input vectors **x** into a state index $s \in \mathcal{S}$. To be concrete we consider a simple example where the state classifier is itself a memoryless VQ having K codewords, one for each state. We will later consider more sophisticated classifiers. Once we have formed a classification of the inputs into K classes, the basic idea of the FSVQ design will resemble that of classified VQ: The FSVQ will attempt to track the input class as its state and have a good state codebook for each class.

A VQ state classifier can be designed using the original generalized Lloyd algorithm for memoryless VQ design. Denote the corresponding codebook $\mathcal{B} = \{c(s); \ s \in \mathcal{S}\}$ where $c(s)$ denotes the code vector associated with state s. For lack of a better name, call this codebook the *classifier codebook* since it can be viewed as a means of classifying the input vectors in order to determine the state codebook (classifying the input by means of finding the nearest neighbor in a codebook). Define the *state classifier* by

$$\mathbf{V}(\mathbf{x}) = \min_{s \in \mathcal{S}}^{-1} d(\mathbf{x}, c(s)).$$

Thus the state classifier produces the state (or, equivalently, the index of the codeword in the state codebook) of the nearest neighbor to **x**. Keep in mind that we are currently using a simple VQ as a state classifier, but more complicated (and possibly better) classification rules can also be used in the design techniques.

Alternatively, we can view $\mathcal{B}$ as the *state label codebook*, a collection of potential state labels (even if the FSVQ is to have labeled-transitions). This initial codebook is the initial super codebook if the FSVQ is to have labeled-states. If the FSVQ is to have labeled-transitions, then the initial codewords can be viewed as the seeds for the development of the individual state codewords (the transition labels). In both cases, encoding the previous input vector using the classifier codebook can be considered as providing a coarse approximation to the previous input vector and hence a form of prediction of the current vector.

We now consider several means of producing a next-state function given the classifier codebook.

Conditional Histogram Design

One of the simplest (and also the oldest) technique for finding a next-state function uses conditional histograms of the classifier codebook in order to determine a next-state function for a labeled-state FSVQ [263] [126]. The classifier codebook is itself the super codebook and the state codebooks will be subsets of the classifier codebook.

This approach was motivated by the observation that in memoryless VQ systems for LPC parameters, each codeword was followed almost always by one of a very small subset of the available codewords [263]. Thus performance should not be much impaired if only these subsets were available as state codebooks. The conditional histogram design accomplishes this by estimating the conditional probabilities of successor codewords if the classifier VQ is used in an ordinary manner, and then forming a labeled state FSVQ by only putting the most probable codeword successors in the super codebook in the state codebook.

Given the classifier $\mathbf{V}$, use it to encode the training sequence and form a conditional histogram giving the relative frequency of occurrence of each codeword given its predecessor. Thus if $\mathbf{x}_l$ is mapped into

$$\hat{\mathbf{x}}_l = \min_{\mathbf{y} \in \mathcal{B}}{}^{-1} d(\mathbf{x}_l, \mathbf{y}),$$

then form the normalized counts

$$p(j|i) = \frac{\text{number of } l \text{ such that } \hat{\mathbf{x}}_{l-1} = c(i) \text{ and } \hat{\mathbf{x}}_l = c(j)}{\text{number of } l \text{ such that } \hat{\mathbf{x}}_{l-1} = c(i)}.$$

Figure 14.5 shows the possible transitions from words in the super codebook to other words in the supercodebook with the transitions labeled by their conditional probability. Once the conditional histograms are determined, form the collections $\mathcal{S}_s$ of possible successor states by picking the $N = 2^R$ most likely successors, that is, $\mathcal{S}_s$ contains the N integers j for which $p(j|s)$ is largest. Thus only those N transitions in 14.5 emanating from codewords having the largest probability would be preserved. The remaining lines would be cut.

Once the collection $\mathcal{S}_s$ is specified for all s, the entire next-state function is specified by the nearest-neighbor encoder rule: $\mathcal{S}_s$ contains all the allowed next states and each state has a label; the actual next state for a given input and current state is the allowed next state whose label is the nearest neighbor to the input vector.

There is an obvious problem with this design. Conditional histograms based on training data can estimate as zero a conditional probability that is in fact small but not zero. This can result in a coding system that when applied to the training sequence is such that the nearest neighbor in the entire super codebook is *always* in the current state codebook; that is, the conditional probabilities of successive super codewords in the training sequence are nonzero for at most K successors, where K is the number of

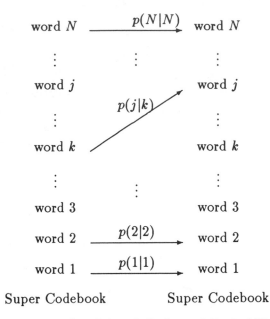

Figure 14.5: Conditional Codeword Probabilities

states. When applied to data outside the training sequence, however, it can happen that if a sequence of input vectors results in the nearest neighbor in the super codebook lying outside the current state codebook. This can result in the encoder having no good choice because of an unlikely change in the input. Once a bad choice is made, the FSVQ no longer produces the same sequence of reproduction vectors as the original memoryless VQ using the super codebook and the coding system "derails" in that its performance degrades from that of the memoryless VQ.

This tendency of some FSVQs to "derail" can be countered by occasional re-initialization, that is, periodically spending the extra bits necessary to perform a full search VQ and force the code into the best possible state. If this is done very infrequently, the extra cost is small.

Nearest Neighbor Design

Another technique for labeled-state FSVQ uses similarity rather than likelihood as a means of selecting the set of allowed new states from a given prior state. In this method, called *nearest neighbor design*, we retain the classifier codebook as a super codebook and as a collection of state labels for the design of a next-state function, but now we use only the distortion (a measure of dissimilarity) between codewords and not relative frequencies. This design technique is limited to distortion measures which are defined for pairs of reproduction vectors as well as for pairs of an input vector and a reproduction vector, e.g., symmetric distortion measures [126]. In this approach we effectively assume that the input vectors are highly correlated and successive source vectors are fairly close to each other. Thus the available state codewords should well approximate the current reproduction codeword.

For each state s find the N nearest neighbor codewords among all the state labels in the entire super codebook and then include these states in the available next state collection $\mathcal{S}_s$. Thus instead of allowing only the most likely successor states, we allow only the closest successor states. This method depends only on the distortion between the codewords while the conditional histogram approach depends on the likelihood of possible successors. The resulting code can be viewed as a form of multidimensional delta modulation since the allowed next reproductions are always close to the current reproduction. The method is clearly inappropriate if the number of codewords K is much larger than the number of possible transitions N since it will likely produce codes that are unable to track rapid variations, a condition analogous to slope overload in delta modulation.

Set Partitioning

Both conditional histogram and nearest-neighbor state codebook design can be considered as variations of *set partitioning* in the sense used by the popular coded modulation schemes for combined channel coding (error correction) and modulation [314] [315], for quantization and vector quantization [232][233][116], and for combined source coding and modulation [115]. One starts with a generally large k-dimensional codebook or super-codebook (a lattice is usually assumed in the coded modulation literature) and partitions the large codebook into several smaller codebooks or sub-codebooks. Each sub-codebook is associated with a transition between states, or equivalently, with a branch in a trellis diagram. While trellises will be considered at some length in the next chapter, for current purposes they can be considered simply as a map of the various time sequence of states, with a branch connecting a particular state at one time to another state at the next time instant. A trellis diagram provides a temporal picture of the possible actions of a finite-state machine in general or an FSVQ in particular.

If the encoder is in a given state, it has available to it those sub-codebooks on branches emanating from that state. If used as an ordinary FSVQ, the encoder would find the best codeword in any of the presented sub-codes. It would then choose the branch corresponding to the sub-code containing the best codeword and advance along that branch to the next state producing a two-part channel codeword: an index indicating the branch taken (i.e., by pointing to the next state) and an index indicating the particular codeword in the corresponding branch sub-code that was selected. Ungerboeck [314] [315] provides a means of generating such a trellis and choosing the sub-codebooks from the super codebook so as to have good properties. When dealing with lattices, the smaller codebooks are typically taken as sublattices.

Omniscient Design

Another FSVQ design technique, called *omniscient design*, is more complicated than the conditional histogram and nearest neighbor methods, but it often gives superior performance, its operation is intuitive, and it can be used with more general classifiers than the simple VQ used so far. The name reflects the fact that we first design state codebooks based on the assumption that the encoder and decoder are omniscient in the sense of knowing a sequence of idealized states classifying the current source behavior. The omniscient encoder is then approximated by a more constrained closed-loop encoder that can be tracked by a nonomniscient labeled-transition FSVQ decoder. If the approximation is reasonably good, then the codebooks designed assuming the extra knowledge should also be fairly good. A variation of the design technique can be used to obtain a labeled-state FSVQ, but we concentrate on the labeled-transition case as it is easier to describe and performs better in simulations.

The omniscient design technique was introduced by Foster et al. [126] and Haoui and Messerschmitt [176] and developed for voice coding by Dunham and Gray [103] and for image coding by Gersho and Aravind [16]. The development here draws on the general classification viewpoint of Aravind and Gersho [16]. The basic idea is simple. We assume that we have a classification rule for the source vectors, that is, a classifier $\mathbf{V}$ that views an input vector $\mathbf{x}$ and assigns it to one of K classes. Without loss of generality we can consider the states to be integers and $\mathbf{V}(\mathbf{x})$ to be an integer from 1 to K. A simple example of such a classifier is a VQ encoder $\mathcal{E}(\mathbf{x})$ with a codebook $\mathcal{B} = \{c(s); s \in \mathcal{S}\}$ where the search is based on a minimum distortion rule. We consider the more general case, however, as more complicated, perhaps heuristically designed, classifiers may yield better performance. The FSVQ will have one state for each class.

We need to design for each state s a state codebook and a next-state selection rule. First pretend that we do not need to select the state according to a next-state function depending on only the channel vector and current state. Assume instead that the next state can be chosen in a memoryless fashion based entirely on the current input vector. A reasonable heuristic for choosing a state would be to simply apply the classifier function to the input vector, that is, find $s_{n+1} = \mathbf{V}(\mathbf{x}_n)$. Unfortunately, however, the decoder cannot track such a next-state rule because it depends on the input rather than only on the encoded input. Hence the state selection is idealized or *omniscient* because of the assumption that the actual input vector can be used to determine the next state when in fact the decoder will not have this information in the final code. Note that, unlike CVQ, if the nth training vector $\mathbf{x}_n$ is classified in state s, then it is the *next* state s_{n+1} that is set to s and the next training vector that should be encoded with $\mathcal{C}_s$.

If we can select a state in this manner, there is a natural state codebook design mechanism. Break the training sequence up into K subsequences T_s; $s = 0, 1, \cdots, K-1$ as follows: The subsequence for state s consists of all of the original training vectors which *follow* a training vector classified as state s. Run the Lloyd algorithm on the subsequence for state s in order to produce the state codebook $\mathcal{C}_s$ having $N = 2^R$ codewords. Note that the state codebooks should be quite good for the words actually in the omniscient state s, that is, which follow a training vector classified as state s. Unfortunately, however, the decoder will not know the omniscient state and we are not permitted to send it extra bits to inform it of the state lest we increase the communication rate. Hence this code does not yet yield an FSVQ. We can, however, approximate the idealized next-state selection and obtain an FSVQ. This is accomplished simply by replacing the actual input vector $\mathbf{x}_n$ used in the next state selection by the current reproduction $\hat{\mathbf{x}}_n$, which is known to the decoder. That is, we keep the codebooks designed for the idealized state, but we actually use a next state rule

$$s_{n+1} = \mathbf{V}(\hat{\mathbf{x}}_n).$$

The resulting FSVQ is a labeled-transition FSVQ since the reproduction vector is completely determined by the current state (which corresponds to a class and has a state codebook associated with it) together with the channel symbol or, equivalently, the next state. The super codebook has NK codewords, where K is the number of classes. (A labeled-state FSVQ would have only K codewords in the super codebook.)

If the reproduction is fairly good, then the classification should be close to the idealized state and hence the actual VQ codebook selected will be good. This method of "closing the loop" so as to form the classification (which can be viewed as a form of prediction) is analogous to the philosophy of predictive quantization where the current reproduction is used in place of the actual input when forming a prediction of the next input.

The following provides a brief alternative synopsis of the omniscient FSVQ design algorithm:

1. Design a classifier for the input vectors.

2. Design a codebook for each class from a training subsequence formed from those vectors whose immediate predecessors are in the given class. This gives a classified VQ coding system.

3. Form a labeled-transition FSVQ from the CVQ by using the decoded reproduction of the previous input vector (instead of the input vector itself) to determine the next state.

The omniscient FSVQ design algorithm is summarized in Table 14.2.

Table 14.2: **Omniscient FSVQ Design Algorithm**

Step 0. *Initialization:*

Given: A classifier **V** (e.g., a VQ) that assigns an integer from 1 to K to each input vector, $N = 2^R$, a threshold $\epsilon > 0$, a training sequence $\mathcal{T} = \{\mathbf{x}_l; \, l = 1, 2, \cdots, L\}$.

Step 1. *State codebook Design:*

For each $s \in S$ use the classifier to partition the training sequence into subsequences

$$\mathcal{T}_s = \{\mathbf{x}_l : \mathbf{x}_l \in \mathcal{T}, \mathbf{V}(\mathbf{x}_{l-1}) = s\}.$$

Use the Lloyd algorithm to design for each subsequence a codebook $\mathcal{C}_s$ using the given threshold.

Step 2. *Next-State Function Design*

Define the next-state function

$$s_{n+1} = \mathbf{V}(\hat{\mathbf{x}}_n),$$

where

$$\hat{\mathbf{x}}_n = \min_{\mathbf{y} \in \mathcal{C}_{s_n}} {}^{-1} d(\mathbf{x}_n, \mathbf{y})$$

is the minimum distortion word in the current state book.

The omniscient FSVQ design has been found to yield the best codes with a reasonable complexity in most applications. It has been particularly successful in coding still images at rates of 1/2 bit per pixel and less.

Stochastic Automaton Design

A final technique which is only mentioned briefly is to use ideas from stochastic automaton theory to design an FSVQ. The basic idea is to allow random state transitions by replacing the next-state transition function $f(u, s)$ by a conditional probability mass function $p(\sigma|u, s)$ for the next state given the current state and the channel symbol, thereby effectively randomizing the next state choice. This converts the FSVQ into a stochastic automaton, where the randomness is in both the driving source and the state conditional probabilities. The automaton then effectively designs itself by making small changes in the conditional state probabilities and tracking the resulting average distortion. Only changes yielding smaller average distortion are pursued. When (or, more correctly, if) the algorithm converges, the final conditional probabilities can approximated by a deterministic next-state function either by simply rounding them to 0 or 1 or by using a pseudo-random sequence to govern a changing, but deterministic, next state assignment. This design technique raises several interesting theoretical questions regarding algorithm convergence and the optimality properties of the final code, but unfortunately FSVQs designed using this approach have yielded disappointing performance [103].

14.6 Design Examples

In this section several simulations using FSVQ are summarized and compared with other systems. The sources considered are the same Gauss Markov, sampled speech, and sampled images considered in the examples of Chapter 16 and first described in Chapter 11. The first two examples are of interest primarily because they indicate the numerical SNR improvements available. FSVQ has proved most successful, however, in the image compression application. For all examples the mean squared error was the distortion measure.

Gauss-Markov Sources

First consider the Gauss-Markov source with the correlation coefficient of 0.9. As before, the training sequence consisted of 128,000 samples as did the test sequence. The first figure compares the performance resulting when the state classifier is itself a VQ and several of the next-state design methods were tried. The vector dimension was $k = 4$ and rates of 1, 2, 3, 4 bits per vector and hence 1/4, 1/2, 3/4, 1 bits per sample were tested. The conditional histogram (CH) and nearest-neighbor (NN) labled-state VQs were designed for for $K = 128$ states. An omniscient labeled-transition FSVQ (O) was designed having the same rates, dimension, and super codebook size as the labeled-state FSVQ, that is, having the same number of states as would the labeled-state FSVQ equivalent to O. The resulting average distortions for the test sequences are depicted in Figure 14.6 together with the performance for an ordinary memoryless VQ (full search) and a lower bound to the optimal performance (the Shannon lower bound to the distortion-rate function, see, e.g., [32] or [159].)

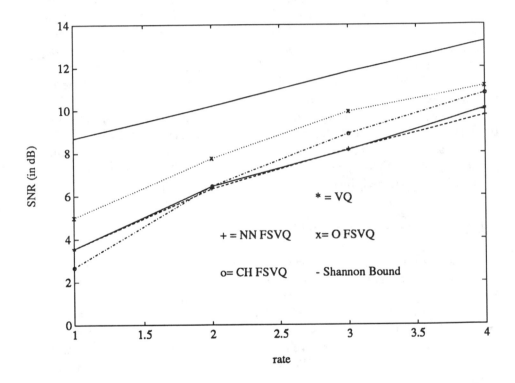

Figure 14.6: Performance Comparison for Next-State Rules: Gauss-Markov Source

A few observations can be drawn from the example. First, the simple NN and CH designs tend to provide performance close to that of the memoryless quantizer in this example, sometimes slightly better and sometimes slightly worse. The methods were included because in some examples they have performed noticeably better than here (see, e.g., [29]) and they are much faster and simpler to design. The omniscient coders do provide consistently superior performance. Remember that the source here is artificial and the state classifier only a simple VQ. As a comparison, an optimized one bit per sample predictive quantizer due to Arnstein [18] achieves an SNR of 10 dB for this source. The omniscient FSVQ achieves the same performance at 3/4 bits per sample.

The next experiment describes a comparison between the performances of an omniscient FSVQ and an ordinary VQ for a fixed rate of 1 bit per sample and a varying dimension of 1 to 5 samples. The number of states was fixed at eight. The Shannon distortion rate function is here 13.2 dB. The results are depicted in Figure 14.7. A dimension 1 FSVQ outperforms

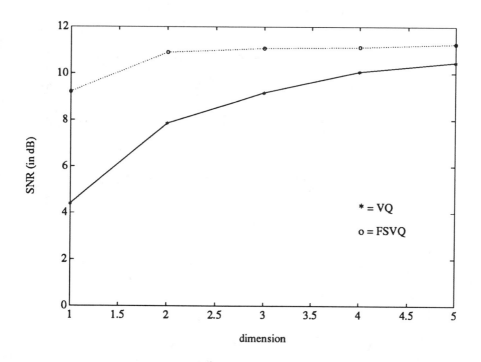

Figure 14.7: FSVQ vs. VQ: Gauss-Markov Source

a scalar quantizer by almost 5 dB and is close in performance to Arnstein's optimum predictive quantizer. This is of interest since the FSVQ has only a finite number of states while the predictive quantizer has an infinite number of states. The performance advantage closes rapidly to about 1 dB and then remains there. The FSVQ of dimension 2, however, outperforms the memoryless VQ of dimension 5. At dimension 5, the eight-state FSVQ was within 2 dB of the Shannon optimum.

Sampled Speech

The experiments were repeated for the previous training sequence of sampled speech consisting of 640,000 samples from five male speakers sampled at 7500 samples per second (about 20 seconds of speech from each speaker). The codes were tested on a test sequence of 78,000 samples from a male speaker not in the training sequence.

The various state selection algorithms exhibit behavior similar to that for the Gauss Markov source and hence we do not detail the results. (Details may be found in [126].) Figure 14.8 provides a comparison between omniscient labeled-transition FSVQ and VQ for a fixed dimension 4 and the same rates as in the Gauss Markov case.

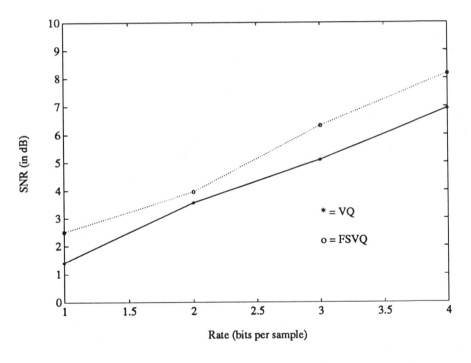

Figure 14.8: FSVQ vs VQ for Sampled Speech: SNR vs. Rate

At the higher rates, the gains of FSVQ over VQ are larger for the sampled speech than for the Gauss-Markov process, reflecting the fact that speech has more complex structure than a Gauss Markov source and the FSVQ is apparently better able to track it. FSVQ consistently provided an improvement of approximately 1.5 dB.

The next figure provides a performance comparison between the labeled-transition omniscient FSVQ and ordinary VQ for a fixed rate of 1 bit per sample and increasing vector dimension.

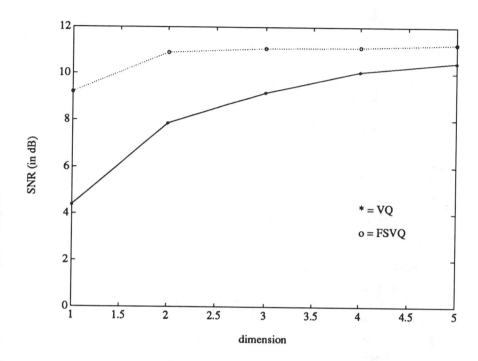

Figure 14.9: FSVQ vs VQ for Sampled Speech: Fixed Rate, Increasing Dimension

Again, the FSVQ improvement at higher rates (here higher dimension) is better than that for the Gauss Markov source.

The coded speech was subjected to the proverbial informal listening tests and the subjective quality reflected the numerical results. FSVQ sounded noticeably better than the memoryless VQ of the same dimension and rate. The improved quality is at the expense of increased complexity and memory.

Image Compression using FSVQ

Aravind and Gersho [16] [17] extended FSVQ to images by extending the notion of a state in a natural manner. With one-dimensional data we considered the collection of states to be identical to a collection of classes which described local behavior, where each class corresponded to a state codebook. We keep the notion of classes, but now we extend the notion of a state to mean a pair of classes, one corresponding to the block above the current block (vector to be coded) and another corresponding to the block to the left of the current block. (In general we might also look at the block to the upper left as well and have states correspond to trios of classes, as is done in [17].) Thus the state of the current block will be implied by the class of the reproduction in the block above together with the class of the reproduction in the block to the left, both blocks having been coded before the current block. If there are M classes, there will be $K = M^2$ states (M^3 if three blocks are used). (Clearly special rules will be needed to handle the top row of blocks and the left column. Special codes can be designed or arbitrary states assigned.) Apart from this difference, the general design algorithm is the same.

State codebooks are designed based on a partition of the training sequence according to an idealized state, that is, according to the true classes of the original input blocks, and codebooks are designed for each of the states. The FSVQ is then formed by closing the loop and classifying the reproductions rather than the original input blocks. Although the general idea is the same, the specific classifier used was much more sophisticated than a simple VQ and yielded much higher quality. They used a perceptually-based edge-classifier similar to that developed by Ramamurthi and Gersho for classified VQ of images [140] [258] [259]. The classifier operates by approximating edges, if present in a block, by a straight line. It distinguishes between four edge orientations and, for each orientation, between the two categories of black-to-white and white-to-black transition. If no edge is detected within a block, a mean classifier classifies the block according to its quantized mean. The classifier had 16 classes (8 for edges, 8 for means) and hence 256 states.

Additional improvement can be obtained by modifying the simple mean squared error distortion measure in a perceptually meaningful way: The distortion between two blocks is measured by squared error, but it is weighted upwards if the two blocks do not fall in the same edge class. This forces a higher penalty for trying to code one edge class into a reproduction of another edge class. This modification is used only in the encoding, however, and not in the centroid computation of the design algorithm. Improvement in the decoding step can be obtained via simple linear interpolation across the block boundaries. Aravind and Gersho reported Peak Signal-to-Noise ratios (PSNR's) in the neighborhood of 28 dB at .32 bpp and 27 dB at .24 bpp on the USC data base using these techniques in [17].

An alternative next-state rule called a *side-match* FSVQ was developed by Kim [204] in an attempt to force the encoder to optimize edge contiguity across block borders without doing explicit edge detection. The state (and hence the appropriate codebook for the current block) is determined by looking only at the pixels adjacent to the current block contained in the block above and to the left of the current block. We refer to these pixels as the *border pixels* of the current block. The basic idea is that the state codebook is selected as the subset of the super codebook (the collection of all codewords in all state codebooks) with the best codewords for matching the border pixels. The collection of states corresponds to the collection of all possible borders. The state selection procedure works as follows: Suppose that C is a super codebook of M codewords. The codebook C can be an ordinary memoryless VQ having a large rate. Given a previous state or border consisting of the pixel values adjacent to the current block and above it and to the left (a row above and a column to the left), select the 2^R codewords from the super codebooks that have the lowest distortion between the border pixels and the adjacent pixels in the codeword. The pixels being matched are pictured in Figure 14.10. The subblock being encoded is the sparsely shaded one. The sixteen squares represent the individual pixels. The pixels in the adjacent blocks used to match the sides are more densely shaded.

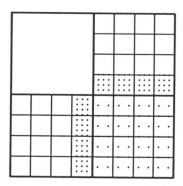

Figure 14.10: Side Match

The words in the state codebook can be listed in order of smallest distortion with the border to largest, an ordering which can be convenient if subsequent noiseless coding is used. Once computed the next-state mapping can of course be implemented by table lookup since it is a purely digital mapping.

As with any FSVQ, the next state selection procedure operates on reproduction blocks available to the decoder and not on the original input blocks. Thus in principle there could be as many as M^2 different borders and hence that many states.

Once the state codebook is determined, the current block is encoded as usual. The resulting codeword will then play its role in determining the new state and codebook for the future blocks to be encoded.

Note that the design is not omniscient and uses a simple nearest neighbor rule to select a state codebook from a large super codebook. The selection procedure explicitly tries to find the subcodebook that best matches edge behavior in previously coded blocks.

This scheme yielded peak signal-to-noise ratios (PSNR) of 30 dB at 1/4 bit per pixel when used in conjunction with Huffman coding, an improvement of 3.5 dB over ordinary VQ. When variable length noiseless coding was not used, the improvement dropped to around 1 dB over ordinary VQ. The lower rate VQ used a super codebook of size 256, a state space of size 65536, and a state codebook size of of 64.

Kim also developed a related, but more complex scheme, which attempts to match overlapping blocks instead of adjacent blocks [204].

FSVQ is still a relatively new approach to image compression, but the high quality of still frame images at 1/2 bit per pixel already obtained shows high promise as a low complexity, high quality image compression algorithm. Because of the simple digital next state selection procedures and the low complexity searches due to the low bit rates and small state codebooks, FSVQ is amenable to VLSI implementation and can do real time image coding at video rates [291].

14.7   Problems

14.1. Show that delta modulation with ideal integration is a special case of a recursive coding system. Define the state and the specific form of the equations (14.1.1) – (14.1.3) for this case.

14.2. Consider a predictive vector quantizer with memory size m and predictor matrices $\mathbf{A}_j$ for $j = 1, 2, 3, \cdots, m$. Show that this structure is a recursive vector quantizer by defining the state and specifying the state-transition function and the specific encoder and decoder mappings as special cases of the general form given in equations (14.1.1) – (14.1.3). Show that the number of distinct states is in general infinite.

14.3. Consider the simple predictive quantizing system where the input sequence is $\mathbf{X}_n$ and reconstructed output sequence is $\hat{\mathbf{X}}_n$, the quantizer has two output levels, ± 1 with threshold value zero, and the predictor is given $\tilde{\mathbf{X}}_n = f(\hat{\mathbf{X}}_{n-1})$ where $f(x) = x$ for $|x| \leq 3$ and $f(x) = 3\,\mathrm{sgn}(x)$ for $|x| > 10$. Assume the initial condition $\hat{\mathbf{X}}_0 = 0$.

Show that this is a finite-state recursive coding system and specify the number of states. Is it an FSVQ? If the input sequence starting at $n = 1$ is $\{1.3, 1.7, 2.1, 1.9, \ldots\}$, compute the first four reconstructed values.

14.4. Consider a PVQ system where the input vectors $\mathbf{X}_n$ have been previously quantized and take on a possibly large but finite set of M values in k-dimensional space. Show that for this case PVQ is a finite-state recursive coding system.

14.5. Given a specified predictor and quantizer for DPCM and viewing DPCM as a particular form of recursive coding system, consider the DPCM decoder structure but discard the standard DPCM encoder. Find an encoder for this decoder that satisfies the nearest neighbor property as defined for an FSVQ, except here it is not necessary to assume that the state space is finite. Specify this encoder algorithmically. *Hint:* the encoder will depend on the quantizer's output levels but not on its decision levels.

14.6. Consider the labeled state VQ derived from a general FSVQ with given next-state function f and decoder β. Draw a block diagram of the decoder using as building blocks unit delays and the mappings f and β, which each have two inputs and one output.

14.7. Is side-matched FSVQ a labeled-state or labeled-transition FSVQ?

14.8. Explain why FSVQs for images typically do not look at the blocks to the right and below the current block when selecting a state.

14.9. The side-match approach is similar to the histogram and nearest neighbor designs in that it begins with a super codebook and then selects state codebooks as subsets. Why do you think the side-match technique works so much better?

14.10. Suppose that a predictive vector quantizer is constrained to produce only a finite number of predicted values of the previous vector (or adjacent blocks in an image coder). For example, the output of the predictor itself might by quantized to one of K possible values. The resulting VQ is a recursive VQ, but is it a finite-state VQ? Is it a finite-state source code in the sense of Shannon? Assuming that the predictor operates on the original source vectors when it is designed, how might you optimize the quantizer used on the predictor so as to obtain a minimum mean squared error predictor? In particular, if you were given a training sequence and told to produce a vector predictor with only 8 possible values, how would you do it?

Chapter 15

Tree and Trellis Encoding

15.1 Delayed Decision Encoder

In this chapter we present the concepts and design algorithms for a class of vector quantization systems with memory that are more general and potentially more powerful than finite-state VQ. The presentation will depend on the previous chapter, so that it is advisable that the reader be familiar with at least the main idea of recursive coding systems and some aspects of FSVQ.

Suppose that we have a recursive vector quantizer as described in Section 14.1, that is, we have a decoder $\hat{\mathbf{x}} = \beta(u, s)$, where s is the decoder state and u the received channel index, an encoder $u = \alpha(\mathbf{x}, s)$, where $\mathbf{x}$ is the input vector and s the encoder state, and a next-state rule $f(u, s)$ giving the next code state if the current state is s and the current channel word is u. It is assumed in a recursive quantizer that there are no channel errors and hence if both encoder and decoder begin in a common state, then the decoder state and encoder state are thereafter identical for the encoding and decoding operation.

In an ordinary recursive quantizer, the encoder mapping α is in fact determined by the decoder mapping: $\alpha(\mathbf{x}, s)$ must produce the channel symbol u which minimizes the distortion $d(\mathbf{x}, \beta(u, s))$. Recall that such a "greedy" algorithm is not inherently optimal because the selection of a low distortion short term path might lead to a bad state and hence higher distortion in the long run. Although a goal in coding system design is to avoid this possibility, there is no guarantee that this will not occur with a greedy encoding algorithm. There is, however, a means of solving the

555

problem and of constructing an encoder that, like the simple memoryless VQ, is optimal for the decoder, or at least nearly so. In addition to the cost of additional complexity, there is another cost for this optimality, and it can be crucial in some applications: coding delay. One means of improving performance over the long run is to permit the encoder to wait longer before producing code vectors. Instead of comparing a single input vector to a single reproduction and producing the channel word, the encoder could simultaneously consider two consecutive input vectors and compare the resulting distortion with two corresponding reproduction vectors resulting from any two channel codewords. This would allow the encoder to ensure a minimum distortion fit for two consecutive input vectors rather than for just one. Clearly this can perform no worse than the original rule, and it could perform better!

More generally, the encoder for a given recursive VQ decoder could view L successive input vectors, then try driving the decoder with every possible sequence of L channel symbols, and find the sequence of channel symbols that produces the best overall fit, that is, the smallest total distortion between the L input and L reproduction vectors. The encoder can now transmit one or more (up to L) of these channel symbols and a new search is begun.

This is the basic idea behind *delayed decision encoding* and the principle can be applied to any recursive decoder. The approach has been called variously *delayed decision encoding, lookahead encoding, multipath search encoding,* and, for reasons to be seen, *tree encoding.* As L gets larger, the technique guarantees that ever longer sequences of input and reproduction vectors will have the minimum possible distortion for the particular decoder and hence one can consider the encoder to be optimal (at least in an asymptotic sense as the encoding block size or lookahead depth gets large) for the decoder, as a simple minimum distortion rule was for a memoryless VQ. Obviously, however, the incurred encoding delay can quickly become intolerable in some applications. For example, if the coding is part of a communication link in a feedback control system, the data would be too stale to be useful. On the other hand, in some applications a delay of many vectors is tolerable, e.g., in speech or image storage images or one-way digital audio or video broadcasting. Also, the encoding complexity grows exponentially with L and can quickly become prohibitive for high or even moderate rates (in bits per input vector).

There are in fact several variations on this concept. One way to assure theoretically better long term performance than memoryless VQ is to have a delayed decision encoder operate essentially as a block code or "giant" vector quantizer: It begins with an initial state, finds an effective sequence of L channel symbols for L input vectors, and then releases the L channel symbols to the channel. It then proceeds to do the same thing for the next group of L input vectors. Typically the scheme will drive the encoder and decoder state back into the initial state s^* at the end of one group of L input vectors. By resetting the state prior to encoding each block, there is no memory of the prior block used in encoding the current block. Hence, in this case the code behaves like a structurally constrained vector quantizer of length Lk where k is the input vector dimension.

An alternative strategy to a blocked delayed decision encoder for improving the greedy encoding algorithm is an *incremental* delayed decision encoder. Here the path map is selected as in the block case, but instead of releasing the entire block of L channel words, only one (or some small number ν) are released and the encoder then again searches ahead L vectors so as to pick a new path and advance further. Such incremental encoding provides a smoother operation and fewer blocking effects in the data than does encoding an entire L-block, but it is not as well understood theoretically and it requires more computation per input vector. All of the delayed decision encoding algorithms can be used in either block or incremental fashion.

Another problem with a delayed decision encoding should also be apparent: At first encounter, the encoding algorithm appears to be computationally horrendous. We have already seen that minimum distortion searches can be very demanding of computation and memory, and now we are proposing that a sequence of several codebooks all be searched before producing any bits for the channel. What makes the approach important for applications is the fact that there exist techniques for accomplishing this search with acceptable complexity. These techniques provide us with a nearly optimal encoder for any recursive decoder. The twin goals of this chapter, then, are the description of such encoding algorithms and the development of design algorithms for good recursive decoders.

15.2 Tree and Trellis Coding

No hint has yet been given as to why this chapter has the words *tree* and *trellis* in its title. We now consider more carefully the operation of a delayed decision encoder and show that a trellis or, more generally, tree, structure arises naturally in the analysis of such codes and plays an important role in their implementation.

Given the recursive VQ previously described, we first focus on the decoder and next-state mapping and abandon the encoder. Let the initial state be s^* and suppose the channel symbols have an alphabet of size $N = 2^R$. For simplicity we shall consider the special case where the channel symbols are binary-valued, so that $N = 2$, that is, the code has a rate of one bit per source vector. While this is clearly inadequate for most applications, it is useful to demonstrate the basic ideas with a minimum of notational clutter. All of the ideas extend in a straightforward manner with the binary trees and trellises becoming more general N-ary trees and trellises.

To depict the operation of the decoder, we use a directed graph called a *tree*, much as we did in the development of entropy codes and tree-searched VQ. Here, however, the tree is used to represent the time evolution of the decoder's operation so that each path through the tree identifies a particular sequence of states taken on by the decoder in successive time instants. Thus, each level of the tree corresponds to the time of arrival of a new input vector, whereas in the case of a TSVQ tree, the entire tree represents the possible ways in which a single input vector is searched. The initial state is the *root node* of the tree and all subsequent nodes will also correspond to decoder states. Suppose the encoder is in the initial state and suppose the channel index is binary valued, either a 0 or 1. If it produces a 0, then the decoder will produce a reproduction $\beta(0, s^*)$ and the decoder will advance to state $s_0 = f(0, s^*)$. If the encoder produces a 1, then the decoder will produce a reproduction $\beta(1, s^*)$ and the decoder will advance to state $s_1 = f(1, s^*)$. This action is depicted in Figure 15.1 by a *branch* of the tree moving upward if the encoder produces a 1 and downward if it produces a 0. The branch is *labeled* by the output produced if that branch is taken and the branch is terminated in a node representing the new state. Thus we have "grown" the first level of a code tree. We now continue by repeating the process for each of the nodes in the first level of the tree. If the encoder is in state s_0 in level 1, then it can produce a 0 with a resulting output $\beta(0, s_0)$ and next state $s_{00} = f(0, s_0)$, or a 1 with resulting output $\beta(1, s_0)$ and next state $s_{01} = f(1, s_0)$. For convenience we subscript the states by the binary channel sequence which led from the initial state to the current state. This sequence is called the *path map* since knowing the sequence one can trace a path through the tree from the root node to the current node. For example, a path map (received sequence) of 1101 will yield a decoded reproduction sequence

$$\beta(1, s*), \ \beta(1, s_1), \ \beta(0, s_{11}), \ \beta(1, s_{110}).$$

Note that with the tree drawn horizontally, the horizontal position of a node corresponds to a particular time, with the root note corresponding to $t = 0$ and time increasing to the right. Table 15.1 summarizes the actions available to the decoder at the first level of the tree. The tree can be extended indefinitely in a similar fashion.

While the tree depicts the possible paths of the decoder, it is the encoder which must make use of this tree in order to search for an effective or

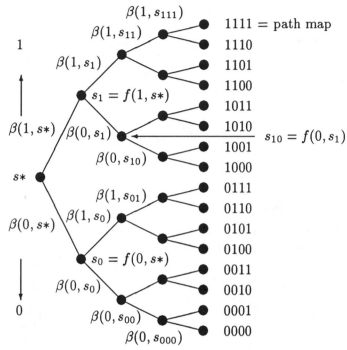

Figure 15.1: Recursive Decoder Tree

Level 1 State	Channel Symbol	Output	Level 2 State
s_0	0	$\beta(0, s_0)$	$s_{00} = f(0, s_0)$
s_0	1	$\beta(1, s_0)$	$s_{01} = f(1, s_0)$
s_1	0	$\beta(0, s_1)$	$s_{10} = f(0, s_1)$
s_1	1	$\beta(1, s_1)$	$s_{11} = f(1, s_1)$

Table 15.1: First Level Decoder Actions

optimal path and thereby determine the channel symbols to be transmitted. If the encoder is to have a delay or *search depth* of L input vectors, then the tree should be extended to L levels. Note that the tree grows exponentially fast with the number of levels since each level has twice as many nodes as the previous level. Given such a code tree, the task of the encoder is now the following: Given an input sequence $\mathbf{x}_0, \mathbf{x}_1, \cdots, \mathbf{x}_{L-1}$, find a path map through the tree (a sequence of symbols from the alphabet of size $N = 2^R$) $u_0, u_1, \cdots, u_{L-1}$ such that the corresponding sequence of states $\sigma_0, \sigma_1, \cdots, \sigma_{L-1}$ (where $\sigma_0 = s^*$) and branch labels (outputs) $(\hat{\mathbf{x}}_0, \cdots, \hat{\mathbf{x}}_{L-1}) = (\beta(\sigma_0, u_0), \cdots, \beta(\sigma_{L-1}, u_{L-1}))$ yields the minimum value of the *path distortion*

$$\Delta(u_0, u_1, \cdots, u_{L-1}) = \sum_{l=0}^{L-1} d(\mathbf{x}_l, \hat{\mathbf{x}}_l), \qquad (15.2.1)$$

where d is the measure of distortion between any pair of vectors ($\mathbf{x}$ and $\hat{\mathbf{x}}$, sometimes called the *per-letter distortion*. Thus the encoder can be thought of as a minimum cost search of a directed graph. This is a much studied problem in operations research, convex programming, and information theory. We shall present two of the most popular algorithms for doing the search. First, however, consider an important special case of the system thus far described.

Suppose that the recursive VQ decoder is in fact a finite state VQ decoder. With only a finite number of states the tree diagram becomes highly redundant. Once one has descended to a sufficient depth of the tree, there are more nodes than there are distinct states of the decoder and hence several of the nodes must represent the same state. The subtree emanating from any node having a given state will be identical for each such node at this level of the tree. No level can have more *distinct* nodes than there are states. Thus we can simplify the picture if we *merge* all those nodes corresponding to the same state to produce a simpler picture of the decoding tree. This leads to a new and simplified map called a *trellis* that more compactly represents of all possible paths through the tree. If the number of states is K, then we need only consider K nodes, one for each state, at each successive time instant after $t = 0$. The trellis consists of the set of all paths starting at the root node and traversing one of the K state nodes at each time instant and terminating at a node at time $K - 1$. The name *trellis* for a merged tree was invented by G. D. Forney, Jr., based on the resemblance of the resulting merged tree to the common garden trellis. As one might suspect, a trellis is easier to search than a general tree because of the additional structure. A *trellis encoding system* is a recursive coding system with a finite-state decoder and a minimum distortion trellis search encoder. It should be pointed out that the goal of searches of trees and trellises is to produce a minimum distortion path, but as with tree structured VQ, simplifications and shortcuts are sometimes taken to speed up a search at the expense of perhaps not finding the true minimum distortion path. The Viterbi algorithm encoder to be considered shortly indeed finds a minimum distortion path to the search depth of the encoder. The M algorithm to be considered finds a good path, but not necessarily the best one.

We shall emphasize trellis coders in this chapter, although we will discuss some aspects of the more general tree encoding systems. As one might suspect, a trellis is easier to search than a general tree because of the additional structure. This class of coding systems can be shown to have the potential for nearly optimal performance based on Shannon theory arguments [150][151][159], but our focus continues on design techniques which produce good, if not optimal, coding schemes. We illustrate the idea of trellis encoding by an example of a decoder structure with a special, but important, form.

A particular type of recursive decoder with a finite number of states, called a *shift register decoder*, consists of a shift register which stores in each stage an R bit binary vector representing a channel symbol and with m stages along with a lookup table having $N^m = 2^{Rm}$ possible reproduction levels as outputs as depicted in Figure 15.2 for the special case of $m = 3$ stages and binary channel symbols, i.e., $R = 1$. The contents of the shift register form an N-ary m-tuple $u = (u_0, u_1, \cdots, u_{K-1})$ (with the lowest index being the most recent entry and hence the leftmost channel symbol in the shift register). If we define the current state to be $s = (u_1, u_2, \cdots, u_{K-1})$ then the decoder output mapping is given by $\mathbf{y}_u = \beta(u_0, s) = g(u)$. Thus $\mathbf{y}_u$ is a code vector the state codebook $\mathcal{C}_s$ and is therefore is an element of the super codebook $\mathcal{C}$ of all possible reproduction vectors. The possible outputs of the shift register decoder are contained in the codebook or lookup table $\mathcal{C}$ and the Nary shift register content, u, forms the index that addresses the codeword $\mathbf{y}_u$ in the table. The functional notation $g(u)$ emphasizes the fact that the decoder can be viewed as a nonlinear filter. Such codes are called *sliding block codes* because a block of input vectors (or in some cases, scalars) is viewed to produce an output symbol and at successive time instants the new block is formed by "sliding" or shifting the block so that the oldest (rightmost) vector is dropped, a new vector is included at the left, and the remainder are just shifted to the right [169]. In the example of Figure 15.2, each symbol is one bit and the decoder is a finite-state VQ decoder with four states. The states are the four possible binary pairs constituting the two oldest channel bits in the register. If the shift register content is $u = b_0 b_1 b_2$, then $u = bs$, where $b = b_0$ is the current

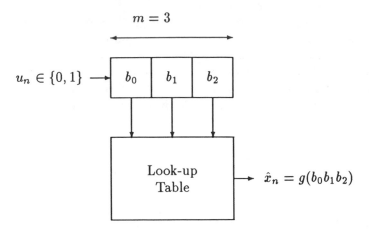

Figure 15.2: Sliding Block Decoder

channel bit (leftmost bit in the register) and $s = b_1b_2$, the state, comprises the remaining two bits. With this convention the reproduction function becomes simply $\beta(u, s) = g(bs) = y_{bs}$. The next-state function is forced by the shift register structure: the next state is always formed by shifting so that the current channel bit becomes the left bit in the state and the previous left state bit becomes the new right state bit. This is summed up by the next-state table for the given example in Table 15.2. Alternatively,

u	v	$s_{\text{next}} = f(u, v)$
0	00	00
0	01	00
0	10	01
0	11	01
1	00	10
1	01	10
1	10	11
1	11	11

Table 15.2: Sliding Block Next State Function

it can be summarized by the formula

$$f(b, b_1b_2) = bb_1.$$

There is nothing unique about this representation of the decoder as a finite-state decoder, it is also common in the literature for authors to reverse the bits describing the state and stick the channel symbol on the right, that is, to consider the state to be b_2b_1 and to write the output function as $g(b_2b_1b_0)$ so that the rightmost bit represents the most recent one. This will result in a different but equivalent next state function.

For this example we can redraw the decoder tree of Figure 15.2 as the trellis diagram shown in Figure 15.3 The initial state is taken as $s* = 00$.

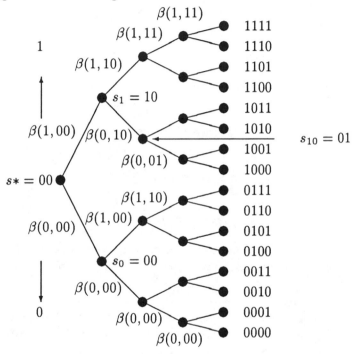

Figure 15.3: Sliding Block Decoder Tree

In some applications of trellis encoding, the original source is actually a sequence of samples of a waveform so that in blocking the waveform into a sequence of vectors, the designer has a choice of the vector dimension k. In memoryless VQ it is clear that the larger is k, the better is the rate-distortion tradeoff that can be achieved; of course, complexity places a limit on k for a given rate. On the other hand, in trellis encoding, the performance measure given by (15.2.1) shows that the distortion to be minimized depends only on the value of the product kL, the total number of samples under consideration rather than on the block size k. This suggests that there might be no sacrifice in performance by using $k = 1$, where the vectors are one-dimensional. This case will arise later in this chapter when we consider a class of trellis encoding systems which combine the use of

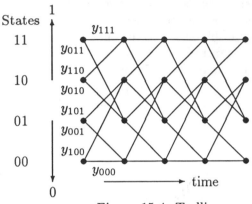

Figure 15.4: Trellis

linear prediction and shift register decoding.

The M Algorithm for Tree Searching

There are a variety of techniques for minimum-cost tree searching and these have been used extensively in error control coding (channel coding) as well as source coding. Many of the algorithms were originally developed for error control coding and many of these have had only limited use in source coding applications. Good surveys may be found in Anderson [10] [11]. The most popular of these algorithms for source coding has proved to be the M algorithm (also called the (M, L) algorithm). This algorithm was originally introduced for source coding by Jelinek and Anderson [196] and it has been extensively studied since, especially for speech coding applications. It is suboptimal in that it may not result in the minimum distortion path through a tree, but it does result in a path through the tree that has small average distortion if not the smallest possible. The algorithm can be used to search any tree and hence can also be used for trellis encoding.

The basic idea of the algorithm is simple: If the encoder starts at an initial node, it does not attempt to find the minimum distortion path for the complete depth L tree. Instead it narrows its search to preserve at most M candidate paths at any time. This is accomplished by effectively pruning the tree as it is searched to permit no more than M pathmaps to be preserved as candidates in any level of the tree. The search begins at the root node. The entire tree is constructed out to level l, where l is the largest integer for which $2^l \leq M$. All 2^l possible path maps are saved

---

Table 15.3: **M Algorithm Tree Encoder**

Step 0. Given: a binary tree described by a collection of nodes $\mathcal{S}$, a collection of pathmaps $\mathcal{M}$ through the tree leading to the terminal nodes, a collection of reproduction vectors produced by the pathmaps according to a decoder $\beta(u, s)$ producing a reproduction given a channel symbol u at node s, a per-letter distortion measure d, and a source input vector $x^L = (x_0, \cdots, x_{L-1})$. For all integers $l = 1, 2, \cdots, L - 1$ let $D_l(u^l)$ denote the distortion for between the first l letters of the input, $(x_0, \cdots, x_{l-1})$, and the reproduction sequence produced by following the tree through a pathmap $u^l = (u_0, \cdots, u_{l-1})$. Set $l = 1$.

Step 1. Grow the tree to levels $l = 1, 2, \cdots, l^*$, where l^* is the largest integer for which $2^{l^*} \leq M$. This requires listing for each l the nodes, the pathmaps into those nodes, and the path distortions between the first l input vectors and the corresponding reproduction vectors along the path. Let $\mathcal{S}_l$ and $\mathcal{M}_l$ denote the collection of all 2^l tree nodes at depth l and all 2^l pathmaps leading into tree nodes of depth l, respectively. Let $\Delta_l = \{D_l(u^l); \ u^l \in \mathcal{M}_l\}$ denote the distortions resulting from encoding the first l input vectors with pathmap u^l through the tree. Set $l^* + 1 \to l$ and go to Step 2.

Step 2. Given the sets $\mathcal{S}_{l-1}$ and $\mathcal{M}_{l-l}$ of surviving nodes and pathmaps at the previous depth, find all depth l nodes and the corresponding pathmaps which are extensions of the surviving nodes and pathmaps and compute the corresponding distortions $D_l(u^l)$. Define the new survivor sets of nodes $\mathcal{S}_l$ and pathmaps $\mathcal{M}_l$ as the M nodes and corresponding pathmaps yielding the smallest distortions.

Step 3. If $l < L-1$, set $l+1 \to l$ and go to Step 2. If $l = L-1$, pick the final pathmap u^L yielding the smallest value of $D_L(u^L)$.

as possible channel codewords. At depth $l + 1$ we first compute the path distortion between the $l + 1$ input vectors and all possible 2^{l+1} length $l + 1$ reproduction sequences corresponding to the nodes at this depth. Now, however, only the M best (in the sense of least path distortion) of these path maps are saved as candidate code sequences. The remaining nodes are removed from the tree and not considered further. The encoder then advances to look at the next level. It now extends all M of the existing candidate paths by one channel word, producing $2M$ length $l + 2$ paths for the one bit per source vector case. Of these, only the M best nodes are saved, the remainder being abandoned forever. The encoder continues in this fashion until the target depth L is reached. At this point the path out of the M candidates having the smallest distortion is selected and the corresponding channel symbols released.

Clearly the larger M gets, the more paths are saved and the more nearly optimal the encoder is. Equally clearly, the larger M implies more computational complexity.

The algorithm is summarized in Table 15.3.

While the M algorithm can be used for trellis encoding, it takes no advantage of the simpler trellis structure so that it is usually used for tree encoding when the number of states is not finite.

The Viterbi Algorithm

The Viterbi algorithm is a minimum-cost search technique specifically suited for a trellis. It is an example of dynamic programming and was originally developed for error control codes. A good general development for this application may be found in Forney [123]. The technique was first applied to source coding by Viterbi and Omura [323].

Suppose that we now have a finite-state recursive decoder and hence a decoder trellis as typified in Figure 15.4. The key idea of the Viterbi algorithm is the following. The minimum distortion path from time 0 to time n must be an extension of one of the minimum distortion paths to a node at time $n - 1$. Thus in order to find the best possible path of length L we compute the best path to each state for each time unit by finding the best extension from the previous states into the current state and we perform this for each time unit up until time L. This idea is known formally as the optimality principle of dynamic programming. The detailed operation is described in Table 15.4.

A key facet above is that we have a block code with block length L, but we do not have to search a block code of 2^L by computing all 2^L possible distortions. Instead we compute a sequence of K distortions, where K is the number of states, and retain the running distortion for each state along with

---

Table 15.4: **Viterbi Algorithm Trellis Encoder**

Step 0. Given a collection of states $S = \sigma_0, \cdots, \sigma_K$, a starting state s^*, a decoder $\beta(u, s)$ producing a reproduction given channel symbol u in state s, a per-letter distortion measure d, a source input vector $x_0, \cdots, x_{L-1}$. Let $D_j(k)$ denote the distortion for state k at time j. Set $D_0(s^*) = 0$, $D_0(s) = \infty$ for $s \neq s^*$. Set $l = 1$.

Step 1. For each current state s find

$$D_l(s) = \min_{\sigma} \left(D_{l-1}(\sigma) + \min_{u: f(u, \sigma) = s} d(x_l, \beta(u, \sigma)) \right)$$

and let s' denote the minimizing value of the previous state and u' the minimizing value of the channel symbol u. Thus the best path into the current state s passes through the previous state s' and is forced by channel symbol u'.

Given the path map $u^{l-1}(s')$ to the best previous state s', form the extended path map $u^l(s) = (u^{l-1}(s'), u')$ giving the best (minimum distortion) path through the trellis from the initial state into the current state s at level l. In other words, for each current state find the distortion resulting from extending the K best paths into the previous states into the current state. The best is picked and the distortion and cumulative path indices to that state saved.

Step 2. If $l < L - 1$, set $l + 1 \rightarrow l$ and go to 1. If $l = L - 1$, pick the final state s_f yielding the minimum of the K values $D_L(s)$. The optimal path map is then $u^L(s_f)$.

the accumulated distortions into each state and the accumulated channel symbol sequence. In particular, the complexity grows with the number of states, not with L.

The algorithm is implemented by keeping track of the following at each level of the trellis: (1) the best path into each node, and (2) the cumulative distortions up to that node. The principal of optimality means that knowing the best path into a node is equivalent to knowing at each state the best possible predecessor state. When the final node is reached, the path map the one which produces the smallest path distortion at the final depth.

15.3   Decoder Design

We have seen that a tree or trellis encoding system is completely described by a recursive VQ decoder together with a search algorithm. If the decoder is a finite-state decoder, then either the M algorithm or the Viterbi algorithm can be used. If it does not have a finite number of states, then the Viterbi algorithm is not applicable. Thus a tree or trellis encoding system can be constructed using any recursive decoder such as a predictive vector quantizer or FSVQ decoder. Any initial decoder immediately gives a complete but likely not optimal system. Therefore, it is natural to try to develop a design algorithm which will take this system and iteratively improve its performance in the same manner that the Lloyd algorithm was used for VQ. It is also natural to seek a means of developing a decoder for a tree or trellis encoding system from scratch in a manner parallel to the splitting algorithm for VQ design. In this we consider both of these issues.

The Trellis Decoder Design Algorithm

Suppose we have an initial finite-state recursive decoder. This means that we have an initial set of state codebooks C_s for each state s. Now we describe a variation of the Lloyd algorithm for FSVQ decoder design that can be used to iteratively improve the decoder for the case of a trellis encoder. To generate training data for the decoder codebooks, simply replace the original minimum distortion single-vector FSVQ encoder by the desired trellis search encoder, either a Viterbi algorithm or an M algorithm of search depth L. Start with a training sequence of input vectors and encode the training sequence with this new encoder, partitioning the input sequence according to state exactly as in the state codebook improvement algorithm for FSVQ as described in Section 14.4. Then, replace all of the reproduction codewords in all of the state codebooks by the appropriate centroids. As in ordinary FSVQ, this replacement necessarily reduces average distortion. Also as in ordinary FSVQ, there is no guarantee that the next step of re-encoding the training sequence using the new decoder will further reduce distortion (except in the limit of large search depth as $L \rightarrow \infty$). Intuitively, however, this problem should be reduced from that of FSVQ because the trellis search encoder is more nearly optimal than is the single vector FSVQ encoder. The trellis decoder iterative improvement algorithm is summarized in Table 15.5.

Table 15.5: **Iterative Decoder Improvement Algorithm**

Step 0. Given: a training sequence $\{\mathbf{x}_l;\ l = 0, 1, \cdots, L - 1\}$, a threshold $\epsilon > 0$, and an FSVQ with super codebook $\mathcal{C}^{(0)} = \{\mathbf{y}_{s,u};\ u \in \mathcal{N},\ s \in \mathcal{S}\}$, where $\mathcal{N}$ is the channel alphabet and $\mathcal{S}$ the state alphabet. Set $m = 1$ and $D_0 = \infty$.

Step 1. Encode the training sequence (using the Viterbi or another selected trellis search algorithm) the training sequence using the super codebook $\mathcal{C}^{(m)}$. The encoding will yield a sequence of code states $\{s_n\}$ and channel symbols $\{u_n\}$. Group the training vectors according to the the channel symbol and code states; that is, partition the training sequence vectors into sets $T_{u,s}$ where $\mathbf{x}_n$ is in $T_{u,s}$ if $(u_n, s_n) = (u, s)$. Let D_m denote the sample average distortion for the entire training sequence.

Step 2. If $m > 0$ and

$$\frac{D_{m-1} - D_m}{D_{m-1}} < \epsilon,$$

then set $\mathcal{C} = \mathcal{C}^{(m)}$ and halt with super codebook $\mathcal{C}^{(m)}$. Otherwise replace every codeword $\mathbf{y}_{u,s}$ by the centroid of the set $T_{u,s}$ for all $u \in \mathcal{N}$ and $s \in \mathcal{S}$ and go to Step 1.

The Lloyd-style trellis code improvement algorithm was developed by Stewart [301] (see also [302]) and it in fact predates the study of FSVQ design, a special case of trellis encoding where $L = 1$.

It is important to note that this algorithm only improves the reproduction levels or state codebooks of the system; it does not cause any changes in the next state function of the decoder. As seen in the FSVQ chapter, next state design is in general a difficult problem. An additional and obvious problem is that the trellis decoder improvement algorithm requires an initial codebook. We consider several ways of generating such codebooks.

Initialization

Any of the FSVQ decoders of the previous chapter can be used as an initial decoder for a trellis encoding system. There is also a variety of other possible choices based on the shift register decoder. If, for example, the channel symbols are binary and the shift-register has m elements, then we can consider the contents of the register to be a binary vector $u = (u_0, u_1, \cdots, u_{m-1})$. We consider bits to be shifted from left to right and we view the oldest $m-1$ bits in the register to be the state of the decoder (and hence there are 2^{m-1} possible states). The next state rule is obvious here: if we are in a state $s = (u_0, \cdots, u_{m-1})$ and receive a channel bit v, then the next state is $(v, u_0, \cdots, u_{m-2})$. To completely describe such a decoder, we need only specify the codebook or contents of a lookup table consisting of the output of the decoder for all possible 2^m contents of the shift register (or all possible combinations of state and channel symbol). Thus given the structure of the finite-state decoder, we need only provide the codebook of outputs $\mathcal{C} = g(u)$ for all binary m-tuples u. We can consider $\beta(v, s)$ to be the *label* of the trellis branch determined by the decoder state s and the channel symbol v.

The oldest technique and one intimately connected with the underlying rate-distortion theory is a randomized codebook selection where the branch labels are selected at random. This is a useful approach for proving theorems and has been successfully used in practice for coding nearly memoryless sources such as speech residuals [196] [336]. It is usually desirable, however, to use knowledge of the source to select an initial codebook in a deterministic fashion. Randomly selected codebooks may work well on the average, but specific realizations can be quite poor.

One of the first techniques for designing codebooks for sources with memory was the intuitively pleasing notion of a plagiarized decoder: Simply use the decoder of some other compression system that is known to be good. The FSVQ decoders are of this type, but the first such codes used DPCM decoders that were truncated to a finite number of states. The decoder was truncated since the usual decoder for a DPCM system has infinite states and does not yield a time-invariant trellis [216].

A somewhat more sophisticated design technique is the *fake process* approach [216]. Here the output of the decoder when driven by a memoryless sequence of channel symbols should resemble statistically the input as much as possible. One can, for example, try to force the marginal probability distribution and the autocorrelation (or power spectrum) of the decoded sequence to closely match those of the input. If the decoder produces sequences which closely resemble the typical sequences of the encoder, then the encoder should be able to find good matches to its typical sequences with the Viterbi or other search algorithm. The "code excited linear predictive" (CELP) speech coders, also known as "vector excitation coders" (VXC), can be viewed as having fake process decoders. See Chapter 16 for a description of this important speech coding technique.

Splitting

All of the initialization techniques described so far begin with a decoder having a fixed structure and then improve on it. We close this section with a means of finding a good shift register decoder of dimension $m + 1$ given a shift register of dimension m. Thus one can begin with a single stage register and iteratively design longer register decoders until some performance or complexity goal is achieved. The technique can be viewed as the trellis encoding extension of the splitting algorithm for ordinary memoryless VQ design since increasing the shift register length by one implies a doubling of the number of reproduction code vectors. This method has also been called the *extension* method by Stewart, who introduced it [301] [302].

Suppose as usual that the channel alphabet is binary and that we have a code $\mathcal{C}_m = \{g(u^m)\}$, where u^m varies over all binary m-tuples. We wish to form a new code $\mathcal{C}_{m+1} = \{g'(u^{m+1})\}$. A simple means of doing this is to replicate the labels in $\mathcal{C}_m$ and define

$$g'(u_0, \cdots, u_m) = g(u_1, \cdots, u_m); \quad u_0 \in (0, 1),$$

that is, given $u_1, \cdots, u_m$, the reproduction has the same value regardless of u_0. This new code has all the same labels available as the old code and hence has all the same available reproduction sequences and can be no worse. It can also be no better for the same reason. Given the new code, however, the iterative improvement algorithm can be run on it to find a better code of the new, larger dimension. The encoding portion of the improvement algorithm will now divide the training sequence into 2^{m+1} groups instead of 2^m and cluster centers will then be found for each of these "finer" groups of input vectors.

Table 15.6 summarizes a complete design algorithm for a trellis coding system for a shift register decoder of length m. The binary channel symbol alphabet is generalized to an alphabet of q symbols.

Table 15.6: **Trellis Decoder Design Algorithm Shift Register Decoder**

Step 0. Given a training sequence $\{\mathbf{x}_l;\ l = 0, 1, \cdots, L-1\}$ and a threshold $\epsilon > 0$. Run the scalar quantizer design algorithm to produce a scalar codebook having q codewords. This is the initial one-dimensional code $\mathcal{C}_1$. Observe that with a decoder of length 1, the resulting vector quantizer of dimension 1 is also a trellis coder of length 1, that is, there is only a single state. Set $i = 1$.

Step 1. *Splitting:* Given the codebook $\mathcal{C}_i$ with q^i codewords $\{g(u)\}$ with u ranging over all q-ary i-tuples, form a new codebook $\mathcal{C}_{i+1}(0)$ with codewords $\{g'(v)\}$, where v ranges over all q^{i+1} q-ary $(i+1)$-tuples with $g'(v_0, v_1, \cdots, v_i) = g(v_1, \cdots, v_i)$ for all q possible values of v_0. Set $i + 1 \rightarrow i$.

Step 2. *Lloyd Improvement Routine:* Run the trellis decoder iterative improvement algorithm of Table 15.5 on $\mathcal{C}_i(0)$ to improve the codebook.

Step 3. If $i = m$, halt with final codebook $\mathcal{C}_m$. Otherwise go to Step 1.

15.4 Predictive Trellis Encoders

Trellis encoded data compression systems can be viewed as a natural extension of ordinary VQ where the decoder is an FSVQ decoder and the encoder is allowed to search several vectors into the future when finding a minimum distortion path instead of being constrained to search only one vector ahead. All of the design algorithms presented so far for trellis coders, however, were based on shift register decoders, a very simple class of FSVQ decoders, and the only design problem is to fill the codebook. One might suspect that better performance could be obtained with more general FSVQ decoders, that is, more complicated finite-state decoders. We have already seen, however, that the design of good next state or state-transition rules for an FSVQ is in general a difficult task and can be strongly dependent on the particular data source to be compressed. Stewart found in several examples, however, that if the signal source is fairly uncorrelated, then the shift register decoders worked quite well and more general finite-state decoders provided little performance improvement [301]. This suggests a natural way to handle highly correlated sources where a simple shift register decoder does not perform as well as desired. First use prediction to "whiten" the source and then use a trellis coding system with a shift register decoder on the whitened source. Finally, reconstruct the final reproduction to the original source by a predictive feedback loop. As in predictive quantization systems studied earlier, the prediction should be done in a closed-loop fashion, where the predictor operates on observations of the decoded reproductions. It should be emphasized that this argument is purely *ad hoc*, there is no theorem guaranteeing that this use of prediction will improve performance. Such a claim requires at least specific experimental evidence to justify it, which is forthcoming.

The basic idea of predictive trellis encoding is very similar to that of predictive VQ with a trellis encoder replacing the memoryless vector quantizer in the transmitter and a shift register decoder replacing the memoryless VQ decoder in the receiver. There are, however, some subtle differences worth noting. In both cases a linear prediction based on past reproductions is subtracted from the input prior to the actual coding operation. In the PVQ case, this prediction involved predicting an entire vector or several samples into the future. In a predictive trellis encoding system, however, the prediction usually does not have to extend far into the future. Also, usually the input source is a scalar valued random sequence and only scalar one-step prediction is performed. For example, for a binary trellis coder with one bit per input sample, the encoder computes the prediction for the next input sample given the reproductions produced by decoding the trellis path map all the way up to the current node; that is, the encoder extends the path one branch at a time (corresponding to one input sample at a time) and it knows the resulting reproduction values up to the current point in the trellis. The actual trellis encoder tries to find a path through the trellis labeled by residuals, but each time it matches a residual for one branch the resulting reproduction is fed back to produce the next residual. The key difference with ordinary trellis encoding is that now it is done within a feedback loop rather than looking only at the input. This makes the prediction more "local" than in a predictive vector quantizer since one is predicting only one sample ahead.

Thus a trellis coder has the potential for producing better predictions and hence whiter residuals. As in the PVQ case, there are a variety of means of designing the predictors, both open loop and closed loop. The structure of a trellis encoder lends itself to a minor variation of the PVQ design techniques that leads to a good joint design of predictor and codebook by iteratively optimizing one for the other. This section follows the presentation in [22] where further details and experiments are described.

Predictive Trellis Encoding System

The decoder and encoder of a predictive trellis encoding system are depicted in Figures 15.5–15.6. For simplicity we will continue to focus on the special case of coding at the rate of 1 bit per input, but the modifications necessary for more general rates are as follows: instead of shifting out one bit for every sample applied to the encoder, simply shift out l bits for every k input samples to achieve a rate of l/k bits per sample. This is simply a vector trellis encoder with vector dimension 3 and channel symbol alphabet size 2^l.

The first part of the decoder resembles an ordinary VQ decoder in the sense that a binary vector (a group of binary channel symbols) is used as an index into a codebook of reproduction values. Here, however, we consider the channel symbols to arrive one bit at a time (rather than one vector at a time) and with the arrival of each new bit a single real number is produced, that is, the codebook contains scalars, not vectors. The binary word that is the input to the lookup table consists of several bits (m in the figure) which arrive one at a time. m is called the *constraint length* of the coding system. As can be seen from Figure 15.5 the scalar output of the table is not the final reproduction, instead it is added to a prediction of the input sample based on p previous reproduction samples. In general the predictor can be nonlinear, but we shall emphasize linear predictors as in the predictive VQ case. Note that the first section of the decoder can be considered as a nonlinear digital filter with binary inputs and a finite collection of real numbers (of high precision) as outputs. This nonlinear filter is cascaded

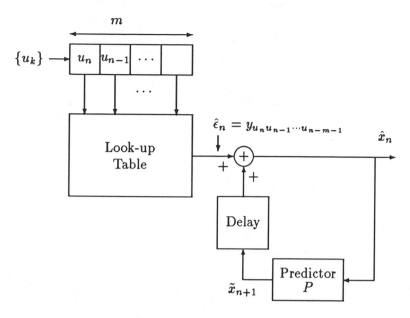

Figure 15.5: Predictive Trellis Decoder

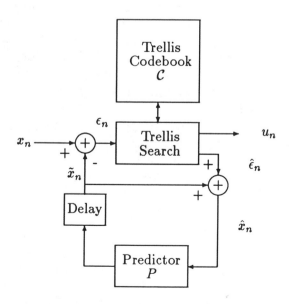

Figure 15.6: Predictive Trellis Encoder

with a linear filter (assuming that the prediction is linear).

To be more precise, if the contents of the shift register at time n are $u_n, u_{n-1}, \cdots, u_{n-m+1}$ and the resulting table output is

$$\beta(u_n, u_{n-1}, \cdots, u_{n-m+1})$$

and if the predictor mapping is P, then the reproduction sequence is defined recursively by

$$\hat{x}_n = \beta(u_n, u_{n-1}, \cdots, u_{n-k+1}) + P(\hat{x}_{n-1}, \cdots, \hat{x}_{n-p+1}).$$

Given an initial state this completely defines the decoder.

As in the nonpredictive trellis encoding system, the action of the decoder can be depicted by a trellis diagram with 2^{m-1} states. Unlike the nonpredictive case, however, the predictive trellis depicting the reproductions of the input signal is not time-invariant, that is, the branch labels change with time because of the feedback loop following the table output. Alternatively, the decoder is still a finite-state machine because the shift register has only a finite number of entries. It is not, however, a finite-state coding system for the original signal as previously defined because the output mapping varies with time, input, that is, the code is not stationary. The Viterbi algorithm can still be used, however, to find a minimum distortion path through this time-varying trellis because the principle of optimality still holds. One would suspect that more memory will be required to store the additional labels, but that the number of states and hence the number of candidate paths that must be retained depends only on the constraint length of the shift register. We shall see, however, that in fact one need only store and use a time-invariant trellis and that the trellis search will itself be done within a feedback loop.

The optimal encoder for a finite-state decoder is, as before, a minimum distortion search matched to the decoder, that is, the encoder should find the binary channel sequence (pathmap) $u_0, \cdots, u_{L-1}$ which produces the output sequence $\hat{x}_0, \cdots, \hat{x}_{L-1}$ for which the average distortion

$$\Delta = \sum_{l=0}^{L-1} d(x_l, \hat{x}_l)$$

is minimized. At first glance, this does not appear to be what the encoder shown in Figure 15.6 is doing. In the figure the trellis search uses a copy of the decoder table to find the minimum distortion fit to the sequence of residuals $\epsilon_n = x_n - \tilde{x}_n$ formed from the difference of the input x_n and the prediction $\tilde{x}_n$ based on the past coded residuals $\hat{\epsilon}_k$; $k < n$. Exactly as in the predictive quantization and predictive vector quantization systems, however, we have that

$$x_n - \hat{x}_n = (x_n - \tilde{x}_n) - (\hat{x}_n - \tilde{x}_n) = \epsilon_n - \hat{\epsilon}_n \qquad (15.4.1)$$

and hence the total overall distortion equals the total distortion in coding the residuals. Hence one can use a time-invariant trellis search algorithm to find the minimum distortion path through the trellis for the residuals and the resulting binary sequence then drives the decoder to produce the minimum distortion reproduction of the original input. Note that as the trellis search proceeds, the input sequence to the trellis search is being modified since it is a residual of the original input and the prediction based on the path selected thus far. Thus the predictive trellis encoder is more complicated than the nonpredictive, but not greatly so. In particular, it can still use the Viterbi algorithm to perform the actual optimal search and it has significantly less complexity than would a tree search algorithm which did not take advantage of the finite-state structure.

We now make the explicit assumption of a pth order linear predictor of the form

$$\tilde{x}_n = P(\hat{x}_{n-1}, \cdots, \hat{x}_{n-p}) = -\sum_{k=1}^{p} a_k \hat{x}_{n-k}, \qquad (15.4.2)$$

where $\{a_k; \ k = 1, \cdots, p\}$ are the linear predictor coefficients.

The Search Algorithm

The trellis search operates on an input sequence $x_0, \cdots, x_{L-1}$, where L is the search depth. When the system is first begun, initial conditions in the form of $\hat{x}_{-p}, \cdots, \hat{x}_{-1}$ are required to form the first prediction $\tilde{x}_0$ and hence the first residual ϵ_0. These past reproductions can be assumed to be all 0 so that $\epsilon_0 = x_0$. At each time n the encoder preserves for each state j the following information:

1. the distortion $D_n(j)$ at time n associated with state j, that is, the distortion of the minimum distortion path from the root node to state j at time n, and

2. the p previous reproduction levels corresponding to this path (for the predictor).

Let $\hat{x}_{n-l}(j); \ l = 1, \cdots, p$ represent the final p symbols of the survivor path ending at state j at time n. Let $y(i, j)$ denote the codeword on the branch connecting nodes i and j. The trellis search algorithm is described in Table 15.7.

Table 15.7: **Predictive Trellis Search Algorithm**

Step 0. Initialization: Set $n = 0$,

$$D_0(0) = 0,$$

$$D_0(j) = \infty; \ 1 \leq j \leq 2^m - 1$$

$$\hat{x}_l = 0; \ l = -p, \cdots, -1.$$

Step 1. For $j = 0, 1, \cdots, 2^{p-1} - 1$ find the following:

$$\tilde{x}_n(j) = -\sum_{l=1}^{p} a_l \hat{x}_{n-l},$$

$$D_{n+1}(j) = \min_i (D_n(i) + d(x_n, \tilde{x}_n(j) + y(i,j))),$$

$$I_n(j) = \min_i^{-1}(D_n(i) + d(x_n, \tilde{x}_n(j) + y(i,j))),$$

where the minimization is over all states i at time n which are connected to state j at time $n+1$. $I_n(j)$ gives the state at time n through which the best path to state j at time $n + 1$ passes. Update the reproduction

$$\hat{x}_{n+1}(j) = \tilde{x}_n(I_n(j)) + y(I_n(j), j).$$

This gives the reproduction when the system is at state j at time $n + 1$. The remaining prediction values are then given by

$$\hat{x}_{n+1-l}(j) = \hat{x}_{n-l+1}(I_n(j)); \ 2 \leq l \leq L.$$

The update for $D_{n+1}(j)$ is the dynamic programming step for the encoder trellis. The predictor updates follow from the linear predictor equation.

Step 3. If $n < L - 1$, set $n + 1 \rightarrow n$ and go to Step 1. Otherwise continue.

Step 4. When the search depth $n = L - 1$ has been reached, find the state (node) j for time $n = L-1$ with the minimum distortion $D_{L-1}(j)$. Release the corresponding pathmap through the trellis to the channel.

A Design Algorithm

The trellis search algorithm completely specifies the encoder given two sets of parameters:

- *the linear prediction coefficients*

$$a_i; \ i = 1, \cdots, p$$

and

- *the prediction error (residual) codewords*

$$y_k; \ k = 0, \cdots, 2^m - 1.$$

A natural choice for $\mathbf{a} = (a_1, \cdots, a_p)^t$ is the same as that made in the PVQ case. Use the optimal linear predictor of Theorem 4.2.2 assuming that the original source inputs rather than the reproductions are the observables. The actual correlation values are, as before, determined from the training sequence. This results in

$$\mathbf{a} = -\mathbf{R}^{-1}\mathbf{v} \qquad (15.4.3)$$

where

$$
\begin{aligned}
\mathbf{R} &= \{R(i-j); \ i = 0, \cdots, p-1; \ j = 0, \cdots, p-1\}, \\
\mathbf{v} &= (R(1), \cdots, R(p)), \\
R(i) &= \frac{1}{L-i} \sum_{k=0}^{L-i-1} x_k x_{k+i}.
\end{aligned}
$$

This selection shares the advantages and disadvantages of the PVQ case. It is simple, intuitive, and likely fairly good when the reproductions are good, that is, when the bit rate is high. On the other hand, it is not the truly optimal predictor since it is based on the perfect originals rather than the coded reproductions and it may be a poor predictor when the rate is low. We shall use this predictor only as an initial guess, however, and will improve on it as part of the design algorithm.

An initial guess for the codewords $\mathcal{C} = \{y_i; \ i = 0, \cdots, 2^{m-1} - 1\}$ can be obtained by the splitting or extension method exactly as in the nonpredictive trellis case. Other initializations are considered in [22].

We assume that the design algorithm has initial values for these parameters. We mention two final points before describing the algorithm. First, keep in mind that for any difference distortion measure $d(x, y) = \rho(x - y)$

(including the squared error distortion emphasized here) we have the property that

$$
\begin{aligned}
d(x_n, \hat{x}_n) &= \rho(x_n - \hat{x}_n) \\
&= \rho(\tilde{x}_n + \epsilon_n - \tilde{x}_n - \hat{\epsilon}_n) \\
&= \rho(\epsilon_n - \hat{\epsilon}_n) = d(\epsilon_n, \hat{\epsilon}_n),
\end{aligned}
$$

where $\hat{x}_n$ is the final reproduction formed if $\tilde{x}_n$ is the prediction and $\hat{\epsilon}_n$ is the selected prediction error codeword. Second, define for each i the set

$$
S_i = \{n : \hat{\epsilon}_n = y_i\}
$$

of all time indices n for which the prediction error ϵ_n is coded into the ith codeword y_i.

For a fixed prediction vector $\mathbf{a}$, the average distortion resulting the given trellis search

$$
\sum_{n=0}^{L-1} d(x_n, \hat{x}_n) = \sum_{n=0}^{L-1} (x_n - \hat{x}_n)^2 = \sum_{n=0}^{L-1} (\epsilon_n - \hat{\epsilon}_n)^2 \qquad (15.4.4)
$$

will be minimized if, instead of using the codewords $\mathcal{C} = \{y_i\}$ originally used in the trellis search, we replace the values (after the search) by the centroids given by

$$
y_i = \frac{1}{\|S_i\|} \sum_{n \in S_i} \epsilon_n; \quad i = 0, 1, \cdots, 2^m - 1, \qquad (15.4.5)
$$

where $\|S_i\|$ is the number of elements in S_i. This operation thus gives the optimal codebook for a given set of linear prediction parameters and a trellis search algorithm. As in the ordinary trellis encoding system, this codebook could now be substituted into the trellis search algorithm to produce an ordinary trellis encoder with a search matched to the decoder.

Given a codebook and a matched trellis search and hence given a reproduction sequence $\hat{x}_n$; $n = 0, \cdots, L-1$, there is a natural method of selecting a new set of linear prediction parameters optimized for this reproduction sequence. The goal is to find the optimal linear prediction coefficients for predicting x_n by

$$
\tilde{x}_n = -\sum_{l=1}^{m} a_l \hat{x}_{n-l}.
$$

From the orthogonality principle (Theorem 4.2.4) applied to the sample distribution, this implies that $\mathbf{a}$ should be such that the prediction errors and the observations are orthogonal in the sense that

$$\sum_{n=0}^{L-1}(x_n - \tilde{x}_n)\hat{x}_{n-i} = 0; \quad i = 1, \cdots, m.$$

Since

$$\hat{x}_n = -\sum_{l=1}^{m} a_l \hat{x}_{n-l} + \hat{\epsilon}_n,$$

this yields the predictor update equation

$$\sum_{l=1}^{m} a_l \sum_{n=0}^{L-1} \hat{x}_{n-l}\hat{x}_{n-i} = -\sum_{n=0}^{L-1}(x_n - q(\epsilon_n))\hat{x}_{n-i}; \quad i = 1, 2, \cdots, m. \quad (15.4.6)$$

If the trellis codebook and predictor parameters are a good match, then both the predictor and quantizer errors should be small and (15.4.6) should be well approximated by (15.4.3), which provides the **a** used for the initial guess.

Having described the basic iterations, we the design algorithm is described in Table 15.8.

The algorithm combines the Lloyd centroid improvement, the minimum distortion trellis search, and the training-sequence-based orthogonality principle in order to iteratively design a complete trellis code. The algorithm is a close relative of the steepest descent or stochastic gradient design algorithm used to design predictive VQ systems in [58]. The similarities and differences of the two design algorithms are discussed in [22].

Design Examples

Predictive trellis encoders are intended for sources with memory. Hence we focus on the principal examples encountered so far: a Gauss-Markov source and sampled speech. The same parameters used previously remain in effect. Figure 15.7 shows the SNR performance for both predictive and nonpredictive trellis encoding as a function of the trellis constraint length m.

The nonpredictive codes are a special case of the predictive codes with the linear predictor coefficients being permanently set to 0 and the predictor update step skipped. Recall that the distortion rate function for this source at 1 bit per sample is 13.2 dB. Both training and test sequences consisted of 30,000 samples and the table and figure report the test results (outside of the training sequence).

Observe that a predictive trellis encoder of constraint length m outperforms a nonpredictive trellis encoder of constraint length $m + 2$, the

Table 15.8: **Trellis Encoding System Design Algorithm**

Step 0. *Initialization:*

Need initial guesses for $C_0 = \{y_i^0; \ i = 0, \cdots, 2^m - 1\}$ and $\mathbf{a}_0 = \{a_l^i; \ l = 1, \cdots, m\}$. Set $\Delta_{-1} = \infty$ and $k = 0$. Fix a threshold $\epsilon > 0$.

Step 1. *Trellis Codebook Update:*

Encode the training sequence with trellis codebook C_k and predictor coefficients $\mathbf{a}_k$ and obtain the resulting average distortion

$$\Delta_k = \sum_{n=0}^{L-1} d(x_n, \hat{x}_n).$$

If $(\Delta_{k-1} - \Delta_k)/\Delta_k < \epsilon$, then stop with codebook $C = C_k$ and predictor coefficients $\mathbf{a} = \mathbf{a}_k$. Otherwise update the trellis codebook using the centroid equation (15.4.5) to obtain C_{k+1}.

Step 2. *Predictor Update:*

Use the new trellis codebook and $\mathbf{a}_k$ to encode the training sequence. Solve (15.4.6) to find the new predictor coefficients $\mathbf{a}_{k+1}$.

Step 3. Set $k + 1 \to k$. Go to Step 1.

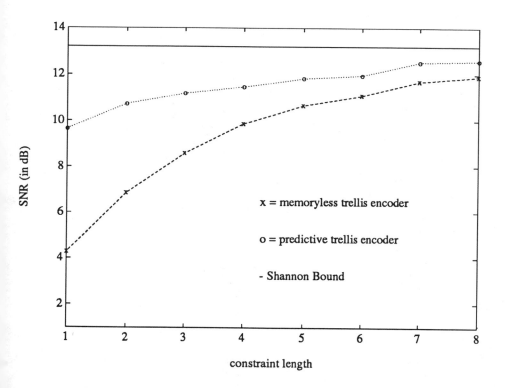

Figure 15.7: SNR for a Gauss-Markov Source

latter code having more than twice the storage and floating point multiply requirements. For a fixed constraint length, the predictive code provides about 4 dB improvement for short constraint lengths, but the improvement diminishes with increasing constraint length.

As a second example, the SNR performance for nonpredictive and predictive trellis encoders applied to the sampled speech training sequence previously described is summarized in Figure 15.8 for a linear predictors of order $m = 2$. The predictive codes provide about 2 dB of improvement for the lower constraint lengths, but the improvement diminishes with increasing constraint length.

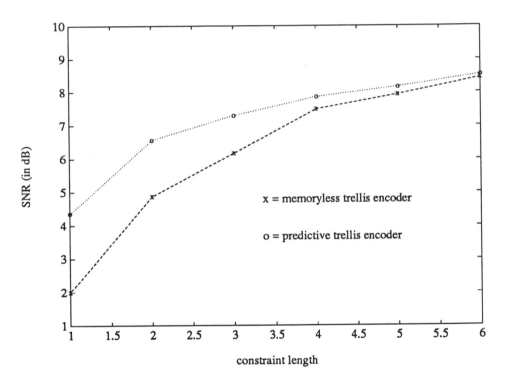

Figure 15.8: SNR for Sampled Speech

15.5 Other Design Techniques

It should be emphasized that the previous s provide only a sampling of approaches to the design of trellis encoding systems. One generalization is to permit finite-state machine codes which, like the predictive trellis encoding system, are not finite-state codes in the strict sense. In particular, one can allow the output of the decoder to depend in a more general way on past reproductions as well as on finite states and one need not restrict the encoder to a matched trellis search. This leads to an approach referred to as *alphabet-constrained data compression* [147]. We also mention two other approaches, one old and one quite new. An approach that has been implicitly discussed is that of designing a decoder using the FSVQ design techniques and then simply replacing the VQ encoder by a trellis search encoder, that is, allow the encoder to search several vectors into the future. This technique provides vector trellis encoding systems using previously developed techniques and can provide good performance in some examples [29]. The primary drawback is that the FSVQ techniques are geared to providing a good decoder for a single vector search, and there are no techniques (yet) for matching the next state design to a trellis search.

A newer technique builds on the duality of source and channel coding and on the recently developed coded modulation schemes originated by Ungerboeck [314][315]. One approach to the construction of these systems is to take good lattice codes (for a channel or a source) and then to partition the lattice into a subset (such as a coset) of the lattice and its translates, thereby obtaining a collection of disjoint codebooks which together give the original lattice. This is the same process as in the discussion of set partitioning in Chapter 14. Here a trellis code is constructed by having each of the disjoint pieces be a state codebook in a lattice and by providing a computationally simple next-state selection procedure. The original code need not be a lattice. For example, it could be a high rate scalar Lloyd-Max quantizer instead of a uniform scalar quantizer.

Systems of this type are called *trellis coded quantizers* and they perform quite well at low complexity on memoryless sources and predictive versions have been developed for sources with memory such as sampled speech. [232] [233].

15.6 Problems

15.1. Suppose that an FSVQ is designed and the encoder is then replaced by a two-vector lookahead to form a (vector) trellis encoded system. Can the resulting system be modeled as an FSVQ? (Either explain why not or provide a description of the encoder and decoder of the new FSVQ including the states, the next-state function, and the output function).

15.2. Develop a variation on the predictive trellis encoder design algorithm that yields a predictive VQ.

15.3. Specialize the description of the Viterbi algorithm in the case of a sliding block decoder.

15.4. Suppose that the state of a sliding block code in the three stage example in the text is considered to be the old shift register contents in reverse order, i.e., if the shift register contains $b_0 b_1 b_2$, then the state is $b_2 b_1$. Draw the resulting trellis.

15.5. Compare the relative advantages and disadvantages of a trellis encoding system of search depth L and a vector quantizer of dimension L. Repeat for the case of a predictive trellis encoding system and a predictive VQ.

15.6. Draw a decoder tree to depth four for a simple delta modulator. Give an example of an input sequence of length 4 for which the tree encoder will produce a lower mean square error than will the usual delta modulator encoder.

Chapter 16

Adaptive Vector Quantization

16.1   Introduction

Typically in VQ applications, source vectors are sequentially extracted from a real signal and are individually coded by a memoryless vector quantizer. Usually, these vectors are not independent and often they are also not identically distributed. For example, the vector statistics may vary slowly in time and hence exhibit nonstationary behavior. Since successive vectors are in general not independent, the conditional distribution of one vector given the observation of some neighboring vectors provides much more information about the vector than does its marginal distribution. The more "information" we have about a vector, the more accurately it can be coded for a given allocation of bits. Consequently, improved coding performance becomes possible if the quantizer can somehow adapt in time or space to suit the local statistical character of the vector source by observing, directly or indirectly, the vectors in some neighborhood of the current vector to be coded. A vector quantizer is *adaptive* if the codebook or the encoding rule is changed in time in order to match observed local statistics of the input sequence. Note that the structurally constrained VQ methods described in Chapter 12 are not adaptive since they focus on techniques for coding of a single vector assuming given statistics for that vector and they make no use of the past or future vectors in a sequence. On the other hand, the predictive and finite-state vector quantizers of Chapters 13 and 14 could be considered to be adaptive by this definition and have sometimes been called such in the literature, but the word "adaptive" is usually associated with systems where the codebook changes slowly with respect to the vector rate

587

rather than changing substantially with each successive vector. The boundary between adaptive vector quantization and recursive vector quantization is not always sharp, however. Usually adaptive schemes are thought of as being more heuristically designed based on intuition to suit particular types of statistical behavior in the vector sequence, while recursive VQ is viewed as a more rigidly structured generic approach. Generally, FSVQ and PVQ are not categorized in the research literature as being adaptive methods.

With significant correlation between successive vectors, it is natural to consider concatenating vectors into "supervectors" and applying fixed (i.e., nonadaptive) VQ to higher dimensional supervectors. In this case, a significant coding gain is indeed likely, but again the price in increased complexity may make this approach infeasible. Adaptive VQ is an alternative solution that can exploit the statistical dependence between vectors while avoiding the complexity that would result from increasing the vector dimensionality. Rather than code each vector independently, we examine the context or local environment of a vector and modify the quantizer to suit our awareness of the "big picture" of how other vectors in the sequence are behaving in order to more efficiently code this particular vector. In the special case where the vector sequence is stationary and the vectors are statistically independent, experience shows that there is usually very little to be gained by using the context to help code an individual vector. In contrast, if there is interdependence between different vectors, observing the context of a vector provides some useful information so that the quantizer can be tuned to suit the conditional distribution of this vector given the observations of its environment. This is indeed adaptation in the intuitive sense of the word, since the way an individual vector is coded is modified according to its environment rather than being fixed in time.

The sequence of vectors will be regarded as a time sequence as in coding of speech or audio signals. In other applications, the successive vectors do not have a direct meaning as a time sequence, but it is still convenient and useful to refer to the sequence of vectors as if it were a time sequence. In many adaptive VQ schemes, the sequence is divided into blocks of fixed length called *frames*. The vectors in each frame are coded by first extracting essential features from the frame that are used by the quantizer to adjust its operation or codebook according to the local character of the current frame. In speech coding, a frame is typically of the order of $20 - 30$ ms. Typically each frame is subdivided into several vectors, where each vector is a block of contiguous waveform samples. In image coding, a frame, or *block*, is typically a square block of pixels, either a sub-block of an image or in some cases a full two-dimensional image. The block is partitioned into vectors that are smaller square blocks of pixels. The successive blocks are normally scanned and represented as a time sequence of vectors, just as conventional raster scanning generates a one-dimensional sequence of individual pixel values from a two-dimensional image.

In this chapter, we describe several different methods for providing some adaptation capability for a vector quantizer. In each case, the quantizer is continually or periodically modified as it observes the local or short-term statistical character of the arriving vectors. The next two sections consider adaptation based on the mean (Section 16.2) and the standard deviation or "gain" (Section 16.3) of a vector. Both are simple but important descriptive features whose statistics might vary with time. The codebook is adapted by shifting or scaling the vectors in a fixed codebook. In the subsequent sections we examine various techniques where the entire codebook is adapted in a more general manner based on acquired information about the local signal statistics.

Forward and Backward Adaptation

There are two fundamentally distinct categories of adaptation that we shall consider. These are *backward* and *forward* adaptation.

In *backward adaptation*, the coding of the current vector makes use of the past of the vector sequence as represented by the quantized data already transmitted to specify prior vectors. In other words, backward adaptation exploits the approximate knowledge of the past to improve the coding of the current vector. Thus backward adaptation limits the way the environment of the current vector can be used to improve its coding. The benefit of this approach is that no additional information is needed by the receiver other than the successive indexes used to specify the selected code vectors.

In *forward adaptation*, information about the environment of a vector to be coded can be extracted from the "future" of the vector sequence (as well as from the past) but this *side information* must be separately coded and transmitted to the decoder so that it can correctly reproduce the current vector. In other words, we can extract information from arriving vectors before they are actually encoded and provide this information to the encoder. Relative to the time at which the encoder sees the vector, the extracted information is in this sense obtained from the "future." A key distinction between forward and backward adaptation is whether or not side information about the adaptation used by the encoder is separately transmitted to the decoder. An important issue in backward adaptation is whether the increase in overall rate due to the use of side information justifies the performance enhancement attained. In many applications, side information can offer a significant performance gain with only a negligible increase in rate.

16.2   Mean Adaptation

Perhaps the simplest useful parameter that describes a signal vector $\mathbf{x} = (x_1, \cdots, x_k)$ is the *mean*, defined simply as the sample average (or *sample mean* or *arithmetic mean*) value of the vector components:

$$m = \frac{1}{k} \sum_{j=1}^{k} x_j.$$

Note that our current use of this term does *not* refer to the expected value of the vector $\mathbf{x}$, which is itself a vector.

Following the notation of Chapter 12, let $\mathbf{1} = (1, 1, \cdots, 1)^t$, the k-dimensional vector with all components equal to unity and $\mathbf{z} = \mathbf{x} - m\mathbf{1}$ denote the residual vector after the mean has been removed. In Chapter 12, we saw that mean removal is the basis of a simple constrained VQ technique. Constrained VQ reduces complexity in coding an individual vector and is concerned only with the internal structure of a vector quantizer for coding a single random vector. In contrast, here we are interested in adaptively estimating the mean of a vector from its context.

Given the vector sequence $\mathbf{X}_n$ consider the scalar sequence of means m_n defined by

$$m_n = \frac{1}{k} \sum_{j=1}^{k} X_{n,j}, \tag{16.2.1}$$

where $X_{n,j}$ denotes the jth component of $\mathbf{X}_n$. The goal is to use observed values of the means to improve the coding of the individual vectors $\mathbf{X}_n$. Let $\mathbf{Z}_n = \mathbf{X}_n - m_n \mathbf{1}$ denote the residual vector after the mean has been removed. Suppose that the sequence of residual vectors can be approximately modeled as independent and identically distributed but the sequence of means m_n are correlated. In this situation, ordinary VQ of the vector $\mathbf{X}_n$ will be enhanced by adding an adaptation technique which examines the mean of neighboring vectors to help encode the current vector.

Block Coding of the Means

In this forward adaptation technique we partition the sequence of vectors into blocks of N vectors. Each vector in the block is then adaptively coded by using information about the means of all the vectors in the block. For notational simplicity, consider only the block of vectors $\mathbf{X}_1, \mathbf{X}_2, \cdots, \mathbf{X}_N$. We compute the *block mean* $M = \frac{1}{N} \sum_{n=1}^{N} m_n$. Then M is scalar quantized and the index, J, identifying the quantized level, $\hat{M}$, is transmitted to the receiver. Now each vector in the block is adaptively coded by subtracting the quantized value, $\hat{M}$, from each vector component forming the *mean-reduced* $\mathbf{Y}_n = \mathbf{X}_n - \hat{M}\mathbf{1}$ vector which is then coded with a fixed (predesigned) vector quantizer. The terminology "mean-reduced" is used here to distinguish from the mean-removed case since only a quantized version of the average mean of the block is removed; this tends to reduce the magnitude of the vector mean but it does not produce a zero mean residual vector. At the receiver, the quantized block mean is added to each component of the reconstructed mean-reduced vectors thereby reproducing the block of vectors. The same procedure is repeated for each successive block of vectors. For simplicity of notation we have omitted a time index from the block means M and $\hat{M}$.

Figure 16.1 shows a block diagram of the mean-adaptive VQ structure. The buffer (operating in first-in first-out or FIFO mode) acquires and stores a block of input vectors for which a block mean value is to be computed and then the vectors are sequentially passed on for encoding. As the vectors are released, the vectors for the next input block are fed into the buffer for the next block computation. We call this buffer a *frame buffer* where a frame refers to the time span corresponding to the number of contiguous vectors stored in the buffer. Note that the figure does not explicitly show the separate encoder and decoder structures, however, by showing the entire quantization system, it is evident that it can be partitioned into an encoder and decoder. Specifically, it can be seen that the reproduced vector $\hat{\mathbf{X}}_n$ is generated in the decoder from the specifications of the VQ index and the index from the block mean quantizer. For notational simplicity, the dependence of the block mean M on time is not indicated, but it is understood that a new block mean index is transmitted once for every block of N vectors. Block mean removal is a forward adaptation technique

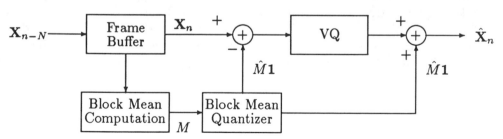

Figure 16.1: Forward Mean-Adaptive VQ with Block Coding of the Means

where side information is needed to specify the quantized block means to the receiver. Typically, the side information for such an application adds a modest increment to the overall bit rate of the coding system. In particular, for a scalar mean quantizer with b bits, the bit rate for the side information is b/M bits per vector. For example, in image coding with VQ each vector might be a 4×4 pixel array and each block a 16×16 array, containing 16 vectors. For a 6 bit block mean quantizer, the side information is 6 bits for every 256 pixels or 0.023 bits per pixel. Image coding is a particularly good application for mean adaptation since images always have a positive mean value that varies spatially with fairly high correlation.

In contrast to mean-adaptive VQ, mean-removed VQ as described in Chapter 12 requires the specification of the quantized mean of each individual vector and is thus not "adaptive," since the quantized mean is directly and exclusively specified to code one particular vector. In this case, the bit allocation for the mean is a substantial fraction of the total number of bits used to code the vector.

Block mean removal is particularly useful when the successive means m_n of the vector sequence vary slowly with the time index, n, so that removal of the quantized block means helps to remove some of the redundancy between vectors. In some applications, mean-adaptive VQ can be advantageously expanded into a hierarchical coding technique where the sequence of block means M_j, where j indicates the jth block in a sequence of blocks, can be viewed as a correlated sequence and is grouped into vectors and each such vector is quantized with VQ [142]. The basic idea of hierarchical VQ was introduced in Chapter 12.

Codebook design for mean-adaptive VQ is straightforward. The scalar quantizer for the block means can be designed using the generalized Lloyd algorithm by first extracting a training set of block means from a given training set of input vectors $\mathbf{X}_n$. As is evident from Figure 16.1, the input-output quantization error is equal to the VQ quantization error, i.e., $\mathbf{X}_n - \hat{\mathbf{X}}_n = \mathbf{Y}_n - \hat{\mathbf{Y}}_n$. Therefore, optimizing the overall performance is achieved by optimizing the vector quantizer for the statistics of its input $\mathbf{Y}_n$. Thus, a training set of $\mathbf{Y}_n$ vectors is generated from the training set of input vectors $\mathbf{X}_n$ by performing block mean quantization and subtracting the quantized block means from each input vector. With the resulting training set, a standard codebook design such as the generalized Lloyd algorithm can be used to design the needed codebook.

It is possible to combine the use of adaptive VQ and constrained VQ into a coding system. In particular, for either of the mean-removed structures depicted in Figures 16.1 and 16.2, the vector quantizer for the mean-reduced vector $\mathbf{Y}_n$ can have any of the constrained structures discussed in Chapter 12.

Predictive Coding of the Means

Another variation of mean-adaptive VQ uses linear prediction to remove redundancy from the sequence of vector means. If the successive vectors are statistically dependent, the sequence of vector means will typically be a correlated scalar valued random sequence. If the vector sequence is stationary, then the mean sequence is also stationary. Hence, it is possible to design an optimal linear predictor that predicts the current mean, m_n from a finite number of prior means m_{n-i} for $i = 1, 2, \cdots, p$. The design of such a predictor has been considered in Chapter 4. The sequence of means can be separately encoded with predictive quantization, i.e., DPCM, and thereby efficiently communicate a quantized approximation of the means to the receiver. The quantized means are subtracted as before from the vector components and then mean-removed vectors are quantized. Figure 16.2 illustrates this version of mean-adaptive VQ. Such a technique was used for image coding by Baker and Gray [23].

Predictive coding of the means still belongs to the category of adaptive VQ because the quantized mean that is subtracted from a vector was obtained using information about the past history of the mean sequence, and therefore of the vector sequence. Thus, the coding of an individual vector is facilitated by making use of the environment (the prior vectors). This scheme eliminates the need for a frame buffer. The memory of this adaptive VQ scheme is contained in the adaptive predictor which stores previously quantized means. This method typically requires more side information than block coding of the means but it leads to much more accurate quantized mean values. In image coding this method can be perceptually advantageous since inadequate coding of the means implies visible blocking effects.

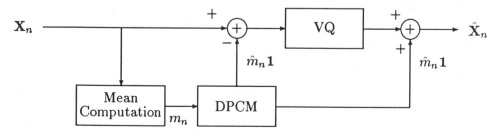

Figure 16.2: Forward Mean-Adaptive VQ with DPCM Coding of the Means

16.3   Gain-Adaptive Vector Quantization

One of the most common ways in which a signal exhibits nonstationarity is in the variation of its short term power with time. This behavior is typical of speech and audio signals which have a wide dynamic range. Suppose we extract a sequence of vectors from such a signal by blocking the scalar sequence into vectors. This produces a sequence of vectors whose energy levels or norms typically vary slowly with time. The norm of a vector can be defined as the rms amplitude of the samples in a vector, although occasionally a somewhat different definition of the norm is of interest. A similar time varying norm may also arise in other signal sources where the vectors have a different physical meaning and are not derived from a blocked scalar sequence. In the short run, we assume the vector gains have a nearly stationary and slowly varying character, but in the long run, they may vary widely with time.

The principle of gain-adaptive quantization is well-known in the case of scalar quantization where an "adaptive quantizer" generally refers to a quantizer that adapts its step size (or amplitude scale) to maintain a suitable loading factor when the input signal power level varies with time. (See the discussion on dynamic range in Section 5.5.) Such gain-adaptive quantizers are particularly effective for coding speech and audio signals because of their wide dynamic range.

Gain-adaptive vector quantizers generalize this concept for the case of a vector sequence with a wide dynamic range of power levels [63]. The basic idea of gain-adaptation is to use the context or environment of the current vector in order to estimate its gain and then scale all the code vectors in a given codebook by this estimate so that their norms are well matched for coding this vector. Of course, an equivalent and simpler way to do this is to normalize the input vector to be coded by the estimated gain and then quantize the normalized vector using a fixed codebook. The decoder must also regenerate the same estimated gain value and then perform a rescaling on the selected code vector to obtain the quantized approximation to the original input vector. Any reference in this discussion to a *normalized* vector means that the vector has been scaled by dividing by an estimate of its norm; it does not imply that the vector has unit norm.

In vector PCM with a fixed (non-adaptive) codebook, the dynamic range of the signal vector norms that can be reasonably handled is severely limited, particularly for speech or audio signals. This problem is somewhat alleviated by the use of shape-gain VQ (see Section 12.10), but this may still prove to be an inadequate solution. Shape-gain VQ is *not* an adaptive technique since it codes each successive vector using exactly the same fixed gain and shape codebooks. If the dynamic range of the vectors varies with time, gain-adaptive VQ is clearly advantageous since it exploits contextual information in coding each vector. It is, however, possible to enhance the performance of shape-gain VQ by adapting the gain codebook while maintaining a fixed shape codebook with unit norm code vectors [277].

Forward Gain Adaptation

Forward adaptation can quite accurately control the gain level of a vector sequence to be quantized, but side information must be transmitted to the receiver. The general structure of forward gain-adaptation VQ is shown in Figure 16.3.

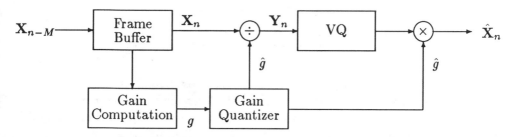

Figure 16.3: Forward Gain-Adaptive VQ

The frame buffer operates in the same way as in Figure 16.1, storing a block of input vectors before they are passed on for encoding. The gain estimator observes all the vectors in the current frame and computes an estimate of the gain value that is in some sense typical of all vectors in the buffer. This gain estimate g is quantized and its index J is transmitted to the decoder. The quantized gain estimate $\hat{g}$ is used to normalize each of the input vectors in the current frame, i.e., the vectors are divided by $\hat{g}$. The normalized vectors are then applied successively to a fixed VQ encoder that has been optimized for coding gain-normalized input vectors, and the encoder generates an index I_n for each vector which is transmitted to the receiver. For simplicity, Figure 16.3 does not explicitly show the separate encoder/decoder parts of the quantizers or the indexes I_n and J.

At the onset of each successive frame, the receiver decodes the gain index J to produce the quantized gain estimate for the current frame. The receiver decodes the successive indexes I_n to produce the code vectors that approximate the normalized input vectors of the current frame. The decoded gain value for the current frame is multiplied by each decoded vector to generate the final quantized approximation to the original signal vectors in the current frame. The same coding and decoding procedure is applied to each successive frame. The operation of the gain estimator will be described later.

Backward Gain Adaptation

The general structure of backward gain-adaptation VQ is shown in Figure 16.4. In this case no buffering is needed. The current input vector is nor-

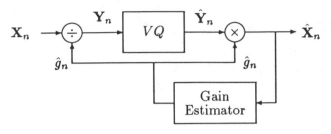

Figure 16.4: Backward Gain-Adaptive VQ

malized by a gain estimate that is an estimate or *prediction* of the current gain value based on quantized information describing the *prior* input vectors. Each input vector $\mathbf{X}_n$ is divided by the gain estimate g_n producing a normalized input vector $\mathbf{Y}_n$ which is then quantized using a codebook that has been optimized for normalized input vectors. The VQ encoder output is the index I_n which is generated for each input vector $\mathbf{X}_n$ and transmitted to the receiver. The receiver has a VQ decoder which maps the indexes I_n into the quantized normalized vectors $\hat{\mathbf{Y}}_n$. Finally, the gain estimate g_n is multiplied with the vector $\hat{\mathbf{Y}}_n$ to obtain the final reconstructed approximation $\hat{\mathbf{X}}_n$ to the original input vector $\mathbf{X}_n$. This structure can be viewed as an extension of a predictive vector quantizer where the prediction is a gain term and the residual formed by division instead of subtraction. As with most adaptive systems, however, the prediction is based on long-term source information instead of only on adjacent vectors.

The gain estimator can be designed to directly operate on the indexes produced by the VQ encoder to generate the gain estimates g_n. Alternatively, equivalent gain estimators can be designed that operate either on the decoded vectors $\hat{\mathbf{Y}}_n$ or on the reconstructed vectors $\hat{\mathbf{X}}_n$. In each case, the same information is available at both transmitter and receiver so that in the absence of transmission errors the receiver reproduces the same gain estimates generated in the transmitter. There are several different embodiments of the gain estimator which will be discussed later.

Gain Estimation

Although the Euclidean or l_2 norm of a vector is often a reasonable and convenient choice for measuring the "size" of a vector, alternative definitions of the norm can be used for defining gain. In the following, we shall assume the *gain* of a vector refers to the vector norm, where the norm is denoted by $||\cdot||$. Usually, the Euclidean norm is assumed, however, the development that follows remains applicable to alternate choices of norm such as the l_1 norm, which is the sum of the magnitudes of the vector components, and the l_∞ norm, which is the maximum magnitude of the vector components.

In forward gain-adaptation, a natural choice of gain estimator is the *average* of the norms of the vectors in a given frame. Thus, if each frame contains M vectors, $\mathbf{X}_{n+1}, \mathbf{X}_{n+2}, \cdots, \mathbf{X}_{n+M}$, then the frame-averaged gain estimate is given by

$$g = \frac{1}{M} \sum_{i=1}^{M} ||\mathbf{X}_{n+i}||. \qquad (16.3.1)$$

The indexing is chosen to indicate that the estimator is looking ahead into the "future" of the input sequence in order to determine the gain value that will begin to be used at time $n + 1$. Note that the gain g will be applied to the coding of the entire block that was stored in the buffer rather than to coding only one vector. This gain estimate is updated once every M time units and is computed after the buffer is reloaded with a completely new set of input vectors. For the case of a Euclidean norm, a reasonable alternative gain estimator is based on the square root of the average of the squared norms. Another option is to use the geometric rather than arithmetic mean of the vector norms:

$$g' = [\prod_{i=1}^{M} ||\mathbf{X}_{n+i}||]^{\frac{1}{M}}.$$

This measure is appropriate in some speech or audio applications where the logarithm of the norm is a more useful measure of the "size" of a vector than the norm itself.

In backward gain-adaptation there are several distinct ways in which a gain estimate can be generated from prior quantized information. Perhaps the simplest choice is to use the block-averaged norm where the norms of M previously reconstructed signal vectors are averaged with a sliding block. Each successive input vector is normalized with a new gain estimate. This sliding block backward gain estimator is

$$g_n = \frac{1}{M} \sum_{i=1}^{M} ||\hat{\mathbf{X}}_{n-i}||. \qquad (16.3.2)$$

This estimate differs from the frame-averaged forward gain estimator of (16.3.1), since we are now averaging over prior data to estimate the gain to be used for the next input vector $\mathbf{X}_n$. Furthermore, it should be noted that each vector $\hat{\mathbf{X}}_{n-i}$ is given by

$$\hat{\mathbf{X}}_{n-i} = \hat{g}_{n-i}\hat{\mathbf{Y}}_{n-i} \qquad (16.3.3)$$

so that the current gain estimate depends indirectly on the previous M gain estimates and the previously quantized normalized signal vectors.

A recursive gain estimate which is computationally simpler to implement is given by

$$g_n = \alpha g_{n-1} + (1 - \alpha)\|\hat{\mathbf{X}}_{n-1}\|. \qquad (16.3.4)$$

This recursion is mathematically equivalent to an exponentially weighted averaging of all prior norms:

$$g_n = \frac{1-\alpha}{\alpha} \sum_{i=1}^{\infty} \alpha^i \|\hat{\mathbf{X}}_{n-i}\|. \qquad (16.3.5)$$

Note that both of the above backward gain estimators are linear time-invariant filtering operations on the sequence of norms $\|\mathbf{X}_{n-i}\|$ of reconstructed signal vectors and both are *ad hoc* estimates. This suggests that an improved linear estimator could be developed by attempting to find an optimal linear filter based on the signal statistics. The statistics of the *unquantized* norms of the original signal vectors are readily accessible in the form of training data, i.e., empirical observations of a long sequence of norms. By using such training data, we can design an optimal finite memory predictor for this sequence based on the linear prediction methods described in Chapter 4. Under the assumption that the norms of the reconstructed vectors $\hat{\mathbf{X}}_n$ are reasonably close to the norms of the original signal vectors $\mathbf{X}_n$, we will obtain a near-optimal gain estimator of the form

$$g_n = \sum_{i=1}^{M} a_i \|\hat{\mathbf{X}}_{n-i}\|. \qquad (16.3.6)$$

g_n is an estimate of $\|\mathbf{X}_n\|$ from observations of M prior values of $\|\hat{\mathbf{X}}_{n-i}\|$.

Typically in speech or other audio waveforms the dynamic range of power levels is very wide and the effect of a gain scaling on the perceived signal level is more effectively measured by the ratio of old and new gains rather than by the difference between the gains. Hence a logarithmic scale is a better measure for the gain in such cases. Consequently a better gain estimation is a linear combination of the logarithm of the gain of previous reconstructed vectors. If we denote $\beta_n = \log\|\hat{\mathbf{X}}_n\|$, then a nonlinear estimate of the gain g_n is obtained from a linear estimate of the log-gain β_n

according to

$$g_n = \exp\left(\sum_{i=1}^{M} a_i \beta_{n-i}\right). \tag{16.3.7}$$

The coefficients a_i can be designed using the linear prediction analysis methods described in Section 4.5 by using a sequence of log-gains, $\log \|\mathbf{X}_n\|$, as training data. Alternatively the simpler estimation methods of (16.3.2) and (16.3.4) can be applied to the log-gain estimation. Thus by setting $a_i = 1/M$, (16.3.7) performs a block-average estimate of the log-gain. Also, by letting $M \to \infty$ and setting

$$a_i = (1 - \alpha)\alpha^{i-1} \tag{16.3.8}$$

(16.3.7) becomes a first order recursive estimator of the log-gain:

$$\log g_n = \alpha \log g_{n-1} + (1 - \alpha)\beta_{n-1}. \tag{16.3.9}$$

Note that (16.3.9) can be expressed directly in terms of the gains as a multiplicative recursion:

$$g_n = \|\hat{\mathbf{X}}_{n-1}\|^{(1-\alpha)} g_{n-1}^{\alpha}.$$

A slight variation of this multiplicative recursion is the vector generalization of the widely used scalar gain-adaptation algorithm of Jayant [193]

$$g_n = m(I_{n-1})g_{n-1}^{\alpha},$$

where the multipliers $m(i)$ are predesigned constants.

Transmission Errors in Gain-Adaptive VQ

It should be noted that a transmission error in the channel between transmitter and receiver may cause an incorrect index to be received. If transmission errors lead to incorrect gain estimates in the receiver, the effect on the reconstructed vectors can be severe and sometimes catastrophic. In forward adaptation, a bit error can alter the index that specifies the gain scaling needed for the next block of vectors. Although this can have a serious effect on the block, subsequent blocks of vector will be unaltered. With sufficiently low error rates, the problem is not serious. High error rates would require some error control techniques to assure reliable recovery of both the gain indexes and the VQ indexes. In backward gain adaptation, a transmission error not only causes the current normalized vector to be incorrectly decoded, but also affects the memory of the gain estimator and therefore may cause incorrect estimates of subsequent gain values. The effect of transmission errors can be much more severe in backward gain adaptation, Since the gain estimator in the receiver operates recursively (in a feedback loop), a bit error that alters an isolated VQ index will have an effect that can be propagated forward in time. When a reproduced vector is incorrectly decoded, it will cause subsequent gain estimates to be incorrectly generated, which then affects subsequent reproduced vectors and so on.

To prevent this propagation of the effect of an error, a "forgetting" capability is introduced to the gain estimator so that the current gain estimate becomes less and less influenced by reproduced vectors from the increasingly remote past. The idea is that a diminishing memory in the gain estimator will cause a gradual reduction of the effect of the transmission error with time. For example, in the recursive gain estimator (16.3.4) or the recursive log-gain estimator (16.3.9), there is an exponentially decaying dependence on the past as long as $|\alpha| < 1$. By choosing α to have a suitable value less than unity the effect of bit errors in transmission will eventually be forgotten. The actual value of α is also important since it determines the time constant of the exponential decay and therefore the duration of the interval in which a bit error will corrupt subsequent performance. Too small a value of α will degrade the ability of the algorithm to track the sequence of vector gains. As always, there are important tradeoffs in practical coder designs.

When an optimal gain estimator is based on linear prediction analysis as in (16.3.6), the resulting prediction error filter, although minimum phase, may have zeros very close to the unit circle. In order to assure robustness to transmission errors, the maximum magnitude of the zeros must be kept below some suitable bound while maintaining a near-optimal prediction filter. The usual solution is to scale down the magnitude of all the zeros of the prediction error filter $A(z)$ by a factor β slightly less than unity, i.e., by replacing $A(z)$ by $A(z/\beta)$. Again a tradeoff exists in choosing the parameter β between the gain-tracking performance and the degree of robustness to transmission errors. This technique, often of value in speech coding, is known as *bandwidth expansion* since it broadens the bandwidths of the spectral resonances or formants in speech processing with linear prediction filters. The use of bandwidth expansion for improving the robustness of backward gain-adaptive VQ is discussed in [61].

Codebook Design for Gain-Adaptive VQ

The performance of gain-adaptive VQ is highly dependent on the effectiveness of the codebook design procedure. We consider first the simpler case of forward adaptation. In this case, as is seen from Figure 16.3, the normalized input y_n to the VQ encoder is completely determined by the sequence of vectors $\{\mathbf{X}_n\}$. The obvious and straightforward procedure for codebook design is to generate a training set of normalized vectors, $\mathbf{y}_n$, that represent the statistics of the input vectors to the encoder and then apply a standard generalized Lloyd algorithm to obtain a codebook. This is in fact the *wrong* procedure to use!

If our objective were to minimize a measure of average distortion (i.e., squared Euclidean distance) between the encoder input $\mathbf{Y}_n$ and the quantized output $\hat{\mathbf{Y}}_n$, then the above procedure would indeed be correct. However, it is really the average distortion between the original signal vector $\mathbf{X}_n$ and its reproduced approximation $\hat{\mathbf{X}}_n$ that we need to minimize in order to optimize the overall coding scheme. Consider specifically the usual squared error distortion measure. Our performance objective is to minimize

$$D = E[||\mathbf{X}_n - \hat{\mathbf{X}}_n||^2]$$

and noting the relationship between $\mathbf{X}_n$ and $\mathbf{Y}_n$ we get

$$D = E[\hat{g}_n^2 ||\mathbf{Y}_n - \hat{\mathbf{Y}}_n||^2]. \tag{16.3.10}$$

Since $\hat{g}_n$ is in general statistically dependent upon $\mathbf{Y}_n$, this average distortion is indeed not an equivalent performance measure to the average distortion between $\mathbf{Y}_n$ and $\hat{\mathbf{Y}}_n$.

To find the optimal quantizer, it is appropriate to consider as the training set the collection of pairs $\{\hat{g}_n, \mathbf{y}_n\}$ since the distortion depends on both the gain and the normalized input vector; also, these pairs can be generated from a sequence of input signal vectors. Now we proceed to find the optimal codebook for a given partition of the training set and the optimal partition for a given codebook.

To determine the optimal partition for a given codebook, we note that given an input vector $\mathbf{Y}_n$ there is a fixed gain $\hat{g}_n$ associated with it. Hence, from (16.3.10), the code vector that minimizes the distortion will be the code vector that minimizes $||\mathbf{Y}_n - \hat{\mathbf{Y}}_n||$. Thus the VQ encoder simply finds the nearest code vector to $\mathbf{Y}_n$ without regard to the value of $\hat{g}_n$ so that the usual nearest neighbor condition is applicable for the VQ encoder.

We next suppose that the VQ encoder is fixed and find the optimal VQ decoder, or in other words find the optimal code vectors for a given cell. The given encoder structure determines a partition of the space of all pairs $\{\hat{g}_n, \mathbf{y}_n\}$ into N regions $\{R_j\}$. Consider a particular region $\{R_i\}$. We seek that vector $\tilde{\mathbf{y}}_{i,opt}$ which minimizes

$$D_i = E[\hat{g}_n^2 ||\mathbf{Y}_n - \hat{\mathbf{y}}||^2 | R_i]$$

over all $\hat{\mathbf{y}}$. The solution can be found directly by differentiating with respect to each component of $\hat{\mathbf{Y}}$ and setting the derivative to zero. The solution obtained indeed yields the minimum of D_i and is given by

$$\mathbf{y}_{i,opt} = \frac{E[\hat{g}_n^2 \mathbf{Y}_n | R_i]}{E[\hat{g}_n^2 | R_i]}. \tag{16.3.11}$$

In implementing this formula using a training set, the expectation is replaced by a summation over the training pairs in the finite cluster corresponding to region R_i. The usual codebook design based on the generalized Lloyd algorithm is therefore applicable with the modified centroid of (16.3.11).

The above discussion was based on the forward adaptation case, where the gain estimator is obtained from the "future" input vectors and it is reasonable to define a training set based on pairs $\{\hat{g}_n, \mathbf{y}_n\}$. In the backward adaptation case, a problem arises with the above codebook design method. To design the codebook using these training pairs, the gain estimator needs the quantized vectors $\{\hat{\mathbf{X}}_n\}$; however, $\{\hat{\mathbf{X}}_n\}$ is not available unless the codebook has already been designed. To circumvent this problem, a two-phase design algorithm was proposed in [63]. In the first phase, actual input vectors rather than quantized ones are used to generate the gain sequence $\{\hat{g}_n\}$ and the corresponding normalized vectors $\{\mathbf{y}_n\}$. The algorithm for forward gain adaptive VQ is then applied to design a codebook based on this training data. In the second phase, the codebook from the first phase is used as an initial codebook and the input training vectors $\{\mathbf{x}_n\}$ are applied to the backward adaptive structure. The estimated gains obtained will differ from those found in the first phase so that a new training set of pairs $\{g_n, \mathbf{x}_n/g_n\}$ is generated. From this training set, the initial codebook is updated using (16.3.11) and then the input sequence $\{\mathbf{x}_n\}$ can be encoded gain to obtain a new training set in the same way and the codebook can be recomputed again. This *closed-loop* design procedure can be repeated until a satisfactory codebook is obtained. Although convergence is not guaranteed with this closed-loop algorithm, a suitable criterion for terminating the algorithm can be based on a comparison between the latest distortion obtained with the best distortion achieved at any prior iteration (16.3.11).

16.4   Switched Codebook Adaptation

A simple and generic technique for forward adaptation is *switched codebook adaptive VQ*. This technique is based on the use of a classifier that looks at the contents of the input frame buffer and decides that the next block of vectors to be encoded belongs to a particular statistical class of vectors from a finite set of K possible classes. The index specifying the class is used to select a particular codebook out of a predesigned set of K codebooks. The index is also transmitted as side information to the receiver. Then each vector in the block is encoded by the VQ encoder which performs a search through the selected codebook using a suitable distortion measure.

The basic scheme of switched codebook adaptation is shown in Figure 16.5. Every M time units, a new frame of M input vectors is stored in

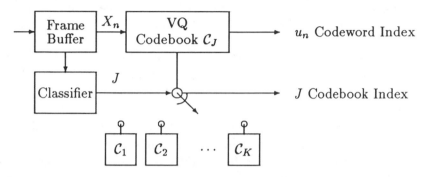

Figure 16.5: Switched Codebook Adaptive VQ

the buffer and a new classification of the frame is made. Thus the side information to identify the class is sent once every M time units while the index to identify the VQ code vector is sent every time unit. This technique can be viewed as a variation of classified VQ, described in Chapter 12, to a forward adaptation scheme. In some applications, it may be useful to have a distortion measure for codebook searching that depends on the specific class of the input block.

If each of the K class-specific codebooks has N code vectors, then the rate needed to specify a code vector is

$$R = \log_2 N + (\log_2 K)/M$$

bits per vector, where the second term is due to the side information. The performance of this scheme is upper bounded by the performance of a non-adaptive vector quantizer whose codebook is the concatenation of the K codebooks. This assumes that the encoder performs an optimal search through the concatenated codebook. However, this non-adaptive quantizer with a concatenated codebook would require a much higher bit rate (per vector); specifically the rate is given by;

$$R = \log_2 N + \log_2 K$$

bits per vector. It should also be noted that direct design of an optimal codebook of size NK may be superior to the concatenated codebook (depending on the design method for the K class-specific codebooks).

The choice of classifier is a design consideration that is often best chosen based on heuristic considerations. Special cases of this structure include the forward adaptation schemes for mean-adaptive VQ and gain-adaptive VQ. In particular, the classifier can be chosen as a VQ encoder that operates on the average of the mean values of each vector in the block (or, in other words, the average of the sum of all samples (vector components) in the buffer). The jth code vector for class i then has the form $c_j + \hat{m}_i$ where the vectors c_j are in a fixed codebook $\mathcal{C}$. This is equivalent to mean-adaptive VQ. Similarly, the classifier can be chosen as a VQ encoder that operates on the average of the norms of the vectors in a block. The jth code vector for class i then has the form $g_i c_j$ so that this is equivalent to gain-adaptive VQ. Of course, in both these examples it is more efficient to use the particular structural forms previously described for mean-adaptive and gain-adaptive VQ for implementing these schemes rather than store separate codebooks for each class as needed for the switched-adaptive structure.

More generally, the classifier might base its decision on some more complex statistical features or on some perceptually based attributes of the buffered vectors. The contents of the frame buffer may be viewed as one supervector of dimension kM (where k is the vector dimension and M the number of vectors in the buffer) and the classifier can then be regarded as a VQ encoder which assigns an index to the supervector. However, a classifier does not need to perform a nearest neighbor partition of the input space as does a VQ encoder. In general, the partition cell (also called a *decision region*) associated with a particular class can be nonconvex and even disconnected.

Codebook design for switched adaptive VQ is performed by sequentially classifying each vector in a training sequence and forming training sets for each class. Thus the classifier would identify a particular frame of training vectors as belonging to class m and then each of these vectors would be assigned to a training set of vectors for class m. Once each class-specific training set is generated, a standard VQ codebook design of any particular size can be used for that class. However, a practical consideration is that the training ratio for each class should be large enough for effective codebook design.

If each codebook has the same number of code vectors, a constant encoding rate will be maintained by the coding scheme. Then the average distortion given the input is in class m will have a particular value that is determined by the statistics of vectors in that class. If the codebook size is allowed to be different for different classes, the resulting quantization scheme will have a variable rate. While variable rate coding is inconvenient for transmission over constant rate channels, it has the advantage that the average distortion of a vector given that it belongs to a particular class can be kept at or below a desired threshold for that class. The threshold can be based on heuristic considerations.

In wideband audio coding, a perceptually based masking model is often used to determine how high a distortion in coding transform coefficients can be tolerated so that the distortion will be inaudible. This effect has been used in a switched adaptive vector quantization scheme where the classifier simply identifies one of a set of distortion levels that can be tolerated in quantizing a vector of DCT coefficients associated with a particular frequency band [54]. The set of codebooks for a given coefficient vector form what is called an *embedded codebook* or a *multiresolution* codebook. Such a codebook is in effect a family of codebooks of increasing size, each of which is designed to code the input source at a given average rate and distortion. For example, the tree structured VQs of Chapter 12 could be used since the nested subtrees form embedded codebooks.

The classifier determines which of the codebooks should be used according to the distortion (or rate) judged appropriate for that class in order to maintain a suitably low level of perceived distortion. Often the family of class-specific codebooks actually consists of one large superset codebook and the class decision determines which subset of this composite codebook will be searched. More generally, in some applications a family of codebooks of different sizes is available for each of a group of statistically distinct source vectors and the total number of bits for coding the group is fixed. This problem can be solved by the method described in the following section.

16.5 Adaptive Bit Allocation for Multiple Vectors

In Chapter 8 we examined the use of bit allocation for quantizing a set of scalar random variables. This technique offered a way to assign a fixed total number of bits among a set of scalar quantizers in a near-optimal manner (subject to a number of assumptions). Now we return to this concept but extend it to the allocation of quantizers to code a set of random vectors subject to a fixed total number of bits for the set. Also, we consider how to make this an adaptive technique when different bit allocations must be dynamically assigned to each of successive sets of vectors. In particular, if a waveform is partitioned into frames and each frame is partioned into vectors, a fixed rate per frame can be maintained while adaptively allocating a different number of bits to each vector in the frame. Of course, allocating a quota b of bits to a particular vector is equivalent to allocating a particular codebook of size 2^b for quantizing that vector so that this could be described as a *codebook allocation* problem. As in the scalar case, there are a variety of ways in which bit allocation can be performed. Here we describe briefly one general approach introduced in [294] [295], that can attain essentially optimal performance under very broad conditions. The presentation here follows the slightly more general development given in [341], [342] where it is applied to image coding. Later in Chapter 17, we shall see another approach to bit allocation that arises in the context of a pruned tree VQ technique. In this section, we first present the main idea of a very general approach to optimal bit allocation and then we examine how this can be applied to adaptively allocate bits (or codebooks, or quantizers) to a set of random variables while maintaining a fixed total quota of bits.

Optimal Bit Allocation

Suppose a signal (or image) is organized into a sequence of frames (or blocks) where each frame consists of a set of random vectors $\mathbf{X}_i$ for $i = 1, 2, \cdots, N$. For each vector $\mathbf{X}_i$ we can design a set of quantizers with

$$\mathbf{Q}_i(\mathbf{X}_i) = \{Q_{i,l}(\mathbf{X}_i) : l = 1, \cdots, M_i\}$$

where $Q_{i,l}(\mathbf{X}_i)$ is the output produced by quantizer l to the ith random vector $\mathbf{X}_i$. The ith input, $\mathbf{X}_i$, is assigned an index L_i to identify which of the N_i quantizers is to be used to code this vector. We call L_i the *coding state* of vector i. Thus, we consider a frame-adaptive coding system where an entire frame is buffered, the vectors in that frame are examined, and the appropriate coding state L_i for each vector in the frame is determined. Usually, the number of possible coding states, M_i for each vector is small and each vector has the same dimension k and the same set of quantizers. In the latter case, $Q_{i,l}$ is independent of i

We define the *rate function* $R_i(L_i)$ and the *distortion function* $D_i(L_i)$ respectively as the number of bits and the resulting distortion when vector $\mathbf{x}_i$ is coded with state L_i. Different coding states can correspond to different ways (or different bit rates) for quantizing a vector. Specifically, $R_i(L_i)$ can be defined as the average word length in bits of variable length channel symbols assigned to the code vectors of the quantizer Q_{i,L_i} or it can be simply the logarithm to base 2 of the number of code vectors in the codebook for Q_{i,L_i}. The choice depends on the particular application of interest but is not relevant to the development that follows. Also, for a distortion measure $d(\cdot, \cdot)$, we can define the distortion function in two ways.

$$D_i'(L_i) = E[d(\mathbf{X}_i, Q_{i,L_i}(\mathbf{X}_i))]$$

or

$$D_i(L_i) = d(\mathbf{x}_i, Q_{i,L_i}(\mathbf{x}_i)).$$

The second case is of interest if we can find the *actual* distortion and rate incurred in coding the observed input vector x_i with each quantizer in the set of quantizers Q_i. This is indeed practical for some applications. In particular, if the coding states for a particular vector correspond to successive levels in a hierarchical quantization scheme (such as a tree structure), then it is not difficult to generate the needed set of distortion-rate pairs. As soon as a frame of vectors is received, the encoder performs each of these quantization operations and then stores the actual distortion-rate pairs $(D_i(L_i), R_i(L_i))$ for each coding state and for each vector in the frame. Note that some side information will be needed to transmit the selected coding state to the decoder and this should be included in the total number of bits allocated per frame.

In the first case, the distortion-rate pairs $(D'_i(L_i), R_i(L_i))$ can be computed in advance using a sequence of frames as training data. This can lead to the off-line design of a fixed coding state (with a corresponding bit allocation) for each vector in the frame in which case the frame is not coded adaptively and no side information is needed. Alternatively, the training vectors can be normalized to have unit gain (norm) and then the average distortion per unit norm vector for a particular state and for a particular bit allocation is found. This then determines the average distortion for all possible vector gains by scaling the unit-norm distortion values. (Specifically for the MSE distortion measure, the scaling factor is the squared norm of the vector.) With this information an adaptive assignment of coding states can be made by simply observing the vector norms in the frame and then invoking an algorithm for optimal state assignment. This method still requires side information to send the coding states for each frame. The second case approach, described below, is more precise since it does not require any approximate (and possibly inadequate) assumptions about the distortion function.

Now that we have set up all this notation, it is time to ask: What, specifically, is the problem to be solved? An optimal coding scheme must solve the problem of finding a state assignment vector

$$\mathbf{L} = [L_1, \cdots, L_N]$$

which minimizes the overall distortion

$$D(\mathbf{L}) = \sum_{i=1}^{N} D_i(L_i).$$

subject to the constraint

$$R(\mathbf{L}) = \sum_{i=1}^{N} R_i(L_i) \leq B$$

where B is the number of bits for coding the entire frame. This is in effect a combinatorics problem since there are only a finite number of possible state assignment vectors. (Note also that there may not exist a solution that gives a total bit allocation exactly equal to the maximum B allowed.) In practice, the number of possible state assigments may be astronomically large and a more efficient solution is needed rather than simply searching through possible state assignments. Note that the state assignment problem is a generalization of the classical bit allocation problem treated in Chapter 8. If each coding state corresponds to a different number of bits, then we would indeed have a bit allocation problem, but in a very general form.

The basic idea underlying the solution to this problem is based on the use of the Lagrange multiplier method for constrained optimization. This method is usually associated with calculus where differentiable functions are assumed, but it also is applicable to discrete optimization problems as is briefly outlined here. The main idea is that we can define an unconstrained optimization problem to minimize the objective function $D(\mathbf{L}) + \lambda R(\mathbf{L})$ over all state vectors $\mathbf{L}$ for each value of the parameter $\lambda \geq 0$. It can easily be shown that the solution to this problem $\mathbf{L}^*(\lambda)$ is identical to the solution to the constrained optimization problem

$$\min_{\mathbf{L}} D(\mathbf{L}) \text{ subject to } R(\mathbf{L}) \leq R(\mathbf{L}^*(\lambda)). \qquad (16.5.1)$$

(See Problem 16.4.) This almost appears to solve our problem, but is not quite the whole story.

The result means that if we pick a value of λ and solve the unconstrained problem, obtaining a value $B^* = R(\mathbf{L}^*(\lambda))$, then we have a solution to the constrained problem for the constraint $R \leq B^*$. But we need to find that λ for which the constraint B^* is equal or nearly equal to the given rate constraint B. Thus, a repeated solution of the unconstrained problem for different but carefully chosen values of λ is needed to locate the best attainable solution for B^* that is less than B but as close to B as possible.

In fact, the unconstrained problem can be efficiently solved for a given value of λ by noting that the objective function can be written as the sum:

$$\sum_{i=1}^{N}[D_i(L_i) + \lambda R_i(L_i)] \qquad (16.5.2)$$

and for each i, the ith term can be separately minimized to find the best L_i. This simplification enormously reduces the complexity of the unconstrained minimization for a given value of λ.

To solve the constrained problem, we need to find the value of λ for which the unconstrained solution will give $(\mathbf{L}^*(\lambda))$ as close as possible to

B but not exceeding B. We wish to do this without laboriously trying an enormous set of possible values for λ. An efficient procedure is indeed available since, as we shall see, there are only a finite set of values of λ, that need be considered. Here we sketch an approach due to Kenneth Rose that leads to the same algorithm originally derived in [295] and later, from a very different perspective, in [268]. The latter approach, to be described in Chapter 17, is of particular interest because it is based on a general technique that also leads to the design of unbalanced tree codebooks for VQ.

Consider first the graph of $D_i(L_i) + \lambda R_i(L_i)$, the ith term of the objective function (16.5.2) plotted versus λ. Recall that we can separately minimize this term to find the optimal L_i. For a particular state $L_i = l$, this gives a straight line starting with height $D_i(l)$ at $\lambda = 0$ rising with positive slope $R_i(l)$ as λ increases. If we superimpose the M_i straight lines for all the coding states (all L_i for $L_i = 1, 2, \cdots, M_i$) on the same plot, it is easy to see that the minimum for each λ of these lines forms a piecewise-linear monotone increasing curve. Each line segment on this curve corresponds to one distortion-rate pair and there is a set of $N \leq M_i$ values of λ at which two segments of the piecewise-linear curve meet. By forming such a curve for each of the i source vectors, we can sum these curves together to obtain a plot of the unconstrained objective function versus λ. This new curve will also be piecewise linear and monotone increasing. Each segment corresponds to a particular coding state $\mathbf{L}$ for the frame with a particular $D(\mathbf{L})$ and $R(\mathbf{L})$. Also, as λ increases, each successive segment will correspond to a unique solution with decreasing rate and increasing distortion value. Every value of λ on a given segment has the same distortion-rate pair. The segment boundaries occur at so-called *singular* values of λ, each of which corresponds to two or more distortion-rate pairs as solutions to the unconstrained problem. Once this curve is obtained, the set of line segments can be mapped to a set of distortion-rate pairs, one for each line segment, which when plotted on a distortion versus rate plane gives the attainable points which lie on the lower convex hull of the set of all possible attainable distortion-rate pairs. The best attainable solution to the constrained problem that lies on this convex hull is then easily found.

These observations lead to an efficient algorithm for searching for a λ which solves (as closely as possible if not exactly) the original constrained problem. The idea is to start with a reasonable initial value of λ, then from examining the unconstrained solutions for this value, a sensible choice of the next singular value to try next can be computed. The algorithm terminates in a finite number of steps at the singular value whose solution rates $R(\mathbf{L}^*(\lambda))$ contain the solution as close as possible to (but not greater than) the constraint B. For a detailed description of the algorithm see [295] or [342]. The latter reference treats the slightly generalized version where the coding states of a quantizer do not necessarily correspond to distinct bit rates.

Adaptive Coding of Multiple Vectors

In many applications a constant bit rate is needed for coding the sequence of vectors representing a signal, yet there may be a large variation from vector to vector in the number of bits needed to code it with adequate fidelity. Based on the preceding bit allocation method, it becomes possible to encode a frame of vectors at a fixed rate while allowing a variable allocation of bits to different vectors in the frame. As discussed above, there are two ways to formulate the problem. One is to assume we know only the average distortion that a given vector in the frame will incur with a particular quantizer (or coding state). The alternative is to code each vector in a given frame with each quantizer and store the actual distortion value for that vector for each coding state. The first method is more of interest if successive frames have a consistent statistical character so that the average distortion is a good indication of what the actual distortion is likely to be for a particular vector. In this method, typically only the vector norms of the current frame can be used to control the final quantizer allocation. The second method is inevitably more powerful since it makes no such assumption of statistical regularity and allocates quantizers to each vector in a particular frame in the best way possible for that particular frame by finding what is the actual performance that will be offered by each quantizer for a particular vector. We focus here on the second method.

Consider, in particular, image coding where each frame represents an image block and each vector is a particular subblock of pixels in that block. The statistics of an image subblock at a fixed location vary from one image to another. Thus the second method is far more powerful for this application.

The main idea of the adaptive coding method is to buffer a frame of input blocks, then compute and store the distortion and associated rate for each vector in the frame for each coding state. This data is then applied to the coding state allocation algorithm described above yielding an optimal coding state **L** for this frame. The coding state is transmitted to the receiver as side information and the indexes produced by the selected quantizers are transmitted to the receiver to specify the code vectors representing each vector in the frame. We do not address here the issue of how to design the family of quantizers to be associated with each vector. A particular approach to this issue is treated in [341] [342]. In this work, each image is partitioned into rectangular blocks (corresponding to frames in our terminology but not related to the use of the word "frame" in video signals) that are horizontal strips each containing many vectors. Each vector is an 8×8 block of pixels. The family of quantizers (coding states) for each vector is organized in a hierarchical manner. The family of quantizers consist of a transform VQ with different bit allocations for vector quantizing subsets of the transform coefficients and for some of the quantizers this is followed by a second stage where the quantization residuals (in the space domain) are partitioned into 4×4 subblocks and each is vector quantized. Not all quantizers operate at different bit rates so that there may be some coding states that operate with a common bit rate. This allows different ways of coding the same vector for a given rate to allow the best choice to be made for possibly quite different types of vectors. High quality image coding has been achieved with this method at relatively low rates. In particular, the image Lenna was coded at 0.3 bpp with a PSNR of 33.1 dB by using 18 coding states for each block [341]. It is interesting to note that in this work the greedy bit allocation algorithm described in Table 8.1 of Chapter 8 was found to give results very close to the optimal algorithm of [295].

16.6 Address VQ

An alternative approach to efficiently coding a frame or block of vectors as one entity will be described in this section in the context of image coding. Suppose one has a VQ of a particular dimension and performance. For simplicity we assume for the moment that the VQ is memoryless, but a recursive VQ such as a predictive or finite-state VQ could also be used. One can effectively increase the vector dimension and, as we shall see, possibly adapt the codebook without changing the basic VQ by taking advantage of the behavior of groups or patterns of contiguous coded vectors. For example, suppose that a VQ has been designed for an image coding application with vectors consisting of 4×4 pixel blocks, that is, dimension 16 vectors. We shall call these 4×4 blocks *small blocks* and their corresponding VQ codewords as *small codewords*. The basic codebook of small codewords contains N_0 words and has a rate $R_0 \equiv (\log N_0)/16$ bits per pixel if used in the ordinary fashion.

If the adjacent pixels groups forming these small blocks are highly correlated, VQ codewords for these adjacent small blocks will tend to occur in patterns. Consider groups or *big blocks* consisting of four contiguous small blocks forming a square; that is, four adjacent 4×4 square pixel blocks which together form a 8×8 square pixel block of dimension 64. This is depicted in Figure 16.6 where the small blocks 1, 2, 3, and 4 make up the single large block. If each of the smaller blocks is encoded using the given

1	2
3	4

Figure 16.6: Small and Big Blocks

VQ, then there will be a total of $N_0^4 = 2^{4R_0}$ possible codeword combinations for the big blocks. If the input blocks are highly correlated, however, only a fraction of these N_0^4 coded big blocks will often appear. The more correlated the adjacent small blocks, the fewer are the regularly appearing patterns of coded big blocks. This suggests an alternative scheme which will have the same performance as the original VQ, but will have a potentially smaller rate. The technique, which is due to Nasrabadi and Feng [110] [241] [242] is called *address-VQ* because of its combined use of a small block VQ and a collection of big block patterns specified by the indexes or addresses of the code vectors representing the small blocks. These two codebooks, the original small codebook and the pattern codebook of combinations of commonly occurring groups of small codewords, are combined into a single extended codebook. The technique can be viewed as simply a combination of VQ and noiseless or entropy coding, but the noiseless coding part involves direct manipulation of the basic VQ codebook.

The extended codebook is constructed and used by the encoder in a manner different from an ordinary VQ. The extended codebook C is the union of the original small block VQ codebook C_0 which has N_0 dimension 16 codewords and a pattern codebook C_1 which has N_1 dimension 64 codewords. The long codewords in C_1 each consist of four of the small codewords from the original VQ concatenated together, but not all such possible combinations are included. Equivalently, we could consider this list to consist of entries which are four indexes or addresses (hence the name of the technique) corresponding to the four small codewords. For pedagogical reasons we consider the list to contain the codeword sequences themselves. This is to avoid possible confusion between addresses in the original VQ codebook C_0 which run from 1 to N_0 and those in the extended codebook $C = C_0 \bigcup C_1$ which run from 1 to $N = N_0 + N_1$.

For the time being we assume that the pattern codebook C_1 contains the N_1 most likely possible patterns of four codewords when the original VQ is applied to the source. This collection can in principle be found by encoding the training sequence with the original VQ and computing the histograms of all N_0^4 possible combinations of four VQ codewords for the small blocks making up a big block. The N_1 most probable combinations are then listed in the pattern codebook C_1. This resembles the conditional histogram construction technique for finite-state VQ in Chapter 14, but there only these probable patterns could be coded and the system could perform quite poorly when faced with unlikely patterns. Here there will be an "escape valve" which permits the correct communication of unlikely patterns at the expense of extra bits.

Note that unlike codebooks considered to this point, the extended codebook contains codewords of two different dimensions, long codewords of dimension 64 and short codewords of dimension 16. Any specific codeword (of either dimension) can be specified to the decoder by sending $\log N$ bits to describe the index of the word in the extended code.

The extended codebook is used to encode the input as follows. The input sequence is parsed into big blocks instead of only small blocks as would be used by the original VQ alone. For a given 8×8 big block the encoder first encodes the big block by using the original VQ on the four subblocks and produces a group of four small VQ codewords. If this particular pattern is in in the pattern codebook C_1, then its index in the extended codebook is sent using $\log N$ bits. If this occurs, then the $\log N$ bits specify a complete big block to the decoder. If the original VQ encodes the input big block into a group of four VQ codewords that is not in C_1, then we essentially return to separate coding of each small block with the original VQ. In this case we have to send four separate indexes, each specifying a small codeword in C_0. Since our extended codebook is used by the encoder and decoder and contains C_0 as a subset of the total code of size N, specifying four small codewords using the extended codebook requires four codewords from the extended codebook or $4 \log N$ bits.

To summarize, if we are lucky enough to encode an input big vector into one of the patterns in C_1, $\log N$ bits suffice to specify the pattern to the decoder. If the input big block does not yield one of these patterns, then we must use the extended code four times and hence send $4 \log N$ bits to the decoder. The basic attribute of a noiseless code operating on the original VQ can be seen here. We have divided the original VQ outputs into two classes, the most probable N_1 possible patterns of four vectors and everything else. The probable big blocks get coded using short words of $\log N$ bits, the improbable big blocks get coded using long words of $4 \log N$ bits. The original VQ would have required $4 \log N_0$ bits to specify a big block. Since $N_1 < N_0^4$ by construction, we use shorter codewords on the probable big blocks of four and longer codewords on the improbable big blocks than would the original VQ. Hopefully the average number of bits per pixel will be reduced from the original VQ.

To determine the rate of address-VQ, suppose that P_1 is the probability that an input big vector of dimension 64 is encoded by the original VQ into a pattern in C_1. If a big block produces a pattern in C_1, then it will take $\log N$ bits to specify the dimension 64 vector. If the pattern is not in C_1, then $4 \log N$ bits will be required. Thus the average number of bits required per big vector will be

$$P_1 \log N + 4(1 - P_1) \log N = 4 \log N - 3P_1 \log N$$

and hence the rate in bits per pixel will be

$$R = \frac{4 - 3P_1}{64} \log(N_0 + N_1). \qquad (16.6.1)$$

By construction this code will yield exactly the same reproduction process as the original VQ, but hopefully the bit rate will be reduced. Since the original bit rate was $R_0 = (\log N_0)/16$, the additional compression factor over the original VQ will be

$$\frac{R_0}{R} = \frac{\log N_0}{(1 - 3P_1/4) \log(N_0 + N_1)}. \qquad (16.6.2)$$

Suppose, for example, that the original VQ has a 7 bit codebook and that the extended codebook has 10 bits, that is, $N_0 = 2^7$ and $N = N_0 + N_1 = 2^{10}$. Then (16.6.1) becomes

$$R = \frac{40 - 30P_1}{64} \qquad (16.6.3)$$

and (16.6.2) becomes

$$\frac{R_0}{R} = \frac{7}{10(1 - 3P_1/4)}. \qquad (16.6.4)$$

For this to be at least 2, for example, we would need to have

$$P_1 \geq \frac{13}{15} \approx .87.$$

In other words, out of 2^{28} possible coded big blocks, the most popular 896 $(2^{10} - 2^7)$ must occur 87in order to have a compression factor of at least 2.

So far no adaptation has been considered, the scheme is effectively a combined VQ and noiseless coder. Adaptation can be accomplished in a variety of ways. One way would be to keep a running estimate of the probabilities of occurrence of big blocks and to adjust the extended codebook accordingly, that is, to add and delete patterns according to the behavior of the source. This procedure can be complicated to compute and implement.

An alternative method for adapting an address-VQ was proposed by Nasrabadi and Feng [241]. Instead of fixing the pattern codebook C_1 as in the original technique or attempting to estimate changing conditional probabilities, the pattern codebook of long codewords is changed depending on the context of the big block using fixed estimates of conditional probabilities based on the original training sequence. As with finite-state VQ, a natural context in the image compression application comprises the adjacent small blocks that have previously been coded. Figure 16.7 shows the big block consisting of the small blocks 1,2,3,4 and the context blocks A, B, C, D, E, F. The small blocks A through F will be used to determine

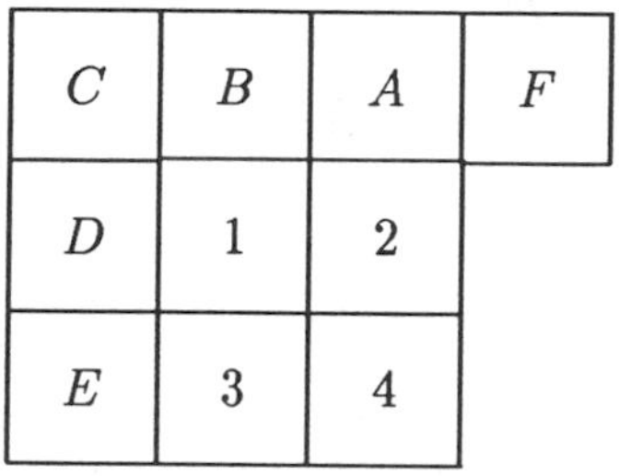

C	B	A	F
D	1	2	
E	3	4	

Figure 16.7: Big Block Context

the code for the current big block consisting of the small blocks 1 through 4. This can be accomplished, for example, by having $\mathcal{C}_1$ contain the N_1 most probable patterns of four small block VQ codewords conditioned on the context blocks A through F. In this case the extended codebook will be modified for each big block depending on the surrounding context, which will be known to the decoder since it depends only on previously sent codewords. This is a natural extension from a fixed codebook having the most probable codeword patterns for big blocks to a variable codebook depending on previously coded small blocks and containing the most probable patterns according to a conditional probability given the previously coded blocks. This is analogous to the side-match finite-state VQ of Chapter 14 where the "border" of the current block was used as context to select a codebook for the current block that matched its surroundings.

Although a natural means of adaptation, this approach has some serious drawbacks. First, there is the extra burden of having to adjust the subcodebook $\mathcal{C}_1$ for each new big block. This involves resorting the collection of all possible patterns of four small block codewords to bring the conditionally most probable N_1 words into the codebook $\mathcal{C}_1$. The second and far more serious problem is the immense computational difficulty of finding a good estimate of the conditional probabilities $P(1, 2, 3, 4 | A, B, C, D, E, F)$. While this can in principle be done from a sufficiently large training sequence, the number of possible given subblocks (N_0^5) and the number of possible big blocks (N_0^4) make this a formidable task. These probabilities can be approximated or estimated, however, by making a Markov-like assumption and using only pairwise conditional probabilities which are relatively easily found. This method is described next.

Following the design of the basic VQ for small blocks, use the training sequence to estimate all conditional probabilities of VQ codewords given adjacent previously coded VQ codewords. These conditional probabilities, which are stored for use during the actual encoding and decoding operations, can be described using Figure 16.7 as $P(1|A)$, $P(1|B)$, $P(1|C)$, and $P(1|D)$. We assume spatial stationarity so that, for example, $P(1|A)$ is the same conditional distribution as $P(2|F)$. In other words, the pairwise conditional probabilities depend only on the relative position of the two small blocks.

Using these conditional probabilities, a score for each big block 1,2,3,4 given a context A, B, C, D, E, F as in Figure 16.7 can be defined by

$$
\begin{aligned}
S(1,2,3,4|A,B,C,D,E,F) = \\
P(1|A) \times P(2|A) \times P(1|B) \times P(2|B) \times P(1|C) \\
\times P(1|D) \times P(3|D) \times P(1|E) \times P(3|E) \\
\times P(2|F) \times P(4|1) \times P(4|2) \times P(4|3).
\end{aligned} \qquad (16.6.5)
$$

This score is interpreted as an estimate of the conditional probability of the big block given the context. The address-VQ now adapts as follows. Given a context of previously coded small blocks (A, B, C, D, E, F), the encoder and decoder first compute the score for all possible patterns and then find the N_1 patterns of four VQ codewords (possibly appearing in 1,2,3,4) having the largest scores according to S and bring these patterns into C_1. Once C_1 is selected, the code works as before. Note that this implies in principle the evaluation and storing of the probabilities $P(n|a)$ for $n = 1, 2, 3, 4$ and $a = A, B, C, D, E$ and for each block coded one must take the current context and compute S for the $N_0^4 = 2^{28}$ possible combinations. This unreasonable computational task can be eased by taking advantage of the structure of S and by suboptimal search. For example, Nasrabadi and Femg did not use all the possible combinations since most of them do not occur. Their address-codebook contained 100,000 address-codevectors which were obtained by finding the most probable ones. Alternatively they could be selected as the words giving minimum distortion. A fast method of resorting is to use a one bit flag for each entry. If the flag is set, then that entry is a part of the addressable region. As a result no reordering is done in hardware.

The four small blocks 1,2,3,4 are encoded using the original VQ. If the resulting pattern is in C_1, the index of the pattern is sent. If the pattern is not in C_1, then the short original codewords in the extended codebook are used four times. Both encoder and decoder independently perform this sorting as both have the decoded context.

The specific score formula given above is that of [241] and others are possible, but this one provides a good compromise between complexity and performance. We now describe a specific design example of [241] by specifying the VQ its performance on a test image from the USC database. They combined their adaptive address-VQ with a simple predictive VQ (which they motivated as a variation on mean-removed VQ). The block predictor is a very simple one: The 9 pixels adjacent to a small block as shown in Figure 16.8 are used to form a simple predictor of the small block by forming the sample mean

$$\overline{m} = \frac{1}{9} \sum_{i=1}^{9} x_i.$$

The sample mean is used as a predictor for all 16 pixels in the small block, that is, the matrix predictor is $\overline{m}$ times the matrix of all 1s. In addition,

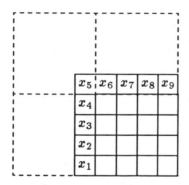

Figure 16.8: Sample Mean Predictor

a classified VQ using the classifier of Ramamurthi and Gersho [259] was used. An adaptive address-VQ was designed using this VQ with $N_0 = 2^7$ and $N_0 + N_1 = 2^{10}$ using a portion of the USC image database not including the Lenna image. The resulting adaptive address-VQ was then applied to the Lenna image as a test. Of a total of 4,096 big blocks, 3222 were encoded using the pattern codebook C_1 and the remaining 874 were encoded using the original VQ codebook C_0 four times. This implies an estimate of finding a big block in the pattern codebook of

$$P_1 = \frac{3222}{4096} \approx .787$$

and hence an average rate from (16.6.1) of $R = 0.256$ bpp. A PSNR of 30.6 dB was obtained for this case. The original VQ for this source was 0.437 bpp, yielding a compression factor of 1.71.

16.7   Progressive Code Vector Updating

A natural approach to adaptive VQ is to continually adapt the code vectors in a codebook based on acquiring updated statistical information about the incoming vectors that are being coded. Often the changes in statistical character of the input vectors are not simply describable as a time variation of some easily identifiable feature of the vectors such as mean or norm. In such cases, the entire codebook can be gradually updated by successively modifying individual code vectors. This can be done in various ways. Three examples are outlined here.

In Paul [253] the distortion incurred in quantizing each input vector is monitored; if it exceeds a threshold, that input vector is added to the codebook as a new code vector. At the same time, the code vector with the largest idle time is deleted in order to maintain a constant codebook size. The idle time is simply the number of time units since that code vector was last selected as a nearest neighbor. This scheme requires the transmission of side information to identify whether or not a codebook modification is made and if so, the quantized value of each component of the new code vector must be transmitted. It is preferable that the quantized input vector, rather than the unquantized version, be added to the encoder's codebook. This scheme could be simplified by having a second fixed codebook (with a much larger size) for quantizing the input vector when a replacement is needed. Then the side information to update the codebook requires a flag indicating that a change is to be made and the index of the new code vector (from the second codebook) that is to replace the dropped code vector. The index of the dropped code vector can be determined by the decoder which also tracks the idle time for each code vector. Note that this scheme implies a variable rate encoder.

An alternate codebook updating method, proposed in [143], is based on monitoring the partial distortion associated with each code vector to determine when a code vector should be updated. As each fixed length block of vectors is encoded, the average partial distortion is accumulated for each code vector. The code vector with largest partial distortion for that block of input vectors is replaced by the centroid (average) of those input vectors coded with that code vector. If this does not sufficiently reduce the partial distortion, the code vector is then split into two code vectors by adding a new code vector displaced from the first one. Then the code vector with the smallest partial distortion is deleted. Side information is needed to communicate the updates to the codebook, but the rate is controlled by the block size.

A third approach to updating code vectors is based on learning algorithms used in neural network applications. The idea is that as each new input vector is applied to the encoder, some new information is obtained about the statistics of the source. This information can then be immediately used to modify a subset of the current code vectors. Let $y_i(n)$ denote the ith code vector at time n in a dynamically changing codebook. Then after the current input vector $X(n)$ arrives and the index j of the nearest neighbor code vector is found, the code vectors are updated according to a rule of the form

$$\mathbf{y}_i(n) = \mathbf{y}_i(n-1) + \alpha_i(n)h_i(j,n)[\mathbf{X}(n) - \mathbf{y}_i(n-1)]$$

where $\alpha_i(n)$ lies between 0 and 1 and is the *step-size* for the ith code vector at time instant n and is assumed to decay to zero as n increases. The function $h_i(j,n)$ is a *neighborhood function* of the jth codevector that is reduced as a function of time and measures how much attention to give to the ith codevector when the jth code vector is the nearest neighbor. Typically $h_i(j,n)$ is nonzero only for indexes i that are "close" to j and decreases with increasing "distance" between i and j where some predefined "distance" measure between indexes is given. Thus, only code vectors that are "close" to the nearest neighbor are updated at each iteration. When the neighborhood of a code vector reduces to include only that code vector itself, the update algorithm becomes

$$\mathbf{y}_i(n) = \mathbf{y}_i(n-1) + \alpha_i(n)S_i(\mathbf{X}(n),n)[\mathbf{X}(n) - \mathbf{y}_i(n-1)]$$

where the selector function $S_i(\mathbf{x},n)$ is one if $i = j$ and zero otherwise. This algorithm is equivalent to an LMS adaptive filtering algorithm for adaptation of the code vectors to minimize mean squared error distortion [144].

The basic idea of this third approach to code vector updating has appeared in several forms, most notably in Kohonen [208] and is sometimes called *Kohonen learning*. This method suffers the serious problem that a grossly excessive amount of side information is required to communicate the codebook updates to the receiver. Consequently, it is primarily of interest as a method for designing VQ codebooks. As such, it has the feature that it may be suited to a one-time pass of a large sequence of training data through the design algorithm without requiring the storage of the training set for repeated iterations as in the generalized Lloyd algorithm.

16.8   Adaptive Codebook Generation

In some cases, the statistics of an ongoing sequence of vectors may undergo a sudden and substantial change so that an entirely new codebook is desirable in order to effectively quantize the subsequent vectors. Several codebook adaptation methods have been proposed to handle these situations.

In one method, a large size "universal" codebook, say of size L code vectors, is designed off-line from training data that attempts to represent the statistical variety of vectors that may be encountered by the coder [250]. The adaptive codebook that is actually used by the encoder is of size N where N is much smaller than L. This codebook is regenerated when needed by buffering a block of input vectors. Using these vectors as training data, a subset of N code vectors from the universal codebook is then selected for the adaptive codebook by a design procedure. If the block of input vectors for training is of size N, one can simply find the closest code vector in the universal codebook to each training vector thereby generating the needed codebook. In another variation of this approach, a larger size block of input vectors (the training set) is used and one Lloyd iteration is performed starting with the previous codebook. After clustering the training vectors into N bins and forming the centroids, the centroids are then quantized with the universal codebook. The new codebook consists of the quantized centroids.

In both variations, the side information needed for identifying the new codebook is at most L bits since this suffices to flag the membership status in the new codebook of each universal code vector. In other words, a binary L-tuple could have a 1 for each of the N codewords in the universal code that are to be included in the current adaptive code and a 0 for all the remaining codewords. Alternatively, we could use $N \log_2 L$ bits to specify the N codewords in the universal code (each described by $\log_2 L$ bits) constituting the adaptive code. Clearly the smaller of the two numbers yields the more suitable indexing. This can indeed be a large amount of overhead, but if the lifetime of the codebook is long enough (before a new codebook generation is required), the overhead can be justified. This approach was specifically designed for image coding where each codebook is used for coding one entire image. For example, suppose the vector is a 4×4 block of pixels, the image is 512×512 block of pixels, the universal codebook has size $4096 = 2^{12}$, and the adaptive codebook has size $N = 64 = 2^6$. Here $L = 4096$ while $N \log_2 L = 768$ and hence it is more efficient to specify the particular subcode using 768 bits per image. Then the overall bit rate is then

$$\frac{6}{16} + \frac{768}{512^2} \approx 0.383 \text{ bpp}$$

including an overhead of less than 1% for side information. The performance is of course upper bounded by that obtained with direct use of the universal codebook at a bit rate of 0.75 bpp.

16.9 Vector Excitation Coding

One of the most important advances in signal compression to emerge in the 1980s is a family of adaptive VQ techniques where the codebook is dynamically updated by filtering a set of fixed code vectors through a time varying spectral shaping filter to generate a new set of code vectors that constitutes an adaptive codebook. In this section, we introduce the principles underlying these powerful and widely used signal coding techniques, which we refer to generically as *vector excitation coding* (VXC) and in a more specific context as code excited linear prediction (CELP). The technique has become widely used in speech coding and more recently it has been applied to image coding.

Direct waveform coding of a signal by quantizing successive blocks of samples with unconstrained VQ will typically require an excessively large dimensions (for example, 40 − 60 samples in speech coding or 64 − 256 pixels in image coding) to adequately exploit the statistical dependence among the samples. For typical rates on the order of 0.25 to 0.50 bits per sample or higher, the complexity prohibits direct use of VQ. However, it is not the dimension that is prohibitive but the corresponding codebook size. One way to approach the coding problem is to retain the vector dimension but drastically reduce the codebook size by making the codebook highly adaptive to local signal characteristics. The reduced codebook size implies that a substantial drop in rate is needed for sending code vector indexes, allowing a relatively large amount of side information to be transmitted. In other words, we partition the data needed to specify the input vector into two parts: the bits needed to specify a suitable code vector from a moderate sized codebook and the bits needed to specify a processing operation that will convert the code vector into a better approximation to the input vector.

Before describing the specifics of VXC, a simple numerical example will be helpful. A recent U.S. government standard for speech coding based on a VXC method, known as the CELP algorithm, uses a vector dimension of 60 samples (for speech sampled at 8 kHz) with a codebook size of 512 or 9 bits [45]. Thus only $9/60 = 0.15$ bit/sample or 1.2 kb/s is used to specify the sequence of code vectors while the side information needed to allow the adaptive codebook to be known to the receiver is approximately 3.6 kb/s.

Let $x(t)$ denote a discrete-time scalar-valued signal. This signal can be directly coded by vector PCM, where blocks of k contiguous samples are formed into vectors and each vector is quantized with k-dimensional VQ having a fixed pre-designed codebook. If the local character of the signal varies widely from one local time epoch to another (as in a speech waveform), it is intuitively clear that an adaptive codebook could substantially improve performance if the adaptation were sufficiently powerful to virtually achieve a codebook that is close to optimal for the local character of the signal.

One way to develop an adaptive codebook for vectors that represent waveform segments is to find a model of the signal whose parameters describes its local statistical character. In Section 4.10 we showed that a signal $x(t)$ can be modeled in a very flexible and general way as the output of a stable linear filter (the *shaping* or *synthesis* filter whose input or *excitation* $e(n)$ is a stationary white noise process). The parameters that specify the transfer function $B(z)$ of this filter serve to characterize the statistical character of $x(n)$. If the excitation is Gaussian, these parameters provide a complete characterization of $x(n)$. Even without a Gaussian excitation, the filter parameters fully determine its spectral density, which is usually sufficient for modeling purposes. By allowing these parameters to vary with time, it is then possible to model a signal whose short-term spectral character changes with time. This way of modeling a signal, often called *source-filter* modeling, is the basis for a family of adaptive VQ techniques described below. It is also the basis for the classical LPC vocoder techniques for speech coding at low rates [189] [190] [19] [234] [247] and for the LPC-VQ system [162] [41] [43] [42] [338] described in Chapter 11.

Non-Adaptive Residual VQ

Suppose for the moment that the signal $x(n)$ is a stationary process that can be modeled by the fixed shaping filter $B(z)$ and a white excitation signal $e(n)$. This model partitions the description of $x(n)$ into a "simpler" residual signal and a set of parameters that specify the filter $B(z)$. For the purpose of coding, it may be more efficient to separately quantize $e(n)$ and the filter parameters rather than directly quantize the waveform $x(n)$. In general, a good model of a signal provides a more parsimonious description or representation of the signal. A block of samples of $x(n)$ may require a far larger codebook for adequate VQ performance than does a block of samples of $e(n)$.

In order to perform model-based coding, the model must either be known in advance or the coding system must first generate the model from the observed signal $x(n)$ to be coded. The source-filter model is found by performing linear prediction analysis on the signal $x(n)$ to determine the optimal prediction error filter $A(z) = 1 - P(z)$. If the predictor order is sufficiently large, the filter acts as a whitening filter producing a white (or nearly white) output $e(n)$. This whitened version of the signal $x(n)$ is often called the *residual* signal since it represents what is left behind after the correlation is removed by the whitening process. Since $A(z)$ is a stable minimum delay filter, its inverse $B(z) = A^{-1}(z)$ exists and is stable. Hence, $B(z)$ can serve as the shaping filter of the model since it maps $e(n)$ back into the original signal $x(n)$. If the process $x(n)$ is stationary, then $B(z)$ is fixed and known by the encoder and decoder so that no side information is needed to specify its parameters. This leads to a simple coding scheme where the input is whitened, the residual samples are blocked into vectors which are then vector quantized, and the quantized vectors are converted back to sequential quantized residual samples which are then filtered with the shaping filter $B(z)$. This scheme is illustrated in Figure 16.9. The

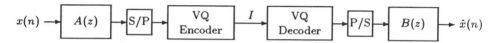

Figure 16.9: Non-adaptive Residual VQ

block labeled "S/P" denotes "serial-to-parallel" and performs the operation of converting a serial sequence of k samples of the incoming waveform into a parallel set of k samples regarded as a vector. The corresponding block labeled "P/S" denotes the reverse operation of taking successive vectors, of k samples each, and producing a serial output sequence of samples.

While non-adaptive residual coding tends to improve coding performance by removing inter-vector redundancy and reducing the size of the codebook needed, it also degrades the VQ coding gain because the intra-vector redundancy is reduced. This coding structure is analogous to transform VQ described in Section 12.6 where we saw that there was no net gain in performing a fixed invertible orthogonal transform to the vectors prior to applying VQ. It is reasonable to expect that although of conceptual interest there is not likely to be any advantage in coding the residual over direct coding of the original signal. On the other hand, when the filters are adapted to track time varying signal statistics, the situation changes and a more interesting coding method results.

Vector Residual Coding

So far we have assumed the shaping and whitening filters are time-invariant. Now we take the next conceptual step toward developing an adaptive codebook technique by allowing the input signal to be modeled by a *time-varying* shaping filter which gradually changes its characteristics to model the differing local statistical character of the signal. Thus we need to determine the evolving character of the required shaping filter and transmit this information in quantized form to the receiver. Both transmitter and receiver will then have the filter parameters needed to model the pair of time-varying shaping and whitening filters. In effect, what we are seeking is a mechanism for periodically sampling the time-varying statistics of the signal to generate a sequence of time invariant filters that collectively model the time-varying whitening and shaping filter pair.

When the short-term spectral character of the input signal $x(n)$ varies with time, an adaptive version of this coding scheme is needed. A general scheme of this type is vector residual coding (VRC). In Figure 16.10, the basic structure of this coding scheme is shown, which can be viewed as a natural evolution from the non-adaptive version of Fig. 16.9. The input

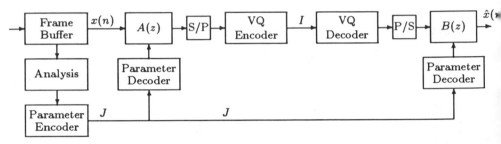

Figure 16.10: Vector Residual Coding (RELP-VQ)

signal is first applied to a frame buffer, which simply stores a set of waveform samples corresponding to a suitable chosen frame size before delivering the successive samples to the whitening filter. The analysis block then computes an optimal estimate of the filter parameters needed to model the current frame of the input signal. Once computed, these parameters are quantized and delivered to the whitening filter $A(z)$. The quantized filter parameters, which specify both $A(z)$ and its inverse $B(z)$ constitute the side information that is transmitted to the receiver once every frame as indicated by the index J for the current frame. A fixed VQ encoder generates an index I to specify the current residual vector. Generally, the signal statistics change very slowly so that the frame duration, M samples, typically consists of several vectors i.e., several vectors are processed with the same filter characteristic before the filter is updated. Thus, if $M = mk$, where m is a positive integer and k is the vector dimension, then the overall rate in bits per frame is $R = mb_v + b_s$ where 2^{b_v} is the VQ codebook size and b_s is the number of side information bits needed to specify the filter parameters in each frame.

Note that the frame buffer effectively allows us to "look ahead" into the future to identify the signal statistics before that segment (frame) of the signal is actually applied to the whitening filter. The indexes that specify the quantized parameters are transmitted to the receiver as "side information," decoded, and then applied to the shaping filter in the receiver. Thus the receiver decodes the residual vectors, converts them into an excitation waveform and feeds this excitation into the time-varying shaping filter, generating the reconstructed approximation $\hat{x}(n)$ of the original waveform $x(n)$.

To fully specify the operation of this adaptive residual coding system several issues must be addressed regarding the filters and their operation. We assume the analysis block has an algorithm that generates a causal, stable, minimum phase recursive (pole-zero) filter of finite order that *approximately* whitens the signal. One of several ways to do this is to use the Levinson-Durbin algorithm to generate a linear prediction filter $P(z)$ as described in Section 4.4. Then we can use $A(z) = 1 - P(z)$ as the (approximate) whitening filter and $B(z) = A^{-1}(z)$ as the corresponding shaping filter. In this case, the shaping filter is an all-pole filter which is widely used in speech coding. Using the Levinson-Durbin algorithm followed by quantization of the chosen filter parameters (e.g., reflection coefficients) yields a coding system sometimes referred to as a *residual excited linear predictive* or *RELP* system. If VQ is also used to represent the filter parameters, then the shaping filter is selected using LPC-VQ from one of a finite set of filters specified by a codebook. In this case, the VRC scheme is called RELP-VQ [264][7].

RELP-VQ has the additional advantage of being well suited to multirate systems. A low rate coding scheme is obtained by using only the LPC-VQ codewords together with voicing and pitch information, as in a classical LPC vocoder [310], where an artificial "buzz-hiss" excitation is generated at the receiver from the voicing and pitch information. The higher rate coder substitutes the waveform coded information as an excitation for the LPC model in place of the vocoder parameters (voicing and pitch).

A second issue not considered so far is the handling of the transition when the filter parameters are updated. Suppose the filter parameters are the coefficients of the autoregressive-moving average difference equation as described in Section 4.10. In each discrete-time unit (i.e., a clock cycle), a new input sample is applied to the filter and a new output sample is generated. The update of the filter parameters can be assumed to take place between two such clock cycles. Specifically, the filter should be updated after the last sample of the last vector in a frame has entered the filter and before the first sample of the first vector in the next frame is applied to the filter. It is reasonable and convenient to assume that the memory in the filter (the internal state of the filter) remains unchanged when the update takes place. This assumption eliminates any ambiguity in specifying the periodically updated filtering operation.

The VRC coding system shown in Figure 16.10 can be regarded as a forward adaptive technique for signal coding since it is looking ahead at the local character of the arriving frame to adapt the filtering operation for coding that frame. The value of VRC or RELP-VQ comes from the adaptive preprocessing of the signal by periodically updating the linear prediction model parameters. To achieve comparable performance by direct VQ of the input signal $x(n)$ in the case of a time varying spectrum would require an enormously large codebook that would be too complex for realistic applications.

Closed-Loop Vector Excitation Coding

One serious weakness of VRC is that there is no guarantee that separate good codings of the shaping filter parameters and the residuals will together produce an accurate coding of the original signal. The problem is that we are coding the residual "open-loop" where a code vector is selected to match the current residual vector as closely as possible and not taking the better course of choosing a code vector that will produce the best possible reproduction of the original input signal. We will accomplish this step by taking a great leap forward from adaptive residual coding to the much more powerful technique of VXC.

The VRC structure structural weakness described above is that the minimum distortion search minimizes a measure of distortion (usually squared Euclidean distance) between the residual and quantized residual vectors. But this is not really the performance measure that matters to the user! Rather it is the error between the original and reconstructed signal that measures the degradation or distortion. One method of causing the search to find the minimum distortion reproduction to the original signal is to find the codevector which *after passing through the shaping filter* produces a reproduced signal vector (block of k samples) that is the best approximation to the current input signal vector (defined as the corresponding block of k samples). Thus we can use any suitable distortion measure and compare the current input vector with the reconstructed signal vector produced from each code vector in the residual (excitation) codebook. Once the best choice is found, the corresponding index is transmitted to the receiver. This gives a so-called *closed-loop* search technique also known as *analysis-by-synthesis*. In general, a closed-loop search technique is a a trial-and-error procedure to determine an excitation signal that after filtering through a synthesis filter best approximates a desired input.

Figure 16.11 illustrates the structure of the closed-loop VXC method, widely known as *code excited linear prediction* or CELP. The codebook

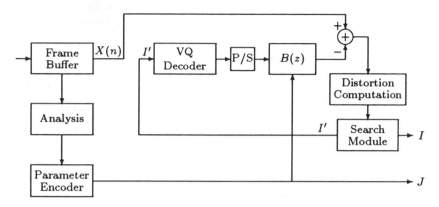

Figure 16.11: CELP (VXC) Encoder

located in the VQ decoder box is called the excitation codebook. The CELP
encoder searches to find that index which when applied to the decoder will
yield the best reconstructed signal vector. Thus the encoder must have a
replica of the decoder and use it to determine the optimal index; this is done
by trying each candidate index I' to generate the reconstructed output that
would result if that index were chosen. The optimal index I is simply the
one which leads to the best reconstructed approximation to the original.
The CELP decoder is identical to that of the VRC decoder shown in the
right side of Figure 16.10. In fact, CELP can be seen to be equivalent
to recursive VQ where a predictive decoder is used and the encoder is a
minimum distortion or nearest neighbor encoder for that decoder. The
state of the recursive VQ decoder is simply the set of past and present
reconstructed samples which constitute the memory of the predictor.

Closed-loop techniques have had widespread applications in speech coding. Closed-loop speech coding where the excitation is generated by a con-
volutional tree code generator was first described by Schroeder and Atal
in [284] for scalar values associated with each tree branch and for the case
of vector values on each branch in [285]. For the case where the shaping
filter selection is accomplished by LPC-VQ, the use of a residual codebook
to drive the filter to produce reproductions for the original input sequence
for a minimum distortion mapping was originated by Stewart [301] and
elaborated in [302] for trellis encoded residuals and in [153] for VQ encoded
residuals. The first reports of a successful speech coder based on closed-
loop searching of a large dimensional VQ residual codebook were due to
Schroeder and Atal [286] and [287], and a closely related speech coding
technique was proposed in [80]. The name "code-excited linear prediction"

or CELP has been used to reflect the fact that the decoder drives a prediction inverse filter with an input residual or excitation selected from a codebook. The name "vector excitation coding" is a slightly more generic name for the large family of closed-loop coding schemes where the excitation is selected by searching a VQ codebook. The speech coder reported in [286] required 1000 times real-time to simulate the coding of one second of speech due to the high complexity of the search process. Subsequently, several papers led to major complexity reductions and modifications that have made possible real-time implementations of CELP coders. See for example, [309], [91], [64], [206]. For a collection containing many recent papers on VXC/CELP coders, see [21].

16.10   Problems

16.1. In an alternative method of forward mean-adaptive VQ, the unquantized block mean is subtracted from each component of the input vector and the resulting mean-reduced vector is vector quantized. The quantized block mean is sent to the receiver as usual. Compare the performance that can be achieved with this alternative to the block mean removal method described in Section 16.2.

16.2. Prove the equivalence of (16.3.4) and (16.3.5).

16.3. Compare a switched-codebook adaptive VQ with a classified VQ on the basis of complexity (search and memory) and bit rate.

16.4. Prove that the solution $\mathbf{L}^*(\lambda)$ to the unconstrained optimization problem to minimize the objective function $D(\mathbf{L}) + \lambda R(\mathbf{L})$ over all state vectors $\mathbf{L}$ for each value of the parameter $\lambda \geq 0$ is identical to the solution of the constrained optimization problem as given in Equation (16.5.1), following the notation of Section 16.5.

16.5. Suppose we have N vector quantizers (not necessarily optimal) available for coding a random vector and a table of the corresponding distortion-rate pairs (D_i, R_i) for $i = 1, 2, \cdots, N$ where the ith pair is the distortion D_i and corresponding number of bits R_i attained with the ith quantizer. Suppose each pair is plotted as a point in the plane of distortion versus rate. Consider the convex upward curve (with positive second derivative) consisting of straight line segments joining a subset of points from the table in such a way that all other points lie above this curve. Show that all solutions to the optimization problem $\min_i [D_i + \lambda R_i]$ for any $\lambda > 0$ are points from the table which lie on this curve.

16.6. Is an adaptive address-VQ a finite-state VQ?

16.7. The progressive codebook updating techniques described are all forward adaptive, that is, they require that the new codewords be communicated via a side-channel. Can the codebook be progressively updated in a backward adaptive fashion? If so, propose a technique for doing so. If not, why not?

16.8. Show that an adaptive codebook with a universal code of L words and a subcode size of N can be indexed by $\log_2 \binom{L}{N}$ bits instead of the L bits mentioned in the text. Show that this is an improvement over L bits. Can you determine if it is an improvement over the $N \log_2 L$ technique?

16.9. Describe clearly an adaptive VQ technique that combines the methods of mean-adaptive and gain-adaptive VQ into a single coding scheme.

16.10. Define the *range* g of a k-dimensional vector $\mathbf{X}$ with components X_i as: $g = \max_i X_i - \min_i X_i$. Suppose a sequence of vectors $\{\mathbf{X}_n\}$ has a corresponding sequence of ranges g_n that are slowly varying with time (highly correlated). Propose an adaptive VQ technique that can take advantage of this characteristic of the vector sequence. Explain at least qualitatively how the adaptive method can achieve enhanced performance over direct use of a fixed vector quantizer on each successive vector.

Chapter 17

Variable Rate Vector Quantization

17.1 Variable Rate Coding

All of the lossy compression schemes considered thus far have been fixed rate schemes, systems that send a fixed number of bits for each input symbol or block or, equivalently, a fixed number of bits per unit time. The only variable rate schemes considered were the noiseless codes of Chapter 9 where channel codewords of differing sizes were assigned to different input vectors. In the noiseless code case the basic strategy is fairly obvious: by assigning short codewords to highly probable input vectors and long codewords to rare input vectors in an intelligent fashion, one can reduce the average number of bits sent. A similar strategy can be applied in vector quantization: instead of sending a fixed number, R, of bits for each input vector, one might have a code that assigns a variable number of bits to different codewords. For example, one could use several bits to describe codewords for active input vectors such as plosives in speech or edges in an image and use fewer bits for codewords describing low activity input vectors such as silence in speech or background in images. Unlike the lossless codes, however, it is not clear that here the best strategy is to assign shorter words to more probable vectors and longer words to less probable since the overall goal is now to minimize average distortion and hence both probability and distortion must be taken into account.

Since variable rate coding systems include fixed rate coding systems as a special case, such schemes are less constrained and are hence capable of better performance at a fixed average number of bits per input vector. Some compression applications, particularly storage and packet switched

631

communication networks are very naturally suited for variable rate coding. For example, pictorial databases, where images are stored in compressed form and retrieved on demand for display, are well-suited for variable rate methods. With variable rate coding, some image regions can be coded with many fewer bits than other regions depending on the amount of detail or complexity in the region. In fast packet networks, speech is typically compressed at variable rates and essentially no data is sent during silent intervals. The use of hardware-based packet switching allows fast routing of the packets so that they can be reconstituted at the receiver without excessive delay. There is considerable interest today in *packet video* where digital video signals are compressed with variable rate coding and packetized for transmission. The irregular distribution of video information with time makes this approach particularly appealing. Scenes with very little motion need only a very low rate while scene changes or scenes with rapid motion require a high rate but occur less often.

For many applications, however, a fixed transmision rate is required. In these cases, there is a cost to be paid for the potential improvement offered by variable-rate coding. The cost is in equipment complexity and implementation: a variable rate coding system requires extra buffering and bit control overhead if it is to be used with a fixed rate communications channel. If a source coder is coding a very active region it will produce more bits than the average rate and hence may produce bits faster than the channel can accept them. These bits must be saved in a buffer until a low activity region of the input permits the rate to reduce (so as to preserve the average) and the channel can catch up. This extra circuitry and overhead can be more than justified by performance improvements, especially if the source exhibits changes in local behavior of which a variable rate system can take advantage. There is another cost that is also of concern in some cases: the one-way end to end transmission delay tends to be much larger with variable length coding. For example, variable length speech coding systems where buffering is used to obtain a constant channel rate typically introduce more than 100 ms of delay, whereas constant rate speech coders can be implemented with delays as low as a few milliseconds, depending on the bit rate.

One approach to constructing a variable rate vector quantizer is simply to combine a fixed rate VQ with a variable rate noiseless code by considering the VQ reproduction vectors to be symbols in an extended source alphabet and constructing a variable length noiseless code for these symbols. This effectively assigns long indexes to the least probable VQ reproduction vectors and short indexes to the most probable. If, for example, Huffman coding is used, the average bit rate can be reduced from the logarithm of the number of VQ reproduction vectors down to the entropy of these reproduction vectors. This can provide a significant improvement if the entropy is much less than the logarithm of the number of reproduction vectors (it can be no greater). In fact, was shown in Chapter 9 by combining arguments from rate-distortion theory and high rate scalar quantization theory that this simple combination of scalar quantization followed by vector entropy coding can achieve performance within approximately one quarter of a bit from the optimal performance given by the rate-distortion function when the bit rate is high. (See also [244] or Chapter 5 of [159].) Ziv [353] has shown that the performance can be made within three quarters of a bit of the optimum even when the bit rate is not asymptotically large, but this bound is loose when one is interested in fractional bit rates.

The simple combination of uniform scalar quantization and entropy coding is the most straightforward and common approach to variable rate vector quantization, but it has its drawbacks. First, it can be quite complicated because good quality codebooks are often large and noiseless codes get quite complicated for large input alphabets. Second, although this approach can yield a good performance at a given average rate (better than the corresponding fixed rate VQ), there is nothing optimal about this factoring of the encoding into two steps. Other coding structures might be much less complex for a given distortion/rate pair. Third, if one has designed a tree-structured VQ (TSVQ) to take advantage of its low complexity and its successive approximation attribute (e.g., for progressive transmission), this structure will be completely lost if the path map through the tree is replaced by a variable length noiseless code.

There are at least two basic approaches to the direct design of variable rate vector quantizers. The first involves varying the dimension or block size of the vector quantizer which is applicable when the signal source is a waveform or image (rather than an intrinsic sequence of vectors as in an LPC parameter vector sequence). In *variable dimension VQ* active vectors are coded with small dimensional codes to permit high rates and inactive vectors are coded with large dimensional codebooks with small rate. In the second approach, the dimension or block size is held constant, but a varying number of bits is assigned to codewords in a convenient manner, possibly with tree-structured VQs having incomplete trees, that is, with leaves or terminal nodes of varying depth, or by another VQ structure that allows variable rate. Unlike the cascade of fixed rate quantizers (scalar or vector) with noiseless codes, in both approaches the VQ is allowed to directly vary its bit rate depending on the local source activity. Most of this chapter is devoted to the second approach and to several relevant ideas from the theory and practice of classification and decision trees. First, however, we describe the general ideas behind variable dimension VQ. A specific implementation incorporating ideas from mean-residual VQ and transform

VQ is described in detail in Vaisey and Gersho [317] [318] [320].

17.2 Variable Dimension VQ

Suppose we have a scalar valued sequence of signal samples and we wish to form a blocked-scalar sequence of vectors from the samples. The usual way is to maintain a constant block size of, say, k samples and obtain a k dimensional vector sequence. Alternatively, we can look ahead at the samples and depending on the local character of the signals, choose a different dimension k_i for the ith vector in the sequence and form a vector $\mathbf{x}_i$ of dimension k_i from the next k samples in the sequence.

In this way a scalar sequence is mapped into a sequence of vectors of varying dimension. To fully specify this sequence, we can describe it as a sequence of pairs, $\{(k_i, \mathbf{x}_i)\}$. The mechanism for choosing the dimension is not of concern to us here but we can assume that we are given a specific set of allowable integer-valued dimensions $\mathcal{K} = \{m_1, m_2, \cdots, m_L\}$. Such a sequence can be coded by the usual nearest neighbor search through a set of VQ codebooks, one for each dimension $m \in \mathcal{K}$ with a distortion measure $d(\mathbf{x}, \hat{\mathbf{x}}, m)$ that depends on the dimension m of the vector $\mathbf{x}$. Thus, we have L codebooks of the form $\mathcal{C}_j = \{\mathbf{c}_{j,i}, i = 1, 2, \cdots, N_j\}$ for $j = 1, 2, \cdots, L$. The encoder must generate an index identifying the selected codevector in the chosen codebook; also, side information must be sent to the receiver to specify the current vector dimension. Note that this is a variable rate coding scheme even in the special case where all codebooks are of the same size, since the rate in bits per input sample will then vary with the dimension of the vector.

The decoder for such a scheme will simply have a copy of each codebook. It must receive both a code vector index and side information that identifies the dimension of the current vector to be decoded.

Codebook design for variable dimension VQ is quite simple. A training sequence of input samples is segmented into a variable dimension vector sequence and all training vectors formed for a particular dimension m_i are formed into a separate training set and then used to design codebook $\mathcal{C}_i$ with any standard design algorithm.

In the above discussion, we have assumed each vector is coded with a simple unstructured VQ. It should be noted that their is no difficulty added in principle if we choose to use a particular constrained VQ method for each vector dimension. A different coding structure can be designed for different dimensions. In particular, each coding structure can itself be a variable rate coding system. A further alternative is to select the variable dimension input vectors in a way to minimize long term average distortion.

By combining optimal segmentation with the Lloyd algorithm and entropy coding one can design good variable-dimension to variable-length codes [67].

Although the variable rate VQ concept has been introduced in the context of a one-dimensional signal, it is equally applicable to multiple dimension signals. In the following section, we describe a particular approach to variable dimension image coding.

Quadtree-Based Variable Block Size VQ

If a region of an image or a period of time of speech consists of fairly homogeneous background information such as the sky surrounding a mountain in a picture or silence in sampled speech, then it is likely possible to code this information with large dimensional vectors because only a fairly small number of such vectors will be needed. Conversely, in periods of high activity such as edges in images and plosives in speech, a rich collection of vectors will be needed to represent well even a small dimensional block. This suggests one basic approach to variable rate VQ: first look at a large block and see if it is fairly inactive. If so, code it with a large dimensional codebook. If not, segment the block into smaller blocks and then continue with the same strategy. Clearly the complexity of the algorithm will depend on the segmentation strategy. One could permit odd block shapes and one could even try to optimize over all possible choices of breaking up an image into smaller pieces, but a simple approach that works quite well is to restrict the allowed block shapes. In particular, one can use *quadtrees* to find a fast segmentation algorithm leading to a variable rate VQ.

A quadtree is a tree representation of an image wherein each nonterminal node has four children. Each node corresponds to a subblock of the image, that is, a subset of the pixels constituting the image. The root node is the entire image, the four nodes at the first level of the tree correspond to four disjoint subblocks. In its simplest form, the image is assumed to be square and the first level nodes correspond to the four equal sized quarter images.

The next layer of the tree corresponds to a partition of the original image into 16 equal parts, and so on. The first layer and the nodes in the second layer descended from one of the first layer nodes are depicted in Figure 17.1. Thus each child represents one quarter of the area (or the number of pixels) of its parent. A general discussion of quadtrees and their applications may be found in Samet [282].

If one has a measure of how complicated or active a node (block) is, this measure can be used to test each node encountered in a VQ encoding algorithm. If a threshold of activity is exceeded, then the node is not yet coded and one proceeds to its children. Otherwise, the node is encoded

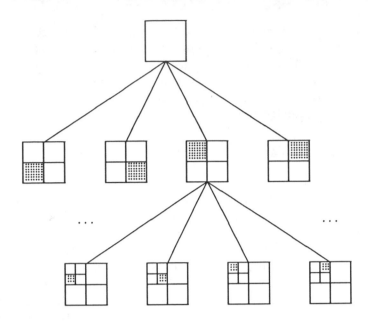

Figure 17.1: Quadtrees

with a VQ designed for the dimension of the block represented by the node. An illustration of a typical variable sized partition based on a quadtree is shown in Figure 17.2. Each of the squares would be coded with a VQ for that dimension. Different VQ structures could be used for the different block sizes. Unlike the fancier type of image segmentation methods used in computer vision applications, the quad tree is attractive for image coding because the partition is readily specified with a modest overhead in bits per pixel. Specifically, for each node above the lowest allowed level only one bit is needed to identify whether that node is split or is a terminal node in the tree. For example, if a a quadtree is based on a 32×32 root node with lowest level consisting of 2×2 blocks, a maximum of $1 + 4 + 16 + 64 = 84$ bits are needed to specify the partition for a worst case side information rate of $85/(32)^2$ or 0.083 bpp. If the lowest level blocks are 4×4 blocks, then the worst case rate is only 0.02 bpp. The total number of bits to specify the complete partition varies with the degree to which the region needs to be decomposed. For example, in Figure 17.2 if the root block is a 32×32 array of pixels, then, exactly 41 (rather than 85) bits are sufficient to specify the entire partition shown in the figure. The overhead for this side information is 0.04 bpp.

There are many heuristic methods for measuring the degree of activity in an image block. For example, if the sample mean of the block differs from

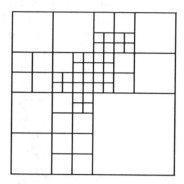

Figure 17.2: Sample Quadtree Segmentation

each of the sample means in the next level subblocks by less than a suitable threshold one might then choose not to subdivide this block. A simple but effective activity measure for a block is the sample variance of the pixels of a block. This quantity actually provide a rough way of estimating the amount of high spatial frequency content in the block. Moreover, to find the sample variance one must first find the sample mean and we found in Section 12.9 that good performance can be achieved by applying a product code to separately code a sample mean and the resulting residual. Hence these two ideas can be combined to form a quadtree based VQ. Thus, for example, one could begin with a large block, e.g., 32×32 pixels, and compute the sample mean and sample variance. If the sample variance is below some threshold, one could then apply a product (mean-residual) VQ to this large block. If the sample variance is large, one could split the block into four equal parts (yielding the four children of the original block) and then treat each one in a similar fashion. Different VQs would be required for each of the allowed block sizes. In [317] [318] [320] the design is somewhat more complicated and specific VQs for the various block sizes are described in detail. In particular, the mean and residual are permitted to have different dimensions and each has its own quadtree. As a first stage of quantization, a "mean image" is formed by segmenting each 32×32 block with a quadtree structure into regions having fairly homogeneous means. Hierarchical VQ is then used to code the mean values by applying a different vector quantizer at each level of the tree.

The result of coding Lenna with the mean quadtree is shown in Figure 17.3 where the bit rate is 0.079 bpp and the PSNR is 24.5 dB. In this example, the smallest subblock size is 4×4 and the 4-dimensional VQ codebooks for the means ranged in size from 64 to 512. An interpolation process is applied to the mean image, enhancing both perceived quality and PSNR, and then the mean image is subtracted from the original to form a residual image. A second quadtree segmentation is performed on the residual image. The large residual blocks are coded using transform VQ (to permit codebooks of reasonable size at the large block sizes) and the small blocks are coded by ordinary memoryless VQ. Further improvement is obtained by permitting a small additional codebook containing codewords peculiar to the specific image, a form of adaptation. The resulting codes gave good performance on the USC database at rates of 0.3 bpp–0.4 bpp. At 0.35 bpp the "Lenna" image yielded a PSNR of 29.8 dB.

Figure 17.3: Quadtree Coded Means at 0.079 bpp

17.3   Alternative Approaches to Variable Rate VQ

A large family of variable rate coding methods can be based on simple modifications to previously described fixed rate VQ techniques. We illustrate this by briefly sketching examples for constrained, adaptive, and finite-state VQ. Later we treat in depth one important variable rate coding technique derived from the constrained VQ technique, TSVQ.

In constrained VQ methods generally a single vector is encoded with a channel symbol that is composed of two or more indexes, identifying component code vectors that are used jointly to reconstruct a reproduction to the source vector. Consider for example shape-gain VQ. Suppose we first encode the gain of a source vector with a scalar gain codebook. In some applications it may happen that the accuracy to which the shape is coded depends on the gain level, e.g., for lower gains fewer bits are adequate to code the shape while for high gain vectors, it is more important to allocate more bits to the shape. In such a case, we can have a *family* of shape codebooks, as many as one for each gain level, and the choice of shape codebook is determined by the index identifying the gain value. The different shape codebooks can have widely different sizes and hence the number of bits to specify the shape will depend on the gain index. Note that no side information is needed. A binary word of fixed length can be used for the gain index to which a variable length binary word is appended to specify the shape code vector. The identity of the shape codebook is known from the gain index.

This is a variable rate coding system, which may not be optimal but it does have some heuristic rationale that suggests it might be effective for some applications. Codebook design for this scheme is readily achieved by first designing the gain codebook from the sequence of training vector gains. Then, the shape vectors from the training set are clustered into separate sets according to the corresponding gain index. Finally a codebook design for the shapes in each cluster is performed by specifying the desired shape codebook size for each gain index.

More generally, any sequential search product code, such as multistage VQ, can be made into a variable rate coding scheme by letting the index of the first feature to be encoded determine which of a set of codebooks of varying sizes is to be used to code the second feature. Similarly the index of the second feature can determine the codebook to be used for the next feature and so on. Thus the generalized sequential search product codes described in Chapter 12 readily provide the basis for variable rate coding systems.

Multistage VQ is sometimes used as a variable rate coding system in a different and simpler way. The actual distortion incurred in encoding the input vector by the first stage is measured. If it exceeds a threshold, a second stage encoding is used; otherwise the encoding is complete. A one bit flag is sent to the receiver to indicate whether or not the second stage encoding is performed. This method can be extended to multiple stages.

Adaptive VQ methods with forward adaptation are also well suited for variable rate coding. By examining a block or frame of successive vectors and encoding a feature such as the frame sample mean or average norm, the index for this feature can control which of a set of codebooks with different rates will be used to encode the individual vectors in the frame. In classified VQ, the classifier can be designed to identify (among other features) how complex the current frame of vectors appears, where "complex"' refers to the amount of information or amount of perceptually important detail is present in the current frame. Thus, the index specifying the class can be an effective indicator of the rate needed for the vectors in the current frame.

Finite-state VQ offers another entry to variable rate coding systems. Each state codebook can have a different rate and since the state is known to the receiver, it also knows how to interpret the current channel symbol which has the form of a varying length binary word.

The extension of TSVQ to variable rate coding has been very extensively studied and has proven to be very effective, particularly in image coding applications. We next examine this topic in depth.

17.4   Pruned Tree-Structured VQ

The successive approximation property of TSVQ suggests a very different structure for a variable rate vector quantizer. Suppose that we have a binary TSVQ with depth L (and hence a rate of L bits/input vector). Call the corresponding tree (or tree-structured codebook, to be precise) $\mathcal{T}$. While there is no method of ensuring that this is an optimal TSVQ, using the design algorithms of Chapter 12 should yield at least a good code. In particular, we assume that the TSVQ tree has been designed from a training set of source vectors as previously described with the splitting method and the generalized Lloyd algorithm so that each node is labeled by the centroid of the training vectors which are assigned to that node. The final codebook is the collection of labels of the terminal nodes or leaves $\tilde{\mathcal{T}}$.

Suppose as indicated in Figure 17.4 we remove or *prune* two of the terminal nodes from the tree descending from a common parent node. This yields a *subtree* S of the original tree which can also be used as a codebook in a tree-structured VQ: the encoder searches the tree as before until it hits a terminal node or leaf. As usual it sends to the receiver the binary word that is the path map through the tree to the terminal node. As before, the corresponding codeword is the label of the terminal node (which is the centroid of all input training vectors which map into that terminal node). What is different is that input vectors mapped into this new terminal node at depth $L - 1$ in the tree will have a path map of length $L - 1$ and not L as do the other terminal nodes. Thus we have an *unbalanced tree* and this encoder produces a variable length binary channel codeword and has a variable rate, which will be strictly less than L if this path to the new

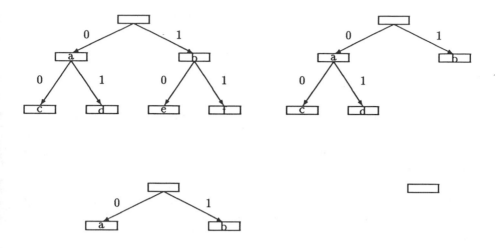

Figure 17.4: Full Tree and Pruned Trees

terminal node is ever taken. On the other hand, the average distortion will increase because we have removed one of the options from the encoder that would have led to lower distortion for certain input vectors. Thus things are better in that we have lowered the average rate, but worse in that we have increased the average distortion. The question is, have we made the tradeoff between average rate and average distortion in a sensible (hopefully optimal) fashion? Likely not yet, but we soon shall.

In exactly the same manner, if we prune a given tree T by removing all branches from any node (terminal or not) within the tree (along with all of the branches and subsequent nodes emanating from the pruned branches), we produce a new subtree which will have a reduced average rate or length and an increased average distortion. If the resulting pruned subtree, S, is used as a tree and codebook for a variable rate TSVQ on a source, then it will have some average distortion and average length, which we denote by $\delta(S)$ and $l(S)$, respectively. Just as we can consider pruned subtrees of the original full tree, we can also consider pruned subtrees of subtrees of the original full tree, as suggested in Figure 17.4. If S' is a pruned subtree of S, then we write $S' \preceq S$ and observe that

$$\text{if } S' \preceq S \text{ then } \delta(S') \geq \delta(S) \text{ and } l(S') \leq l(S).$$

The average distortion cannot increase from the smaller subtree S' to the tree S that contains it because of the way the original TSVQ was designed. The input vectors encoded into a particular node were used to design the codebook (consisting of two codewords in the binary case) represented by the children nodes and this new codebook must be better than the single codeword labeling the parent node (since the codebook was obtained by running the Lloyd algorithm on an initial code which included the parent label). The average length must decrease since all of the path maps including this node have been reduced to the depth of this node.

We can now formulate one approach to designing a variable rate TSVQ. Given a complete TSVQ with codebook tree T designed as in Chapter 12 with depth L and given a target rate constraint R, what is the *optimal* subtree S of T in the sense of providing the smallest possible average distortion while having average rate or codeword length (the number of bits per input vector) no greater than R? We can formalize this notion using an *operational distortion-rate function* defined by

$$\hat{D}(R) = \min_{S \preceq T:\, l(S) \leq R} \delta(S).$$

In principle this problem can be solved by brute force exhaustion. For example, Figure 17.5 plots as dots a typical collection of distortion/rate pairs for subtrees of a depth 3 TSVQ codebook. The operational distortion-rate function is given by the staircase. It can happen that more than one dot will lie on the straight line segments connecting the lowest points, but the figure reflects the simpler (and more common) behavior where the dots are the breakpoints.

In this simple example the minimum average distortion subtree with rate no greater than R corresponds to the dot on or just to the left of the vertical line drawn at R bits. Solution by exhaustion is clearly impractical if the original full tree is not small, but the figure does provide some useful insight. One can argue that it is not in fact the operational distortion-rate function that is most important here. Specifically, if you want a source code of a given average rate, instead of just picking one of the pruned TSVQs from the figure, you could *time share* two of the codebooks. That is, you could use one TSVQ a fraction λ of the time and the other a fraction $1 - \lambda$ of the time. In this manner you could, in principle, achieve an average distortion anywhere on the convex line in the figure connecting the dots on the operational distortion-rate curve, the *convex hull* of the distortion-rate curve or the lower boundary of the convex hull of the set of achievable distortion/rate pairs. These points always lie on or below the operational distortion-rate curve in that they yield strictly smaller average distortion for a fixed average rate. (This of course assumes that the time-sharing of two trees can be implemented without sending side information to the receiver.) Note that in this manner one can achieve the *optimal* performance for any rate between 0 and 3 bits per input vector in the sense that one has the

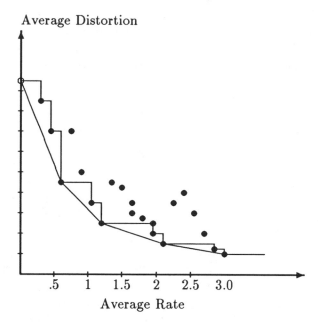

Figure 17.5: Distortion/Rate Points of Subtrees

minimum distortion code given that the coders are pruned subtrees of the original complete tree. The original tree, unfortunately, may not be optimal in the sense of yielding the smallest possible distortion for all trees of the same rate and depth.

One can argue that even if time-sharing is not practicable, the only pruned subtrees that should be used as codes are those lying on the convex hull since the others do not trade off rate and distortion as efficiently since distortion/rate pairs on the operational distortion-rate curve but not on the convex hull will by dominated by a point on the convex hull giving smaller average distortion for the same average rate.

This argument suggests that it is of interest to determine the set of pruned subtrees whose distortion/rate pairs lie on the convex hull. This is a reduced goal from that of determining all distortion/rate pairs on the operational distortion-rate curve, but we shall see that it is a much simpler task and that the resulting trees are embedded, that is, a sequence of subtrees–a property not shared by the trees yielding the complete operational distortion-rate curve. In the next section we shall see that the problem of finding the subtrees yielding the convex hull of the distortion-rate curve is solvable in a straightforward and efficient manner. Specifically, we will develop an algorithm which finds the lower boundary of the convex hull of the rate/distortion pairs of pruned subtrees of a given tree and which specifies the subtrees that lie on this convex hull. This provides a means of designing variable rate TSVQs which will be seen to provide better performance (in the sense of trading off rate and distortion) than both fixed rate TSVQ and fixed rate full-search VQ. The problem of finding all of the trees on the distortion-rate curve is far more difficult and has been considered by Lin, Storer, and Cohn [213][212].

The algorithm to be presented is a generalization of an algorithm of Breiman, Friedman, Olshen, and Stone [36] and we will refer to it as the generalized *BFOS* algorithm. Breiman et al. developed an algorithm for optimally pruning a tree trading off the number of leaves with probability of error for a classification tree and mean squared error for a regression tree. Their pruning algorithm, part of a program called CART^{TM} for "classification and regression trees," extends to more general complexity and cost criteria in a straightforward manner [69]. The algorithm is also closely related to the bit allocation algorithm in [294] [295] which was discussed briefly in Chapter 16, although the latter approach is couched in somewhat different language and was not applied to tree structures. The algorithm has a Lagrangian flavor in that it minimizes the functional $J(\mathcal{S}) = \delta(\mathcal{S}) + \lambda l(\mathcal{S})$ over all pruned subtrees $\mathcal{S}$ of an initial tree $\mathcal{T}$. By varying the parameter λ (which can be interpreted as a Lagrange multiplier and as the slope of the distortion vs. rate curve), all of the distortion/rate pairs on the convex hull of the operational distortion-rate curve can be found. The extension of the BFOS algorithm and its application to pruned TSVQ closely follows the development in [69]. The algorithm is presented in some generality since, as will be seen in later sections, it can be used for other resource allocation problems as well as for pruned TSVQ design. For example, the BFOS algorithm can be used to design an optimal bit allocation in transform VQ systems. The algorithm is not, however, developed in as general a form as [69] and this does result in some simplification.

Before formalizing the procedure, however, the basic idea can be informally described with reference to the example of Figure 17.5. Suppose that one begins with the full tree, which is the rightmost dot in the figure. One can compute the slope to any other distortion/rate pair, that is, the slope of the line connecting the rightmost dot with any other dot. To find the subtree corresponding to the adjacent point on the convex hull of all such pairs it suffices to find the pair which yields the *smallest* magnitude slope. Any dot having a slope of larger magnitude will necessarily lie above the line connecting the rightmost dot with the dot yielding the smallest magnitude slope. What makes the algorithm work is that it turns out that when computing these slopes one need only consider further subtrees of the subtree corresponding to the current dot on the convex hull. As a result, once the second dot is selected, one can then find the next dot to the left on the convex hull by repeating the procedure: compute the slopes from the current dot to all other dots and select the dot having the lowest magnitude slope.

17.5 The Generalized BFOS Algorithm

Real-valued functions on trees and their subtrees, such as average length $l(\mathcal{S})$ and average distortion $\delta(\mathcal{S})$, are known as *tree functionals*. Strictly speaking, however, we have not defined l and δ on *all* subtrees of a tree, only on the pruned subtrees, that is, only on subtrees with the same root node. It will prove convenient to consider tree functionals as being defined for all subtrees, including the singleton subtrees defined as subtrees consisting of only one node. This can be done in a natural manner that will be described later.

These two tree functionals and many other interesting tree functionals share certain properties that play a key role in the optimization algorithm to be described. We begin this section by introducing some notation and defining these properties. We will focus on tree functionals that are *linear* and *monotonic* in a sense to be defined shortly. The algorithm works for more general *affine* functionals, but we focus on the linear case for simplicity since it is enough for the applications considered here.

A tree $\mathcal{T}$ will be thought of as a set of nodes, $\mathcal{T} = \{t_0, t_1, \cdots\}$, where t_0 will always mean the root node. The set of leaves or terminal nodes of the tree will be denoted $\tilde{\mathcal{T}}$. A subtree $\mathcal{S}$ of a tree $\mathcal{T}$ is itself a tree that is rooted at some node $t \in \mathcal{T}$, where it is not required that the leaves $\tilde{\mathcal{S}}$ of the subtree $\mathcal{S}$ also be leaves of $\mathcal{T}$. If it is true that $\tilde{\mathcal{S}} \subset \tilde{\mathcal{T}}$, then if t is the root node of $\mathcal{S}$ we refer to $\mathcal{S}$ as a *branch* of $\mathcal{T}$ from t and write $\mathcal{S} = \mathcal{T}_t$. Thus, a branch from t contains the entire set of nodes in $\mathcal{T}$ that descend from t. A *pruned subtree* $\mathcal{S}$ of $\mathcal{T}$ is a subtree with the same root node t_0 as the full tree, a relation which we denote by $\mathcal{S} \preceq \mathcal{T}$. We will also consider a single node t as being a subtree with only a single node.

A tree functional $u(\mathcal{S})$ is an assignment of a real number to all subtrees of a given tree $\mathcal{T}$, including the complete tree $\mathcal{T}$ and all of the individual nodes. The tree functional is said to be *monotonically increasing* if $\mathcal{S}' \preceq \mathcal{S}$ implies that $u(\mathcal{S}) \geq u(\mathcal{S}')$. (The functional is bigger for the bigger tree.) It is called *monotonically decreasing* if $\mathcal{S}' \preceq \mathcal{S}$ implies that $u(\mathcal{S}) \leq u(\mathcal{S}')$. (The functional is smaller for the bigger tree.) It is called simply *monotonic* if it is either monotonically increasing or decreasing. In our application of a TSVQ the average length $l(\mathcal{S})$ is monotonically increasing and the average distortion $\delta(\mathcal{S})$ is monotonically decreasing.

A tree functional $u(\mathcal{S})$ is said to be *linear* if it has the property that its value is given by the sum of its values at the leaves:

$$u(\mathcal{S}) = \sum_{t \in \tilde{\mathcal{S}}} u(t). \tag{17.5.1}$$

This implies that if $\mathcal{R}$ is any pruned subtree of $\mathcal{S}$, then we have the decomposition

$$u(\mathcal{S}) = u(\mathcal{R}) + \sum_{t \in \tilde{\mathcal{R}}} \Delta u(\mathcal{S}_t) \tag{17.5.2}$$

where $\Delta u(\mathcal{S}_t) = u(\mathcal{S}_t) - u(t)$ is the change in the tree functional from the root node of the subtree to its terminal leaves. Thus the tree functional of a subtree $\mathcal{S}$ can be found from the tree functional of a further subtree $\mathcal{R}$ by adding the change resulting from all branches $\mathcal{S}_t$ for $t \in \tilde{\mathcal{R}}$ emanating from the smaller subtree. Note in particular that this provides a means of testing the change in value of the tree functional that would result from pruning any of the branches emanating from the leaves of the smaller subtree. Given a subtree $\mathcal{S}$, we shall be interested in examining the change in value of a functional on $\mathcal{S}$ if a single branch $\mathcal{S}_t$ rooted at a node $t \in \mathcal{S}$ is pruned. By knowing this change for every t in $\mathcal{S}$, we can then choose the best possible branch to prune.

To consider $l(\mathcal{S})$ and $\delta(\mathcal{S})$ in this framework, observe that given a source (and hence a probability distribution) and a TSVQ we can associate with each node t in the tree a probability $P(t)$ of reaching that node when encoding the given source with that tree-structured codebook. These probability values can be determined from a training sequence as part of the TSVQ design. Note that for any pruned subtree $\mathcal{S}$ of $\mathcal{T}$ we must have that

$$\sum_{t \in \tilde{\mathcal{S}}} P(t) = 1.$$

If $l(t)$ is the length of the path from the root node to node t, then the average length of a pruned TSVQ corresponding to a pruned subtree $\mathcal{S}$ will be

$$l(\mathcal{S}) = \sum_{t \in \tilde{\mathcal{S}}} P(t)l(t). \tag{17.5.3}$$

Thus if we define $u(t) = P(t)l(t)$, then

$$u(\mathcal{S}) = \sum_{t \in \tilde{\mathcal{S}}} u(t)$$

is a linear tree functional and $u(\mathcal{S}) = l(\mathcal{S})$ for all pruned subtrees. Hence we can extend the definition of l to all subtrees via (17.5.3). This definition has a natural interpretation. Given an arbitrary (not necessarily pruned) subtree S of T having root node t, then $l(S)$ is the probability that the input is mapped into node t multiplied by the conditional expected length of the path maps (resulting from encoding input sequences given that they are mapped into t by the original TSVQ). If the root node of S is at depth m in the tree, then $l(t) = m$ and $\Delta l(S) = l(S) - l(t) = l(S) - m$ is the portion of the average length due to S.

Similarly, if $d(t)$ is the average distortion at node t (the expected distortion resulting from encoding an input vector into the reproduction label corresponding to that node, then

$$\delta(S) = \sum_{t \in \tilde{S}} P(t)d(t)$$

is also a linear tree functional which agrees with the original definition on pruned subtrees and has a similar interpretation on other subtrees: it is the probability of mapping into the root node t of the subtree times the conditional expected distortion resulting from the subtree, where the conditioning is on being mapped into t. The value of the functional for each node can be determined as as offshoot of the tree design procedure based on a training set.

The linear functionals, now defined on all subtrees, are still monotonic for essentially the same reasons they were monotonic on pruned subtrees: a smaller tree must have a smaller average length and the complete TSVQ tree design ensures that conditional average distortion (conditioned on reaching any node) must be decreased by the codewords designed for that node.

Let u_1 and u_2 be two monotonic linear tree functionals, with u_1 monotonically increasing and u_2 decreasing. A typical example is $u_1(S) = l(S)$ and $u_2(S) = \delta(S)$. Let $\mathbf{u}(S) = (u_1(S), u_2(S))$ be the vector-valued function on the set of all subtrees of T, with u_1 and u_2 as components. Then consider the set of points $\{\mathbf{u}(S) : S \preceq T\}$, and its convex hull. When $u_1(S) = l(S)$ and $u_2(S) = \delta(S)$, this is just the set of points for all pruned subtrees of T, plotted in the distortion-rate plane and its convex hull. By monotonicity, the singleton tree consisting of just the root t_0 has the smallest u_1 and the largest u_2; the full tree T itself has the largest u_1 and the smallest u_2. Therefore, $\mathbf{u}(t_0)$ is the upper left corner of the convex hull; $\mathbf{u}(T)$ is the lower right corner. This is the case shown in Figure 17.5 where u_1 represents the average rate and u_2 represents the distortion.

It is a remarkable fact and the key to the generalized BFOS algorithm that the optimal subtrees are nested in the sense that if $(\mathbf{u}(T), \mathbf{u}(S_1), \mathbf{u}(S_2), \cdots, \mathbf{u}(S_n), \mathbf{u}(t_0))$ is the list of vertices clockwise around (the lower boundary of) the convex hull of all distortion/rate pairs, then $t_0 \preceq S_n \preceq \cdots \preceq S_2 \preceq S_1 \preceq T$. Hence it is possible to start with the full tree at $\mathbf{u}(T)$

and prune back to the root at $\mathbf{u}(t_0)$, producing a list of nested subtrees which trace out the vertices of the lower boundary of the convex hull. This fact is proved in [69].

It remains to show how to obtain $\mathcal{S}_{i+1}$ by pruning back $\mathcal{S}_i$. Actually, we will show only how to obtain $\mathcal{S}'$ by pruning back $\mathcal{S}$, where $\mathcal{S}$ and $\mathcal{S}'$ are subtrees on the convex hull with

$$\mathcal{S}_{i+1} \preceq \mathcal{S}' \preceq \mathcal{S} \preceq \mathcal{S}_i$$

and $\mathcal{S}' \neq \mathcal{S}$. This distinction is made because it turns out that the pruned subtree $\mathcal{S}_{i+1}$ of $\mathcal{S}_i$ can always be found by removing a sequence of single branches from the latter and each of these interim pruned subtrees will lie on the straight line connecting the vertices corresponding to $\mathcal{S}_{i+1}$ and $\mathcal{S}_i$. Thus each face of the convex hull may contain several nested subtrees, each formed by pruning one branch from the next larger subtree. The algorithm will use this procedure repeatedly, starting with $\mathcal{T}$, until t_0 is reached, which occurs after a finite number of steps since there are only a finite number of subtrees.

Suppose we are at a point $\mathbf{u}(\mathcal{S})$ that is an internal point of a face F of the convex hull. Let λ be the magnitude of the slope of F. Consider the set of all pruned subtrees of $\mathcal{S}$ obtained by pruning off a single branch $\mathcal{S}_t$ from some interior node $t \in \mathcal{S}$. Each interior node $t \in \mathcal{S}$ corresponds to a pruned subtree $\mathcal{R}(t)$ in this set. Then $\mathbf{u}(\mathcal{S}) = \mathbf{u}(\mathcal{R}(t)) + \mathbf{\Delta u}(\mathcal{S}_t)$ for each $\mathcal{R}(t)$, by Eq. (17.5.2). Hence as illustrated in Figure 17.6, each vector $\mathbf{\Delta u}(\mathcal{S}_t)$ must have slope no smaller in magnitude than λ, else some $\mathbf{u}(\mathcal{R}(t))$ would lie outside the convex hull.   For each t, the slope of the vector $\mathbf{\Delta u}(\mathcal{S}_t)$ must be greater in magnitude than (or equal to) the slope of the face F. In fact it can be shown that there exists at least one $\mathbf{\Delta u}(\mathcal{S}_t)$ with slope precisely equal to λ in magnitude. In particular, $\mathbf{u}(\mathcal{R}(t)) \in F$ if t is a leaf of the nested subtree $\mathcal{S}_{i+1}$ (corresponding to the next vertex), i.e., if $t \in \tilde{\mathcal{S}}_{i+1} \preceq \mathcal{S}_i$. Hence the algorithm only needs to find the interior node $t \in \mathcal{S}$ for which the magnitude of the slope of $\mathbf{\Delta u}(\mathcal{S}_t)$ is a minimum and if more than once such node exists, the one for which $u_1(\mathcal{S}_t)$ is smallest should be selected. For this t, the slope of $\mathbf{\Delta u}(\mathcal{S}_t)$ coincides with the slope of F, and the subtree $\mathcal{R}$ obtained by pruning off the branch $\mathcal{S}_t$ lies on F.

The above reasoning describes the heart of the algorithm. In essence, we need an efficient way to examine a particular pruned subtree $\mathcal{S}$ that has already been obtained from the original tree $\mathcal{T}$ and find which internal node to use as the root of the branch to be pruned next. To do this efficiently, we need to compute and store a data structure that gives some key numerical values associated with each node of the subtree $\mathcal{S}$. We assume that for each node t of the original tree $\mathcal{T}$ we have available the values $l(t)$, $\delta(t)$, and $P(t)$ which were computed as part of the original TSVQ design. Therefore, for

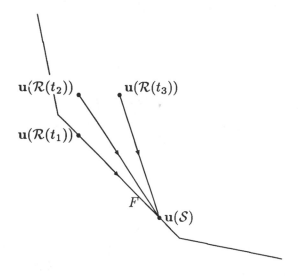

Figure 17.6: Some of the points $\mathbf{u}(\mathcal{R}(t))$.

each t in $\mathcal{S}_t$ we compute $\Delta\mathbf{u}(\mathcal{S}_t) = (\Delta u_1(\mathcal{S}_t), \Delta u_2(\mathcal{S}_t))$, which shows the effect of pruning off the branch $\mathcal{S}_t$ from node t. We also compute the magnitude $\lambda(t)$ of the slope of $\Delta\mathbf{u}(\mathcal{S}_t)$, namely, $\lambda(t) = -\Delta u_2(\mathcal{S}_t)/\Delta u_1(\mathcal{S}_t)$. In order to simplify the task of finding that node in $\mathcal{S}$ for which $\lambda(t)$ is smallest, for each node t we can also store $\lambda_{\min}(t)$, the minimum such magnitude for all descendent nodes of t, that is, for all interior nodes of $\mathcal{S}_t$. This value can be computed recursively, by starting at the leaves of $\mathcal{S}$ and moving upwards, at each step finding the minimum of $\lambda(t)$ and $\lambda_{\min}(t')$ for all children nodes t' of the current node t. For example, in the case of a binary tree, $\lambda_{\min}(t) = \min\{\lambda(t), \lambda_{\min}(t_L), \lambda_{\min}(t_R)\}$, where $t_L = \text{left}(t)$ and $t_R = \text{right}(t)$ are the left and right children, respectively, of t. Pointers specifying the addresses of the data structures containing $\text{left}(t)$, $\text{right}(t)$ and $\text{parent}(t)$ are usually also necessary unless they can be computed from t, e.g., $\text{left}(t_i) = t_{2i+1}$, $\text{right}(t_i) = t_{2i+2}$ and $\text{parent}(t_i) = t_{\lfloor (i-1)/2 \rfloor}$. Of course the actual tree must also specify the label or codeword associated with each node, but for the purpose of tree pruning we can ignore this aspect of the tree specification.

To simplify the description of the algorithm, we continue to assume that the tree $\mathcal{T}$ is binary. This, however, is by no means necessary. From the preceding discussion, we see that the tree can be stored as an array of nodes, each containing the following information:

$$node(t) = \begin{bmatrix} \Delta\mathbf{u}(\mathcal{S}_t) \\ \lambda(t) \\ \lambda_{\min}(t) \\ \text{left}(t) \\ \text{right}(t) \end{bmatrix}.$$

are the left and right children, respectively, of t. The one-time initialization of Table 17.1 is first performed on the original tree $\mathcal{T}$.

Table 17.1: Initialization

For each leaf node t:
  $\Delta\mathbf{u}(\mathcal{S}_t) \leftarrow \mathbf{0}$
  $\lambda_{\min}(t) \leftarrow \infty$
For each interior node t:
  $\Delta\mathbf{u}(\mathcal{S}_t) \leftarrow \mathbf{u}(\mathcal{S}_t) - \mathbf{u}(t)$
  $\lambda(t) \leftarrow -\Delta u_2(\mathcal{S}_t)/\Delta u_1(\mathcal{S}_t)$
  $\lambda_{\min}(t) \leftarrow \min\{\lambda(t), \lambda_{\min}(t_L), \lambda_{\min}(t_R)\}$

The minimum $\lambda(t)$, over all interior nodes of the tree, is maintained in $\lambda_{\min}(t_0)$. As the algorithm proceeds, the tree is replaced by successive nested pruned subtrees and at each step the data structure is updated to contain the correct entries for the current subtree. So, starting with the full tree $\mathcal{T}$, branches are successively pruned off until $\lambda_{\min}(t_0) = \infty$, indicating that only the root node, t_0, remains. The complete algorithm is given in Table 17.2.

The vector variable $\mathbf{u}$ maintains a running account of $\mathbf{u}(\cdot)$ for the tree as it is pruned. Thus, as each new pruned subtree is determined, the variable $\mathbf{u}$ gives the point on the convex hull for this subtree. Step 2 examines the current subtree and searches for the best branch to be pruned off next. It does this by beginning at the root and tracing a path down the tree searching for the node with the minimum $\lambda(t)$. This node will have its current magnitude slope equal to the minimum magnitude slope of all its descendents. Since the latter value is stored in the data structure for this node, this condition can be determined as soon as this node is reached without continuing the search any deeper into the tree. When it is found, the branch from this node is pruned off, the change in $\mathbf{u}(\cdot)$ is recorded in

Table 17.2: BFOS Pruning Algorithm

Step 0. Set $u(t_0) + \Delta u(S_{t_0}) \to u$.
Start with the entire tree as the "subtree".

Step 1. If $\lambda_{\min}(t_0) = \infty$, quit. (Stop if the minimum magnitude slope over all interior nodes of the current subtree is infinite, which happens when the remaining pruned subtree consists only of the root node.)
Set $t_0 \to t$ (set the node variable t to be the root node.)

Step 2. Find the best subtree of the current subtree that is formed by pruning a branch.

   Step 2a: If $\lambda(t) = \lambda_{\min}(t_0)$, go to Step 3. (This happens when node t is found whose magnitude slope equals the smallest possible over all interior nodes in the branch S_t emanating from t. This is the node at which one should prune.)
   Otherwise we must have $\lambda(t) > \lambda_{\min}(t_0)$, so continue to Step 2b to find a better node.

   Step 2b: Set t equal to the descendant which has the smallest $\lambda_{\min}$. This value must equal $\lambda_{\min}(t_0)$.
   If $\lambda_{\min}(t_L) = \lambda_{\min}(t_0)$, then set $t_L \to t$.
   Otherwise $t_R \to t$.

   Step 2c: Go to Step 2a.

Step 3. At this point, node t has been found from which the branch S_t is to be pruned so that the resulting pruned subtree is best.
Set $\Delta u(S_t) \to \Delta$.
Set $\infty \to \lambda_{\min}(t)$.

Step 4. Update all node entries to account for the pruned branch by moving up from t to t_0

   Step 4a: If $t = t_0$, go to Step 5. (Updating has proceeded all the way to the root.)

   Step 4b: Set $\text{parent}(t) \to t$.
   Set $\Delta u(S_t) - \Delta \to \Delta u(S_t)$.
   Set $-\Delta u_2(S_t)/\Delta u_1(S_t) \to \lambda(t)$.
   Set $\min\{\lambda(t), \lambda_{\min}(t_L), \lambda_{\min}(t_R)\} \to \lambda_{\min}(t)$.

   Step 4c: Go to Step 4a.

Step 5. Set $u - \Delta \to u$.
The best subtree, the slope $\lambda_{\min}(t_0)$, and the tree functional vector u have been found. Go to Step 1.

the vector variable Δ, and since the node now becomes a leaf in the newly formed subtree, its value of $\lambda_{\min}(t)$ is reset to infinity (or a very large number). The second inner loop (Step 4) retraces the path back to the root, updating the data entries for all ancestor nodes to reflect the pruning. Finally, $\mathbf{u}$ is updated and printed out, along with the slope $\lambda_{\min}(t_0)$ if desired.

The main loop of the algorithm is performed somewhere between M and N times, where N is the number of nodes in the tree and M is the number of extreme points around the lower boundary of the convex hull.

17.6   Pruned Tree-Structured VQ

As previously discussed, complete tree-structured quantizers can be pruned back to obtain variable rate tree-structured quantizers. To perform the pruning optimally in a distortion-rate sense, the generalized BFOS algorithm can be used with u_1 as the expected length and u_2 as the expected distortion. The resulting sequence of pruned subtrees will lie on the convex hull of the operational distortion-rate function. In the vector case, with binary trees, optimal pruning can provide a significant improvement in the signal to quantization noise ratio (SNR), as shall be seen. Such variable rate tree-structured quantizers have the desirable feature, not found in variable rate quantizers based on entropy coding, of graceful degradation in the face of buffer overflow. For example, if the transmit buffer begins to approach its capacity, the path map from the root to a leaf could be truncated at some ancestor to save bits at a cost of increased distortion. This is possible because of the successive approximation nature of tree-structured quantization. In addition, as with fixed rate TSVQ, pruned TSVQ is amenable to progressive transmission applications. For simplicity we consider only binary trees.

A pruned TSVQ (PTSVQ) is designed by applying the pruning algorithm on a TSVQ to optimally tradeoff average distortion and average length. In this case the tree functionals reduce to

$$u_2(t) = P(t)d(t) \tag{17.6.1}$$

$$u_1(t) = P(t)l(t) \tag{17.6.2}$$

so that

$$u_2(\mathcal{S}) = \delta(\mathcal{S}) = \sum_{t \in \tilde{\mathcal{S}}} P(t)d(t) \tag{17.6.3}$$

and

$$u_1(\mathcal{S}) = l(\mathcal{S}) = \sum_{t \in \tilde{\mathcal{S}}} P(t)l(t). \tag{17.6.4}$$

Pseudo-code for the complete PTSVQ design is given in Section 17.11.

17.7 Entropy Coded VQ

The most immediate variation of the PTSVQs is to apply the same technique to trade off average distortion with average entropy instead of average length. Minimizing the entropy instead of the average length is a good idea if the VQ output is to be followed by an entropy coder in an attempt to achieve additional compression by noiseless coding. Since entropy codes can effectively reduce the average length of the final code to the entropy of the vector, the VQ should be designed to minimize its output entropy. These codes, called entropy-pruned TSVQ (EPTSVQ), lose the successive approximation property of PTSVQ (unless special entropy codes are used), but they retain the rapid search property of trees. The reward for the loss of the successive approximation property is the gain of further compression[68]. This is accomplished simply by replacing (17.6.2) and (17.6.4) by

$$u_1(t) = -P(t) \log P(t) \qquad (17.7.1)$$

$$u_1(\mathcal{S}) = l(\mathcal{S}) = -\sum_{t \in \tilde{\mathcal{S}}} P(t) \log P(t). \qquad (17.7.2)$$

If one is able to tolerate even more complexity, then one can design a full search VQ to minimize average distortion for a given output entropy rather than for a given number of reproduction vectors. This can be accomplished by a variation of the Lloyd algorithm which replaces the minimization of average distortion

$$\bar{\rho} = E(\rho(\mathbf{X}, \hat{\mathbf{X}}))$$

by the minimization of a modified functional

$$\bar{d} = E(\rho(\mathbf{X}, \hat{\mathbf{X}})) + \lambda H(\hat{\mathbf{X}}),$$

where $H(\hat{\mathbf{X}})$ is the entropy of the VQ output and λ can be viewed as a Lagrange multiplier. If we let $i(\hat{\mathbf{x}})$ denote the index (typically a binary word), of the VQ reproduction vector $\hat{\mathbf{x}}$, then $H(\hat{\mathbf{X}}) = H(i(\hat{\mathbf{X}}))$ since the index and the reproduction vector exactly determine each other. Suppose now that the length of the noiselessly encoded reproduction (or reproduction index) is $l(i(\hat{\mathbf{x}}))$. We assume that the noiseless code is optimal for the VQ output vector in the sense that

$$l(i) = -\log P(i); \qquad (17.7.3)$$

that is, the length of the noiseless codeword is equal to the negative logarithm of the probability of the word (which can be estimated given a VQ and a training sequence). This assumption is accurate if the probabilities of the codewords are powers of two and Huffman coding is used on the reproduction vectors. If the probabilities are not powers of two, (17.7.3) can be considered as an approximation since a code approximately having this property exists (remember that we are now dealing with noiseless codes applied to vectors, the reproduction vectors produced by the codebook, as in Section 9.5). With this assumption the goal is now to minimize

$$\bar{d} = E(d(\mathbf{X}, \hat{\mathbf{X}}) + \lambda l(i(\hat{\mathbf{X}}))) = E(\rho(\mathbf{X}, \hat{\mathbf{X}})) \tag{17.7.4}$$

where the modified distortion measure ρ is defined by

$$\rho(\mathbf{x}, \hat{\mathbf{x}}) = d(\mathbf{x}, \hat{\mathbf{x}}) + \lambda l(i(\hat{\mathbf{x}})).$$

This provides a form amenable to minimization with descent techniques.

The algorithm resembles a Lloyd algorithm with some important differences: The modified distortion measures is used for encoding given a reproduction codebook and entropy coder, that is, one uses a nearest neighbor mapping that balances distortion with codeword probability. Then the reproduction codebook is updated by a centroid with respect to the squared error alone, as this provides the best label for all input vectors mapping into this word. Lastly, the probabilities of the codewords are estimated and used to form a new entropy coder. The process then iterates to convergence. Note that this design produces a variable rate full search VQ, there is no tree structure and hence the encoding is quite complex.

The algorithm proceeds as in Table 17.3. (A more detailed version can be found in Section 17.11.)

For a given value of λ, the final code will have some average distortion and entropy values. One can either experimentally find the λ yielding a desired entropy or one can step through possible values to plot the entire distortion-rate curve. If arbitrarily large codebooks are allowed, then the resulting entropy-constrained VQ (ECVQ) will have better performance than either PTSVQ or EPTSVQ. If the codebook size is constrained for complexity or memory reasons, however, it may turn out that EPTSVQ is superior.

17.8   Greedy Tree Growing

It is not necessary to use an ordinary complete TSVQ designed by the splitting Lloyd algorithm as the starting point for pruning. One can use any tree-structured code as a starting point, including a variable length tree. Thus, for example, the individual node-splitting algorithm of Makhoul et al. [229] can be used to grow a tree which can then be pruned back. In

---

Table 17.3: **Entropy-Constrained VQ Design**

Step 0: Given an initial PTSVQ codebook, a noiseless coding of the PTSVQ codebook, a training sequence, and a threshold.

Step 1: Encode the training sequence with the minimum distortion rule with the modified distortion measure ρ. During this computation find the centroids corresponding to each codeword index and find the probability (relative frequency) of each codeword index. If the average modified distortion is small enough, quit. Otherwise continue.

Step 2: Replace the reproduction vectors in the PTSVQ codebook by their centroids under the original distortion measure d. Based on the index probabilities $P(i)$ found in Step 1, update the list of codeword lengths $l(i)$. Go to Step 1.

---

their approach, the terminal node contributing the most average distortion is split until the desired number of terminal nodes is reached. Application of this splitting rule to the design of variable length codes was suggested in unpublished work of Robert Lindsay of Unisys Corporation at a NASA Data Compression Conference. This algorithm is greedy in that it only considers the short term effects of extending the tree, that is, it only considers what happens with the addition of a single new pair of leaves grown from a current leaf. A variation of this algorithm which was introduced by Riskin [267][269] more closely resembles the classification tree growing algorithms of [36] and also provides another greedy algorithm that strongly resembles the pruning algorithm. Instead of splitting the node that contributes the most distortion, split the node that provides the best tradeoff between total average distortion over all terminal nodes and average rate, that is, that provides the largest possible magnitude slope $|\Delta d/\Delta l|$, where Δd is the decrease in average distortion resulting from a specific node split and Δl is the increase in average length resulting from the split. This can be viewed as a myopic inverse to the pruning rule, which chooses a subtree so as to *minimize* the magnitude slope so that the average distortion is increased as little as possible for a decrease in average length. Instead, we choose a supertree (formed by splitting one leaf) so that the average distortion is decreased as much as possible for an increase in average length and hence the magnitude slope is *maximized*.

Because this tree growing algorithm is greedy, only considering the effect of adding a single leaf, one can produce better trees in general by running the pruning algorithm on the final grown tree. Lindsay considered a simple form of pruning by exactly inverting his growing procedure, he simply removed leaves in the inverse order to which they were added (a "last in first out" or LIFO rule). The BFOS algorithm can do better, however, since it finds the best of all pruned subtrees in the sense of trading off distortion and rate. The Lindsay algorithm is, however, simple and works quite well [269].

Greedy growing provides a means of directly growing a variable length tree without first growing a complete tree. This has one side benefit in that leaves having a small number of training vectors can be frozen and not split further.

17.9   Design Examples

In this section we describe a variety of design simulations for PTSVQ. Further details and experiments may be found in [69] [68] [220] [270].

Sampled Speech

These experiments used a locally generated 2 minute training sequence of 8 kHz sampled speech from 3 males and 3 females. The test sequence was 40 seconds of speech from a male and a female not represented in the training sequence. Our distortion measure was mean squared error. All vectors had 8 dimensions and consisted of 8 contiguous samples of the waveform so that the training sequence was composed of 120,000 vectors. The speech was high-frequency pre-emphasized with a filter $1 - 0.9z^{-1}$, divided into 128 sample frames, and Hamming windowed.

The starting (complete) TSVQ had depth 12 and hence the resolution of the full tree is 1.5 bits/sample. We chose this experiment as a good example of a reasonably complex source that could demonstrate the effects of pruning. As we have discussed in Chapter 11, however, direct waveform vector quantization of speech (especially at these rates) has had limited success, and we have found more profitable uses of pruned TSVQ in other VQ applications. Figure 17.7 demonstrates test sequence performance of variable rate pruned TSVQ compared against full search VQ and fixed rate complete tree-structured VQs designed by the generalized Lloyd algorithm.

To generate the pruned trees, we applied the generalized BFOS algorithm with the pruning criteria given by u_1 = average length and $u_2 =$

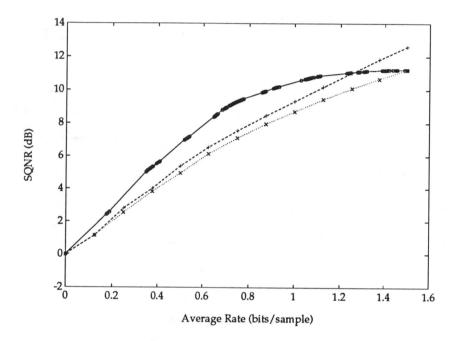

Figure 17.7: Pruned TSVQ performance: Resolution vs. distortion. •= pruned TSVQ, x = TSVQ, + = full search VQ.

average distortion. Note that, providing the additional complexity and delay inherent in the buffering of a variable rate system is acceptable, the
pruned TSVQs outperform both alternative systems for a significant range
of average rates. In addition, at most rates, our systems require significantly fewer computational resources than the same rate full search VQ,
since the number of distortion calculations is bounded by twice the depth
of the starting tree for the pruned TSVQ. Memory requirements can be
significantly greater, however, although they are also (obviously) bounded
by those of the starting tree.

Figure 17.8 plots the per-vector entropy of the pruned TSVQ and the
full search fixed rate VQ, which shows how much further compression could
be achieved if the VQ is followed by an entropy coder with nearly-optimal
performance.

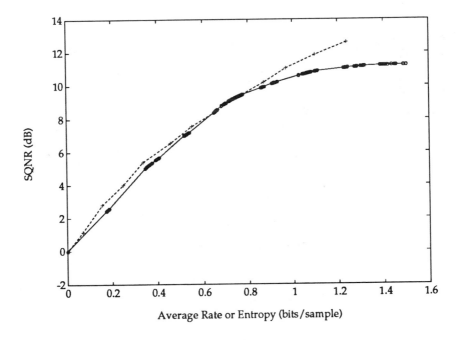

Figure 17.8: Pruned TSVQ peformance:Entropy vs. distortion. •= pruned
TSVQ, + = full search VQ.

Performance is comparable for several rates. If entropy coding is to
be used, however, then the performance of the PTSVQ can be improved
by pruning using entropy as the measure of rate instead of average length.
Figure 17.9 shows the performance of entropy-pruned TSVQ along with full
search VQ and ordinary complete TSVQ.

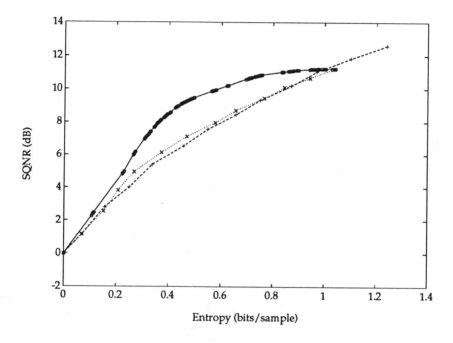

Figure 17.9: Entropy-Pruned TSVQ. •= pruned TSVQ, x = TSVQ, + = full search VQ.

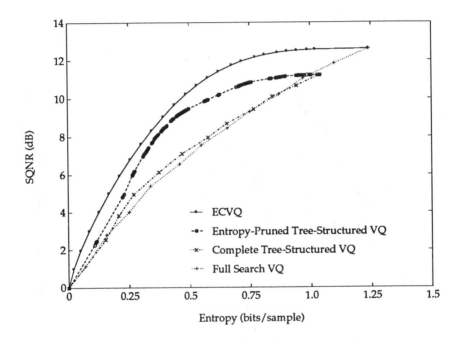

Figure 17.10: Entropy Constrained VQ

If one wishes to extract the maximum possible performance without regard to complexity, then an entropy constrained VQ can be designed, yielding the performance shown in Figure 17.10 in comparison with both entropy-pruned TSVQ, complete TSVQ, and full search VQ.

LPC-PTSVQ

A PTSVQ can be used to vector quantize LPC coefficients as in Section 11.4. The setup is as before and the (modified) Itakura-Saito distortion of (11.4.3)–(11.4.4) is used.

For this experiment the prediction filters were chosen to have 11 coefficients. The training and test sequences described at the beginning of this section were used. Figures 17.11 and 17.12 show the distortion vs.

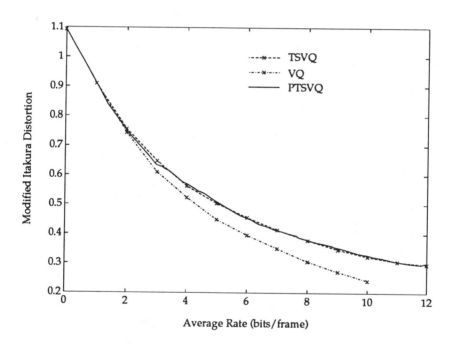

Figure 17.11: LPC Vector Quantization: Distortion vs. Rate.

rate performance for full search VQ, complete tree-structured VQ, PTSVQ, EPTSVQ, and ECVQ systems for the modified Itakura-Saito distortion measure. The PTSVQ and EPTSVQ were initialized with a depth 12 tree and ECVQ was initialized with a size 1024 codebook. The variable rate vector quantization systems provide little or no improvement over the fixed rate systems. In fact, performance was so close that in some cases training

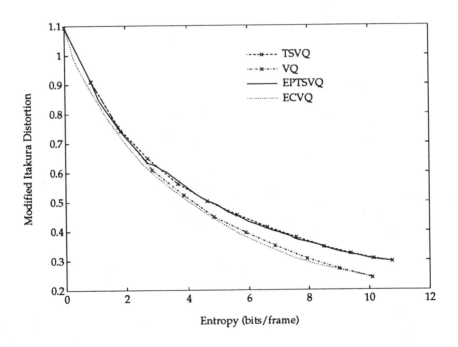

Figure 17.12: LPC Vector Quantization: Distortion vs. Entropy.

sequence dependencies caused slightly worse performance for the variable rate VQ systems when compared to the fixed rate systems. In the figure, average rate is $\log_2$(codebook size) for TSVQ and VQ and average length for PTSVQ.

Note that the measured sample entropy of a full search LPC-VQ is quite close to its rate (as previously observed by Juang, et al. [199]): for our sequences, an 8-bit codebook for the modified Itakura-Saito distortion measure had an empirical index entropy of 7.87 bits/vector on the training sequence (or 98% of fixed rate), and an 8-bit codebook for the modified Itakura-Saito distortion measure had an empirical index entropy of 7.88 bits/vector. This compares with an empirical index entropy of 7.02 bits/vector (88% of fixed rate) for an 8-bit full search VQ designed for the speech waveform under the mean square error distortion measure. Given these facts, one would predict little gain for the variable rate VQ techniques over standard VQ designs, and this was indeed the case.

Switched Gauss-Markov Sources

It is interesting to evaluate pruned TSVQ on a source to which we can apply distortion-rate theory. For simple sources such as the first order Gauss-Markov source, however, it has been found experimentally that full search VQ, complete TSVQ, and pruned TSVQ all have very similar performance (the closeness of full search VQ and complete TSVQ for this source was noted in [168]). A source that is complex enough to demonstrate a significant discrepancy between system performances, as waveform coding of speech does, is the switched source that, at the beginning of each switching period, chooses randomly between two memoryless Gaussian sources with different variances. We plot the performance of full search VQ, complete TSVQ, and pruned TSVQ as well as the theoretical performance bound given by the distortion-rate function of the composite source (due to Fontana [121]) in Figure 17.13. The switched source has a long switching period (required by the theoretical bounds) and subsource variances 1 and 1000.

Image Coding

The extension of the TSVQ image compression simulations to pruned trees is straightforward. PTSVQ codes were designed for both the USC database and on the magnetic resonance brain training sequence. In this section we describe both experiments.

The code vectors for the USC data base were 4×4 pixel blocks. The full tree was designed for rates of $n/16$ bits/pixel for $n = 0, 1, \cdots, 12$. The

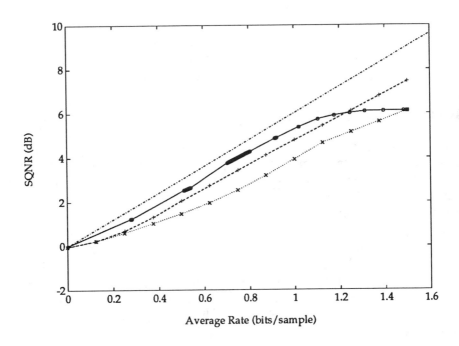

Figure 17.13: Composite Source: PTSVQ, full search, and TSVQ

largest complete TSVQ was pruned using the training sequence. The resulting peak SNR is plotted against rate for both complete and pruned TSVQ in Figure 17.14. Figure 17.15 shows the test image Lenna at a rate

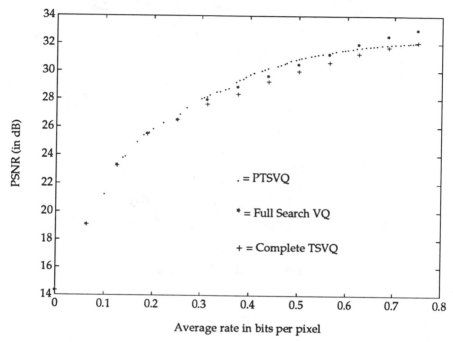

Figure 17.14: Peak SNR vs. Average Rate: USC database

of 0.3125 bpp for full search memoryless VQ (SNR = 27.92dB) and PTSVQ (SNR = 28.16dB). Although the PTSVQ is only slightly better, it is simpler in the sense that it is a tree-structured code (and more complicated in that it has a variable rate).

The performance improvement becomes more significant if one combines PTSVQ with predictive VQ. Figure 13.5 shows the pixels in the adjacent blocks used to form a linear prediction of the current block. The code is designed forming the linear predictors based on the training sequence and uncoded inputs as in Chapter 13 and then designing the codebooks open loop (based on the residuals between the prediction based on the true inputs). The resulting prediction residuals proved to have a more highly skewed (less uniform) distribution than did the original directly coded vectors, suggesting that subsequent entropy coding would yield further compression. (The two sequences have equal entropy, but a uniformly distributed random vector cannot be further compressed in the sense of reducing its average length.) The performance of the resulting predictive PTSVQ is

Figure 17.15: Test Image for VQ and PTSVQ at 0.3125 bpp

plotted along with ordinary PTSVQ, predictive VQ, and memoryless VQ and TSVQ in Figure 17.16.   Figure 17.17 shows Lenna coded by both predictive VQ and predictive PTSVQ at 0.3125 bpp. The full search predictive VQ yields 29.29 dB and the predictive PTSVQ yields 31.02 dB. Figure 17.18 shows Lenna's shoulder using memoryless VQ at 0.3125 bpp, predictive PTSVQ at 0.3125 bpp, predictive PTSVQ at 0.5 bpp, and the original rate of 8 bpp.   Figures 17.19 and 17.20 show the corresponding results for entropy pruning and constrained entropy codes without and with prediction.

Next we consider the analogous results for the MRI sagittal brain images. Once again both memoryless and predictive PTSVQ are considered, although now the predictions are based on the smaller group of pixels shown in Figure 13.6. Figure 17.21 shows the SNR for memoryless PTSVQ along with that for full search VQ and complete TSVQ.

Figure 17.22 shows the SNR for predictive and memoryless PTSVQ along with that for predictive full search VQ and memoryless full search VQ for Eve. Figure 17.23 shows a full search VQ at 1.5 bpp alongside a PTSVQ for comparison. Figure 17.24 compares the same PTSVQ with the original (reproduced at 8 bpp). Figure 17.25 shows both the error signal between the 1.5 bpp coded image and the original image (on the left) and

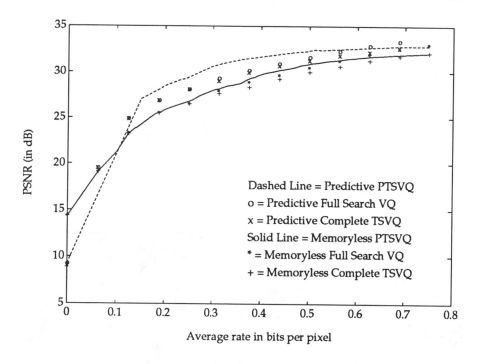

Figure 17.16: Predictive PTSVQ: USC database

Figure 17.17: Predictive VQ and PTSVQ: Lenna at 0.3125 bpp

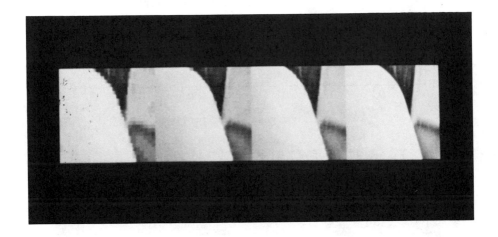

Figure 17.18: Four Shoulders of Lenna

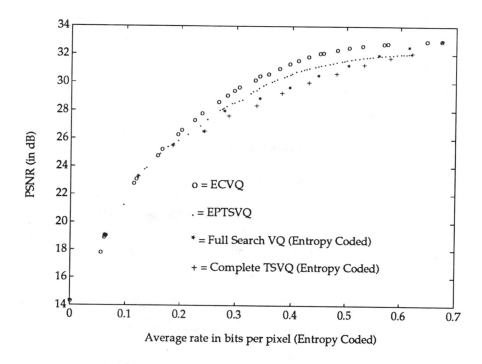

Figure 17.19: PSNR vs. average rate for entropy-coded USC database.

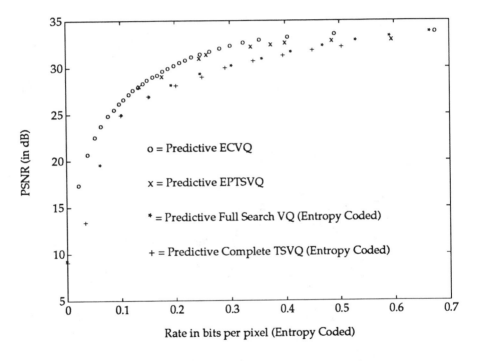

Figure 17.20: PSNR vs. average rate for entropy-coded USC database prediction residuals.

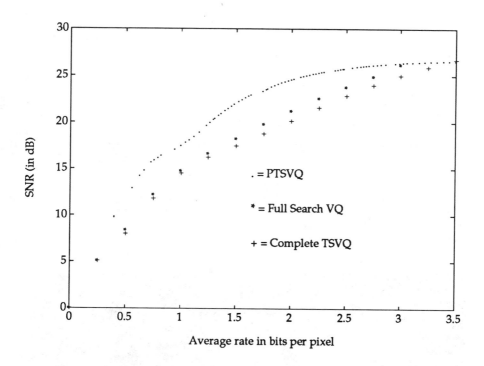

Figure 17.21: Performance of memoryless PTSVQ on MRI

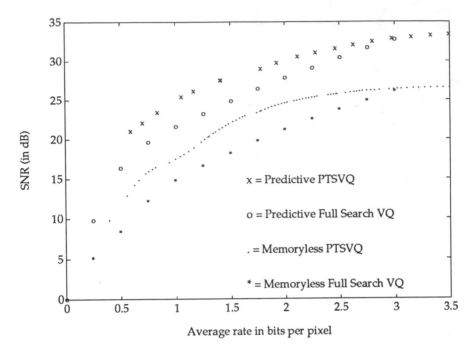

Figure 17.22: Performance of predictive PTSVQ on MRI

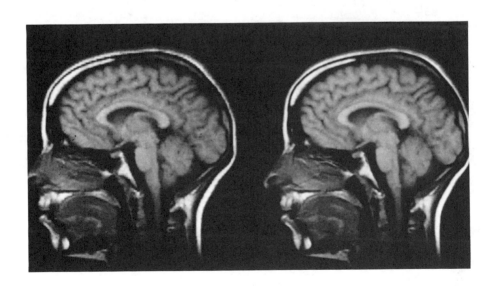

Figure 17.23: Full Search VQ vs. PTSVQ: MRI at 1.5 bpp

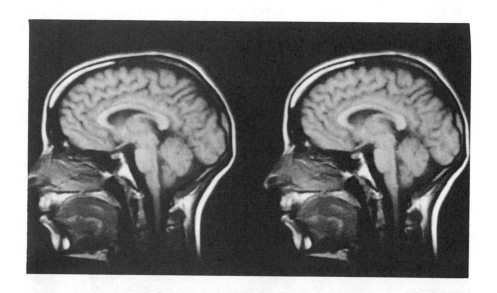

Figure 17.24: Original vs. PTSVQ: MRI

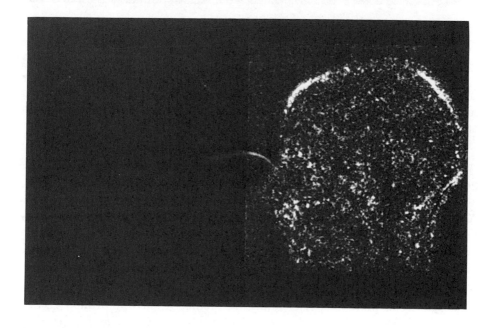

Figure 17.25: Error Signal

the same signal amplified by 10 (on the right). Note that without the amplification the error signal is uniformly dark.

Figure 17.26 shows a plot of the instantaneous rate of the PTSVQ along

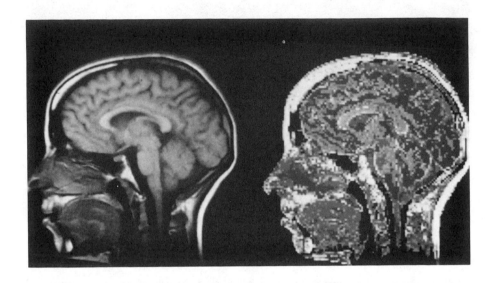

Figure 17.26: Instantaneous Rate

side of the encoded image, with higher rate (length) being whiter and lower rate being darker. The figure demonstrates that, as one would expect, most of the high rate words are devoted to edges.

Figure 17.27 shows the coded test image using predictive PTSVQ at rates of 0.63 bpp, 0.76 bpp, and 1.16 bpp along with the original for comparison. In general the pruned TSVQ outperforms both complete TSVQ and full searched VQ and the same remains true when prediction is added. Also in general, the prediction tends to improve performance, especially if it yields a highly skewed residual distribution. If, on the other hand, prediction yields a highly uniform residual distribution, then performance may actually suffer since the VQ will have small advantage over a simple scalar quantizer.

Figure 17.28 plots the performance for predictive entropy-pruned TSVQ, predictive full search VQ, and predictive complete TSVQ. The gain pruning now provides over full search and TSVQ is larger than in the nonpredictive codes, as was the case for the USC database.

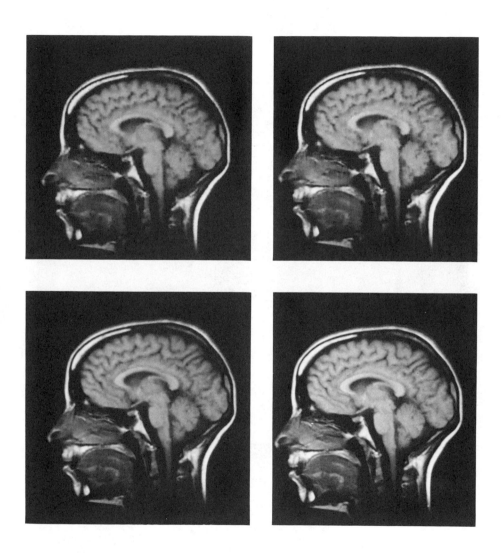

Figure 17.27: Test MR Image using predictive PTSVQ: 0.63 bpp, 0.76 bpp, 1.16 bpp, original

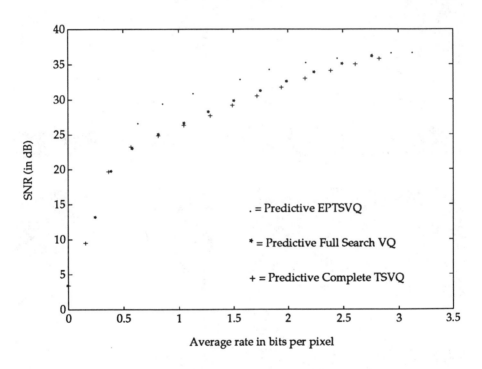

Figure 17.28: Predictive PTSVQ performance for MRI

17.10 Bit Allocation Revisited

We have seen that transform coding provides a means of transforming an input vector into the frequency domain wherein a vector quantizer can be used to separately quantize separate groups of coefficients using different codes. If the quantization is scalar, each coefficient is coded separately. In both cases, given a division of the coefficients into groups, how should a given total number of bits be allocated to each group? We have already considered this problem in Chapter 8 and again in Chapter 16, but we now show that it can also be solved by the BFOS algorithm using the approach described in [69]. More generally this approach can determine the optimal bit allocation for any VQ involving multiple codebooks, including sub-band coding and classified VQ as well as transform coding. The application of the generalized BFOS algorithm to bit allocation is due to Riskin [267][268].

Suppose the ith class (or frequency band or coefficient group or subband) has its own sequence of n_i codebooks $C_{i,1}, C_{i,2}, \cdots, C_{i,n_i}$, where, along the sequence, the distortion is decreasing and the bit rate is increasing. For example, these codebooks can be designed using the generalized BFOS algorithm to be nested trees tracing the convex hull of the operational distortion-rate function for the ith coefficient group or class. This could be a sequence of full search codebooks designed by the splitting Lloyd algorithm.

Construct the following tree: the root node has one child per class, and the subtree rooted at each child is the degenerate, unary tree of length n corresponding to the linearly linked list of codebooks for that band. A typical tree is depicted in Figure 17.29.

With u_1 as the sum of bit rates over all the leaves and u_2 as the sum of distortions, the generalized BFOS algorithm will find the optimal tradeoff between the overall rate and distortion, and will furthermore specify the codebooks to be used at each frequency. In this application, the generalized BFOS algorithm is equivalent to the algorithm proposed by Shoham and Gersho [294] which also uses Lagrangian methods to search the convex hull of the achievable distortion-rate region. (See Chapter 12.) In addition, since the generalized BFOS algorithm generates a set of good allocations that cover a wide range of total available bits, it provides a solution to the "soft" bit allocation problem of Bruckstein [37], in which the total number of bits available is allowed to vary based on the resulting performance attainable (leading to a variable rate system). He provides an analytic solution when quantizer error cost and bit usage penalties are monotonic, smooth (differentiable) functions; the generalized BFOS algorithm is applicable when these quantities are discrete or otherwise not well modeled by such functions.

The BFOS bit allocation algorithm can be restated the form of an earlier

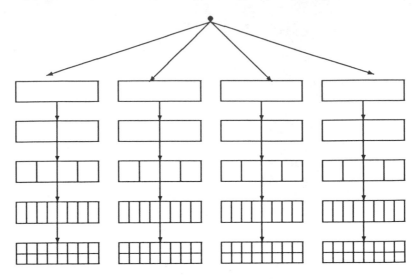

Figure 17.29: Bit Allocation Tree

bit allocation algorithm of Westerink, Biemond, and Boekee [329] as in Table 17.4.

In Step 2 we pick the maximum of the slope function $s(v_n, k_n)$ rather than the minimum because we are working with the reciprocal.

One benefit of the algorithm is that one can find the entire convex hull of the distortion-rate curve very quickly. The complexity of the algorithm can be measured as follows. Let M be the number of classes and q be the maximum allowable number of bits that can be allocated to a particular class. First, $(q - 1)M$ subtractions are performed to calculate the differences in distortion for the quantizer. This assumes that all of the quantizers' distortion-rate curves are convex. (If they are not convex, we will have a small amount of added complexity in order to find the lower convex hull of the curve.) Now, we have M ordered lists of (at most) $q - 1$ numbers to merge. It is known that this can be done with at most $qM \log_2 M$ comparisons, by repeatedly merging two lists at a time.

For sub-band and transform coding, the bit allocation algorithm will yield a fixed rate code. For classified VQ, it allocates different numbers of bits to the different classes; hence, it yields a variable rate system.

As an example of the algorithm, Figure 17.30 is the distortion-rate curve for a classified VQ with four codebooks for an image coding application. The experiment and analysis are from Riskin [267] [268]. The training sequence is a set of five 256 by 256 Magnetic Resonance (MR) brain scans that were locally generated using a General Electric Signa machine. Each codebook can have 1, 2, 4, 8, or 16 codewords (0, 1, 2, 3, or 4 bits) and

Table 17.4: **Bit-Allocation Algorithm**

Step 1. Determine the initial bit allocation by assigning to each class the admissible quantizer that has the highest bit rate.

Step 2. Calculate for each class n all possible values of the slope function

$$s(v_n, k_n) \equiv \frac{r_n(v_n) - r_n(k_n)}{d_n(v_n) - d_n(k_n)}$$

with $d_n(v_n) < d_n(k_n)$, where $r_n(v_n)$ and $d_n(v_n)$ are the rate and distortion, respectively, of the nth quantizer under the bit assignment v_n, and find for each class the quantizer for which $s(v_n, k_n)$ is maximal. Note that $s(v_n, k_n)$ is the ratio of the change in rate to distortion between two bit assignments, v_n and k_n, which is exactly the reciprocal of λ in the BFOS algorithm.

Step 3. Determine the class for which the maximum $s(v_n, k_n)$ is the largest and assign to this class the quantizer for which this maximum is obtained.

Step 4. Calculate the new rate R and the new distortion D and check if they are sufficiently close to the desired bit rate or the desired distortion; if so, then stop the algorithm.

Step 5. Repeat 2, 3, and 4, but do 2 only for the class to which the new quantizer is assigned (the other values have not changed).

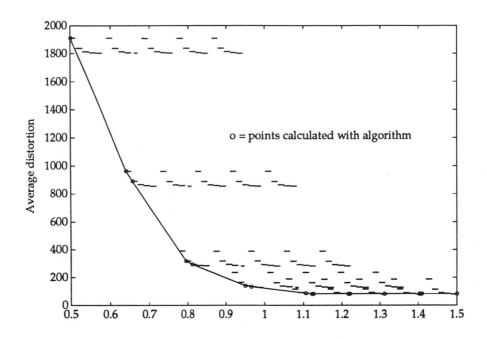

Figure 17.30: Distortion-Rate Curve with Optimal Bit Allocation

the (rate, distortion) pair for every possible bit allocation scheme is plotted here. The "o" marks were calculated by hand with the method described in this paper and form the set of extreme points on the convex hull, which is the solid line.

In Figure 17.31, the signal-to-noise ratio (SNR), defined as

$$\text{SNR} = 10.0 \log_{10} \frac{D(0)}{(\frac{1}{256})^2 \sum_{i=1}^{256} \sum_{j=1}^{256} (x_{ji} - \hat{x}_{ji})^2}$$

appears for this same classified VQ. Here x_{ji} is the input pixel and $\hat{x}_{ji}$ is

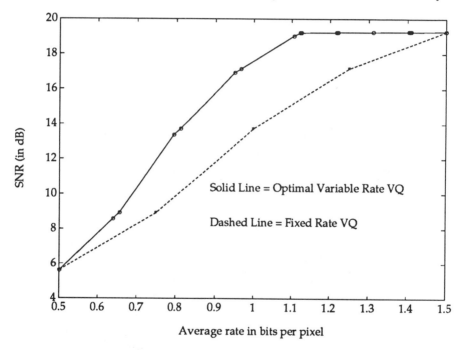

Figure 17.31: SNR for Classified VQ

the quantized pixel. Because we have taken the log of $\hat{D}_T(R)$, the graph is no longer convex. At the same time, however, we can see that the SNR will be highest for the bit allocations found with this algorithm.

As an example, the bit allocation algorithm led to a gain of 3.4 dB in the SNR for the variable rate classified VQ over fixed rate classified VQ at 1 bit per pixel. The chosen image was a Magnetic Resonance Image (MRI) of a brain in the above training sequence. More importantly, though, the subjective image quality of the variable rate image was much higher.

17.11   Design Algorithms

This section provides details on the design algorithms used in this chapter. The pseudo-code is taken from [220] and [269]. Except for the general pruning algorithm described first, the code is restricted to the mean squared error distortion measure for simplicity. More general algorithms may be found in [68,69].

BASIC BFOS PRUNING ALGORITM

Initialization:

For each leaf node t:
$\quad \Delta \mathbf{u}(\mathcal{S}_t) \leftarrow \mathbf{0}$
$\quad \lambda_{\min}(t) \leftarrow \infty$
For each interior node t:
$\quad \Delta \mathbf{u}(\mathcal{S}_t) \leftarrow \mathbf{u}(\mathcal{S}_t) - \mathbf{u}(t)$
$\quad \lambda(t) \leftarrow -\Delta u_2(\mathcal{S}_t)/\Delta u_1(\mathcal{S}_t)$
$\quad \lambda_{\min}(t) \leftarrow \min\{\lambda(t), \lambda_{\min}(t_L), \lambda_{\min}(t_R)\}$

$\mathbf{u} \leftarrow \mathbf{u}(t_0) + \Delta \mathbf{u}(\mathcal{S}_{t_0})$
while $(\lambda_{\min}(t_0) < \infty)$ **do**
$\quad t \leftarrow t_0$
$\quad$ **while** $(\lambda(t) > \lambda_{\min}(t_0))$ **do**
$\quad\quad$ **if** $(\lambda_{\min}(t_L) = \lambda_{\min}(t_0))$
$\quad\quad$ **then** $t \leftarrow t_L$
$\quad\quad$ **else** $t \leftarrow t_R$
$\quad\quad$ **end**
$\quad \Delta \leftarrow \Delta \mathbf{u}(\mathcal{S}_t)$
$\quad \lambda_{\min}(t) \leftarrow \infty$
$\quad$ **while** $(t \neq t_0)$ **do**
$\quad\quad t \leftarrow \text{parent}(t)$
$\quad\quad \Delta \mathbf{u}(\mathcal{S}_t) \leftarrow \Delta \mathbf{u}(\mathcal{S}_t) - \Delta$
$\quad\quad \lambda(t) \leftarrow -\Delta u_2(\mathcal{S}_t)/\Delta u_1(\mathcal{S}_t)$
$\quad\quad \lambda_{\min}(t) \leftarrow \min\{\lambda(t), \lambda_{\min}(t_L), \lambda_{\min}(t_R)\}$
$\quad\quad$ **end**
$\quad \mathbf{u} \leftarrow \mathbf{u} - \Delta$
$\quad$ **print** $\lambda_{\min}(t_0), \mathbf{u}$
$\quad$ **end**.

The vector variable $\mathbf{u}$ maintains a running account of $\mathbf{u}(\cdot)$ for the tree as it is pruned. The first inner **while** loop searches a path beginning at the root for the node with the minimum $\lambda(t)$. When it is found, the branch from this node is pruned off, and the change in $\mathbf{u}(\cdot)$ is recorded in the vector variable Δ. The second inner **while** loop retraces the path back to the root, updating all ancestors to reflect the pruning. Finally, $\mathbf{u}$ is updated and printed out, along with the slope $\lambda_{\min}(t_0)$ if desired.

PRUNED TREE-STRUCTURED VECTOR QUANTIZER DESIGN

The following algorithm is for PTSVQ design. The definition and manipulation of tree structures are described in the accompanying notes, as is the modification required for EPTSVQ design.

(0) Initialization:

Given:

a training sequence of K-dimensional vectors, $X^{(0)}, X^{(1)}, \ldots, X^{(N-1)}$,

a complete binary tree-structured codebook $c(\cdot)$, L layers deep, with K-dimensional vectors at each node,

a complete binary tree structure of L layers with node contents:

$(d(t), count(t))$ set to zero at each node,

$(\Delta d(t), \Delta r(t))$ set to zero at each node,

and $(\Delta d(t), \Delta r(t), \lambda(t), \lambda_{min}(t))$ at each node with $\Delta d(t)$ and $\Delta r(t)$ set to zero and $\lambda_{min}(t)$ set to infinity at all terminal nodes.

(1) Encode each training sequence vector, updating $d(t)$ and $count(t)$:

For $n = 0, \ldots, N - 1$

$d(root) \leftarrow d(root) + (\frac{1}{K}) \sum_{k=0}^{K-1} (x_k - c_k(root))^2$

$count(root) \leftarrow count(root) + 1$

$t \leftarrow root$

while (t is an internal node)

$d_l \leftarrow (\frac{1}{K}) \sum_{k=0}^{K-1} (x_k - c_k(left(t)))^2$

$d_r \leftarrow (\frac{1}{K}) \sum_{k=0}^{K-1} (x_k - c_k(right(t)))^2$

if $(d_l < d_r)$

$\quad t \leftarrow left(t)$

$\quad d(t) \leftarrow d(t) + d_l$

else

$\quad t \leftarrow right(t)$

$\quad d(t) \leftarrow d(t) + d_r$

$count(t) \leftarrow count(t) + 1$

(2) Calculate $\Delta d(t)$ and $\Delta r(t)$ for all internal nodes:

For all internal t,

$\Delta d(t) \leftarrow \frac{d(left(t)) + d(right(t)) - d(t)}{N}$

$\Delta r(t) \leftarrow \frac{count(t)}{N}$

(3) Evaluate $d(root), r(root)$:

$\quad d(root) \leftarrow \frac{d(root)}{N}$

$\quad r(root) \leftarrow 0$

(4) Use a post-ordered traversal to calculate $\Delta d(t)$ and $\Delta r(t)$ from $\Delta d(t)$ and $\Delta r(t)$:

$\quad t \leftarrow root$

$\quad$ repeat

$\qquad$ while (t is an internal node) do

$\qquad\quad t \leftarrow left(t)$

$\qquad$ while ($t \neq root$ and t is a right child) do

$\qquad\quad t \leftarrow parent(t)$

$\qquad\quad \Delta d(t) \leftarrow \Delta d(t) + \Delta d(left(t)) + \Delta d(right(t))$

$\qquad\quad \Delta r(t) \leftarrow \Delta r(t) + \Delta r(left(t)) + \Delta r(right(t))$

$\qquad\quad \lambda(t) \leftarrow -\frac{\Delta d(t)}{\Delta r(t)}$

$\qquad\quad \lambda_{min}(t) \leftarrow \min(\lambda(t), \lambda_{min}(left(t)), \lambda_{min}(right(t)))$

$\qquad$ if ($t \neq root$)

$\qquad\quad t \leftarrow right(parent(t))$

$\quad$ until ($t = root$)

(5) Prune:

$\quad r \leftarrow r(root) + \Delta r(root)$

$\quad d \leftarrow d(root) + \Delta d(root)$

$\quad$ while ($\lambda_{min}(root) < \infty$) do

$\qquad t \leftarrow root$

$\qquad$ while ($\lambda(t) > \lambda_{min}(root)$) do

$\qquad\quad$ if ($\lambda_{min}(left(t)) = \lambda_{min}(root)$)

$\qquad\qquad t \leftarrow left(t)$

$\qquad\quad$ else

$\qquad\qquad t \leftarrow right(t)$

$\qquad$ delete all descendents of t from the tree structure

$\qquad \Delta d \leftarrow \Delta d(t)$

$\qquad \Delta r \leftarrow \Delta r(t)$

$\qquad \lambda_{min}(t) \leftarrow \infty$

$\qquad$ while ($t \neq root$) do

$\qquad\quad t \leftarrow parent(t)$

$\qquad\quad \Delta d(t) \leftarrow \Delta d(t) - \Delta d$

$\qquad\quad \Delta r(t) \leftarrow \Delta r(t) - \Delta r$

$\qquad\quad \lambda(t) \leftarrow -\frac{\Delta d(t)}{\Delta r(t)}$

$\qquad\quad \lambda_{min}(t) \leftarrow \min(\lambda(t), \lambda_{min}(left(t)), \lambda_{min}(right(t)))$

$\qquad d \leftarrow d - \Delta d$

$\qquad r \leftarrow r - \Delta r$

$\qquad$ write tree structure, d, r.

Notes:

(A) A binary tree structure is a set of nodes t, each containing information. The *root* node has an externally known address. A node either has a left and right child node (in which case it is internal) or it has no children (in which case it is terminal). The functions $left(t)$ and $right(t)$ are defined for all internal nodes t; they return the addresses of the left and right children of t, respectively. The function $parent(t)$ is defined for all nodes except the root; it returns the address of the node t' for which either $t = left(t')$ or $t = right(t')$. The depth of a node t is given by the number of applications of $parent(t)$ required to return *root*. A complete tree has all its terminal nodes at the same depth.

(B) In order to convert the above algorithm to EPTSVQ design, the step

$$\Delta r(t) \leftarrow \frac{count(t)}{N}$$

in (2) should be changed to

$$\Delta r(t) \leftarrow [count(t)\log_2(count(t)) - count(left(t))$$
$$\times \log_2(count(left(t))) -$$
$$count(right(t))\log_2(count(right(t)))]/N.$$

(C) This algorithm can be broken into two sections. Steps (1), (2), and (3) can be considered an initialization stage and depend on the definitions of rate and distortion. Steps (4) and (5) use information generated by the first three steps but are themselves independent of the definitions of rate and distortion. The numbers $\Delta d(t)$ and $\Delta r(t)$ calculated in steps (1), (2), and (3) correspond to the quantities $u_1(\mathcal{R}_t)$ and $u_2(\mathcal{R}_t)$ for height-1 trees $\mathcal{R}_t$ rooted at each internal node in [68].

(D) Each tree structure written in step (5) represents the pruned subtree of the initial TSVQ codebook that will yield the corresponding distortion d and rate r on the training sequence used in step (1). This training sequence does not have to be the same as the training sequence used to design the initial complete TSVQ codebook.

ENTROPY-CONSTRAINED VECTOR QUANTIZER DESIGN

(0) Initialization:

Given:

a training sequence of K-dimensional vectors, $X^{(0)}, X^{(1)}, \ldots X^{(N-1)}$,

a Lagrange multiplier, λ,

a convergence threshold, ϵ,

an index set, I,

an initial reproduction codebook of K-dimensional vectors,

$\{C_k(i), k = 0, \ldots, K - 1\}_{i \in I}$

(perhaps designed using the generalized Lloyd algorithm),

and initial codeword lengths $\{l(i)\}_{i \in I}$

(perhaps estimated from the initial codebook as

$l(i) = -\log_2(\text{card}(\{n : i = \alpha(X^{(n)})\})/N)$ where the function

α is defined in step (1) below and is used for this purpose with $\lambda = 0$

set $t \leftarrow 0, J^{(0)} \leftarrow \infty$.

(1) Find the nearest neighbor of each training sequence vector:

$$\alpha(X^{(n)}) \leftarrow \min_{i \in I}^{-1}\{[\textstyle\sum_{k=0}^{K-1}(X_k^{(n)} - C_k^{(i)})^2] + \lambda l(i)\}$$

(2) Calculate the encoding distortion:

$$J^{(t+1)} \leftarrow (\tfrac{1}{NK})\textstyle\sum_{n=0}^{N-1}\{[\textstyle\sum_{k=0}^{K-1}(X_k^{(n)} - C_k^{(\alpha(X^{(n)}))})^2] + \lambda l(\alpha(X^{(n)}))\}$$

(3) Find a new set of reproductions (centroid):

for all $i \in I$,

if $(\text{card}(\{n : i = \alpha(X^{(n)})\})) = 0$

remove i from I

else

$$C_k(i) \leftarrow \textstyle\sum_{n:i=\alpha(X^{(n)})} \frac{X_k^{(n)}}{\text{card}(\{n:i=\alpha(X^{(n)})\})}$$

(4) Find new codeword lengths:

$$l(i) \leftarrow -\log_2(\text{card}(\{n : i = \alpha(X^{(n)})\})/N), \text{ for all } i \in I,$$

(5) Check for convergence:

if $(\frac{J^{(t)} - J^{(t+1)}}{J^{(t+1)}} > \epsilon)$

$t \leftarrow t + 1$, go to (1)

(6) Quit with reproduction codebook $C(i)$ and codeword lengths $\{l(i)\}_{i \in I}$.

Notes

(A) The function card gives the number of elements (cardinality) in the set that is its argument. The function $\min_{s \in S}^{-1}(f(s))$ returns the element s in S that minimizes the function $f(s)$ over all elements in S. The function is defined for a finite set S so that there always exists at least one such s. If there are more than one, select one arbitrarily.

(B) The index set I is typically $\{0, \ldots, M-1\}$ initially.

(C) The algorithm is usually run for different values of λ, corresponding to different average rates $R = \frac{1}{NK} \sum_{n=0}^{N-1} l(\alpha(X^{(n)}))$.

GREEDY TREE GROWING ALGORITHM

The following is the complete unbalanced tree design algorithm in pseudo-code. It is taken from [269]. A node's left and right child are designed ahead of time so that its $\Delta i(s, t)$ can be determined before it is split. The nodes t_L and t_R refer to the left and right child, respectively, of the node that has just been split. A node t's left and right child are returned by left(t) and right(t). The RATE is the current average rate of the tree as measured on the training sequence. The parameter SLOPELIST is the list of candidate slopes for the nodes that have been designed but not split. The node that would measure the largest $\Delta i(s, t)$ when split is $t_{\max}$. The TargetRate is the desired average rate of the final tree. The tree is grown until it measures this rate on the training sequence.

```
Design root node (Centroid of the entire training sequence)
RATE = 0.0
Split root node using GLA
RATE = 1.0
t_L ← left(root)
t_R ← right(root)
while (RATE < TargetRate)

        Design t_L's 2 children using GLA but DON'T split it
        Calculate Δi(s, t_L)
        INSERT Δi(s, t_L) onto SLOPELIST
        Design t_R's 2 children using GLA but DON'T split it
        Calculate Δi(s, t_R)
        INSERT Δi(s, t_R) onto SLOPELIST
        Search SLOPELIST for t_max
        DELETE Δi(s, t_max) off of SLOPELIST
        Split t_max
        RATE += p(t_max)
        t_L ← left(t_max)
        t_R ← right(tmax)
```

Calculate distortion of the final tree
Print RATE and distortion of the final tree
end

17.12   Problems

17.1. (a) Use the high resolution quantization approximations of Chapter 5 to find an approximate value for the entropy of a uniformly scalar quantized process which has a smooth probability density function. Is it worth the effort to entropy code the quantizer output? In particular, what additional compression is obtainable?

   (b) Next use the high resolution approximation to approximate the entropy when a nonuniform quantizer is used. Your answer should depend on the point density function of the quantizer.

   (c) Show that if the approximations are accurate, then the uniform quantizer entropy is lower than that of the nonuniform quantizer. (*Hint:* Use Jensen's inequality.) What are the implications of this fact?

17.2. Determine whether the following tree functionals are linear or not:

   (a) The number of leaves.

   (b) The number of nodes.

   (c) The entropy of the leaves

17.3. (Problem courtesy W. Y. Chan.) A TSVQ design has produced the following 2-level binary tree with four terminal nodes. The nodes are numbered so that node 1 is the root and nodes 4-7 are terminal nodes. The probability and average distortion (P, D) for each (numbered) node are: 1: (1, 3.0), 2: (0.57, 2.0), 3: (0.43, 2.5), 4: (0.17, 0.4), 5: (0.4, 1.0), 6: (0.33, 1.2), 7: (0.1, 2.0). Using the BFOS algorithm, find and sketch all pruned subtrees lying on the convex hull of the distortion-rate curve. Trace each iteration of the algorithm by tabulating Δ_U, λ, and λ_{min} for each node. Plot the convex hull and show all attainable points that are above the convex hull.

17.4. Give examples of sources where pruned TSVQ might be expected to provide little gain over complete TSVQ. Give examples where pruned TSVQ would be expected to provide large gains over complete TSVQ.

17.5. Suppose that one needs a fixed rate code, but in order to have a simple encoder a TSVQ (not necessarily complete) is used and the leaves are indexed from 1 to 2^{kR} and these indices form a fixed rate code. Suppose now that you must design a new code having only $2^{k(R-1)}$ words from this code. How would you do it by pruning so that the new code would be as good as possible? Can you think of a way that these two codes could be combined to form a progressive transmission system?

17.6. Describe how the pseudo-code for the design algorithms of Section 17.11 would have to be modified for discrete data with an average Hamming distortion.

17.7. (Problem courtesy E. Riskin.) You are given a classified vector quantizer and you wish to design a variable rate code for it. You are given the distortion-rate performance of each of the 4 classes in Table 7c, where R denotes rate and D distortion. Use the generalized BFOS algorithm to determine the set of optimal bit allocations. In a case where two or more slopes of magnitude of increase in distortion to decrease are tied, deallocate bits from all the classes with that slope.

   (a) How many possible bit allocations are there and how many lie on the convex hull of the distortion-rate plane (i.e., are optimal)?

   (b) Assuming that all four classes are equally likely, list the (distortion,rate) pairs of the optimal quantizers and the bit allocations in terms of how many bits are assigned to each class for each bit allocation. You can neglect the contribution to the bit rate due to sending side information. The maximum number of bits that can be assigned to a given class is 3 bits.

   (c) Repeat part b but let p(class 0) = 0.5, p(class 1) = 0.25, p(class 2) = 0.15, and p(class 3) = 0.1.

Class 0		Class 1		Class 2		Class 3	
R	D	R	D	R	D	R	D
0	10,000	0	1000	0	500	0	500
1	4000	1	700	1	300	1	250
2	2000	2	300	2	200	2	150
3	1000	3	75	3	150	3	75

Table 17.5: Distortion and rate values for the four classes

Bibliography

[1] N. M. Abramson. *Information Theory and Coding.* McGraw-Hill, New York, 1963.

[2] H. Abut, editor. *Vector Quantization.* IEEE Reprint Collection. IEEE Press, Piscataway, New Jersey, May 1990.

[3] H. Abut, R. M. Gray, and G. Rebolledo. Vector quantization of speech and speech-like waveforms. *IEEE Trans. Acoust. Speech Signal Process.*, ASSP-30:423–435, June 1982.

[4] H. Abut and S. A. Luse. Vector quantizers for subband coded waveforms. In *International Conference on Acoustics, Speech, and Signal Processing*, volume 1, pages 10.6.1–10.6.4, San Diego,Calif., March 1984.

[5] S. Adlersberg and V. Cuperman. Transform domain vector quantization for speech signals. In *International Conference on Acoustics, Speech, and Signal Processing*, volume 4, pages 45.44.1–45.4.4, Dallas,Texas, April 1987.

[6] J. P. Adoul and M. Barth. Nearest neighbor algorithm for spherical codes from the Leech lattice. *IEEE Trans. Inform. Theory*, IT-34:1188–1202, 1988.

[7] J. P. Adoul and P. Mabilleau. 4800 bps RELP vocoder using vector quantization for both filter and residual representation. In *International Conference on Acoustics, Speech, and Signal Processing*, volume 1, page 601, Paris, April 1982.

[8] K. Aizawa, H. Harashimia, and H. Miyakawa. Adaptive vector quantization of picture signals in discrete cosine transform domain. *Electronics and Communications in Japan,Part I*, 70, 1987.

[9] M. R. Anderberg. *Cluster Analysis for Applications.* Academic Press, San Diego, 1973.

[10] J. B. Anderson. Recent advances in sequential encoding of analog waveforms. In *1978 National Telecommunications Conference Record*, pages 19.4.1–19.4.5, Birmingham. Ala., December 1978.

[11] J. B. Anderson. Effectiveness of sequential tree search algorithms in speech digitization. In *1979 International Conference on Communications Conference Record*, pages 8.1.1–8.1.6, Boston. Mass., June 1979.

[12] J. B. A. Anderson and J. B. Bodie. Tree encoding of speech. *IEEE Trans. Inform. Theory*, IT-20:379–387, 1975.

[13] M. Antonini, M. Barlaud, and P. Mathieu. Image coding using lattice vector quantization of wavelet coefficients. In *Proceedings of the International Conference on Acoustics, Speech, and Signal Processing*, Toronto, Canada, May 1991.

[14] M. Antonini, M. Barlaud, P. Mathieu, and I. Daubechies. Image coding using wavelet transform. Submitted to *IEEE Trans. SP*.

[15] M. Antonini, M. Barlaud, P. Mathieu, and I. Daubechies. Image coding using vector quantization in the wavelet transform domain. In *Proceedings of the International Conference on Acoustics, Speech, and Signal Processing*, pages 2297–2300, Albuiquerque, April 1990.

[16] R. Aravind and A. Gersho. Low-rate image coding with finite-state vector quantization. In *Proceedings ICASSP*, pages 137–140, Tokyo, 1986.

[17] R. Aravind and A. Gersho. Image compression based on vector quantization with finite memory. *Optical Engineering*, 26:570–580, July 1987.

[18] D. S. Arnstein. Quantization error in predictive coders. *IEEE Trans. Comm.*, COM-23:423–429, April 1975.

[19] B. S. Atal and S. L. Hanauer. Speech analysis and synthesis by linear prediction of the speech wave. *J. Acoust. Soc. Am.*, 50:637–655, 1971.

[20] B. S. Atal and M. R. Schroeder. Predictive coding of speech signals and subjective error criteria. *IEEE Trans. Acoust. Speech Signal Process.*, ASSP-27:247–254, 1979.

[21] B.S. Atal, V. Cuperman, and A. Gersho, editors. *Advances in Speech Coding*. Kluwer Academic Publishers, July 1991.

[22] E. Ayanŏglu and R. M. Gray. The design of predictive trellis wave-form coders using the generalized Lloyd algorithm. *IEEE Trans. Comm.*, COM-34:1073–1080, November 1986.

[23] R. L. Baker and R. M. Gray. Image compression using non-adaptive spatial vector quantization. In *Conference Record of the Sixteenth Asilomar Conference on Circuits Systems and Computers*, Asilomar Calif., October 1982.

[24] R. L. Baker and R. M. Gray. Differential vector quantization of achromatic imagery. In *Proceedings of the International Picture Coding Symposium*, March 1983.

[25] R. L. Baker and J. L. Salinas. A motion compensated vector quantizer with filtered prediction. In *International Conference on Acoustics, Speech, and Signal Processing*, volume 2, pages 1324–1327, New York, NY, April 1988.

[26] R. L. Baker and H.-H. Shen. A finite-state vector quantizer for image sequence coding. In *International Conference on Acoustics, Speech, and Signal Processing*, volume 2, pages 760–763, Dallas,Texas, April, 1987.

[27] C. F. Barnes and R. L. Frost. Necessary conditions for the optimality of residual vector quantizers. In *Abstracts of the 1990 IEEE International Symposium on Information Theory*, page 34, San Diego, Calif., January 1990.

[28] A. R. Barron. The strong ergodic theorem for densities: generalized Shannon-McMillan-Breiman theorem. *Ann. Probab.*, 13:1292–1303, 1985.

[29] C.-D. Bei and R. M. Gray. Simulation of vector trellis encoding systems. *IEEE Trans. Comm.*, COM-34:214–218, March 1986.

[30] W. R. Bennett. Spectra of quantized signals. *Bell Systems Technical Journal*, 27:446–472, July 1948.

[31] J. L. Bentley. Multidimensional binary search trees used for associative searching. *Comm. ACM*, pages 209–226, September 1975.

[32] T. Berger. *Rate Distortion Theory*. Prentice-Hall Inc., Englewood Cliffs,New Jersey, 1971.

[33] J. C. Bezdek. A convergence theorem for the fuzzy ISODATA cluser-ing algorithms. *IEEE Trans. Pattern Anal. and Mach. Int.*, 3:1–8, 1980.

[34] R. E. Blahut. *Principles and Practice of Information Theory.* Addison-Wesley, Reading, Mass., 1987.

[35] R. N. Bracewell. *The Hartley Transform.* Oxford University Press, New York, NY, 1986.

[36] L. Breiman, J. H. Friedman, R. A. Olshen, and C. J. Stone. *Classification and Regression Trees.* Wadsworth, Belmont,California, 1984.

[37] A. M. Bruckstein. On 'soft' bit allocation. *IEEE Trans. Acoust. Speech Signal Process.*, 35:614–617, May 1987.

[38] J. A. Bucklew and G. L. Wise. Multidimensional asymptotic quantization theory with rth power distortion measures. *IEEE Trans. Inform. Theory*, IT-28:239–247, March 1982.

[39] S. Bunton and G. Borriello. Practical dictionary management for hardware data compression: *Development of a theme by Ziv and Lempel.* In *Proceedings of the Sixth MIT Conference on Advanced Research in VLSI*, pages 33–50, March 1990. To appear, *Communications of the ACM*, June, 1991.

[40] P. J. Burt and E. H. Adelson. The Laplacian pyramid as a compact image code. *IEEE Trans. Comm.*, COM-31:552–540, April 1983.

[41] A. Buzo, A. H. Gray, Jr., R. M. Gray, and J. D. Markel. Optimal quantizations of coefficient vectors in LPC speech. In *International Conference on Acoustics, Speech, and Signal Processing*, pages 52–55, Washington,D. C., April 1979.

[42] A. Buzo, A. H. Gray, Jr., R. M. Gray, and J. D. Markel. Speech coding based upon vector quantization. *IEEE Trans. Acoust. Speech Signal Process.*, ASSP-28:562–574, October 1980.

[43] A. Buzo, A. H. Gray, Jr., R. M. Gray, and J. D. Markel. Speech coding based upon vector quantization. In *International Conference on Acoustics, Speech, and Signal Processing*, Denver CO, April 1980.

[44] S. J. Campanella and G. S. Robinson. A comparison of orthogonal transformations for digital speech processing. *IEEE Transactions Commun. Technol.*, COM-19:1045–1049, 1971.

[45] J. P. Campbell, Jr., T.E. Tremain, and V.C. Welch. The DoD 4.8 kbps standard (proposed federal standard 1016). In B.S. Atal, V. Cuperman, and A. Gersho, editors, *Advances in Speech Coding*. Kluwer Academic Publishers, 1991.

[46] J. C. Candy. A use of limit cycle oscillations to obtain robust analog-to-digital converters. *IEEE Trans. Comm.*, COM-22:298–305, March 1974.

[47] J. C. Candy. A use of double integration in sigma delta modulation. *IEEE Trans. Comm.*, COM-33:249–258, March 1985.

[48] J. C. Candy. Decimation for sigma delta modulation. *IEEE Trans. Comm.*, COM-34:72–76, January 1986.

[49] J. C. Candy and O. J. Benjamin. The structure of quantization noise from sigma-delta modulation. *IEEE Trans. Comm.*, COM-29:1316–1323, Sept. 1981.

[50] J. C. Candy, Y. C. Ching, and D. S. Alexander. Using triangularly weighted interpolation to get 13-bit PCM from a sigma delta modulator. *IEEE Trans. Comm.*, pages 1268–1275, November 1976.

[51] P. Cappello, G. Davidson, A. Gersho, C. Koc, and V. Somayazulu. A systolic vector quantization processor for real-time speech coding. In *Proceedings of the International Conference on Acoustics, Speech, and Signal Processing*, volume 3, pages 2143–2146, Tokyo, April 1986.

[52] A. E. Cetin and V. Weerackody. Design vector quantizers using simulated annealing. In *Proceedings of the International Conference on Acoustics, Speech, and Signal Processing*, page 1550, December 1988.

[53] W.-Y. Chan and A. Gersho. Enhanced multistage vector quantization with constrained storage. In *Proc. 24th Asilomar Conf. Circuits, Systems, and Computers*, November 1990.

[54] W.-Y. Chan and A. Gersho. High fidelity audio transform coding with vector quantization. In *Proceedings of the International Conference on Acoustics, Speech, and Signal Processing*, pages 1109–1112, April 1990.

[55] W. Y. Chan and A. Gersho. Constrained constrained storage quantization of multiple vector sources by codebook sharing. *IEEE Trans. Comm.*, COM-38(12):11–13, January 1991.

[56] W. Y. Chan and A. Gersho. Constrained-storage vector quantization in high fidelity audio transform coding. In *Proceedings of the International Conference on Acoustics, Speech, and Signal Processing*, Toronto, Canada, May 1991.

[57] W.Y. Chan, S. Gupta, and A. Gersho. Enhanced multistage vector quantization by joint codebook design. Submitted for Publication, 1991.

[58] P. C. Chang and R. M. Gray. Gradient algorithms for designing predictive vector quantizers. *IEEE Trans. Acoust. Speech Signal Process.*, ASSP-34:679–690, August 1986.

[59] P. C. Chang, R. M. Gray, and J. May. Fourier transform vector quantization for speech coding. *IEEE Trans. Comm.*, COM-35:1059–1068, October 1987.

[60] D. T. S. Chen. On two or more dimensional optimum quantizers. In *International Conference on Acoustics, Speech, and Signal Processing*, pages 640–643, Hartford,CT, 1977.

[61] J.-H. Chen. A robust low-delay CELP speech coder at 16 kb/s. In B.S. Atal, V. Cuperman, and A. Gersho, editors, *Advances in Speech Coding*, pages 25–35. Kluwer Academic Publishers, 1991.

[62] J-H. Chen and A. Gersho. Covariance and autocorrelation methods for vector linear prediction. In *Proceedings of the International Conference on Acoustics, Speech, and Signal Processing*, pages 1545–1548, Dallas, Texas, April 1986.

[63] J. H. Chen and A. Gersho. Gain-adaptive vector quantization with application to speech coding. *IEEE Trans. on Communications*, COM-35:918–930, Sept. 1987.

[64] J. H. Chen and A. Gersho. Real-time vector APC speech coding at 4800 bps with adaptive postfiltering. In *Proceedings of the International Conference on Acoustics, Speech, and Signal Processing*, pages 2185–2188, Dallas, April 1987.

[65] D.-Y. Cheng and A. Gersho. A fast codebook search algorithm for nearest-neighbor pattern matching. In *Proceedings of the International Conference on Acoustics, Speech, and Signal Processing*, volume 1, pages 265–268, Tokyo, April 1986.

[66] D. Y. Cheng, A. Gersho, B. Ramamurthi, and Y. Shoham. Fast search algorithms for vector quantization and pattern matching. In *Proceedings of the International Conference on Acoustics, Speech, and Signal Processing*, pages 911.1–911.4, San Diego, March 1984.

[67] P. A. Chou and T. Lookabaugh. Locally optimal variable-to-variable length source coding with respect to a fidelity criterion. In *Proceedings 1991 IEEE International Symposium on Information Theory*, page 238, Budapest, Hungary, June 1991. IEEE Press.

[68] P. A. Chou, T. Lookabaugh, and R. M. Gray. Entropy-constrained vector quantization. *IEEE Trans. Acoust. Speech Signal Process.*, pages 31–42, January 1989.

[69] P. A. Chou, T. Lookabaugh, and R. M. Gray. Optimal pruning with applications to tree-structured source coding and modeling. *IEEE Trans. Inform. Theory*, pages 299–315, March 1989.

[70] Barry A. Cipra. A new wave in applied mathematics. *Science*, 240:858–859, August 1990.

[71] R. J. Clarke. *Transform Coding of Images*. Academic Press, Orlando, Fla., 1985.

[72] A. G. Clavier, P. F. Panter, and D. D. Grieg. Distortion in a pulse count modulation system. *AIEE Transactions*, 66:989–1005, 1947.

[73] A. G. Clavier, P. F. Panter, and D. D. Grieg. PCM distortion analysis. *Electrical Engineering*, pages 1110–1122, November 1947.

[74] J.-M. Combes, A. Grossmann, and Ph. Tchamitchian, editors. *Wavelets*. Springer-Verlag, Berlin, second edition, 1990.

[75] J. H. Conway and N. J. A. Sloane. Fast quantizing and decoding algorithms for lattice quantizers and codes. *IEEE Trans. Inform. Theory*, IT-28:227–232, March 1982.

[76] J. H. Conway and N. J. A. Sloane. Voronoi regions of lattices,second moments of polytopes,and quantization. *IEEE Trans. Inform. Theory*, IT-28:211–226, March 1982.

[77] J. H. Conway and N. J. A. Sloane. A fast encoding method for lattice codes and quantizers. *IEEE Trans. Inform. Theory*, IT-29:820–824, 1983.

[78] J. H. Conway and N. J. A. Sloane. On the Voronoi regions of certain lattices. *SIAM Journal of Alg. Disc. Math.*, 1983.

[79] J. H. Conway and N. J. A. Sloane. *Sphere Packings,Lattices and Groups*. Springer-Verlag, New York, 1988.

[80] M. Copperi and D. Sereno. ImprovedLPC excitation based on pattern classification and perceptual criteria. In *Proc. of 7th Int. Conf. on Pattern Recongition*, pages 860–862, Montreal, 1984.

[81] T. M. Cover. Enumerative source coding. *IEEE Trans. Inform. Theory*, IT-17:73–77, June 1973.

[82] H. M. S. Coxeter. *Regular Polytopes*. Dover, New York, 1973. Third edition.

[83] R. E. Crochiere, S. M. Webber, and J. K. L. Flanagan. Digital coding of speech in sub-bands. *Bell Syst. Tech. J.*, 55:1069–1086, October 1976.

[84] A. Croisier, D. Esteban, and C. Galand. Perfect channel splitting by use of interpolation/decimation/tree decomposition techniques. In *Proceedings Inter. Conf. on Information Science and Systems*, Patras, Greece, 1986.

[85] V. Cuperman. *Efficient waveform coding of speech using vector quantization*. Ph. d. dissertation, University of California, Santa Barbara, February 1983.

[86] V. Cuperman and A. Gersho. Adaptive differential vector coding of speech. In *Conference Record GlobeCom 82*, pages 1092–1096, December 1982.

[87] V. Cuperman and A. Gersho. Vector predictive coding of speech at 16 kb/s. *IEEE Trans. Comm.*, COM-33:685–696, July 1985.

[88] C. C. Cutler. Transmission systems employing quantization, 1960. U. S. Patent No. 2,927,962.

[89] G. Davidson, R. Aravind, T. Stanhope, and A. Gersho. Real-time speech compression with a VLSI vector quantization processor. In *Proceedings of the International Conference on Acoustics, Speech, and Signal Processing*, pages 37.6.1–37.6.4, Tampa, Florida, March 1985.

[90] G. Davidson, P. Cappello, and A. Gersho. Systolic architectures for vector quantization. *IEEE Trans. Acoust. Speech Signal Process.*, 36:1651–1664, October 1988.

[91] G. Davidson and Allen Gersho. Complexity reduction methods for vector excitation coding. In *Proceedings of the International Conference on Acoustics, Speech, and Signal Processing*, volume 4, pages 3055–3058, Tokyo, April 1986.

[92] Grant Davidson and Allen Gersho. Application of a VLSI vector quantization processor to real-time speech coding. *IEEE Journal on Selected Areas in Commun.*, SAC-4(1):112–124, January 1986.

[93] L. D. Davisson. Comments on 'sequence time coding for data compression'. *Proc. IEEE*, 54:2010, 1966.

[94] L. D. Davisson. Universal noiseless coding. *IEEE Trans. Inform. Theory*, IT-19:783–795, 1973.

[95] L. D. Davisson and R. M. Gray, editors. *Data Compression*, volume 14 of *Benchmark Papers in Electrical Engineering and Computer Science*. Dowden,Hutchinson,& Ross, Stroudsburg,Penn., 1976.

[96] L. D. Davisson, R. J. McEliece, M. B. Pursley, and M. S. Wallace. Efficient universal noiseless source codes. *IEEE Trans. Inform. Theory*, IT-27:269–279, 1981.

[97] R. A. DeVore, B. Jawerth, and B. Lucier. Image coding through wavelet transform coding. Preprint.

[98] R. A. DeVore, B. Jawerth, and B. Lucier. Data compression using wavelets: error, smoothness, and quantization. In J. A. Storer and J. H. Reif, editors, *Proceedings Data Compression Conference*, April 1991.

[99] R. Dianysian and R. L. Baker. A VLSI chip set for real time vector quantization of image sequences. In *Proc. IEEE Int. Symp. Circuits and Systems*, volume 1, pages 221–224, May 1987.

[100] B. W. Dickinson. Autoregressive estimation using residual energy ratios. *IEEE Transactions on Information Theory*, IT-24:503–506, July 1978.

[101] E. Diday and J. C. Simon. Clustering analysis. In K. S. Fu and K. S. Fu, editors, *Digital Pattern Recognition*. Springer-Verlag, NY, 1976.

[102] D.E. Dudgeon and R. M. Mersereau. *Multidimensional Digital Signal Processing*. Prentice-Hall, Inc., Englewood Cliffs, New Jersey, 1984.

[103] M. Ostendorf Dunham and R. M. Gray. An algorithm for the design of labeled-transition finite-state vector quantizers. *IEEE Trans. Comm.*, COM-33:83–89, January 1985.

[104] J. C. Dunn. A fuzzy relative of the ISODATA process and its use in detecting well-separated clusters. *Journal of Cybernetics*, 3:32–57, 1974.

[105] J. Durbin. The fitting of time series models. *Rev. Inst. Inter. Statist.*, 28:233–243, 1960.

[106] W. H. Equitz. Fast algorithms for vector quantization picture coding. In *International Conference on Acoustics, Speech, and Signal Processing*, pages 725–728, 1987.

[107] W. H. Equitz. A new vector quantization clustering algorithm. *IEEE Trans. Acoust. Speech Signal Process.*, pages 1568–1575, October 1989.

[108] M. Vedat Eyuboğlu and Jr. G. David Forney. Lattice and trellis quantization with lattice- and trellis-bounded codebooks–Part I: high-rate theory for memoryless sources. Submitted for possible publication.

[109] N. Farvardin and J.W. Modestino. Optimum quantizer performance for a class of non-gaussian memoryless sources. *IEEE Trans. Inform. Theory*, pages 485–497, May 1984.

[110] Y. Feng and N. M. Nasrabadi. A dynamic address-vector quantization algorithm based on interblock and intercolor correlation for color image coding. In *Proceedings of the International Conference on Acoustics, Speech, and Signal Processing*, pages 1755–1758, May 1989.

[111] T. Fine. Properties of an optimal digital system and applications. *IEEE Trans. Inform. Theory*, IT-10:287–296, Oct 1964.

[112] T. R. Fischer. Quantized control with data compression constraints. *Optimal Control Applications and Methods*, 5:39–55, January 1984.

[113] T. R. Fischer. A pyramid vector quantizer. *IEEE Trans. Inform. Theory*, IT-32:568–583, July 1986.

[114] T. R. Fischer. Geometric source coding and vector quantization. *IEEE Trans. Inform. Theory*, IT-35:137–145, January 1989.

[115] T. R. Fischer and M. W. Marcellin. Joint source/channel trellis coding. *IEEE Trans. Comm.*, 1990. To appear.

[116] T. R. Fischer, M. W. Marcellin, and M. Wang. Trellis coded vector quantization. *IEEE Trans. Inform. Theory*, 1991. Submitted for possible publication.

[117] T. R. Fischer and D. J. Tinnen. Quantized control with differential pulse code modulation. In *Conf. Proceedings, 21st Conf. on Decision and Control*, pages 1222–1227, December 1982.

[118] T. R. Fischer and D. J. Tinnen. Quantized control using differential encoding. *Optimal Control Applications and Methods*, pages 69–83, 1984.

[119] J. K. Flanagan, D. R. Morrell, R.L. Frost, C.J. Read, and B. E. Nelson. Vector quantization codebook generation using simulated annealing. In *Proceedings of the International Conference on Acoustics, Speech, and Signal Processing*, pages 1759–1762, Glasgow, Scotland, May 1989.

[120] P. E. Fleischer. Sufficient conditions for achieving minimum distortion in a quantizer. In *IEEE Int. Conv. Rec.,Part 1*, pages 104–111, 1964.

[121] R. J. Fontana. *A Class of Composite Sources and their Ergodic and Information Theoretic Properties*. Ph.d. Dissertation, Stanford University, 1978.

[122] E. Forgey. Cluster analysis of multivariate data: efficiency vs. interpretability of classification. *Biometrics*, 21:768, 1965. (Abstract).

[123] G. D. Forney, Jr. The Viterbi algorithm. *Proc. IEEE*, 61:268–278, March 1973.

[124] G. D. Forney, Jr. Coset codes–Part II: Introduction and geometrical classification. *IEEE Trans. Inform. Theory*, IT-34:1152–1187, 1989.

[125] G. D. Forney, Jr. Multidimensional constellations Part II: Voronoi constellations. *Journal of Selected Areas in Communications*, 7:941–958, 1989.

[126] J. Foster, R. M. Gray, and M. Ostendorf Dunham. Finite-state vector quantization for waveform coding. *IEEE Trans. Inform. Theory*, IT-31:348–359, May 1985.

[127] B. Fox. Discrete optimization via marginal analysis. *Management Science*, 13:210–216, November 1966.

[128] J. H. Friedman, F. Baskett, and L.J. Shustek. An algorithm for finding nearest neighbors. *IEEE Trans. on Comput*, C-24(10):1000–1006, October 1975.

[129] F. N. Fritsch and R. E. Carlson. Monotone piecewise cubic interpolation. *SIAM J. Numer. Anal.*, 17:238–246, April 1980.

[130] R. L. Frost, C. F. Barnes, and F. Xu. Design and performance of residual quantizers. In J. A. Storer and J. H. Reif, editors, *Proceedings Data Compression Converence*, pages 129–138, Snowbird, Utah, April 1991. IEEE Computer Society Press.

[131] K. E. Fultz and D. B. Penick. The T1 carrier system. *Bell Sys. Tech. Journal*, 44:1405–1452, September 1965.

[132] N. T. Gaarder and D. Slepian. On optimal finite-state digital transmission systems. *IEEE Trans. Inform. Theory*, IT-28:167–186, March 1982.

[133] G. Gabor and Z. Gyorfi. *Recursive Source Coding*. Springer-Verlag, New York, 1986.

[134] R. G. Gallager. *Information Theory and Reliable Communication*. John Wiley & Sons, New York, 1968.

[135] R. G. Gallager. Variations on a theme by Huffman. *IEEE Trans. Inform. Theory*, IT-24:668–674, Nov. 1978.

[136] S. Geman and D. Geman. Stochastic relaxation, Gibbs distribution, and the Bayesian restoration of images. *IEEE Trans. Pattern Anal. and Mach. Int.*, 11(6):689–691, 1984.

[137] A. Gersho. Asymptotically optimal block quantization. *IEEE Trans. Inform. Theory*, IT-25:373–380, July 1979.

[138] A. Gersho. On the structure of vector quantizers. *IEEE Trans. Inform. Theory*, IT-28:157–166, March 1982.

[139] A. Gersho and V. Cuperman. Vector quantization: A pattern-matching technique for speech coding. *IEEE Communications Magazine*, 21:15–21, December 1983.

[140] A. Gersho and B. Ramamurthi. Image coding using vector quantization. In *International Conference on Acoustics, Speech, and Signal Processing*, volume 1, pages 428–431, Paris, April 1982.

[141] A. Gersho, T. Ramstad, and I. Versvik. Fully vector-quantized subband coding with adaptive codebook allocation. In *International Conference on Acoustics, Speech, and Signal Processing*, volume 1, pages 10. 7. 1–4, March 1984.

[142] A. Gersho and Y. Shoham. Hierarchical vector quantization of speech with dynamic codebook allocation. In *Proceedings of the International Conference on Acoustics, Speech, and Signal Processing*, pages 10.9.1–10.9.4, San Diego, March 1984.

[143] A. Gersho and M. Yano. Adaptive vector quantization by progressive codevector replacement. In *Proceedings of the International Conference on Acoustics, Speech, and Signal Processing*, pages 4.6.1–4.6.4, 1985.

[144] Allen Gersho. Adaptive vector quantization. *Annales des Télécommunications*, 41(9-10):470–480, September-October 1986.

[145] Allen Gersho. Optimal nonlinear interpolative vector quantization. *IEEE Trans. Comm.*, COM-38(9-10):1285–1287, September 1990.

[146] J. D. Gibson. Adaptive prediction in speech differential encoding systems. *Proc. IEEE*, 68:1789–1797, November 1974.

[147] J. D. Gibson and T. R. Fischer. Alphabet-constrained data compression. *IEEE Trans. Inform. Theory*, IT-28:443–457, May 1982.

[148] J. D. Gibson and K. Sayood. Lattice quantization. In *Advances in Electronics and Electron Physics*, volume 72, pages 259–330. Academic Press, New York, NY, 1988.

[149] R. M. Gray. On the asymptotic eigenvalue distribution of Toeplitz matrices. *IEEE Trans. Inform. Theory*, IT-18(6):725–730, November 1972.

[150] R. M. Gray. Sliding-block source coding. *IEEE Trans. Inform. Theory*, IT-21(4):357–368, July 1975.

[151] R. M. Gray. Time-invariant trellis encoding of ergodic discrete-time sources with a fidelity criterion. *IEEE Trans. Inform. Theory*, IT-23:71–83, 1977.

[152] R. M. Gray. Toeplitz and circulent matrices: II. ISL technical report no. 6504–1, Stanford University Information Systems Laboratory, April 1977. (Available on request from author.).

[153] R. M. Gray. Vector quantization. *IEEE ASSP Magazine*, 1,No. 2:4–29, April 1984.

[154] R. M. Gray. Oversampled sigma-delta modulation. *IEEE Trans. Comm.*, COM-35:481–489, April 1987.

[155] R. M. Gray. *Probability, Random Processes, and Ergodic Properties.* Springer-Verlag, New York, 1988.

[156] R. M. Gray. Spectral analysis of quantization noise in a single-loop sigma-delta modulator with dc input. *IEEE Trans. Comm.*, COM-37:588–599, 1989.

[157] R. M. Gray. *Entropy and Information Theory.* Springer-Verlag, New York, 1990.

[158] R. M. Gray. Quantization noise spectra. *IEEE Trans. Inform. Theory*, IT-36:1220–1244, November 1990.

[159] R. M. Gray. *Source Coding Theory.* Kluwer Academic Press, Boston, 1990.

[160] R. M. Gray and H. Abut. Full search and tree searched vector quantization of speech waveforms. In *International Conference on Acoustics, Speech, and Signal Processing*, pages 593–596, Paris, May 1982.

[161] R. M. Gray, A. Buzo, A. H. Gray, Jr., and Y. Matsuyama. Distortion measures for speech processing. *IEEE Trans. Acoust. Speech Signal Process.*, ASSP-28:367–376, August 1980.

[162] R. M. Gray, A. Buzo, Y. Matsuyama, A. H. Gray, Jr., and J. D. Markel. Source coding and speech compression. In *Proceedings of the International Telemetering Converence*, volume XIV, pages 871–878, Los Angeles, Calif., Nov. 1978.

[163] R. M. Gray, W. Chou, and P. W. Wong. Quantization noise in single-loop sigma-delta modulation with sinusoidal inputs. *IEEE Trans. Inform. Theory*, IT-35:956–968, 1989.

[164] R. M. Gray and L. D. Davisson. *Random Processes: A Mathematical Approach for Engineers.* Prentice-Hall, Englewood Cliffs,New Jersey, 1986.

[165] R. M. Gray, A. H. Gray, Jr., G. Rebolledo, and J. E. Shore. Rate distortion speech coding with a minimum discrimination information distortion measure. *IEEE Trans. Inform. Theory*, IT-27(6):708–721, Nov. 1981.

[166] R. M. Gray and E. Karnin. Multiple local optima in vector quantizers. *IEEE Trans. Inform. Theory*, IT-28:256–261, March 1982.

[167] R. M. Gray, J. C. Kieffer, and Y. Linde. Locally optimal block quantizer design. *Inform. and Control*, 45:178–198, May 1980.

[168] R. M. Gray and Y. Linde. Vector quantizers and predictive quantizers for Gauss-Markov sources. *IEEE Trans. Comm.*, COM-30:381 – 389, February 1982.

[169] R. M. Gray and D. S. Ornstein. Sliding-block joint source/noisy-channel coding theorems. *IEEE Trans. Inform. Theory*, IT-22:682–690, 1976.

[170] R. M. Gray, M. Ostendorf, and R. Gobbi. Ergodicity of Markov channels. *IEEE Trans. Inform. Theory*, 33:656–664, September 1987.

[171] A. H. Gray, Jr., R. M. Gray, and J. D. Markel. Comparison of optimal quantizations of speech reflection coefficients. *IEEE Trans. Acoust. Speech Signal Process.*, ASSP-25:9–23, February 1977.

[172] U. Grenander and G. Szego. *Toeplitz Forms and Their Applications.* University of California Press, Berkeley and Los Angeles, 1958.

[173] J. M. Hammersley and D. C. Handscomb. *Monte Carlo Methods.* Monographs on Applied Probability and Statistics. Chapman and Hall, New York, 1979.

[174] H.-M. Hang and B. Haskell. Interpolative vector quantization of color images. *TCOM*, COM-36:465–470, 1987.

[175] H.-M. Hang and J. W. Woods. Predictive vector quantization of images. *IEEE Trans. Comm.*, COM-33:1208–1219, November 1985.

[176] A. Haoui and D. G. Messerschmitt. Predictive vector quantization. In *International Conference on Acoustics, Speech, and Signal Processing*, volume 1, pages 10. 10. 1–10. 10. 4, San Diego, March 1984.

[177] S. Haykin. *Adaptive Filter Theory.* Prentice-Hall, Englewood Cliffs, New Jersey, 1986.

[178] E. E. Hilbert. Cluster compression algorithm: a joint clustering/data compression concept. Publication 77-43, Jet Propulsion Lab, Pasadena, Calif., December 1977.

[179] Y.-S. Ho and A. Gersho. Variable-rate contour-based interpolative vector quantization for image coding. In *Conference Record: IEEE Global Commun. Conf.*, pages 1890–1893, November 1988.

[180] Y.-S. Ho and A. Gersho. Variable-rate multi-stage vector quantization for image coding. In *Proceedings of the International Conference on Acoustics, Speech, and Signal Processing*, pages 1156–1159, 1988.

[181] Y. S. Ho and A. Gersho. Classified transform coding of image using interpolative vector quantization. In *Proceedings of the International Conference on Acoustics, Speech, and Signal Processing*, pages 1890–1893, May 1989.

[182] Y. S. Ho and A. Gersho. A pyramidal image coder using contour-based interpolative vector quantization. In *Proc. SPIE Int. Symp. on Visual Commun. and Image Processing*, volume 1199, pages 733–740, November 1989.

[183] Y. S. Ho and A. Gersho. A variable rate image coding scheme with vector quantization and clustering interpolation. In *Proceedings 1989 Globecom*, pages 25.5.1–25.5.1, 1989.

[184] J. E. Hopcroft and J. D. Ullman. *Introduction to Automata Theory, Language, and Computation*. Addison-Wesley, Reading, Mass., 1979.

[185] J.-Y. Huang. *Quantization of correlated random variables*. Ph.D. Dissertation, Yale University, New Haven, Conn., 1962.

[186] J.-Y. Huang and P. M. Schultheiss. Block quantization of correlated Gaussian random variables. *IEEE Trans. Comm.*, CS-11:289–296, September 1963.

[187] D. A. Huffman. A method for the construction of minimum redundancy codes. *Proceedings of the IRE*, 40:1098–1101, 1952.

[188] H. Inose and Y. Yasuda. A unity bit coding method by negative feedback. *Proc. IEEE*, 51:1524–1535, November 1963.

[189] F. Itakura and S. Saito. Analysis synthesis telephony based on the maximum likelihood method. In *Proceedings of the 6th International Congress of Acoustics*, pages C–17–C–20, Tokyo, Japan, August 1968.

[190] F. Itakura and S. Saito. A statistical method for estimation of speech spectral density and formant frequencies. *Electron. Commun. Japan*, 53-A:36–43, 1970.

[191] J. E. Iwersen. Calculated quantizing noise of single-integration delta-modulation coders. *Bell Syst. Tech. J.*, pages 2359–2389, September 1969.

[192] Jr. J. B. O'Neil. A bound on signal-to-quantizing noise ratios for digital encoding systems. *Proc. IEEE*, 55:2887–2892, March 1967.

[193] N. S. Jayant. Adaptive quantization with one word memory. *Bell Syst. Tech. J.*, pages 1119–1144, September 1973.

[194] N. S. Jayant, editor. *Waveform Quantization and Coding*. IEEE Press, Piscataway, NJ, 1976.

[195] N. S. Jayant and P. Noll. *Digital Coding of Waveforms*. Prentice-Hall, Englewood Cliffs,New Jersey, 1984.

[196] F. Jelinek and J. B. Anderson. Instrumentable tree encoding of information sources. *IEEE Trans. Inform. Theory*, IT-17:118–119, Jan 1971.

[197] C. B. Jones. An efficient coding system for long source sequences. *IEEE Trans. Inform. Theory*, IT-27:280–291, 1981.

[198] B.-H. Juang and A. H. Gray, Jr. Multiple stage vector quantization for speech coding. In *International Conference on Acoustics, Speech, and Signal Processing*, volume 1, pages 597–600, Paris, April 1982.

[199] B.-H. Juang, D. Y. Wong, and A. H. Gray, Jr. Distortion performance of vector quantization for LPC voice coding. *IEEE Trans. Acoust. Speech Signal Process.*, ASSP-30:294–304, April 1982.

[200] J. C. Kieffer. Sliding-block coding for weakly continuous channels. *IEEE Trans. Inform. Theory*, IT-28:2–10, 1982.

[201] J. C. Kieffer. Stochastic stability for feedback quantization schemes. *IEEE Trans. Inform. Theory*, IT-28:248–254, March 1982.

[202] J. C. Kieffer and J. G. Dunham. On a type of stochastic stability for a class of encoding schemes. *IEEE Trans. Inform. Theory*, 1983.

[203] J. C. Kieffer and M. Rahe. Markov channels are asymptotically mean stationary. *Siam Journal of Mathematical Analysis*, 12:293–305, 1980.

[204] T. Kim. New finite state vector quantizers for images. In *Proceedings of the International Conference on Acoustics, Speech, and Signal Processing*, pages 1180–1183, 1988.

[205] R. A. King and N. M. Nasrabadi. Image coding using vector quantization in the transform domain. *Pattern Recognition Letters*, 1:323–329, 1983.

[206] W.B. Kleijn, D.J. Krasinski, and R.H. Ketchum. Improved speech quality and efficient vector quantization in selp. In *Proceedings of the International Conference on Acoustics, Speech, and Signal Processing*, volume 1, pages 155–158, New York City, April 1988.

[207] D. E. Knuth. *The art of computer programming.* Addison-Wesley, Reading, Mass., 1981.

[208] T. Kohonen. *Self-organization and associative memory.* Springer-Verlag, Berlin, third edition, 1989.

[209] H. P. Kramer and M. V. Mathews. A linear coding for transmitting a set of correlated signals. *IRE Transactions on Information Theory,* IT-23:41–46, Sept. 1956.

[210] P. Van Laarhoven and E. Aarts. *Simulated Annealing: Theory and Applications.* D. Reidel Publishing Company, Dordrecht, Holland, 1987.

[211] N. Levinson. The Wiener RMS (root mean squared) error criterion in filter design and prediction. *J. Math. Phys.,* 25:261–278, 1947.

[212] J. Lin and J. A. Storer. Resolution-constrained tree-structured vector quantization for image compression. In *Proceedings 1991 IEEE International Symposium on Information Theory,* page 251, Piscataway, New Jersey, June 1991. IEEE Press.

[213] J. Lin, J. A. Storer, and M. Cohn. On the complexity of optimal tree pruning for source coding. In J. A. Storer and J. H. Reif, editors, *Proceedings Data Compression Converence,* pages 63–72, Snowbird, Utah, April 1991. IEEE Computer Society Press.

[214] S. Lin. *Introduction to Error Correcting Codes.* Prentice-Hall, Englewood Cliffs,NJ, 1970.

[215] Y. Linde, A. Buzo, and R. M. Gray. An algorithm for vector quantizer design. *IEEE Trans. Comm.,* COM-28:84–95, January 1980.

[216] Y. Linde and R. M. Gray. A fake process approach to data compression. *IEEE Trans. Comm.,* COM-26:702–710, April 1978.

[217] B. Liu and T. P. Stanley. Error bounds for jittered sampling. *IEEE Trans. Automat. Control,* AC-10:449–454, Oct. 1965.

[218] S. P. Lloyd. Least squares quantization in PCM. Unpublished Bell Laboratories Technical Note. Portions presented at the Institute of Mathematical Statistics Meeting Atlantic City New Jersey September 1957. Published in the March 1982 special issue on quantization of the *IEEE Transactions on Information Theory,* 1957.

[219] S. P. Lloyd. Least squares quantization in PCM. *IEEE Trans. Inform. Theory,* IT-28:127–135, March 1982.

[220] T. Lookabaugh, E. A. Riskin, P. A. Chou, and R. M. Gray. Variable rate vector quantization for speech, image, and video compression. *IEEE Transactions on Communications*, 1991. To appear.

[221] T. D. Lookabaugh and R. M. Gray. High-resolution quantization theory and the vector quantizer advantage. *IEEE Trans. Inform. Theory*, 35:1020–1033, September 1989.

[222] T. L. Lookabaugh. *Variable Rate and Adaptive Frequency Domain Vector Quantization of Speech*. Ph.D. Dissertation, Stanford University, 1988.

[223] A. Lowry, S. Hossain, and W. Millar. Binary search trees for vector quantization. In *Proceedings of the International Conference on Acoustics, Speech, and Signal Processing*, pages 2206–2208, Dallas, 1987.

[224] D. G. Luenberger. *Optimization by Vector Space Methods*. Wiley, New York, 1969.

[225] J. Lukaszewicz and H. Steinhaus. On measuring by comparison. *Zastos. Mat.*, 2:225–231, 1955. (In Polish.).

[226] T. J. Lynch. Sequence time coding for data compression. *Proc. IEEE*, 54:1490–1491, 1966.

[227] J. MacQueen. Some methods for classification and analysis of multivariate observations. In *Proc. of the Fifth Berkeley Symposium on Math. Stat. and Prob.*, volume 1, pages 281–296, 1967.

[228] J. Makhoul. Linear prediction: A tutorial review. *Proc. IEEE*, 63:561–580, 1975.

[229] J. Makhoul, S. Roucos, and H. Gish. Vector quantization in speech coding. *Proc. IEEE*, 73. No. 11:1551–1587, November 1985.

[230] S. G. Mallat. A theory for multiresolution signal decomposition: The wavelet representation. *IEEE Trans. Pattern Anal. and Mach. Int.*, 11(7):674–693, July 1989.

[231] S. G. Mallat. A compact multiresolution representation: The wavelet model. Preprint, 1990.

[232] M. W. Marcellin and T. R. Fischer. Trellis coded quantization of memoryless and Gauss-Markov sources. *IEEE Trans. Comm.*, pages 82–93, 1990.

[233] M. W. Marcellin, T. R. Fischer, and J. D. Gibson. Predictive trellis coded quantization of speech. *IEEE Trans. Acoust. Speech Signal Process.*, ASSP-38:46–55, January 1990.

[234] J. D. Markel and A. H. Gray, Jr. *Linear Prediction of Speech*. Springer-Verlag, New York, 1976.

[235] J. Max. Quantizing for minimum distortion. *IEEE Trans. Inform. Theory*, pages 7–12, March 1960.

[236] B. McMillan. Communication systems which minimize coding noise. *Bell Syst. Tech. J.*, 48(9):3091–3112, Nov 1969.

[237] N. Metropolis, A. W. Rosenbluth, M.N. Rosenbluth, A. H. Teller, and E. Teller. Equations of state calculations by fast computing machines. *J. Chem. Phys.*, 21:1087–1091, 1953.

[238] K. Motoishi and T. Misumi. On a fast vector quantization algorithm. In *Proceedings of the VIIth Symp. on Inform. Theory and its Applications*, 1984.

[239] K. Motoishi and T. Misumi. Fast vector quantization algorithm by using an adaptive searching technique. In *Abstracts of the 1990 IEEE International Symposium on Information Theory*, San Diego, Calif., January 1990.

[240] T. Murakami, K. Asai, and E. Yamazaki. Vector quantizer of video signals. *Electronics Letters*, 7:1005–1006, Nov. 1982.

[241] N. M. Nasrabadi and Y. Feng. Image compression using address-vector quantization. *IEEE Trans. Comm.*, 38:2166–2173, December 1990.

[242] N. M. Nasrabadi and Y. Feng. A multi-layer address-vector quantization. *IEEE Trans. Circuits and Systems*, CAS-37:912–921, July 1990.

[243] N. M. Nasrabadi and R. A. King. Image coding using vector quantization: A review. *IEEE Transactions on Communications*, COM-36:957–971, August 1988.

[244] D. L. Neuhoff. Source coding strategies: simple quantizers vs. simple noiseless codes. In *Proceedings 1986 Conf. on Information Sciences and Systems*, volume 1, pages 267–271, March 1986.

[245] D. L. Neuhoff, R. M. Gray, and L. D. Davisson. Fixed rate universal block source coding with a fidelity criterion. *IEEE Trans. Inform. Theory*, 21:511–523, 1975.

[246] M. Orchard. A fast nearest neighbor search algorithm. In *Proceedings of the International Conference on Acoustics, Speech, and Signal Processing*, Toronto, Canada, May 1991.

[247] D. O'Shaughnessy. *Speech Communication*. Addison-Wesley, Reading, Mass., 1987.

[248] Y. Özturk and H. Abut. Multichannel linear prediction and applications to image coding. *Archiv Elect. Übertrag.*, 43(5):312–320, 1989.

[249] M. D. Paez and T. H. Glisson. Minimum mean squared error quantization in speech PCM and DPCM systems. *IEEE Trans. Comm.*, COM-20:225–230, April 1972.

[250] S. Panchanathan and M. Goldberg. Algorithms and architecture for image adaptive vector quantization. In *Proc. SPIE Conf. on Visual Communications and Image Processing '88*, volume 1001, pages 336–344., Cambridge, Mass., November 1988.

[251] P. Papantoni-Kazakos and R. M. Gray. Robustness of estimators on stationary observations. *Ann. Probab.*, 7:989–1002, Dec. 1979.

[252] R. Pasco. *Source coding algorithms for fast data compression*. Ph. D. Dissertation, Stanford University, 1976.

[253] D. Paul. A 500-800 bps adaptive vector quantization vocoder using a perceptually motivated distance measure. In *Conf. Record, IEEE Globecom*, pages 1079–1082, 1982.

[254] D. P. Peterson and D. Middleton. Sampling and reconstruction of wave-number limited functions in n-dimensional euclidean spaces. *Inform. and Control*, 5:279–323, 1962.

[255] W. K. Pratt. *Image Transmission Techniques*. Academic Press, New York, 1979.

[256] M. Rabbani and P. W. Jones. *Digital Image Compression Techniques*, volume TT7 of *Tutorial Texts in Optical Engineering*. SPIE Optical Engineering Press, Bellingham, Washington, 1991.

[257] V. Ramamoorthy and T. Ericson. Modulo-PCM with side information. In *European Signal Proc. Conf., EUSIPCO-80*, pages 395–398, Erlangen, September 1980.

[258] B. Ramamurthi and A. Gersho. Image coding using segmented codebooks. In *Proceedings International Picture Coding Symposium*, Mar. 1983.

[259] B. Ramamurthi and A. Gersho. Classified vector quantization of images. *IEEE Trans. Comm.*, COM-34:1105–1115, November 1986.

[260] V. Ramasubramanian and K.K. Paliwal. An optimized k-d tree algorithm for fast vector quantization of speech. In *Proc. European Signal Processing Conf.*, pages 875–878, Grenoble, 1988.

[261] V. Ramasubramanian and K.K. Paliwal. An efficient approximation-elimination algorithm for fast nearest-neighbor search based on a spherical distance coordinate formulation. In *Proc. European Signal Processing Conf.*, Barcelona, September 1990.

[262] T. A. Ramstad. Sub-band coder with a simple bit allocation algorithm: a possible candidate for digital mobile telephony. In *International Conference on Acoustics, Speech, and Signal Processing*, pages 203–207. IEEE Press, May 1982.

[263] G. Rebolledo. *Speech and waveform coding based on vector quantization*. Ph.D. Dissertation, Stanford Univ., Aug. 1982.

[264] G. Rebolledo, R. M. Gray, and J. P. Burg. A multirate voice digitizer based upon vector quantization. *IEEE Trans. Comm.*, COM-30:721–727, April 1982.

[265] R. F. Rice and J. R. Plaunt. Adaptive variable-lentgh coding for efficient compression of spacecraft television data. *IEEE Transactions on Comm. Tech.*, COM-19:889–897, December 1971.

[266] Robert F. Rice. Some practical universal noiseless coding techniques, March 1979.

[267] E. A. Riskin. *Variable Rate Vector Quantization of Images*. Ph. D. Dissertation, Stanford University, 1990.

[268] E. A. Riskin. Optimal bit allocation via the generalized BFOS algorithm. *IEEE Trans. Inform. Theory*, 37:400–402, March 1991.

[269] E. A. Riskin and R. M. Gray. A greedy tree growing algorithm for the design of variable rate vector quantizers. *IEEE Trans. Signal Process.*, November 1991. To appear.

[270] E. A. Riskin, T. Lookabaugh, P. A. Chou, and R. M. Gray. Variable rate vector quantization for medical image compression. *IEEE Transactions on Medical Imaging*, 9:290–298, September 1990.

[271] J. Rissanen. Generalized Kraft inequality and arithmetic coding. *IBM J. Res. Develop.*, 20:198–203, 1976.

[272] J. Rissanen. Optimum block models with fixed-length coding. IBM research center technical report, IBM, San Jose,California, 1982.

[273] J. Rissanen and G. G. Langdon. Universal modeling and coding. *IEEE Trans. Inform. Theory*, IT-27, 1981.

[274] K. Rose. *Deterministic annealing , clustering, and optimization*. Ph. D. Dissertation, California Institute of Technology, 1991.

[275] K. Rose, E. Gurewitz, and G. C. Fox. A deterministic annealing approach to clustering. *Pattern Recognition Letters*, 11:589–594, 1990.

[276] K. Rose, E. Gurewitz, and G. C. Fox. Statistical mechanics and phase transitions in clustering. *Physical Review Letters*, 65:945–948, 1990.

[277] M. J. Sabin. Fixed-shape adaptive-gain vector quantization for speech waveform coding. *Speech Communication*, 8:177–183, 1989.

[278] M. J. Sabin and R. M. Gray. Product code vector quantizers for waveform and voice coding. *IEEE Trans. Acoust. Speech Signal Process.*, ASSP-32:474–488, June 1984.

[279] M. J. Sabin and R. M. Gray. Global convergence and empirical consistency of the generalized Lloyd algorithm. *IEEE Trans. Inform. Theory*, IT-32:148–155, March 1986.

[280] D. J. Sakrison. *Communication Theory: Transmission of Waveforms and Digital Information*. Wiley, New York, 1968.

[281] D. J. Sakrison. A geometric treatment of the source encoding of a Gaussian random variable. *IEEE Trans. Inform. Theory*, IT-14:481–486, May 1968.

[282] H. Samet. The quad tree and related hierarchical data structures. *ACM Computing Surveys*, 16:188–260, June 1984.

[283] K. Sayood, J. D. Gibson, and M. C. Rost. An algorithm for uniform vector quantizer design. *IEEE Trans. Inform. Theory*, IT-30:805–814, 1984.

[284] M. R. Schroeder and B. S. Atal. Rate distortion theory and predictive coding. In *Proceedings of the International Conference on Acoustics, Speech, and Signal Processing*, pages 201–204, Atlanta, Goergia, March 1981.

[285] M. R. Schroeder and B. S. Atal. Speech coding using efficient block codes. In *Proceedings of the International Conference on Acoustics, Speech, and Signal Processing*, pages 1668–1671, March 1982.

[286] M. R. Schroeder and B. S. Atal. Stochastic coding of speech signals at very low bit rates. In *Conf. Record, IEEE Inter. Conf. Commun.*, pages 1610–1613, Amsterdam, May 1984.

[287] M. R. Schroeder and B. S. Atal. Code-excited linear prediction (CELP): High-quality speech at very low bit rates. In *Proceedings of the International Conference on Acoustics, Speech, and Signal Processing*, pages 937–940, Tampa, Florida, March 1985.

[288] A. Segall. Bit allocation and encoding for vector sources. *IEEE Trans. Inform. Theory*, IT-22:162–169, March 1976.

[289] C. E. Shannon. A mathematical theory of communication. *Bell Syst. Tech. J.*, 27:379–423,623–656, 1948.

[290] C. E. Shannon. Coding theorems for a discrete source with a fidelity criterion. In *IRE National Convention Record, Part 4*, pages 142–163, 1959.

[291] H.-H. Shen and R. L. Baker. A finite state/frame difference interpolative vector quantizer for low rate image sequence coding. In *International Conference on Acoustics, Speech, and Signal Processing*, pages 1188–1191, 1988.

[292] L. Shepp, D. Slepian, and A. D. Wyner. On prediction of moving average processes. *Bell Syst. Tech. J.*, pages 367–415, 1980.

[293] Y. Shoham and A. Gersho. Pitch synchronous transform coding of speech at 9.6kb/s based on vector quantization. In *Conference Record: IEEE International Conference on Communications*, pages 1179–1182, Amsterdam, 1984.

[294] Y. Shoham and A. Gersho. Efficient codebook allocation for an arbitrary set of vector quantizers. In *International Conference on Acoustics, Speech, and Signal Processing*, pages 1696–1699, 1985.

[295] Y. Shoham and A. Gersho. Efficient bit allocation for an arbitrary set of quantizers. *IEEE Trans. Acoust. Speech Signal Process.*, ASSP-36(9):1445–1453, September 1988.

[296] J. E. Shore and R. M. Gray. Minimum-cross-entropy pattern classification and cluster analysis. *IEEE Transactions on Pattern Analysis and Machine Intellegence*, PAMI-4:11–17, Jan. 1982.

[297] G. L. Sicuranza and G. Ramponi. Adaptive nonlinear prediction of TV image sequences. *Electronics Letters*, 25:526–527, 13 April 1989.

[298] M. J. T. Smith and T.P. Barnwell, III. Exact reconstruction techniques for tree-structured subband coders. *IEEE Trans. Acoust. Speech Signal Process.*, ASSP-34:434–441, June 1986.

[299] M. R. Soleymani and S. D. Morgera. An efficient nearest neighbor search method. *IEEE Trans. Comm.*, COM-35:677–679, 1987.

[300] R. Steele. *Delta Modulation Systems*. Pentech Press, London, 1975.

[301] L. C. Stewart. Trellis data compression. Stanford university information systems lab technical report 1905–1, Stanford University, July 1981. Stanford University Ph.D. thesis.

[302] L. C. Stewart, R. M. Gray, and Y. Linde. The design of trellis waveform coders. *IEEE Trans. Comm.*, COM-30:702–710, April 1982.

[303] J. Storer. *Data Compression*. Computer Science Press, Rockville, Maryland, 1988.

[304] P. F. Swaszek, editor. *Quantization*, volume 29 of *Benchmark Papers in Electrical Engineering and Computer Science*. Van Nostrand Reinhold Company,Inc., New York, NY, 1985.

[305] D. W. Tank and J. J. Hopfield. Simple 'neural' optimization networks: an A/D converter, signal decision circuit, and a linear programming circuit. *IEEE Trans. Circuits and Systems*, CAS-33:533–541, May 1986.

[306] B. P. M. Tao and R. M. Gray H. Abut. Hardware realization of waveform vector quantizers. *IEEE J. Selected Areas in Commun.*, SAC-2:343–352, March 1984.

[307] N. Tishby. A dynamical systems approach to speech processing. In *Proceedings of the International Conference on Acoustics, Speech, and Signal Processing*, pages 365–368, Albuquerque, New Mexico, April 1990.

[308] J. T. Tou and R. C. Gonzales. *Pattern Recognition Principles.* Addison-Wesley, Reading, Mass., 1974.

[309] I. Trancoso and B. S. Atal. Efficient procedures for finding the optimum innovation in stochastic coders. In *Proceedings of the International Conference on Acoustics, Speech, and Signal Processing*, pages 1681–1684, Tokyo, Japan, April 1986.

[310] T. Tremain. The government standard linear predictive coding algorithm:LPC-10. *IEEE Trans. Acoust. Speech Signal Process.*, ASSP-36(9):40–49, April 1982.

[311] A. V. Trushkin. Sufficient conditions for uniqueness of a locally optimal quantizer for a class of convex error weighting functions. *IEEE Trans. Inform. Theory*, IT-28:187–198, March 1982.

[312] C. Tsao and R. M. Gray. Shape-gain matrix quantizers for LPC speech. *IEEE Trans. Acoust. Speech Signal Process.*, ASSP-34:1427–1439, December 1986.

[313] B. P. Tunstall. *Synthesis of noiseless compression codes.* Ph. D. Thesis, Georgia Institute of Technology, 1968.

[314] G. Ungerboeck. Channel coding with multilevel/phase signals. *IEEE Trans. Inform. Theory*, IT-28:55–67, January 1982.

[315] G. Ungerboeck. Trellis-coded modulation with redundant signal sets, Parts I and II. *IEEE Communications Magazine*, 25:5–21, February 1987.

[316] P.P. Vaidyanathan and P. Q. Hoang. Lattice structures for optimal design and robust implementation of two-channel perfect reconstruction QMF banks. *IEEE Trans. Acoust. Speech Signal Process.*, ASSP-36:81–94, January 1988.

[317] D. J. Vaisey and A. Gersho. Variable block-size image coding. In *International Conference on Acoustics, Speech, and Signal Processing*, pages 1051–1054, 1987.

[318] D. J. Vaisey and A. Gersho. Variable rate image coding using quad trees and vector quantization. *Signal Processing IV: Theories and Applications*, pages 1133–1136, 1988.

[319] J. Vaisey and A. Gersho. Simulated annealing and codebook design. In *Proceedings of the International Conference on Acoustics, Speech, and Signal Processing*, pages 1176–1179, New York, April 1988.

[320] J. Vaisey and A. Gersho. Image compression with variable block size segmentation. *IEEE Trans. Signal Process.*, ASSP-40, 1992. to appear.

[321] M. Vetterli. Multi-dimensional sub-band coding: some theory and algorithms. *Signal Processing*, 6:97–112, April 1984.

[322] E. Vidal. An algorithm for finding nearest neighbors in (approximately) constant average time complexity. *Pattern Recognition Letters*, 4:145–157, 1986.

[323] A. J. Viterbi and J. K. Omura. Trellis encoding of memoryless discrete-time sources with a fidelity criterion. *IEEE Trans. Inform. Theory*, IT-20:325–332, May 1974.

[324] S. Wang, E. Paksoy, and A. Gersho. Nonlinear prediction of speech with vector quantization. In *Proc. Int. Conf. Spoken Language Processing*, pages 29–32, Kobe, Japan, November 1990.

[325] Shihua Wang and Allen Gersho. Interframe coding of LPC parameters with band-splitting vq. In *Conf. Rec., 1991 IEEE Workshop on Speech Coding for Telecommunications*, September 1991.

[326] J. Ward. Hierarchical grouping to optimize an objective function. *J. Amer. Stat. Assoc.*, 37:236–244, March 1963.

[327] T. A. Welch. A technique for high-performance data compression. *Computer*, pages 8–18, 1984.

[328] E. J. Weldon, Jr. and W. W. Peterson. *Error Correcting Codes*. MIT Press, Cambridge, Mass., 1971. Second Ed.

[329] P. H. Westerink, J. Biemond, and D. E. Boekee. An optimal bit allocation algorithm for sub-band coding. In *International Conference on Acoustics, Speech, and Signal Processing*, pages 757–760, 1988.

[330] P. H. Westerink, D. E. Boekee, J. Biemond, and J. W. Woods. Subband coding of images using vector quantization. *IEEE Trans. Comm.*, COM-36:713–719, June 1988.

[331] P. Whittle. On the fitting of multivariable autoregressions and the approximate cononical factorization of a spectral density matrix. *Biometrika*, 50:129–134, 1963.

[332] B. Widrow. A study of rough amplitude quantization by means of Nyquist sampling theory. *IRE Transactions Circuit Theory*, CT-3:266–276, 1956.

[333] B. Widrow. Statistical analysis of amplitude quantized sampled data systems. *Transactions Amer. Inst. Elec. Eng.,Pt. II: Applications and Industry*, 79:555–568, 1960.

[334] B. Widrow and S. Stearns. *Adaptive signal processing.* Prentice Hall, Englewood Cliffs,New Jersey, 1985.

[335] R. A. Wiggins and E. A. Robinson. Recursive solution to the multichannel filtering problem. *J. Geophys. Res.*, 70:247–254,1885–1891, April 1965.

[336] S. G. Wilson and D. W. Lytle. Trellis encoding of continuous-amplitude memoryless sources. *IEEE Trans. Inform. Theory*, IT-28:211–226, March 1982.

[337] I. H. Witten, R. M. Neal, and J. G. Cleary. Arithmetic coding for data compression. *Communications of the ACM*, 30:520–540, 1987.

[338] D. Wong, B.-H. Juang, and A. H. Gray, Jr. An 800 bit/s vector quantization LPC vocoder. *IEEE Trans. Acoust. Speech Signal Process.*, ASSP-30:770–779, October 1982.

[339] J. W. Woods and S. D. O'Neil. Subband coding of images. *IEEE Trans. Acoust. Speech Signal Process.*, ASSP-34:1278–1288, October 1986.

[340] John W. Woods, editor. *Subband Image Coding.* Kluwer Academic Publishers, Boston, 1991.

[341] S.W. Wu and A. Gersho. Optimal block-adaptive image coding with constrained bit-rate. In *Proc. 24th Asilomar Conf. Circuits, Systems,and Computers*, pages 336–340, November 1990.

[342] S.W. Wu and A. Gersho. Rate-constrained optimal block-adaptive coding for digital tape recording of hdtv. *IEEE Trans. Video Technology in Circuits and Systems*, 1(1):100–112, March 1991.

[343] X. Wu and K. Zhang. A better tree-structured vector quantizer. In J. A. Storer and J. H. Reif, editors, *Proceedings Data Compression Converence*, pages 392–401, Snowbird, Utah, April 1991. IEEE Computer Society Press.

[344] Y. Yamada, K. Fujita, and S. Tazaki. Vector quantization of video signals. In *Proceedings of Annual Conference of IECE*, page 1031, Japan, 1980.

[345] Y. Yamada, S. Tazaki, and R. M. Gray. Asymptotic performance of block quantizers with a difference distortion measure. *IEEE Trans. Inform. Theory*, IT-26:6–14, Jan. 1980.

[346] Mei Yong, Grant Davidson, and Allen Gersho. Encoding of LPC spectral parameters using switched-adaptive interframe vector prediction. In *Proceedings of the International Conference on Acoustics, Speech, and Signal Processing*, pages 402–405, New York, April 1988.

[347] P. L. Zador. *Development and evaluation of procedures for quantizing multivariate distributions.* Ph. D. Dissertation, Stanford University, 1963.

[348] P. L. Zador. Topics in the asymptotic quantization of continuous random variables, 1966. Unpublished Bell Laboratories Memorandum.

[349] P. L. Zador. Asymptotic quantization error of continuous signals and the quantization dimension. *IEEE Trans. Inform. Theory*, IT-28:139–148, Mar. 1982.

[350] K. Zeger and A. Gersho. A stochastic relaxation algorithm for improved vector quantiser design. *Electronics Letters*, 25:896–898, July 1989.

[351] K. Zeger, J. Vaisey, and A. Gersho. Globally optimal vector quantizer design by stochastic relaxation. *IEEE Trans. Signal Process.*, 1992. to appear.

[352] J. Ziv. Distortion-rate theory for individual sequences. *IEEE Trans. Inform. Theory*, IT-26:137–143, Mar. 1980.

[353] J. Ziv. Universal quantization. *IEEE Trans. Inform. Theory*, IT-31:344–347, 1985.

[354] J. Ziv and A. Lempel. A universal algorithm for sequential data compression. *IEEE Trans. Inform. Theory*, IT-23:337–343, 1977.

[355] J. Ziv and A. Lempel. Compression of individual sequences via variable-rate coding. *IEEE Trans. Inform. Theory*, IT-24:530–536, 1978.

Index